Observational Cosmology

Observational cosmology is a rapidly developing field and this book covers some of the breadth of recent developments, such as precision cosmology and the concordance cosmological model, inflation, gravitational lensing and shear, the extragalactic far-infrared and X-ray backgrounds, downsizing and baryon wiggles. Forthcoming major facilities are covered, including radio, X-ray, submm-wave and gravitational wave astronomy. Suggestions for further reading provide accessible and approachable jumping off points for students aiming to further their studies. Produced by Open University academics and drawing on decades of Open University experience in supported open learning, the book is completely self-contained with numerous exercises (with full solutions provided). Designed to be worked through sequentially by a self-guided student, it also includes clearly identified key facts and equations as well as informative chapter summaries.

Stephen Serjeant is a Reader in Cosmology at The Open University. He led the extragalactic science case of the SCUBA-2 All Sky Survey, and co-led the active galaxies science theme of the ATLAS Key Project on the Herschel Space Observatory. Stephen also coordinates the science faculty's broadcasting at The Open University and is the lead science academic for the BBC1 science show *Bang Goes the Theory*.

Cover image: A multi-wavelength, false-colour view of the M82 galaxy. X-ray data recorded by Chandra appears in blue; infrared light recorded by Spitzer appears in red; Hubble's observations of hydrogen emission appear in orange, and the bluest visible light appears in yellow-green.

Observational Cosmology

Author:

Stephen Serjeant

CAMBRIDGE UNIVERSITY PRESS

Cambridge, New York, Melbourne, Madrid, Cape Town, Singapore, São Paulo, Delhi, Dubai, Tokyo

Cambridge University Press
The Edinburgh Building, Cambridge CB2 8RU, UK

In association with THE OPEN UNIVERSITY

The Open University, Walton Hall, Milton Keynes MK7 6AA, UK

Published in the United States of America by Cambridge University Press, New York.

www.cambridge.org
Information on this title: www.cambridge.org/9780521157155

First published 2010.

Edited and designed by The Open University.

Typeset by The Open University.

Printed and bound in the United Kingdom by Latimer Trend and Company Ltd, Plymouth.

This book forms part of an Open University course S383 *The Relativistic Universe*. Details of this and other Open University courses can be obtained from the Student Registration and Enquiry Service, The Open University, PO Box 197, Milton Keynes MK7 6BJ, United Kingdom: tel. +44 (0)845 300 60 90, email general-enquiries@open.ac.uk

http://www.open.ac.uk

British Library Cataloguing in Publication Data available on request.

Library of Congress Cataloguing in Publication Data available on request.

ISBN 978-0-521-19231-6 Hardback
ISBN 978-0-521-15715-5 Paperback

Additional resources for this publication at www.cambridge.org/9780521157155

1.1

OBSERVATIONAL COSMOLOGY

Introduction

> I have gathered a posie of other men's flowers, and nothing but the thread that binds them is my own.
>
> Montaigne

Observational cosmology is in a tremendously exciting time of rapid discovery. Cosmology can be enriching and enjoyable at this level no matter what your aims are, but my guiding principle for the topics in this book has been: what would I ideally like a person finishing an undergraduate degree and starting a PhD in observational cosmology to know? What would represent a balanced undergraduate introduction that I would like them to have had?

Throughout this book, I've tried to give readers enough grounding to appreciate the current topics in this enormously active and exciting field, and to give some sense of the gaps — and in some cases chasms — in our understanding. I haven't forgotten that the step up to third-level undergraduate study can be difficult and daunting, so I've included further reading sections. Some of the items in these lists will take you to more leisured introductions and backgrounds to some of the material that we shall cover. Nevertheless, this book is intended to be fully self-contained.

I've also given some jumping-off points if readers want to go into more depth. You'll find these mostly in the further reading sections, but also in some footnotes and figure captions. There are some references to journal articles, such as 'Hughes et al., 1998, *Nature*, **394**, 241'. The first number is a volume number, and the second is a page number. Many of these can currently be read online, either in preprint form or as published papers, at http://adsabs.harvard.edu/abstract_service.html. Some of the further reading is most easily found on the internet, but internet addresses are transitory so I've tried to keep these to a minimum. References to arXiv or astro-ph reference numbers are to the preprint server, currently at http://arxiv.org or various worldwide mirrors. Entering the article identification in the search there usually results in the paper. Sometimes the further reading section will point to more advanced material, beyond the normal scope of an undergraduate degree. I've chosen to do this partly in order to ease the transition from undergraduate to postgraduate level for those of you who are on that track. The online abstract service also has the facility to list later papers that have cited any given paper, so it's very useful for literature reviews. At every stage, each level is a big step up from the previous one and the transition can be difficult. I don't intend this book to be a postgraduate textbook, but if I can ease the transition to that level, then all to the good.

Inevitably the selection of topics betrays my own biases and interests, and there are undoubtedly many exciting areas not covered. But the biggest problem is that this is a fast-paced field with lots of exciting and rapid developments. Some future advances are foreseeable, such as gravitational wave astronomy or the Square Kilometre Array, and I can give tasters for what these fabulous new facilities promise, so this book should keep its relevance for a few years at least. As I write this, the Herschel and Planck satellites are waiting to be launched in French Guyana. However, the 'unknown unknowns' I can do nothing about. This is the mixed blessing of writing a book during the golden age of cosmology.

Finally, I would like to thank David Broadhurst, Mattia Negrello, Andrew Norton, Robert Lambourne, Jim Hague and Carolyn Crawford for their critical readings of early drafts of this book. Any errors that I somehow managed to sneak through their careful ministrations are down to me alone. I would also like to thank the editors and artists at The Open University for turning my scribbles into something beautiful.

Chapter 1 Space and time

God does not care about our mathematical difficulties. He integrates empirically.

Albert Einstein

Introduction

How did the Universe begin? How big is the observable Universe? Why is the night sky dark? What will the Universe be like in the year one trillion? What is the ultimate fate of the Universe? This chapter will answer these questions and more, and give you the tools that you need for understanding modern precision cosmology.

Although it's not necessary for you to have met special relativity and the Robertson–Walker metric before, you may find that we take these subjects at a fast pace in this chapter if these are new topics to you. If so, you may find Appendix B on special relativity helpful, or you might try a more comprehensive introduction to expanding spacetime metrics such as that in Robert Lambourne's *Relativity, Gravitation and Cosmology* (see the further reading section).

1.1 Olbers' paradox

In 1823, the Dutch astronomer Heinrich Wilhelm Olbers asked a profound question: if the Universe is infinite, then every line of sight should end on a star, so why isn't the night sky as bright as the Sun? The fact that these stars are further away doesn't help, as we'll show.

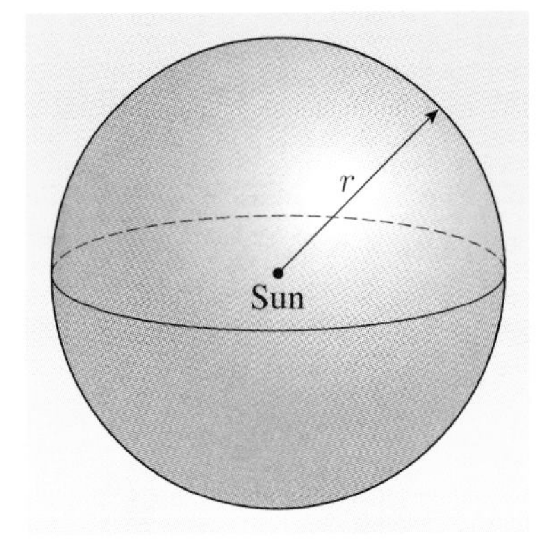

Figure 1.1 A sphere of radius r surrounding the Sun.

At a distance r from the Sun, its light is spread evenly over a sphere with an area $4\pi r^2$, as shown in Figure 1.1. If the Sun has luminosity L, then the energy flux S from the Sun must be

$$S = \frac{L}{4\pi r^2}, \tag{1.1}$$

i.e. $S \propto L/r^2$. Meanwhile, if the diameter of the Sun is D, then the angular diameter θ of the Sun (see Figure 1.2) will be given by

$$\theta \simeq \tan\theta = \frac{D}{r},$$

where the approximation $\theta \simeq \tan\theta$ is valid for small angles measured in radians. Therefore we have

$$\theta \propto \frac{D}{r}.$$

So the Sun's angular area on the sky (in, for example, square degrees or steradians) must be proportional to $(D/r)^2$. Therefore the **surface brightness** (flux per unit area on the sky, e.g. per square degree) is proportional to $(L/r^2)/(D/r)^2 = L/D^2$, which is a constant independent of r. So if all stars are like the Sun (i.e. similar luminosities and diameters), then all stars should have surface brightnesses similar to that of the Sun. If every line of sight ends on a star, then the whole sky should be about as bright as the Sun.

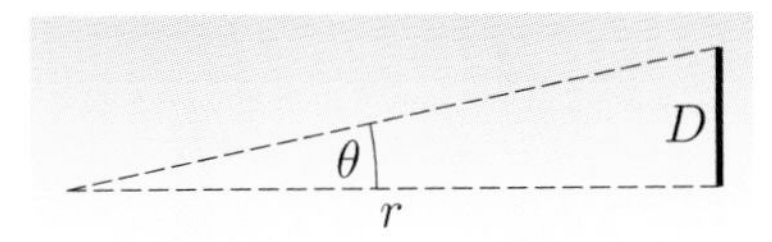

Figure 1.2 The angle θ varies approximately as D/r.

1.2 Olbers' paradox in a different way

Here is a different approach to the same problem. Suppose that there are ρ stars per unit volume, in a Universe that's homogeneous (the same seen from every point) and isotropic (no preferred direction). How many stars have fluxes in the range S to $S + \mathrm{d}S$? (Here, $\mathrm{d}S$ can be thought of as a limitingly-small[1] increment of S, which we use in preference to δS.) First, let's assume for now that all stars are identical, and have luminosity L. Consider a radial shell around the Earth, with radius r and thickness $\mathrm{d}r$ (Figure 1.3). The volume of this shell is the area times the thickness, or $4\pi r^2\,\mathrm{d}r$. (Another way of finding the volume of the shell is to subtract $\frac{4}{3}\pi r^3$ from $\frac{4}{3}\pi(r + \mathrm{d}r)^3$, and neglect terms of the order $(\mathrm{d}r)^2$.) The number of stars in this shell is ρ times the volume of the shell:

$$\mathrm{d}N = \rho \times 4\pi r^2\,\mathrm{d}r \propto \rho r^2\,\mathrm{d}r. \tag{1.2}$$

(We're assuming a flat space for now, known as **Euclidean** space — curved spaces will come later.) The flux S of a star varies with distance according to Equation 1.1, which implies that

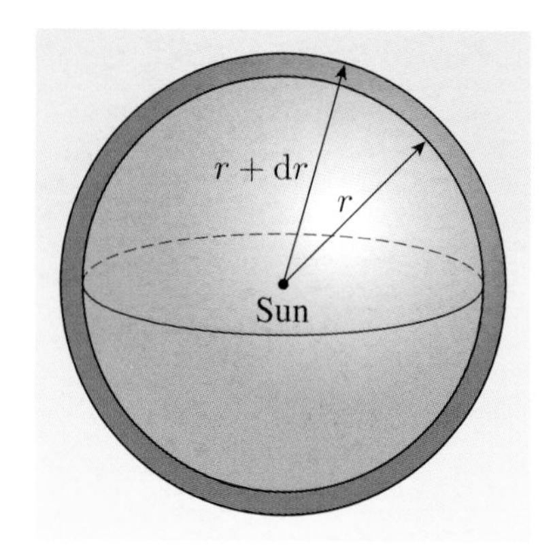

Figure 1.3 A shell of radius r and thickness $\mathrm{d}r$ around the Sun.

$$\frac{\mathrm{d}S}{\mathrm{d}r} \propto Lr^{-3}.$$

So the number of stars with fluxes between S and $S + \mathrm{d}S$ is

$$\begin{aligned}\mathrm{d}N = \frac{\mathrm{d}N}{\mathrm{d}S}\,\mathrm{d}S &= \frac{\mathrm{d}N}{\mathrm{d}r}\,\frac{\mathrm{d}r}{\mathrm{d}S}\,\mathrm{d}S \\ &\propto \rho r^2 \frac{r^3}{L}\,\mathrm{d}S \propto r^5\,\mathrm{d}S.\end{aligned} \tag{1.3}$$

This $\mathrm{d}N$ is the same as in Equation 1.2; all stars have the same luminosity L, so Equation 1.1 gives an exact one-to-one correspondence between S and r, so the interval $(r, r + \mathrm{d}r)$ corresponds exactly to an interval $(S, S + \mathrm{d}S)$.

We can also write the last result as $\mathrm{d}N/\mathrm{d}S \propto r^5$. Now, by rearranging Equation 1.1 we have that

$$r = \left(\frac{L}{4\pi S}\right)^{1/2} \propto S^{-1/2}$$

and therefore

$$\frac{\mathrm{d}N}{\mathrm{d}S}\,\mathrm{d}S \propto S^{-5/2}\,\mathrm{d}S.$$

So we find that

$$\frac{\mathrm{d}N}{\mathrm{d}S} \propto S^{-5/2}. \tag{1.4}$$

Relations of this kind are known in cosmology as **number counts** or **source counts**, and play an important role, as we'll see. $\mathrm{d}N/\mathrm{d}S$ is the number of stars $\mathrm{d}N$ in a flux interval $\mathrm{d}S$, which is a slightly different way of regarding the rate of increase of N with respect to S. The general form $y \propto x^a$ is sometimes

[1] The infinitesimal quantity $\mathrm{d}S$, which we refer to vaguely as a 'limitingly-small' version of δS, can be defined more rigorously using a mathematical discipline called *non-standard analysis*. This takes us beyond the scope of this book, but if you are concerned by manipulating infinitesimals no differently to other algebraic symbols, try, for example, H. Jerome Keisler's book *Elementary Calculus: An Approach Using Infinitesimals*, which is available online.

called a **power law**, with a in this case being the **power law index**. Here, $\mathrm{d}N/\mathrm{d}S$ is a power law function of S, with a power law index of $-5/2$.

Now, we've assumed that all the stars are identical, but suppose instead that there are several types of star, each with a different luminosity L_i and number density ρ_i, with $i = 1, 2, 3, \ldots$. Each type of star will have its own number counts $\mathrm{d}N_i/\mathrm{d}S = k_i S^{-5/2}$, where k_i is some constant specific to type i. The total number counts will still obey a $-5/2$ power law:

$$\frac{\mathrm{d}N}{\mathrm{d}S} = \sum \frac{\mathrm{d}N_i}{\mathrm{d}S} = \sum \left(k_i S^{-5/2}\right) = S^{-5/2} \sum k_i \propto S^{-5/2}.$$

So any homogeneous, isotropic population of stars produces a $-5/2$ power law for number counts. But this leads to a profound problem: the total flux of stars brighter than S_0 is

$$S_{\text{total}} = \int_{S_0}^{\infty} S \frac{\mathrm{d}N}{\mathrm{d}S}\, \mathrm{d}S \propto \int_{S_0}^{\infty} S^{-3/2}\, \mathrm{d}S \propto S_0^{-1/2},$$

which diverges as S_0 tends to zero. So the sky should be infinitely bright!

Exercise 1.1 First, we've argued that a homogeneous, isotropic Universe gives you a sky as bright as the Sun. Next, we've argued that the sky is infinitely bright in a homogeneous, isotropic Universe. They can't both be true, and it's not a mistake in the algebra, so what's different in our assumptions? ■

The night sky is a long way from being as bright as the Sun, and is certainly not infinitely bright. So what is the answer to Olbers' profound question? It's not that the Universe is opaque — in fact, as we shall see later in this book, the Universe is surprisingly transparent at optical wavelengths. It's also not that stars have finite lifetimes, because that doesn't stop lines of sight ending on a star eventually and inevitably.

Part of the answer is that the Universe is only finitely old. Edgar Allan Poe was the first to point out this solution, in his 1848 book *Eureka: a Prose Poem*. But another part of the answer is that we don't live in a static, flat space. Rather, we live in a curved, expanding spacetime, which we shall meet in the next section.

1.3 Metrics: the Universe in a nutshell

We're about to commit possibly the greatest ever act of hubris: to describe the geometry of the Universe in a single equation. To do this, we'll need to remind you of a few preliminaries and notations. Pythagoras's theorem is

$$H^2 = x^2 + y^2$$

for a right-angled triangle with length x, height y and hypotenuse H. In three dimensions this is just

$$H^2 = x^2 + y^2 + z^2.$$

So, if two points in space are separated by $(\delta x, \delta y, \delta z)$, their separation δL is given by

$$(\delta L)^2 = (\delta x)^2 + (\delta y)^2 + (\delta z)^2.$$

In pre-relativistic physics, if one observer measures the separation to be δL, then all observers measure the same δL regardless of how they are moving. This also has all observers agreeing over the passage of time: a separation in time of δt of two events is the same for all observers. (**Event** is used to mean a point in *both* space and time.) The whole aim of fundamental physics is to describe the workings of the Universe in an observer-independent way, so these observer-independent quantities, called **invariants**, often play a central role. This is why it's meaningless to say that the laws of physics are different somewhere else in the Universe: if they're different somewhere else, they weren't fundamental laws in the first place. Many quantities in physics owe almost all their interest to the fact that they are conserved in all (or nearly all) situations: energy, momentum, angular momentum, baryon number, lepton number, strangeness, isospin.

Having said that, there is currently no consistent theory of everything. The best description of the very small, quantum physics, contradicts the best theory of the very large, general relativity. We proceed in the hope that the apparently-invariant quantities discovered so far will lead us closer to the underlying workings of the Universe ...

In Einstein's special relativity, neither spatial separations nor time intervals are invariant, but there is a combined spacetime interval that *is* invariant:

$$(\delta s)^2 = (c\,\delta t)^2 - (\delta x)^2 - (\delta y)^2 - (\delta z)^2, \tag{1.5}$$

where c is the speed of light in a vacuum. The coefficients on the right-hand side (in this case $+1, -1, -1, -1$) are known as the **metric coefficients**. (Note that some books choose to use a $(-1, 1, 1, 1)$ metric instead.) Together, these coefficients make up the **metric tensor** (see Appendix B); we shall discuss tensors later in this book.

The metric of special relativity has many consequences with which you should be familiar, such as time dilation, Lorentz contraction, Lorentz transformations and the non-universality of simultaneity. If you need reminders, Appendix B gives a very brief summary of special relativity. Freely falling particles move on paths for which the total interval s is a maximum along that path, similarly to Fermat's principle in optics. These optimal paths are known as **geodesics**.

Worked Example 1.1

Two ticks of a watch are separated by $(\delta t, 0, 0, 0)$ in the frame of the watch, with $\delta t = 1$ second. The watch is moving at a constant velocity v relative to an observer, for whom the ticks are separated by $(\delta t', \delta x', \delta y', \delta z')$. Using Equation 1.5 or otherwise, show that $\delta t' = \gamma(v)\,\delta t$ with $\gamma = (1 - v^2/c^2)^{-1/2}$, and calculate the spacetime interval δs between the ticks.

If I put the watch on a piece of string and whirl it around my head, and it takes half a second (according to the watch) to go round once, what is the total spacetime interval of the watch's world-line of one orbit? (A world-line is the set of events that trace the path of an object through spacetime.)

Solution

Without loss of generality, we can choose coordinates in which the observer is moving along the x-axis, so $\delta y' = \delta z' = 0$. Now $v = \mathrm{d}x'/\mathrm{d}t'$, and from $\delta s' = \delta s$ we have $(c\,\delta t)^2 = (c\,\delta t')^2 - (\delta x')^2$. If we divide by $(c\,\delta t')^2$, we find

$$\frac{c^2(\delta t)^2}{c^2(\delta t')^2} = 1 - \frac{(\delta x')^2}{c^2(\delta t')^2} = 1 - \frac{(\mathrm{d}x')^2}{c^2(\mathrm{d}t')^2},$$

so

$$\left(\frac{\delta t}{\delta t'}\right)^2 = 1 - \left(\frac{\mathrm{d}x'}{c\,\mathrm{d}t'}\right)^2 = 1 - \frac{v^2}{c^2}$$

hence

$$\delta t' = \gamma(v)\,\delta t.$$

For the second part, $\delta s = c\,\delta\tau$, where τ is the proper time, i.e. the time measured by the watch. Here, we have $\delta\tau = \delta t$, so the total interval is $0.5c$ metres, or 0.5 light-seconds. When the watch is being whirled around, τ is in an accelerating frame, but *it's always true that* $\delta s = c\,\delta\tau$, so the total interval is 0.5 light-seconds.

Worked Example 1.2

What is the total spacetime interval between any two points on a light ray, and how does the interval relate to causality in the Universe? (*Hint*: If events can be connected only by a signal travelling faster than light, they cannot be in causal contact with each other.)

Solution

The spacetime interval between any two points on a light ray is zero. The connection to causality is best illustrated in the lightcone diagram shown in Figure 1.4. An event at the origin can send a message at light speed or slower to any event in the future lightcone. Similarly, any event in its past lightcone could have affected it. The spacetime intervals between the origin and these events are **time-like** intervals, $(\delta s)^2 > 0$. Events outside the lightcone cannot affect, or be affected by, the event at the origin. The spacetime intervals between the origin and these events are **space-like**, i.e. $(\delta s)^2 < 0$. Points on the lightcone have exactly zero spacetime interval between one another. The $\delta s = 0$ intervals are sometimes referred to as **null**.

Exercise 1.2 The highest-energy cosmic rays have energies of 10^{20} eV and above. Most cosmic rays are protons, with rest masses of $938.28\,\mathrm{MeV}/c^2$. The diameter of our Galaxy is roughly 100 000 light-years. Calculate how long it would take to cross the Galaxy, according to the highest-energy cosmic rays. (*Hint*: You don't need to know the conversion between eV and joules, nor do you need the conversion between light-years and metres.) ■

We can also describe the invariant spacetime interval in spherical coordinates, in infinitesimals:

$$ds^2 = (c\,dt)^2 - dr^2 - (r\,d\theta)^2 - (r\sin\theta\,d\phi)^2,$$

where ds^2 means $(ds)^2$. These coordinates are shown in Figure 1.5.

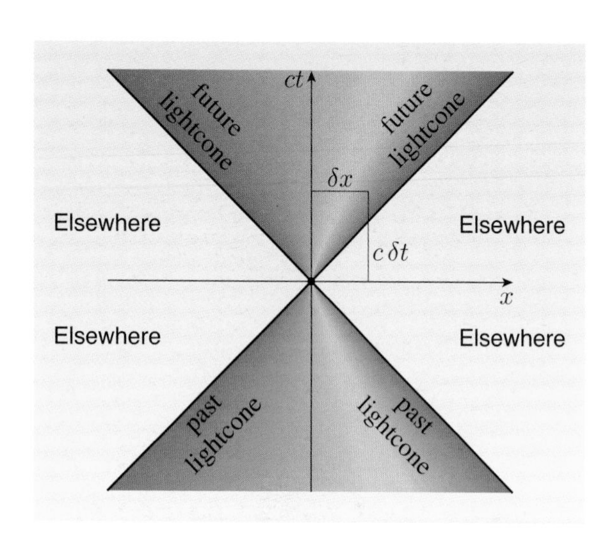

Figure 1.4 The lightcone in special relativity. The point at position $(\delta x, c\,\delta t)$ has $c\,\delta t > \delta x$, so $(c\,\delta t)^2 - (\delta x)^2 > 0$, implying that $(\delta s)^2 > 0$, meaning that the invariant interval between that point and the origin is time-like.

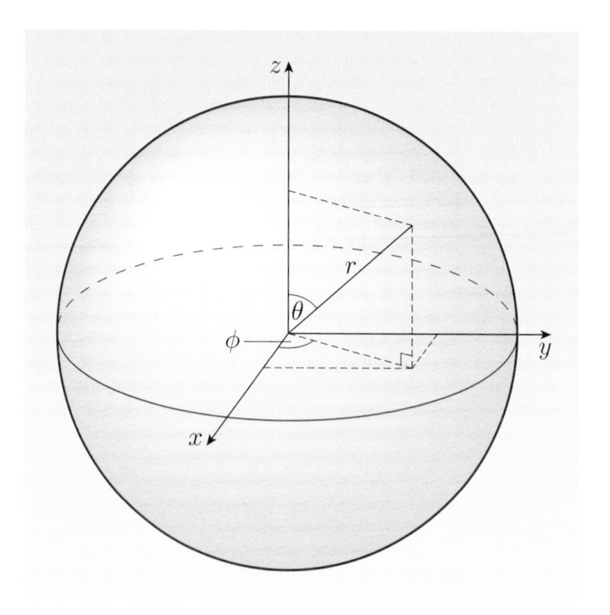

Figure 1.5 The position of a point can be specified in terms of Cartesian coordinates (x, y, z) or in terms of spherical coordinates (r, θ, ϕ).

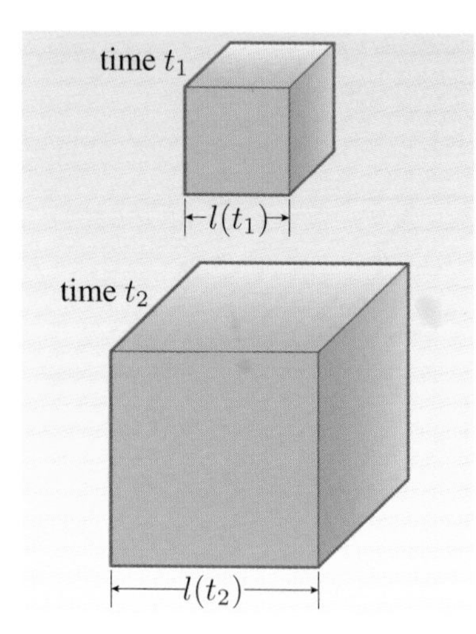

Figure 1.6 A cubical volume of the Universe. The length of the side l expands with the scale factor of the Universe, so $l(t_2) = (R(t_2)/R(t_1))\,l(t_1)$.

To describe an expanding Universe, we could modify the metric by multiplying the spatial parts with a time-dependent expansion factor:

$$ds^2 = c^2\,dt^2 - R(t)\left(dr^2 + r^2\,d\theta^2 + r^2\sin^2\theta\,d\phi^2\right),$$

where $R(t)$ is called the **scale factor** of the Universe. A schematic representation of this is shown in Figure 1.6.

In fact, the most general homogeneous, isotropic metric is

$$ds^2 = c^2\,dt^2 - R^2(t)\left(\frac{dr^2}{1-kr^2} + r^2\,d\theta^2 + r^2\sin^2\theta\,d\phi^2\right), \tag{1.6}$$

where the constant k determines whether the Universe is spatially flat ($k = 0$), spherical ($k = +1$) or hyperbolic ($k = -1$). (We use only these three values of k because other values can be found by rescaling R and r; for example, if $k = -3$, then the substitutions $r' = r\sqrt{3}$ and $R' = R/\sqrt{3}$ give an equation of the same form as Equation 1.6 for $k = -1$.) Figure 1.7 illustrates some two-dimensional surfaces in which $k = +1$, 0 or -1, to give you some intuition (if not actually a visualization) of the three-dimensional counterparts. You might reasonably object that these two-dimensional representations oversimplify the situation. In physics (and general relativity especially) it's often easier to describe something mathematically than it is to visualize it; physics makes tremendous demands on the imagination. Perhaps the human brain doesn't have the cognitive machinery to be able to visualize curved expanding spacetimes.

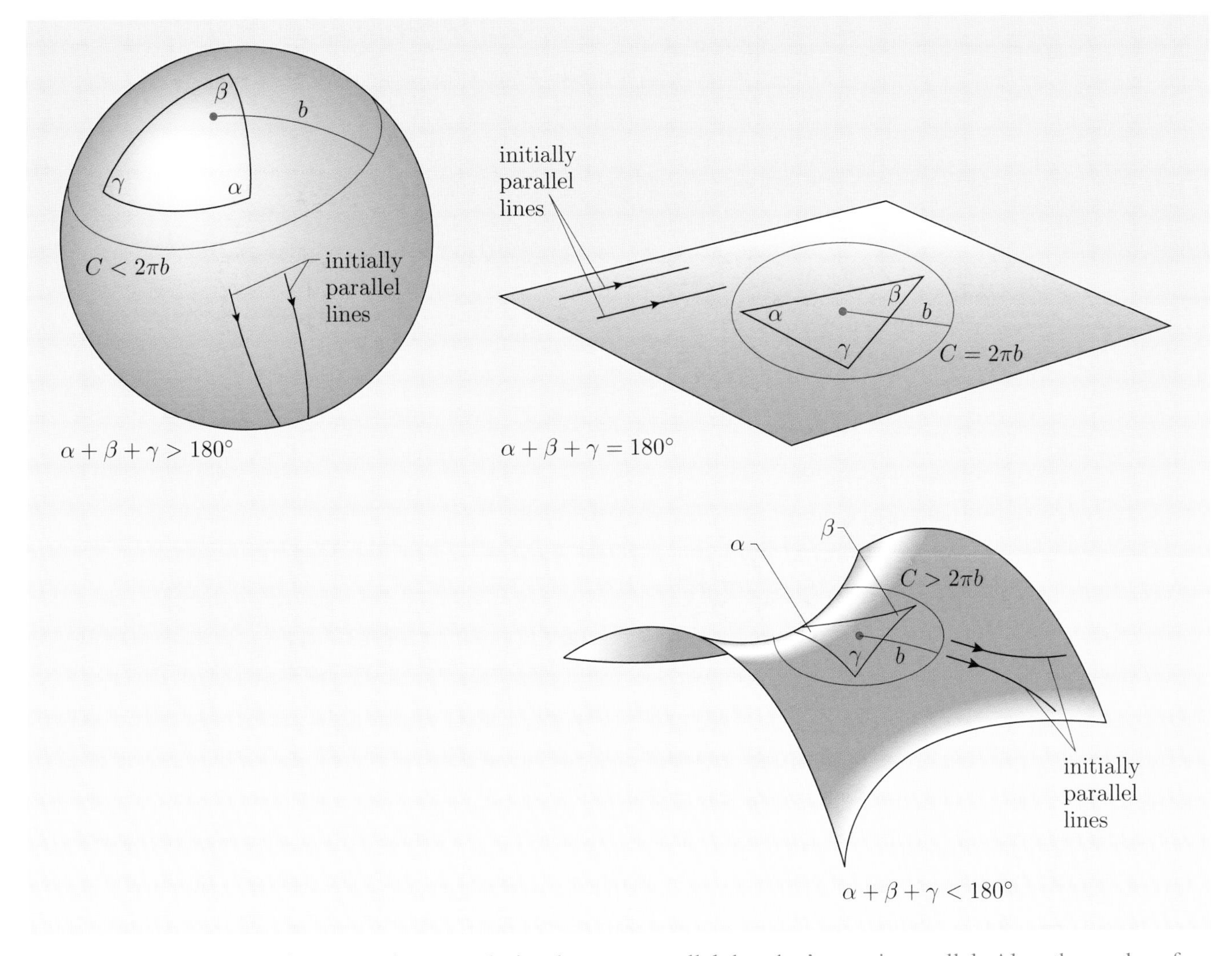

Figure 1.7 Curved surfaces may have geodesics that start parallel, but don't remain parallel. Also, the angles of a triangle need not add up to $180°$, nor is the circumference of a circle necessarily 2π times the radius. The spherical model has $k = +1$, the flat model has $k = 0$ and the saddle-shaped model has $k = -1$.

Equation 1.6 is known as the **Robertson–Walker metric** or sometimes as the **Friedmann–Robertson–Walker metric**. Because this metric is spatially homogeneous and isotropic, any point can be chosen for the origin $r = 0$. We sometimes refer to t as coordinate time or **cosmic time**. This metric has preferred inertial reference frames in which the expansion of the Universe is isotropic. It's in such a frame that the coordinates t, r, θ, ϕ are measured. We'll refer to this reference frame as the **cosmic rest frame**, and it's assumed to be equivalent to the frame of the cosmic microwave background, of which more later. If we took an observer in the cosmic rest frame and applied a velocity boost (a Lorentz transformation) to see what the expanding Universe would look like from a moving observer's point of view, we'd find that it no longer looked isotropic.

Equation 1.6 is our hubristic attempt to describe the Universe in one line, and we shall meet this equation many times in this book.

It may now be worth discussing some frequent misconceptions about the Robertson–Walker metric.

- If the Universe is expanding, why am I not getting taller?
- Your head is not free-floating from your feet; your body is bound by chemical bonds.
- Why does the Earth not drift away from the Sun as the Universe expands?
- Equation 1.6 describes a Universe that is exactly homogeneous and isotropic. Locally, though, that's obviously not right. Within the Solar System the local metric is not the Robertson–Walker metric, because the gravitational field of the Sun dominates. Equation 1.6 is an approximation that becomes increasingly good at larger and larger scales, but on small scales spacetime clearly has much more structure. When we consider the collapse of density perturbations in Chapter 4, we'll see that it's wrong to think of the Earth feeling a gentle tug from the expansion of the Universe, which is overwhelmed by the attraction from the Sun.
- What is the Universe expanding into?
- In general relativity, the intrinsic curvature of spacetime (including the expansion of space) can be specified using measurements *within* the spacetime, using geodesics. This means that if we want to describe the curvature of spacetime, we don't need to embed spacetime in a higher-dimensional space. But if we don't need to refer to higher-dimensional space, and it doesn't affect anything that we can measure, do we need to hypothesize its existence at all? In any case, we would have no reason to believe that this higher-dimensional space is flat anyway. So, we don't have evidence that the Universe is expanding 'into' anything at all — it is simply expanding.
- Where is the middle of the Universe, which everything exploded out of?
- This is a widely-held misconception that perhaps dates back to Lemaître's phrase 'the primeval atom'. We frequently illustrate the expansion of space with an analogy of an expanding balloon (such as in Figure 1.8), but what is rarely pointed out is that the radial coordinate is *time-like*, not space-like. The explosion started at the centre, but this was at the beginning of *time* in the Robertson–Walker universe, not in a particular location. In fact, one might reasonably say that it occurred everywhere, at every point in space. Also, if the Universe is hyperbolic ($k = -1$), then the Universe may be infinite. If so, then even in the earliest moments of the history of the Universe, there would still be infinitely more matter outside any given volume than inside that volume.

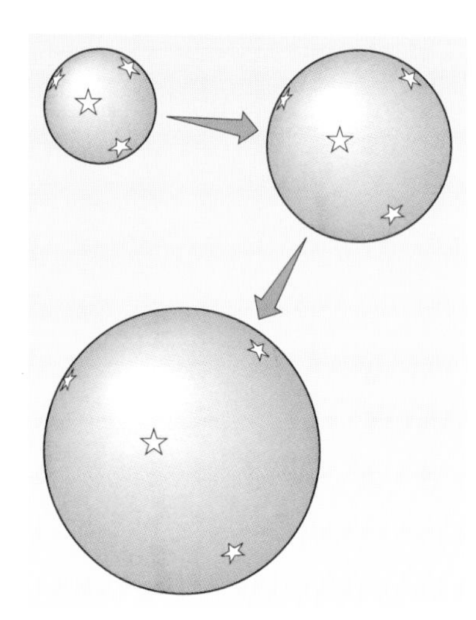

Figure 1.8 The balloon metaphor for the expanding Universe. Note that the objects on the balloon don't themselves expand.

- What caused the Universe to fling itself apart in the Big Bang?
- This question suggests an input of energy, flinging matter from an initial state of rest, but this is not how the field equations of general relativity are formulated. Classically at least, the answer is that these are just the initial conditions to Einstein's field equations. But just saying 'that's how it started' is clearly a very unsatisfactory answer, and arguably avoiding the question. One attraction of the theory of inflation is that it gives a mechanism for this initial expansion; we shall meet inflation in later chapters.

Note that Equation 1.6 defines a preferred reference frame, in which the expansion is isotropic. As we shall see, this is well-supported by observations both of the large-scale structure of the galaxy distribution, and of the cosmic microwave background. Nevertheless, is it possible to conceive of a universe consistent with Einstein's field equations in which there are no preferred reference frames? One possibility is a fractal structure, and we shall meet this in later chapters when discussing inflation.

The field equations of Einstein's general relativity determine both k and $R(t)$. These equations can be shown to yield

$$\left(\frac{\mathrm{d}R}{\mathrm{d}t}\right)^2 = \dot{R}^2 = \frac{8\pi G(\rho_\mathrm{m} + \rho_\mathrm{r})R^2}{3} - kc^2 + \frac{\Lambda c^2 R^2}{3}, \tag{1.7}$$

$$\frac{\mathrm{d}^2 R}{\mathrm{d}t^2} = \frac{\mathrm{d}}{\mathrm{d}t}\left(\frac{\mathrm{d}R}{\mathrm{d}t}\right) = \ddot{R} = -4\pi G\left(\rho_\mathrm{m} + \rho_\mathrm{r} + \frac{3p}{c^2}\right)\frac{R}{3} + \frac{\Lambda c^2 R}{3}, \tag{1.8}$$

Proving these equations would take us a long way outside the scope of this book into what is usually graduate-level physics, but if you wish to pursue this rewarding path you might try, for example, *Relativity, Gravitation and Cosmology* by R. Lambourne for an advanced undergraduate-level introduction, or *General Relativity: An Introduction for Physicists* by M.P. Hobson, G.P. Efstathiou and A.N. Lasenby.

where ρ_m is the average matter density of the matter in the Universe, ρ_r is an equivalent matter density for radiation (derived using $E = mc^2$), G is Newton's gravitational constant, p is the pressure of the matter and radiation, and R is a function of time, $R = R(t)$, though we drop the function notation for clarity and brevity. These equations are known as the **Friedmann equations**. Both the densities and p also vary with time. Λ is known as the **cosmological constant**, and features in Einstein's field equations for general relativity. Physically, it represents an in-built tendency of space to expand (or, for $\Lambda < 0$, contract). Some special cases are fairly simple: for example, if $k = \Lambda = 0$, then $R(t) \propto t^{2/3}$ in a matter-dominated universe, or $R(t) \propto t^{1/2}$ in a radiation-dominated universe.

We can derive Equation 1.8 from Equation 1.7 by differentiation. This will give us a term involving $\mathrm{d}(\rho_\mathrm{m} + \rho_\mathrm{r})/\mathrm{d}t$, but we could treat a part of the Universe as a box of gas, and use the conservation of energy to show that the change in energy density equals the $p\,\mathrm{d}V$ work, i.e. $\mathrm{d}((\rho_\mathrm{m} + \rho_\mathrm{r})c^2 R^3) = -p\,\mathrm{d}(R^3)$. Therefore $\mathrm{d}((\rho_\mathrm{m} + \rho_\mathrm{r})c^2 R^3)/\mathrm{d}t = -p\,\mathrm{d}(R^3)/\mathrm{d}t$.

Exercise 1.3 Derive Equation 1.8 from Equation 1.7, using the conservation of energy. ■

(The issue of $p\,\mathrm{d}V$ work is slightly more subtle in general relativity, since it's not immediately clear what the work is done against, but the full relativistic calculation gives the same result.)

In the next few sections, we shall explore some of the surprising aspects of this expanding spacetime model, before returning to Olbers' profound paradox in the next chapter.

1.4 Redshift and time dilation

Imagine two photons being emitted at an epoch t_1 when the scale factor of the Universe was R_1. Suppose also that the photons were emitted a short time δt apart. They arrive at the Earth now at time t_0 when the scale factor is R_0. When the photons were emitted at t_1, the distance between them was $c\,\delta t$. Now, the distance between them is $(R_0/R_1)c\,\delta t$, i.e. stretched by a factor R_0/R_1 due to the expansion of the Universe. The second photon will arrive $(R_0/R_1)\,\delta t$ later than the first. This implies that distant clocks in the Robertson–Walker universe appear time-dilated by a factor R_0/R_1, sometimes called **cosmological time dilation**. We'll find it useful to define the **dimensionless scale factor** a as

$$a = \frac{R_1}{R_0}, \tag{1.9}$$

so $a = 1$ today and $a < 1$ in the past.

A similar argument applies to the photons themselves. Treating them this time as waves, the distance between two peaks of a light wave will be expanded by the same factor R_0/R_1, i.e. the wavelength is longer, and the light is shifted to the red. We define **redshift** (symbol z) using

$$1 + z = \frac{R_0}{R_1} = \frac{1}{a} = \frac{\lambda_{\text{observed}}}{\lambda_{\text{emitted}}}, \tag{1.10}$$

where $\lambda_{\text{observed}}$ is the observed wavelength of the photon, and λ_{emitted} is the original photon wavelength when the light was emitted. Sometimes this is written as

$$z = \frac{\lambda_{\text{observed}} - \lambda_{\text{emitted}}}{\lambda_{\text{emitted}}}. \tag{1.11}$$

A high redshift means that there has been a big increase in the expansion factor since the light was emitted. Redshift is sometimes misleadingly referred to as 'recession', since a receding object would have a Doppler shift to the red. Indeed, galaxies are not stationary relative to each other, but have relative velocities up to even $1000\,\text{km}\,\text{s}^{-1}$. In astronomy these are usually known as **peculiar velocities**, and these will indeed contribute both blue and red Doppler shifts. However, cosmological redshift swamps these effects at distances beyond about $100\,\text{Mpc}$, and you should not confuse Doppler shifts with the redshift from cosmological expansion. The distance between us and a distant galaxy is getting bigger because of the expansion of the Universe, but this is a physically distinct situation to a galaxy moving away in a flat, non-expanding spacetime.

One alternative to the Robertson–Walker model is the 'tired light' universe, proposed by Fritz Zwicky in 1929. In this model, redshift is due to photons gradually losing energy during their passage through the universe, due to some interaction with intervening matter. There are many observations that are difficult to reproduce in this model, but in particular, the experimental detection of cosmological time dilation has made this interpretation untenable. Figure 1.9 shows the decay times of supernovae as a function of redshift, which show exactly the $(1 + z)$ time dilation predicted by theory.

But to measure redshifts, we need to know λ_{emitted}. This can be done using atomic or molecular transitions that occur at particular quantized energies, and so involve the emission or absorption of photons with particular quantized wavelengths. If we can identify the transition in the distant object, we know λ_{emitted}, provided that atoms behaved in the same way in the early Universe.

But did they? And if not, how could we tell? It turns out that many characteristic atomic and molecular transitions can easily be recognized at high redshifts (see, for example, Figure 1.10), so any differences must be fairly subtle. If the strength of the electromagnetic interaction were different at early cosmic epochs, this would change the atomic fine structure constant $\alpha = e^2/(4\pi\varepsilon_0\hbar c) \simeq 1/137$. The fractional difference in wavelength $(\delta\lambda/\lambda)$ between a pair of relativistic fine structure lines is proportional to α^2, so changes in α should lead to wavelength shifts between some emission lines in distant cosmological objects.

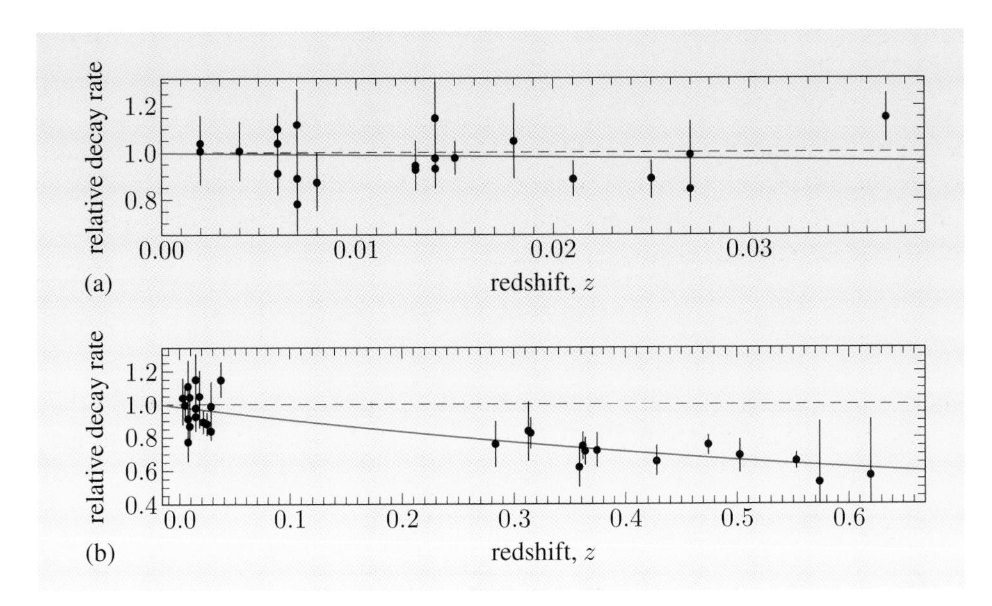

Figure 1.9 Supernova decay rates in the nearby Universe (top), and in the high-redshift Universe (bottom). The dashed line shows the tired light prediction of no time dilation, and the red line shows the $(1+z)^{-1}$ time dilation expected in the Robertson–Walker metric. The high-redshift data strongly support the expanding universe model; the measured variation is $(1+z)^{-0.97\pm0.10}$.

So far, comparisons of the atomic and molecular transitions in the early Universe with laboratory experiments have not yielded any uncontested evidence for α being any different in the early Universe; some claimed detections of changes in α have not been corroborated by other experiments, and it is clear that the experiments are both very difficult and prone to systematic errors. In terrestrial laboratories, $\dot{\alpha}/\alpha = (-2.6 \pm 3.9) \times 10^{-16}$ per year, i.e. consistent with no change. However, it remains possible that ongoing cosmological experiments will make a ground-breaking detection of a change in α. Some speculative theories allow for possible changes in α, such as supersymmetry or M-theory. However, these theories do not (or don't yet) predict the specific variations in α with redshift that some groups have claimed, in any unique and unforced way.

1.5 Cosmological parameters

How fast is the Universe expanding?

Suppose that the distance between us and a distant galaxy is $\ell = D \times R$, where R is the current scale factor of the Universe, and D is some constant. By differentiating this, we find that ℓ is increasing at a rate

$$\frac{\mathrm{d}\ell}{\mathrm{d}t} = D\,\frac{\mathrm{d}R}{\mathrm{d}t}$$

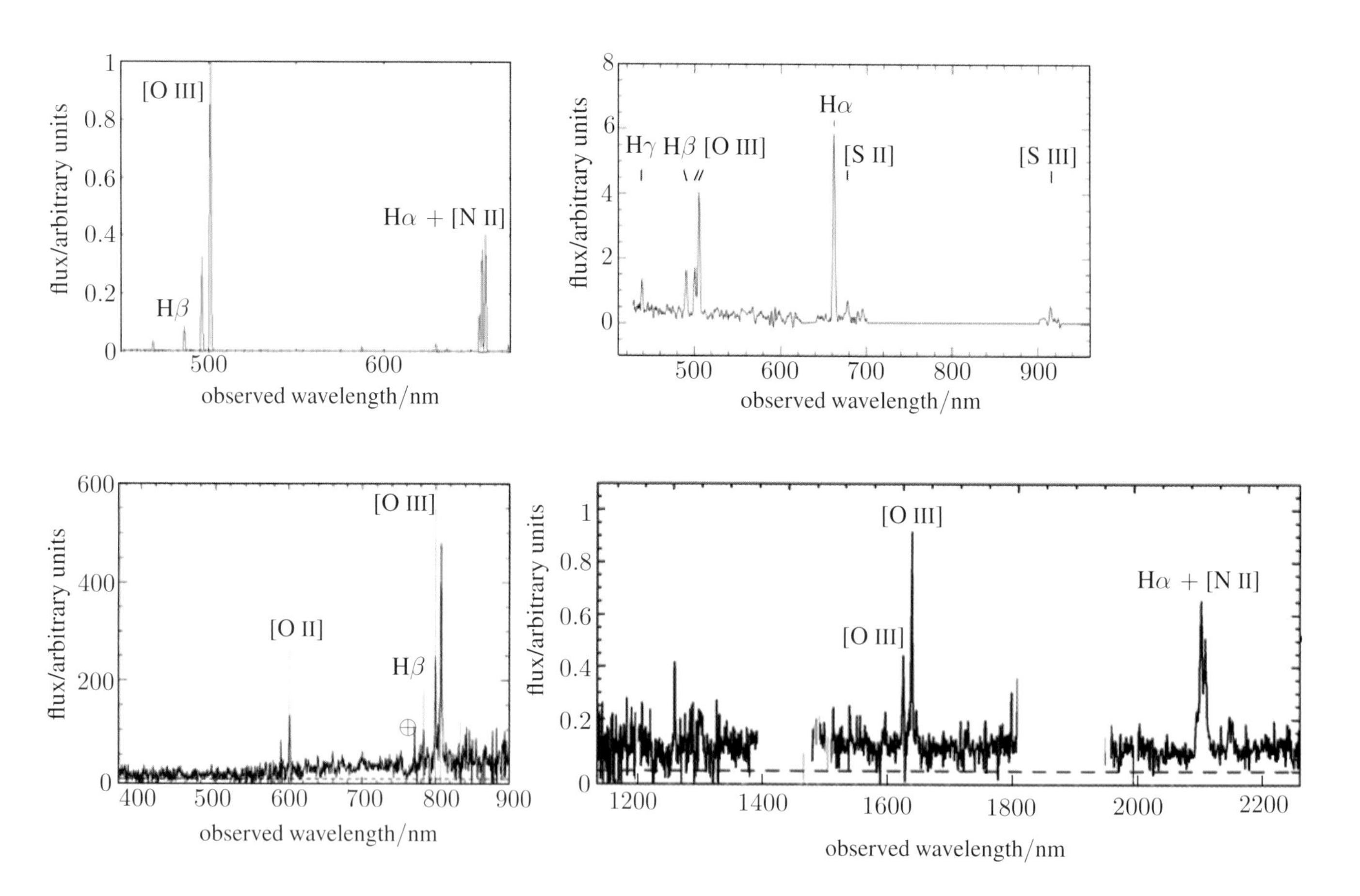

Figure 1.10 Spectra of objects in the local Universe and in the high-redshift Universe, showing many of the same characteristic spectral features. The y-axes in all the spectra are relative flux. The top left panel is a planetary nebula in our Galaxy, M57. The top right panel is an H II region in the Virgo cluster of galaxies. The bottom left panel shows the spectrum of a star-forming galaxy at a redshift of $z = 0.612$ (the $\oplus$ symbol marks absorption from the Earth's atmosphere), and the bottom right panel shows another star-forming galaxy at a redshift of $z = 2.225$. In all cases there is the characteristic [O III] emission line doublet at a rest-frame wavelength of 495.9, 500.7 nm, as well as other emission lines such as Hα 656.3 nm, Hβ 486.1 nm, [N II] 654.8, 658.4 nm. The bottom right panel has the emission lines redshifted into the near-infrared range, in which only certain regions of the spectrum are available, for reasons of atmospheric transparency.

because of the expansion of the Universe. But $D = \ell/R$, so

$$\frac{\mathrm{d}\ell}{\mathrm{d}t} = \frac{\ell}{R}\frac{\mathrm{d}R}{\mathrm{d}t} = \ell \times \left(\frac{1}{R}\frac{\mathrm{d}R}{\mathrm{d}t}\right) = \ell \times H,$$

where

$$H = \dot{R}/R \tag{1.12}$$

is known as the **Hubble parameter**, whose current value is known as H_0. If we regard $\mathrm{d}\ell/\mathrm{d}t$ as an *apparent* recession velocity v, then we have

$$v = \ell \times H, \tag{1.13}$$

i.e. the apparent recession velocity v is proportional to distance ℓ, but recall the warnings in Section 1.4. Sometimes this apparent flow is called the **Hubble flow**.

Beware: H is frequently (but misleadingly) known as the **Hubble constant** (Hubble constant is a fair description of H_0, however). Although it's virtually constant over our lifetimes, it certainly isn't constant over the history of the Universe. In some sense, H_0 is a measure of the current expansion rate of the Universe, and it has the value $72 \pm 3\,\mathrm{km\,s^{-1}\,Mpc^{-1}}$, or about $2 \times 10^{-18}\,\mathrm{s^{-1}}$. This is also sometimes written as $H_0 = 100h\,\mathrm{km\,s^{-1}\,Mpc^{-1}}$, with $h = 0.72 \pm 0.03$. This may seem deliberately obtuse, but the Hubble parameter is so fundamental that it affects many other cosmological measurements, so some observational cosmologists opt to quote their results in terms of h.

If we divide Equation 1.7 by R^2, we obtain

$$H^2 = \left(\frac{\dot{R}}{R}\right)^2 = \frac{8\pi G(\rho_\mathrm{m} + \rho_\mathrm{r})}{3} + \frac{\Lambda c^2}{3} - \frac{kc^2}{R^2}. \tag{1.14}$$

The terms on the right-hand side drive the expansion of the Universe. It's common in cosmology to define their fractional contributions:

$$\Omega_\mathrm{m} = \frac{8\pi G\rho_\mathrm{m}}{3H^2}, \tag{1.15}$$

$$\Omega_\mathrm{r} = \frac{8\pi G\rho_\mathrm{r}}{3H^2}, \tag{1.16}$$

$$\Omega_\Lambda = \frac{\Lambda c^2}{3H^2}, \tag{1.17}$$

$$\Omega_k = \frac{-kc^2}{R^2H^2}, \tag{1.18}$$

where the subscript 'm' stands for 'matter' and the subscript 'r' stands for 'radiation'. Equation 1.14 then implies that

$$\Omega_\mathrm{m} + \Omega_\mathrm{r} + \Omega_\Lambda + \Omega_k = 1. \tag{1.19}$$

We can also define $\Omega_\mathrm{total} = \Omega_\mathrm{m} + \Omega_\mathrm{r} + \Omega_\Lambda = 1 - \Omega_k$. All these Ω parameters, known as **density parameters**, are functions of time, except in a few special cases. Figure 1.11 shows how the density parameters varied with the scale factor of the Universe. As with the Hubble parameter, we shall use a subscript 0 for the present-day value, e.g. $\Omega_{\Lambda,0} = \Lambda c^2/(3H_0^2)$. However, be warned that many textbooks omit 0 subscripts for the present-day Ω values.

Historically, another notation has been used, which is now out of favour: $q_0 = -(R\ddot{R}/\dot{R}^2)_0 = \frac{1}{2}\Omega_{\mathrm{m},0} - \Omega_{\Lambda,0}$ and $\sigma_0 = \Omega_{\mathrm{m},0}/2$. Besides mentioning them here, we won't use this notation in this book.

It's also useful to define a critical density ρ_crit such that

$$\Omega_\mathrm{m} = \frac{\rho_\mathrm{m}}{\rho_\mathrm{crit}}, \tag{1.20}$$

so that

$$\rho_\mathrm{crit} = \frac{3H^2}{8\pi G}, \tag{1.21}$$

(we'll explain below why this is 'critical').

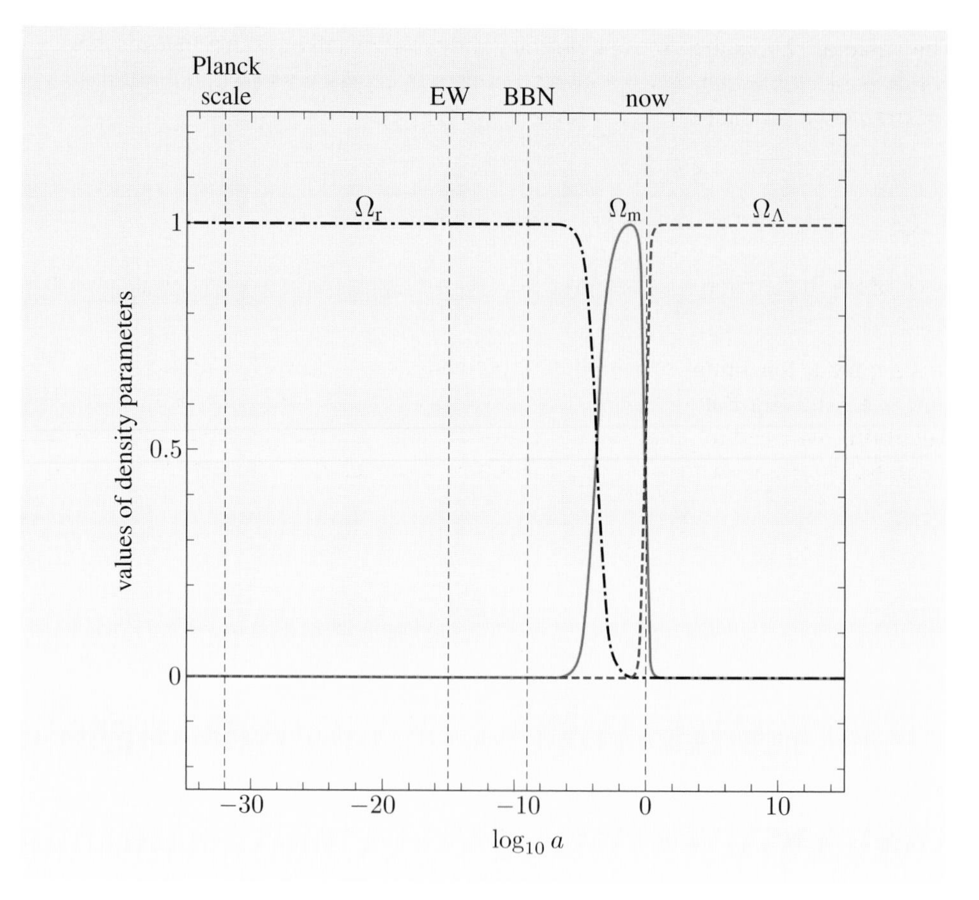

Figure 1.11 Past and future variations in the density parameters given the present-day WMAP cosmological parameters, with dimensionless scale factor $a = R/R_0$. Some key times are marked: the Planck scale, the epoch of electroweak symmetry breaking (EW), Big Bang nucleosynthesis (BBN) and the present.

Matter densities are often expressed relative to this critical density. For example, the baryon density of the Universe, ρ_b, is sometimes written as

$$\Omega_b = \frac{\rho_b}{\rho_{crit}}. \tag{1.22}$$

The matter density of the Universe can be expressed as

$$\begin{aligned}\rho_m &= \Omega_m\,\rho_{crit} = 1.8789 \times 10^{-26}\,\Omega_m h^2\,\text{kg m}^{-3} \\ &= 2.7752 \times 10^{11}\,\Omega_m h^2\,\text{M}_\odot\,\text{Mpc}^{-3},\end{aligned} \tag{1.23}$$

where $1\,\text{M}_\odot$ is the mass of the Sun.

We shall see in later chapters that most of the matter content of the Universe is **dark matter** that neither absorbs nor emits light. Dark matter appears to interact only (or very nearly only) through gravitation, and its only observational consequences so far have been via its gravitational effects.

The current experimental values from the WMAP satellite (which we shall meet later) are

$$\Omega_{m,0}\,h^2 = 0.1326 \pm 0.0063, \tag{1.24}$$

$$\Omega_{\Lambda,0} = 0.742 \pm 0.030, \tag{1.25}$$

$$\Omega_{b,0}\,h^2 = (2.273 \pm 0.062) \times 10^{-2}. \tag{1.26}$$

We'll show in Chapter 2 that the contribution from radiation and neutrinos gives

$$\Omega_{r,0}\,h^2 \simeq 4.2 \times 10^{-5}. \tag{1.27}$$

The value of $\Omega_{r,0}$ is therefore negligible, and we'll usually assume that it's zero in this book. Note that WMAP doesn't constrain $\Omega_{m,0}$ on its own, but rather constrains the product of $\Omega_{m,0}$ with the Hubble parameter squared.

1.6 The age of the Universe

How old is the Universe? And how long does it take light from distant galaxies to reach us?

It is a quite astonishing feat of modern precision cosmology that we know the time since the Big Bang to better than a few per cent accuracy. To see how this is calculated, we need to relate the redshift z to the age of the Universe at that epoch, using only present-day observable quantities. The Hubble parameter H is closely related to $\mathrm{d}z/\mathrm{d}t$:

$$\begin{aligned} H = \frac{1}{R}\frac{\mathrm{d}R}{\mathrm{d}t} = \frac{R_0}{R}\frac{\mathrm{d}(R/R_0)}{\mathrm{d}t} &= (1+z)\,\frac{\mathrm{d}(1/(1+z))}{\mathrm{d}t} \\ &= (1+z)\,\frac{\mathrm{d}(1/(1+z))}{\mathrm{d}z}\frac{\mathrm{d}z}{\mathrm{d}t} \\ &= \frac{-1}{1+z}\frac{\mathrm{d}z}{\mathrm{d}t}. \end{aligned} \quad (1.28)$$

Next, we re-cast Equation 1.14 in terms of $-kc^2$ (assuming $\Omega_r = 0$):

$$-kc^2 = H^2R^2 - \frac{8\pi G\rho_m R^2}{3} - \frac{\Lambda c^2 R^2}{3}. \quad (1.29)$$

In particular, at the present time we have

$$-kc^2 = H_0^2R_0^2 - \frac{8\pi G\rho_{m,0} R_0^2}{3} - \frac{\Lambda c^2 R_0^2}{3}, \quad (1.30)$$

where $\rho_{m,0}$ is the present-day matter density. Clearly, the right-hand sides of Equations 1.29 and 1.30 must be equal. But $\rho_m = \rho_{m,0} \times R_0^3/R^3$, so

$$H^2R^2 - \frac{8\pi G\rho_{m,0} R_0^3}{3R} - \frac{\Lambda c^2 R^2}{3} = H_0^2R_0^2 - \frac{8\pi G\rho_{m,0} R_0^2}{3} - \frac{\Lambda c^2 R_0^2}{3}.$$

We can express this in terms of the present-day density parameters $\Omega_{m,0}$ and $\Omega_{\Lambda,0}$ using Equations 1.15 and 1.17. After rearranging, this gives

$$\left(\frac{H}{H_0}\right)^2 = \frac{R_0^2}{R^2} + \Omega_{m,0}\left(\frac{R_0^3}{R^3} - \frac{R_0^2}{R^2}\right) + \Omega_{\Lambda,0}\left(1 - \frac{R_0^2}{R^2}\right). \quad (1.31)$$

We can simplify this using $1 + z = R_0/R$, which gives

$$\begin{aligned} \left(\frac{H}{H_0}\right)^2 = (1+z)^2 &+ \Omega_{m,0}\left\{(1+z)^3 - (1+z)^2\right\} \\ &+ \Omega_{\Lambda,0}\left\{1 - (1+z)^2\right\}, \end{aligned} \quad (1.32)$$

which can be rearranged to give

$$\left(\frac{H}{H_0}\right)^2 = (1+z)^2(1 + z\,\Omega_{m,0}) - z(2+z)\,\Omega_{\Lambda,0}. \quad (1.33)$$

Finally, using Equation 1.28, we reach

$$\left(\frac{\mathrm{d}z}{\mathrm{d}t}\right)^2 = H_0^2(1+z)^2\left\{(1+z)^2(1+z\,\Omega_{\mathrm{m},0}) - z(2+z)\,\Omega_{\Lambda,0}\right\}, \quad (1.34)$$

from which we can easily find $\mathrm{d}t/\mathrm{d}z$. Although admittedly pretty ghastly, this equation does have the advantage of using only present-day observable quantities, and we'll be referring to it several times in this book.

In general, $\mathrm{d}t/\mathrm{d}z$ can't be integrated analytically, so $t(z)$ can be calculated only by numerically integrating $\mathrm{d}t/\mathrm{d}z$. The time to $z = \infty$ is the age of the Universe, and this is shown in Figure 1.12. Equation 1.34 can also be integrated to give the time taken for light to reach us from redshift z. This is known as the **lookback time**.

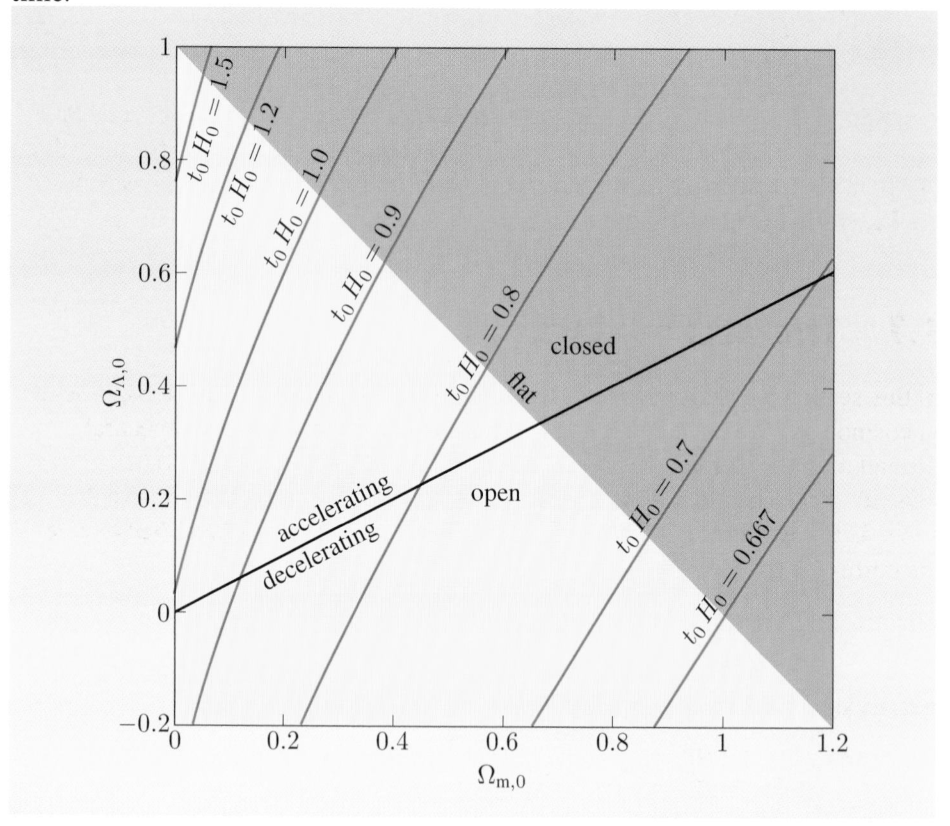

Figure 1.12 The age of the Universe times the Hubble parameter, for several cosmological models. Also shown is the spatial geometry (open, flat and closed) and whether the present-day expansion of the Universe would be accelerating or decelerating.

How does this estimate of the age of the Universe compare to the ages of the oldest objects in the Universe? There is now a well-developed theory for main sequence stellar evolution that can be used to find the ages of stars. Particularly useful are globular clusters (e.g. Figure 1.13), which are some of the oldest gravitationally-bound objects in the Universe. The stars that comprise any given globular cluster are believed to have formed at about the same time (though the ages of globular clusters vary). More luminous stars spend less time on the main sequence in the colour–magnitude diagram, so if one can find the luminosity of the stars in a globular cluster that are just leaving the main sequence, one can infer an age for the globular cluster.

The oldest known globular cluster appears to be 12.7 ± 0.7 Gyr old. In the 1990s it was recognized that globular cluster ages are an important constraint on the age of the Universe, and therefore on the cosmological parameters that control the geometry and fate of the Universe (Figure 1.12). But as we shall see in the next section, there seemed to be very good reasons to expect that we live in a Universe with $\Omega_{\mathrm{m},0} = 1$ and $\Lambda = 0$, which as you will see turns out to be significantly younger, so these stars appeared to be older than the Universe.

Figure 1.13 The globular star cluster M80. Most of its stars are older and redder than our Sun.

Exercise 1.4 Starting from Equation 1.33, or otherwise, show that in an $\Omega_{\mathrm{m}} = 1$, $\Lambda = 0$ universe,

$$\frac{R}{R_0} = \left(\frac{t}{t_0}\right)^{2/3}. \tag{1.35}$$

Exercise 1.5 Using Equation 1.35, show that the age of the Universe in an $\Omega_{\mathrm{m}} = 1$, $\Lambda = 0$ model is $t_0 = 2/(3H_0)$, and evaluate the age in Gyr for the value of H_0 in Section 1.5. A spacetime that expands in this way is sometimes called the **Einstein–de Sitter model**. ■

1.7 The flatness problem

In this section we shall introduce you to a profound, and as yet unsolved, problem in cosmology. We have already noted that the density parameters in general depend on time. For example, what was Ω_{m} at a redshift of $z = 1000$ (about the redshift of the cosmic microwave background)? Let's assume for now that $\Lambda = 0$, so Ω_Λ is always zero, and $\Omega_{\mathrm{m}} = 1 - \Omega_k$. Equation 1.18 can be modified to give the current value of Ω_k,

$$\Omega_{k,0} = \frac{-kc^2}{R_0^2 H_0^2},$$

and dividing this into Equation 1.18 gives

$$\frac{\Omega_k(z)}{\Omega_{k,0}} = \frac{R_0^2 H_0^2}{R^2 H^2} = (1+z)^2 \left(\frac{H_0}{H}\right)^2.$$

Now we know how to relate H to z (Equation 1.33), so we can find how Ω_k and Ω_{m} evolve.

First, if $\Lambda = 0$ and $\Omega_{\mathrm{m},0} = 1$, then $\Omega_{k,0} = 0$. But this can happen only if $k = 0$, so Ω_k must always be zero, and $\Omega_{\mathrm{m}} = 1$ at all times.

But if $\Lambda = 0$ and $\Omega_{\mathrm{m},0} = 0.7$, then at $z = 1000$, $\Omega_{\mathrm{m}} = 0.9996$. As redshift increases, Ω_k tends to zero and Ω_{m} tends to 1. About one second after the Big Bang, $\Omega_{\mathrm{m}} = 1 - 10^{-15}$. So if the present-day value of Ω_{m} is not 1, it must have had only a very tiny offset from 1 in the early Universe. What could cause it to be so close to 1, but not quite equal to 1?

This fine-tuning problem is worse if we include a non-zero Λ. If $\Omega_{\Lambda,0} = 0.7$ and $\Omega_{\mathrm{m},0} = 0.3$, then at $z = 1000$ their values were $\Omega_\Lambda = 3.3 \times 10^{-9}$ and $\Omega_{\mathrm{m}} = 0.999\,999\,996$.

Figure 1.14 shows how the Ω parameters depend on redshift, for various cosmological models. If $\Omega_\mathrm{m} = 1$ and $\Omega_\Lambda = 0$, then they keep these values throughout the history of the Universe. This would remove the need to explain the cosmological fine-tuning in the Ω parameters, and led to the expectation among at least some astronomers that the likely values are $\Omega_{\mathrm{m},0} = 1$ and $\Omega_{\Lambda,0} = 0$. However, there is now good experimental evidence to reject this particular cosmological model, so we are left with the problem: what caused the fine-tuning in the early Universe?

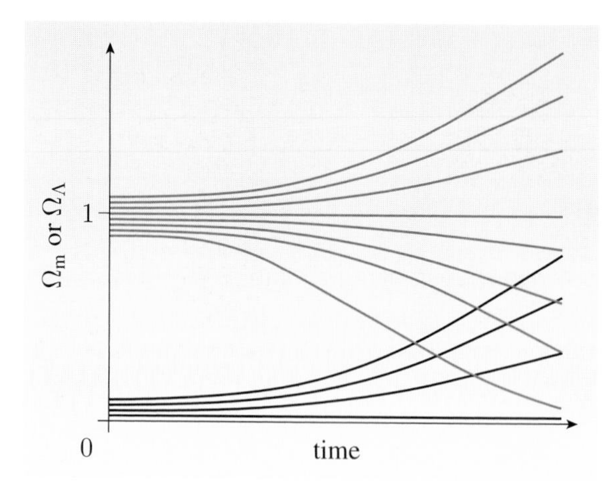

Figure 1.14 Schematic illustration of the variation of the density parameters Ω_m (upper, red lines) and Ω_Λ (lower, black lines) with time, for various cosmological models. Note that the time-dependence of a density parameter for any given starting point also depends on the starting values of the other density parameters.

One possible solution for the smallness of Ω_k, or rather a class of possible solutions, is inflation, which we shall meet in later chapters. But it is by no means certain that this is the correct solution. Inflation also does not offer any clear explanation for the fine-tuning of Ω_Λ. There may be Nobel prizes to be had for successful insights into the origin of Ω_Λ.

1.8 Distance in a warped spacetime

In 1963, Maarten Schmidt and Bev Oke made an astonishing discovery: the extraterrestrial radio source 3C273 is at what was then an unprecedentedly vast distance, a redshift of $z = 0.158$. Suddenly astronomers realized that telescopes could explore a much bigger volume of the Universe than had been supposed, and see objects as they were much earlier in the history of the Universe. This produced great excitement and optimism in cosmology. This also meant that 3C273 must be extremely luminous. We shall explore the causes of this prodigious luminosity in later chapters.

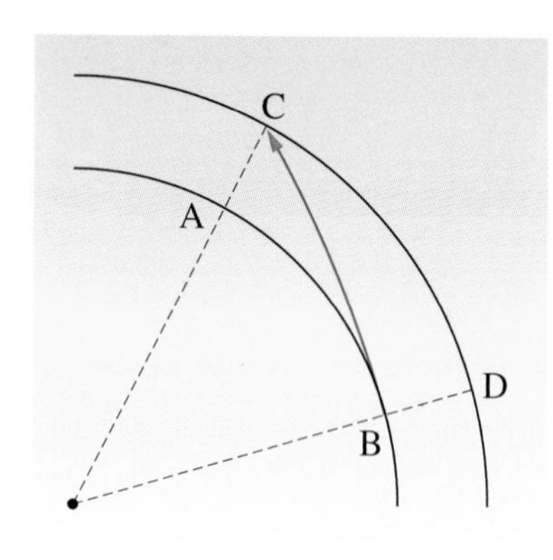

Figure 1.15 Different options for measuring cosmological distances in an expanding universe.

Using $\mathrm{d}z/\mathrm{d}t$ from Section 1.6, then integrating numerically with the cosmological parameters from Section 1.5, it turns out that the light from 3C273 has taken 1.9 billion years to reach us, which is about 14% of the age of the Universe. Does this mean that 3C273 is 1.9 billion light-years away?

It turns out that 'distance' is a surprisingly tricky concept to define in an expanding spacetime. Figure 1.15 shows some of the problems. Do we mean the distance that the light has travelled (B to C)? Or do we mean the distance that 3C273 was at when the light was emitted (B to A)? Or do we mean the distance that 3C273 is at now (D to C)?

Cosmologists have settled on a convention to use D to C (neglecting any peculiar velocities), in the preferred reference frame of the Robertson–Walker metric. This distance is known as the **comoving distance**. The comoving distance to 3C273 is about 637 Mpc, or about 2.1 billion light-years. This is longer than the light-travel distance, because the space has been expanding since the light was emitted, and 3C273 is now further away than when it emitted the light. Figure 1.16 illustrates how the Universe looks in **proper coordinates** (proper distances are those that would be measured by a tape measure at a fixed time in the cosmic rest frame), and in comoving coordinates.

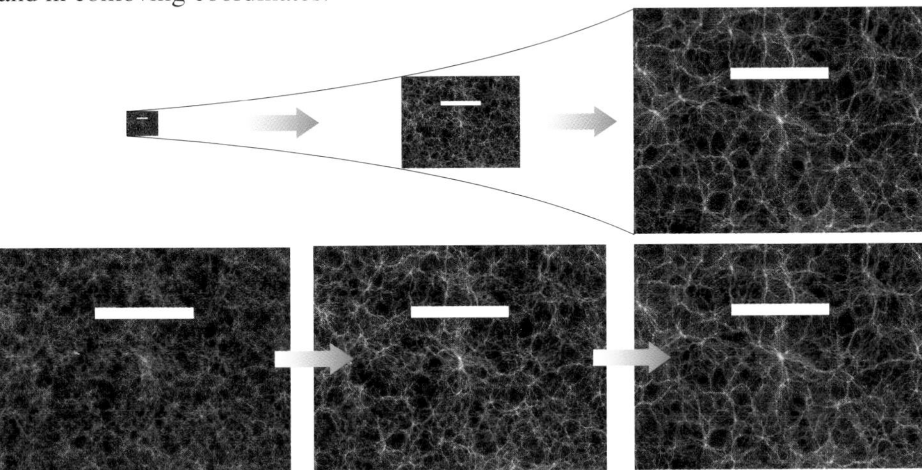

Figure 1.16 A simulation of the expanding Universe as seen in proper coordinates and in comoving coordinates. The panels show a simulation of a segment of the Universe at redshifts of $z = 5.7$ (left), $z = 1.4$ (centre) and $z = 0$ (right). The upper panels are shown in proper coordinates, while the lower panels are shown in comoving coordinates. A white bar shows a comoving length of $125/h$ Mpc. Note also the gradual increase in large-scale structure in this simulation with time, which we shall return to in later chapters.

To calculate the comoving distance to a cosmological object, we can use the Robertson–Walker metric, Equation 1.6. We want to know the radial distance between us and a distant object, at a fixed coordinate time $t = t_0$ (i.e. the present), perhaps imagining a tape measure stretched between us and it, which we read at the time $t = t_0$. Therefore $\mathrm{d}t = 0$ and $\mathrm{d}\theta = \mathrm{d}\phi = 0$. The remaining non-zero terms of Equation 1.6 are

$$\mathrm{d}s^2 = \frac{-R^2(t_0)\,\mathrm{d}r^2}{1 - kr^2}.$$

This is -1 times the square of a spatial separation. We define the comoving distance d_{comoving} via

$$\mathrm{d}d_{\text{comoving}} = \frac{R(t_0)\,\mathrm{d}r}{\sqrt{1 - kr^2}} = \frac{R_0\,\mathrm{d}r}{\sqrt{1 - kr^2}} \tag{1.36}$$

(with apologies for the profusion of the letter d) so that

$$d_{\text{comoving}} = R_0 \int_0^r \frac{\mathrm{d}r'}{\sqrt{1 - kr'^2}}. \tag{1.37}$$

This integral has a standard solution, depending on the value of k:

$$d_{\text{comoving}} = \begin{cases} R_0 \sin^{-1} r & \text{if } k = +1, \\ R_0 r & \text{if } k = 0, \\ R_0 \sinh^{-1} r & \text{if } k = -1. \end{cases} \tag{1.38}$$

Now, it's all very well imagining imaginary tape measures, but how can we relate this to real observable quantities? To find out, consider the light ray arriving at the Earth from the distant object. Light rays have $\mathrm{d}s = 0$, and the motion of this light ray is purely radial, so $\mathrm{d}\theta = \mathrm{d}\phi = 0$. The motion of the light ray is therefore just

$$\frac{R(t)\,\mathrm{d}r}{\sqrt{1-kr^2}} = c\,\mathrm{d}t. \tag{1.39}$$

Our aim is to calculate d_{comoving} following Equation 1.38.

In general, $R(t)$ can't be expressed analytically, though there are a few analytic special cases (e.g. $R(t) \propto t^{2/3}$ for a matter-dominated universe with $\Omega_{\text{m}} = 1$ and $\Lambda = 0$). But it's more helpful to express the right-hand side of Equation 1.39 in terms of redshift, which (unlike lookback time) is directly observable. To do this, we start with the chain rule for differentiation:

$$c\,\mathrm{d}t = c\,\mathrm{d}R\,\frac{\mathrm{d}t}{\mathrm{d}R} = \frac{c\,\mathrm{d}R}{\mathrm{d}R/\mathrm{d}t} = \frac{c\,\mathrm{d}R}{RH}, \tag{1.40}$$

where we have used $H = \dot{R}/R$. Next, $R = R_0/(1+z)$, which we can differentiate to find

$$\mathrm{d}R = \frac{-R_0}{(1+z)^2}\,\mathrm{d}z.$$

Putting this into Equation 1.40 gives

$$c\,\mathrm{d}t = \frac{-c}{HR}\,\frac{R_0}{(1+z)^2}\,\mathrm{d}z = \frac{-c}{(1+z)H}\,\mathrm{d}z$$

and therefore

$$\frac{R\,\mathrm{d}r}{\sqrt{1-kr^2}} = \frac{-c\,\mathrm{d}z}{(1+z)H}. \tag{1.41}$$

The comoving distance is $\int_0^r R_0\,(1-kr'^2)^{-1/2}\,\mathrm{d}r'$ (Equation 1.37), and this is almost in the right form. $R_0 = (1+z)R$, so $R_0\,\mathrm{d}r = (1+z)R\,\mathrm{d}r$, and therefore

$$\frac{R_0\,\mathrm{d}r}{\sqrt{1-kr^2}} = \mathrm{d}d_{\text{comoving}} = \frac{-c\,\mathrm{d}z}{H}, \tag{1.42}$$

where H is a function of redshift. Therefore the comoving distance is just

$$d_{\text{comoving}} = c\int_0^z \frac{\mathrm{d}z'}{H(z')}. \tag{1.43}$$

But we know H in terms of z and the observed cosmological parameters — we found this back in Equation 1.33. Putting this in here, and using $R_0/R = 1+z$, we get

$$\mathrm{d}d_{\text{comoving}} = \frac{-c}{H_0}\,\frac{\mathrm{d}z}{\sqrt{(1+z)^2(1+z\,\Omega_{\text{m},0}) - z(2+z)\,\Omega_{\Lambda,0}}},$$

thus

$$d_{\text{comoving}} = \frac{c}{H_0} \int_0^z \frac{\mathrm{d}z'}{\sqrt{(1+z')^2(1+z'\,\Omega_{\text{m},0}) - z'(2+z')\,\Omega_{\Lambda,0}}}. \tag{1.44}$$

(We have integrated the previous differential from z to 0, but used its minus sign to swap the limits and obtain a positive integral.) There are a few special cases where this integral comes out as a relatively simple expression, such as when $\Lambda = 0$ and $\Omega_\text{m} = 1$:

$$d_{\text{comoving}} = \frac{2c}{H_0}\left(1 - (1+z)^{-1/2}\right) \textit{ only for } \Omega_\text{m} = 1,\ \Lambda = 0. \tag{1.45}$$

1.9 The edge of the observable Universe

How big is the observable Universe today? If we lived in a flat, unexpanding space, the observable Universe would be the region of the Universe that could have sent us a light signal. The radius would be ct_0, where t_0 is the age of the Universe. This radius would be increasing at a speed c.

In the Robertson–Walker expanding universe, this is different in a rather astonishing way. First, we'll calculate the size of the observable Universe. This size is the same in comoving coordinates and in proper coordinates, provided that we're referring to the size at time $t = t_0$ (i.e. now) when the scale factor is $R = R_0$.

The size of the observable Universe is therefore the comoving radius in Equation 1.44 as redshift z tends to infinity. In general, this comes out as a number times c/H_0. For an $\Omega_\text{m} = 1$ and $\Lambda = 0$ universe, you can see from Equation 1.45 that the radius of the observable Universe is $2c/H_0$, or about 8300 Mpc for the currently-accepted value of H_0 in Section 1.5. For the currently-accepted values of the density parameters in Section 1.5, the radius of the observable Universe comes out at about $3.53c/H_0$. The volume enclosed is sometimes referred to as the **Hubble volume**.

Next, we'll calculate how fast this observable Universe is growing, in proper distances rather than comoving distances. Again, we take the time now to be t_0 and the current scale factor to be R_0. Suppose that we see a distant object at a redshift z and a proper distance $R_0 r$. After a time δt, the time will be $t_1 = t_0 + \delta t$, and the new scale factor of the Universe will be R_1. Therefore the new proper distance to this distant object will be

$$R_1 r = R_0 r + \left(\frac{\mathrm{d}R}{\mathrm{d}t}\,\delta t\right) r \simeq R_0 r + (R_0 H_0\,\delta t) r.$$

Therefore the rate of change of proper distance is

$$\frac{\mathrm{d}(Rr)}{\mathrm{d}t} \simeq \frac{R_1 r - R_0 r}{\delta t} = R_0 H_0 r, \tag{1.46}$$

i.e. just H_0 times the current proper distance. This is one sense in which the Hubble parameter is measuring the rate of the expansion of the Universe.

So, for the currently-accepted values of the density parameters in Section 1.5, the observable Universe is getting bigger at an astonishing rate of $(3.53c/H_0) \times H_0 = 3.53c$. In Section 1.5 we met Equation 1.46 in a different guise (Equation 1.13), but warned that the left-hand side is not a *recession* velocity, but rather an *apparent* recession. Here, you see our reason for this warning! An object moving through a flat, unexpanding space has a maximum speed of c, but an expanding spacetime is a very different physical situation, and the maximum cosmological apparent 'recession' speed in our Universe is currently about $3.53c$.

1.10 Measuring distances and volumes

Redshift is one way of measuring distances, but to convert redshift into a distance, we need to know the cosmological parameters, especially H_0. If we want to measure these fundamental cosmological parameters, we need independent ways of measuring distances.

One useful distance measure is the **angular diameter distance**. If an object has a known size D, and subtends an angle θ in radians, then the angular diameter distance is

$$d_\mathrm{A} = \frac{D}{\theta}. \qquad (1.47)$$

(Compare Figure 1.2, in which r takes the place of d_A.)

Another useful measure is the **proper motion distance**. If an object has a known transverse velocity u (i.e. motion in the plane of the sky), and has an observed angular motion of $\mathrm{d}\theta/\mathrm{d}t$, then the proper motion distance is defined as

$$d_\mathrm{M} = \frac{u}{\mathrm{d}\theta/\mathrm{d}t}. \qquad (1.48)$$

Finally, we can also define the **luminosity distance**. If an object has a known luminosity L, and the observed flux is S, then the luminosity distance is

$$d_\mathrm{L} = \left(\frac{L}{4\pi S}\right)^{1/2}. \qquad (1.49)$$

These three approaches to measuring distance give the same answer in a flat, unexpanding space, but they are surprisingly different in the Robertson–Walker metric. Figure 1.17 illustrates how these distances are constructed in Robertson–Walker universes. As usual, we use t_0 and R_0 for the current time and current scale factor, respectively. Suppose that photons are emitted from a distant object of size D at a time t_1, when the scale factor was R_1. From the figure, we see that $D = R_1 r\theta$, so $d_\mathrm{A} = R_1 r$.

Next, suppose that another object at the same redshift is moving with a proper transverse velocity u, and is seen to move at an angular speed $\mathrm{d}\theta/\mathrm{d}t$, as in Equation 1.48. Here, the cosmological time dilation first noted in Section 1.4 comes into play. If t' is the time measured when the photons are emitted, then $\mathrm{d}t'/\mathrm{d}t = R_1/R_0$. The proper velocity is

$$u = \frac{\mathrm{d}D}{\mathrm{d}t'} = \frac{\mathrm{d}(R_1 r\theta)}{\mathrm{d}t'},$$

and substituting this into Equation 1.48, we find $d_\mathrm{M} = R_0 r$. Note that when $k = 0$, this equals the comoving distance (compare Equation 1.38).

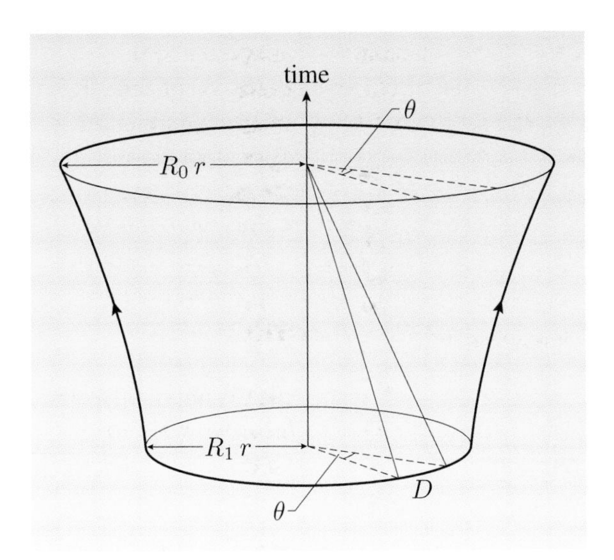

Figure 1.17 Light rays in the Robertson–Walker metric, for illustrating distance measures.

Finally, suppose that an object at this redshift has a bolometric luminosity L (where 'bolometric' means the total over all wavelengths). The photons are distributed over a sphere with a proper area of $4\pi(R_0 r)^2$ (see Figure 1.17). The energy emitted in a time $\mathrm{d}t'$ will be $L\,\mathrm{d}t'$, but the redshifting will reduce the energy received by a factor of R_1/R_0. Therefore the flux received will be

$$S = \frac{L\,\mathrm{d}t'\,R_1/R_0}{\mathrm{d}t} \times \frac{1}{4\pi(R_0 r)^2} = L\frac{\mathrm{d}t'}{\mathrm{d}t}\frac{R_1}{R_0}\frac{1}{4\pi}\frac{1}{R_0^2 r^2}.$$

But because of cosmological time dilation, $\mathrm{d}t'/\mathrm{d}t = R_1/R_0$, so

$$S = L\frac{R_1}{R_0}\frac{R_1}{R_0}\frac{1}{4\pi}\frac{1}{R_0^2 r^2} = \frac{LR_1^2}{4\pi r^2 R_0^4} = \frac{L}{4\pi(R_0^2 r/R_1)^2}.$$

Comparing this to Equation 1.49, we see that $d_\mathrm{L} = R_0^2 r/R_1$. It would be wonderful if comparing these three distance measures for a single object gave us constraints on the cosmological parameters. It's perhaps a little disappointing, then, that they are all closely related. Using $1 + z = R_0/R_1$, we find that

$$d_\mathrm{L} = (1+z)d_\mathrm{M} = (1+z)^2 d_\mathrm{A}, \qquad (1.50)$$

independent of the cosmological parameters. The constraints on the cosmological parameters can instead be gleaned from how these distance measures vary with redshift. Figure 1.18 shows how the angular diameter distance varies with redshift, for various cosmological models.

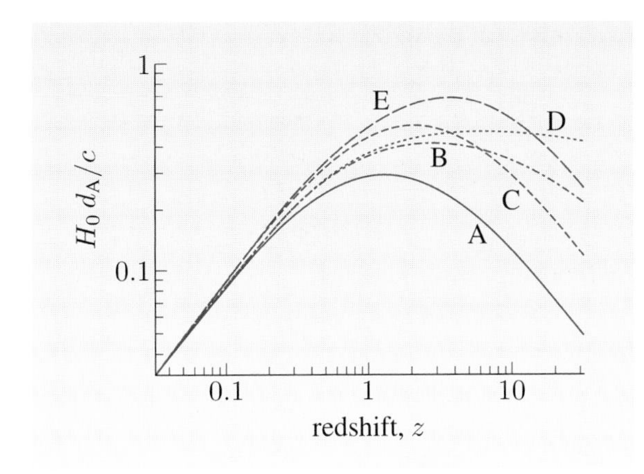

Figure 1.18 The variation of angular diameter distance with redshift, for the following cosmological models. A: $\Omega_\mathrm{m} = 1$, $\Lambda = 0$. B: $\Omega_{\mathrm{m},0} = 0.1$, $\Lambda = 0$. C: $\Omega_{\mathrm{m},0} = 0.1$, $\Omega_{\Lambda,0} = 0.9$. D: $\Omega_{\mathrm{m},0} = 0.01$, $\Lambda = 0$. E: $\Omega_{\mathrm{m},0} = 0.01$, $\Omega_{\Lambda,0} = 0.99$.

Again, the Robertson–Walker metric holds another surprise: the angular diameter distance d_A has a maximum value, as you can see in Figure 1.18. What does this mean? In flat unexpanding space, objects always appear smaller when they are placed further away, but in the Robertson–Walker spacetime, an object that is placed further away can appear *larger*. This is partly because the Universe was smaller when the light was emitted, so the object was then nearer to us. It's partly also to do with the geometry of the space. For example, light rays emitted on a two-dimensional spherical surface at the South pole will initially diverge, but will eventually converge again as they approach the North pole. A spherical unexpanding space would therefore still have a maximum angular diameter distance.

We shall see how these distances affect Olbers' paradox later in this book. In the meantime, the following exercise will give you a clue to how Olbers' paradox is resolved in a Robertson–Walker universe.

Exercise 1.6 How does surface brightness (flux per square degree) vary with redshift? ■

In observational astronomy we rarely measure the total luminosities of distant objects; instead, we tend to measure the redshift and the flux in a particular wavelength interval $\Delta\lambda_{\rm obs}$. Two effects change the observed flux: first, the observed wavelength interval $\Delta\lambda_{\rm obs}$ corresponds to a smaller wavelength interval in the emitted frame, because $\Delta\lambda_{\rm obs} = (1+z)\Delta\lambda_{\rm em}$; second, the distant object may emit different amounts of light at *rest* wavelengths of $\lambda_{\rm em}$ and $\lambda_{\rm obs}$. This latter effect is known as the **K-correction** for historical reasons, and we shall meet it later in this book. If the underlying spectrum is a 'power law', i.e. if the flux per unit frequency is $S_\nu \propto \nu^{-\alpha}$, then a useful expression for the luminosity is

$$\frac{L_\nu}{10^{26}\,{\rm W\,Hz^{-1}\,sr^{-1}}} = \frac{S_\nu}{10^{-26}\,{\rm W\,Hz^{-1}\,m^{-2}}}\left(\frac{d_{\rm M}}{3241\,{\rm Mpc}}\right)^2 (1+z)^{1+\alpha}. \quad (1.51)$$

Finally, cosmologists often use the term **comoving volume** to describe volumes with the expansion factor divided out (Figure 1.16). We shall use this many times throughout this book. Imagine a patch of sky with an angular area $\delta\Omega$ (in units, for example, of square degrees). We can convert this to a proper area at any redshift using the angular diameter distance: $\delta A = d_{\rm A}^2\,\delta\Omega$. Now imagine that we are observing a slab at redshift z with a proper area δA and proper thickness $R(1-kr^2)^{-1/2}\,{\rm d}r$. The proper volume will therefore be ${\rm d}V_{\rm proper} = \delta A \times (1-kr^2)^{-1/2} R\,{\rm d}r$, or

$${\rm d}V_{\rm proper} = d_{\rm A}^2(z)\,\delta\Omega \times \frac{R\,{\rm d}r}{\sqrt{1-kr^2}}. \quad (1.52)$$

Now the comoving volume is just ${\rm d}V_{\rm comoving} = (1+z)^3 \times {\rm d}V_{\rm proper}$, so

$${\rm d}V_{\rm comoving} = d_{\rm A}^2(z)\,\delta\Omega\,\frac{R\,{\rm d}r}{\sqrt{1-kr^2}} \times (1+z)^3. \quad (1.53)$$

There are many ways of integrating this, but one approach is to express it in terms of the proper motion distance $d_{\rm M} = R_0 r$. (Recall that in a flat universe this is equivalent to the comoving distance.) Then

$${\rm d}V_{\rm comoving} = \frac{d_{\rm M}^2(z)}{\sqrt{1+\Omega_{k,0}H_0^2 d_{\rm M}^2/c^2}}\,{\rm d}(d_{\rm M})\,\delta\Omega. \quad (1.54)$$

We'll express this in another useful way in Chapter 6, Equation 6.26.

The function $\mathrm{d}V_{\text{comoving}}/\mathrm{d}z$ is plotted in Figure 1.19. This can be integrated to give $V_{\text{comoving}}(z)$, the total volume enclosed by a sphere with radius z centred on us. The form of the comoving volume equation depends on whether $k = 0$, $k = 1$ or $k = -1$. For reference, there are analytic solutions for the comoving volume over the whole sky, in terms of the proper motion distance d_{M}:

$$V(d_{\text{M}}) = \begin{cases} 4\pi D_{\text{H}}^3 (2\Omega_{k,0})^{-1}[(d_{\text{M}}/D_{\text{H}})\sqrt{1+\Omega_{k,0}d_{\text{M}}^2/D_{\text{H}}^2} - |\Omega_{k,0}|^{-1/2}\sin^{-1}((d_{\text{M}}/D_{\text{H}})|\Omega_{k,0}|^{1/2})] & \text{if } k=+1, \\ \frac{4}{3}\pi d_{\text{M}}^3 & \text{if } k=0, \\ 4\pi D_{\text{H}}^3 (2\Omega_{k,0})^{-1}[(d_{\text{M}}/D_{\text{H}})\sqrt{1+\Omega_{k,0}d_{\text{M}}^2/D_{\text{H}}^2} - |\Omega_{k,0}|^{-1/2}\sinh^{-1}((d_{\text{M}}/D_{\text{H}})|\Omega_{k,0}|^{1/2})] & \text{if } k=-1, \end{cases} \tag{1.55}$$

where $D_{\text{H}} = c/H_0$ is sometimes known as the **Hubble distance**. Also, for reference, the proper motion distance can be expressed as

$$d_{\text{M}} = \begin{cases} D_{\text{H}} \dfrac{1}{|\Omega_{k,0}|^{1/2}} \sin\left[|\Omega_{k,0}|^{1/2} \displaystyle\int_0^z \left\{(1+z)^2(1+\Omega_{\text{m},0}z) - z(2+z)\Omega_{\Lambda,0}\right\}^{-1/2} \mathrm{d}z\right] & \text{if } k=+1, \\ d_{\text{comoving}} & \text{if } k=0, \\ D_{\text{H}} \dfrac{1}{|\Omega_{k,0}|^{1/2}} \sinh\left[|\Omega_{k,0}|^{1/2} \displaystyle\int_0^z \left\{(1+z)^2(1+\Omega_{\text{m},0}z) - z(2+z)\Omega_{\Lambda,0}\right\}^{-1/2} \mathrm{d}z\right] & \text{if } k=-1. \end{cases} \tag{1.56}$$

Equation 1.44 gives the expression for d_{comoving}. Remember that proper motion distance d_{M} is equal to comoving distance d_{comoving} *only when* $k = 0$.

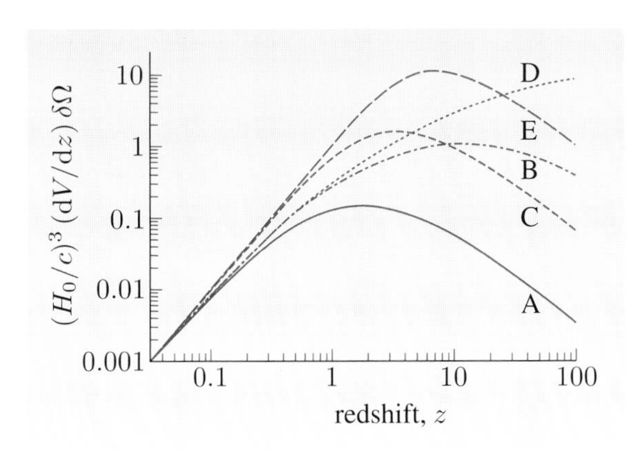

Figure 1.19 The variation of the comoving volume derivative $\mathrm{d}V/\mathrm{d}z$ with redshift, for the various cosmological models in Figure 1.18, and for an angular area on the sky of $\delta\Omega$.

1.11 The fate of the Universe

The cosmological parameters in Section 1.5 can also tell us the fate of the Universe. An extraordinary consequence of the success of the Robertson–Walker model is that we know the ultimate fate of the atoms in our bodies, at least up to about the year AD 10^{35}.

We can again re-cast Equation 1.14, this time by changing the variables to $a = 1/(1+z) = R/R_0$ and $\tau = H_0 t$. This comes out as

$$\left(\frac{\mathrm{d}a}{\mathrm{d}\tau}\right)^2 = 1 + \Omega_{\text{m},0}\left(\frac{1}{a} - 1\right) + \Omega_{\Lambda,0}(a^2 - 1).$$

This differential equation can be solved numerically, and the predicted fate of the Universe is shown in Figure 1.20 as a function of $\Omega_{m,0}$ and $\Omega_{\Lambda,0}$. The cosmological parameters in Section 1.5 are very clearly in the regime of expanding forever.

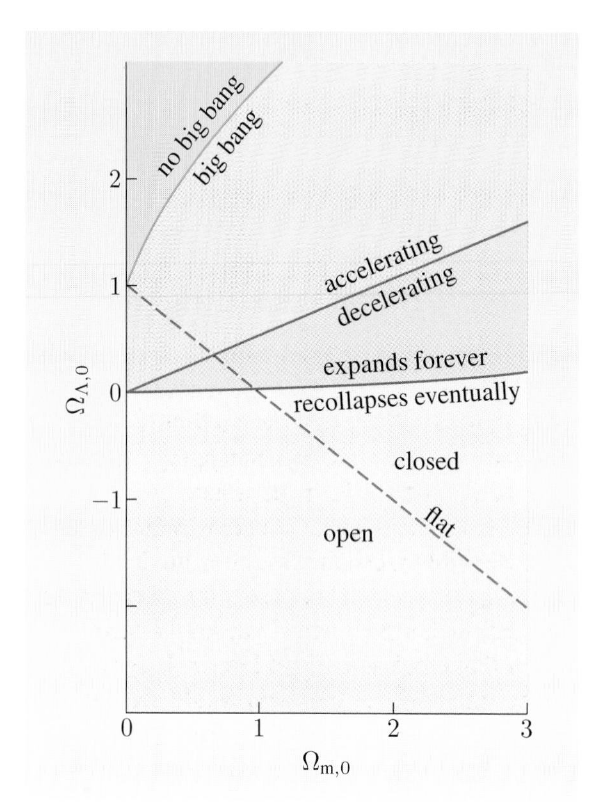

Figure 1.20 The predicted fate of the Universe as a function of the present-day cosmological density parameters.

What will this be like? The Universe will become increasingly sparse, as the matter density decreases and Ω_m tends to zero. The cosmological constant will then dominate, and the Universe will tend to the $\Omega_\Lambda = 1$ model. Equation 1.14 will reduce to

$$H^2 = \frac{\Lambda c^2}{3}$$

so the Hubble constant will, finally, be truly a constant. The expansion will be exponential, as you can see from substituting $H = \dot{R}/R$ into the equation above:

$$\frac{1}{R}\frac{\mathrm{d}R}{\mathrm{d}t} = \sqrt{\frac{\Lambda c^2}{3}},$$

which has the solution $R \propto \mathrm{e}^{ct\sqrt{\Lambda/3}}$. This is sometimes known as **de Sitter spacetime**.

This exponential expansion has a curious consequence. Regions of the Universe that were once in causal contact eventually lose contact with each other, as the rapidly expanding space makes it impossible for even light signals to pass between them. To show this, imagine a light signal being sent out in this universe. How far can it get? The light signal will be in an exponentially expanding universe, so in a sense it will go infinitely far, but if we normalize our distances by the scale factor, then we'll see that the light signal reaches only a finite *comoving* region.

Light signals still satisfy Equation 1.39, i.e. $R(t)\,\mathrm{d}r = c\,\mathrm{d}t$ (note that $k = 0$ because $\Omega_k = 1 - \Omega_\mathrm{m} - \Omega_\Lambda = 1 - 0 - 1 = 0$), but this time $R(t)$ is very different. If we set $R(t) = R_1$ at a time $t = t_1$, then

$$\frac{R(t)}{R(t_1)} = \frac{R(t)}{R_1} = \frac{\mathrm{e}^{ct\sqrt{\Lambda/3}}}{\mathrm{e}^{ct_1\sqrt{\Lambda/3}}} = \mathrm{e}^{c(t-t_1)\sqrt{\Lambda/3}}.$$

If we treat $R_1 r$ as our choice of comoving distance, then we have

$$R_1\,\mathrm{d}r = c\,\mathrm{e}^{-c(t-t_1)\sqrt{\Lambda/3}}\,\mathrm{d}t$$

so

$$R_1 r = \int_{t=t_1}^{\infty} c\,\mathrm{e}^{-c(t-t_1)\sqrt{\Lambda/3}}\,\mathrm{d}t = \frac{c}{c\sqrt{\Lambda/3}} = \sqrt{\frac{3}{\Lambda}} = \frac{c}{H}.$$

So as time tends to infinity, the light signal penetrates a comoving distance of only $R_1 r = \sqrt{3/\Lambda}$. Objects beyond a comoving distance of $\sqrt{3/\Lambda}$ cannot be seen, because the intervening space is expanding so quickly that even a light signal cannot cross it. But H is constant and the expansion rate is unchanging, so this is true at *any* time. What this would look like is a fixed horizon around you at a distance of $\sqrt{3/\Lambda}$, and neighbouring galaxies being accelerated away from you towards this horizon and out of your observable Universe, which is gradually being emptied out. However, you would never see a galaxy *cross* this horizon: its redshift would get larger as it approached the horizon, and if you could watch a clock in that galaxy, the time dilation of that clock would get longer. If t_2 is the coordinate time when the galaxy reaches the horizon, then you would see the clock slow at it approached t_2, but it would never quite reach t_2 from your point of view. However, from the galaxy's point of view, the passage of time is unaffected. There, they would see *your* clocks running slowly, as you passed out of *their* observable Universe.

You may recognize this redshifting and time dilation from descriptions of objects falling into the event horizon of a black hole (which you will also meet in Chapter 6). Indeed, the horizon at $\sqrt{3/\Lambda}$ is a **cosmological event horizon**. The Universe in the far future will look like a black hole, but inside-out.

Exercise 1.7 How big, in megaparsecs and in metres, will the cosmological event horizon be? You will need the cosmological parameters in Section 1.5. How does this compare to the current size of the observable Universe? ■

How far ahead can we look? When Ralph Alpher and George Gamow realized that the early Universe was hot and dense enough for nuclear reactions, and calculated the amount of heavy elements production, they were condemned by some physicists for their rashness. What grounds do we have, the critics argued, for believing that the same physical theories applied three minutes after the Big Bang? The Universe provided the rebuttal: the predictions of primordial nucleosynthesis have been very extensively confirmed, as we shall see in later chapters. Nevertheless, the words of warning from these critics should still ring in our ears as we extrapolate to the distant future.

At the moment all the baryons in the Universe are either involved in the cycle of star birth and death, or could potentially take part. However, by about the year one trillion (10^{12} years), the Universe will be too sparse to support more star formation. At that point, baryons will either be in degenerate matter (in white dwarfs or neutron stars), or be locked up in brown dwarfs, or have fallen into black holes, or just be atoms or molecules too sparsely distributed to form new stars. Looking further ahead, we eventually reach the epoch of possible proton decay.

In the standard model of particle physics, the proton is stable and does not decay. However, some 'grand unified theories' in particle physics predict eventual proton decay. The best current limit on proton half-life $t_{1/2}$ comes from the Super-Kamiokande experiment in Japan, which found $t_{1/2} > 10^{35}$ years. Perhaps 10^{35} years is the furthest ahead that one might venture to predict the contents of the Universe. But who would be around to contradict you if you got it wrong?

Summary of Chapter 1

1. In a flat Euclidean space, the number counts of a homogeneous and isotropic distribution of objects vary as $\mathrm{d}N/\mathrm{d}S \propto S^{-5/2}$, but the total flux diverges.
2. In special relativity, lengths and times are not observer-independent, but the relativistic interval s is invariant.
3. $\delta s = 0$ always for light rays, and $\delta s = c\,\delta\tau$ always for massive particles, where τ is proper time.
4. Any homogeneous, isotropic expanding Universe consistent with special relativity can be described by the Robertson–Walker metric
$$\mathrm{d}s^2 = c^2\,\mathrm{d}t^2 - R^2(t)\left(\frac{\mathrm{d}r^2}{1-kr^2} + r^2\,\mathrm{d}\theta^2 + r^2\sin^2\theta\,\mathrm{d}\phi^2\right). \quad \text{(Eqn 1.6)}$$
5. Cosmological redshift z, given by
$$1+z = \frac{R_0}{R_1} = \frac{1}{a} = \frac{\lambda_{\text{observed}}}{\lambda_{\text{emitted}}}, \quad \text{(Eqn 1.10)}$$
is caused by the expansion of the Universe, not by the Doppler effect. Random galaxy motions (known as proper motions) can contribute additional red or blue shifts from the relativistic Doppler effect.
6. Nevertheless, if one regards cosmological redshift as an *apparent* recession velocity, then the apparent velocity is proportional to distance from the observer, with the constant of proportionality known as the Hubble parameter.
7. The contributions to the energy density of the Universe from matter and the cosmological constant are denoted as Ω_m and Ω_Λ, respectively, and are defined by
$$\Omega_\mathrm{m} = \frac{8\pi G\rho_\mathrm{m}}{3H^2}, \quad \text{(Eqn 1.15)}$$
$$\Omega_\Lambda = \frac{\Lambda c^2}{3H^2}. \quad \text{(Eqn 1.17)}$$
These determine the age and fate of the Universe.
8. Neglecting radiation, if $\Omega_\mathrm{m} + \Omega_\Lambda = 1$ at any time, then this is true at all times. Also, if either Ω_m or Ω_Λ is zero, then this is also true at all times. In all other situations, there is a fine-tuning problem in the early Universe for the values of Ω_m and Ω_Λ.

9. 'Distance' can have several meanings in a Robertson–Walker metric. We have defined the angular diameter distance, proper motion distance and luminosity distance:

$$d_{\mathrm{A}} = \frac{D}{\theta}, \quad \text{(Eqn 1.47)}$$

$$d_{\mathrm{M}} = \frac{u}{\mathrm{d}\theta/\mathrm{d}t}, \quad \text{(Eqn 1.48)}$$

$$d_{\mathrm{L}} = \left(\frac{L}{4\pi S}\right)^{1/2}. \quad \text{(Eqn 1.49)}$$

These give different but related values for the distance. Angular diameter distance has a maximum value, so objects placed further away could sometimes appear larger.

10. The comoving distance to any distant object is the current proper distance, neglecting peculiar velocities. (This turns out to be equal to the proper motion distance when $k = 0$.)

11. The proper size of the observable Universe is increasing at a rate much larger than the speed of light, c. An object moving in a flat spacetime is a very different physical situation to free-floating objects in an expanding space.

12. The currently-accepted cosmological parameters imply that the foreseeable fate of the Universe is exponential expansion.

Further reading

- For a more leisurely introduction to the Robertson–Walker metric, see Lambourne, R., 2010, *Relativity, Gravitation and Cosmology*, Cambridge University Press.
- For a useful review of distance measures in cosmology (though pre-dating dark energy, which we shall meet in later chapters, and the observation that $\Omega_\Lambda \simeq 0.7$), see Carroll, S.M., Press, W.H. and Turner, E.L., 1992, 'The cosmological constant', *Annual Review of Astronomy and Astrophysics*, **30**, 499.

Chapter 2 The cosmic microwave background

I would rather live in a world where my life is surrounded by mystery than live in a world so small that my mind could comprehend it.

Harry Emerson Fosdick

Introduction

We effectively answer Olbers' paradox in this chapter, with the first and most famous cosmic background light. You will also find out how quantitative cosmology is done using this background, which has already resulted in two Nobel prizes.

2.1 The discovery of the cosmic microwave background

The Big Bang theory is supported by three major observations, one of which you have already met, and the other two you will meet in this chapter: the expansion of the Universe, the cosmic microwave background (CMB), and primordial nucleosynthesis. Some would add the large-scale structure of matter in the Universe, which we shall also meet later in this book in various guises.

Perhaps the most powerful recent development is the fact that many disparate observations all converge on the same cosmological model, described in Chapter 1. Our cosmological model is *over*-determined, in the sense that there are more independent experimental constraints than there are parameters to constrain. Perhaps this is a sign of a mature scientific discipline; the field is now described as 'precision cosmology', of which more later. This expression came into use about the time of the microwave background measurements from the Wilkinson Microwave Anisotropy Probe (WMAP), for reasons that will become clear in this chapter.

The microwave background is the redshifted light from when the Universe was last opaque. Because of the finite speed of light, this light must appear to any observer as a spherical, receding surface, with the observer at the centre. For this reason the microwave background is sometimes called the 'surface of last scattering'. In this sense, Olbers was exactly right — the sky is indeed uniformly bright. It has a black body spectrum to an excellent approximation — in fact, it is the most perfect black body spectrum known in existence (Figure 2.1). From calculations of the probability of photon–electron collisions as a function of the ionization of the primordial gas, the redshift of the last scattering surface can be estimated as $z \simeq 1000$.

Around the time of last scattering is also the epoch of **recombination**. Before that time, photons suffered many Thomson-scattering collisions with electrons, and would ionize any atoms that tried to form. Once the density dropped enough for the Universe to become transparent, the electrons and nuclei were free to combine to form atoms, and the photons were free to travel in the newly-transparent

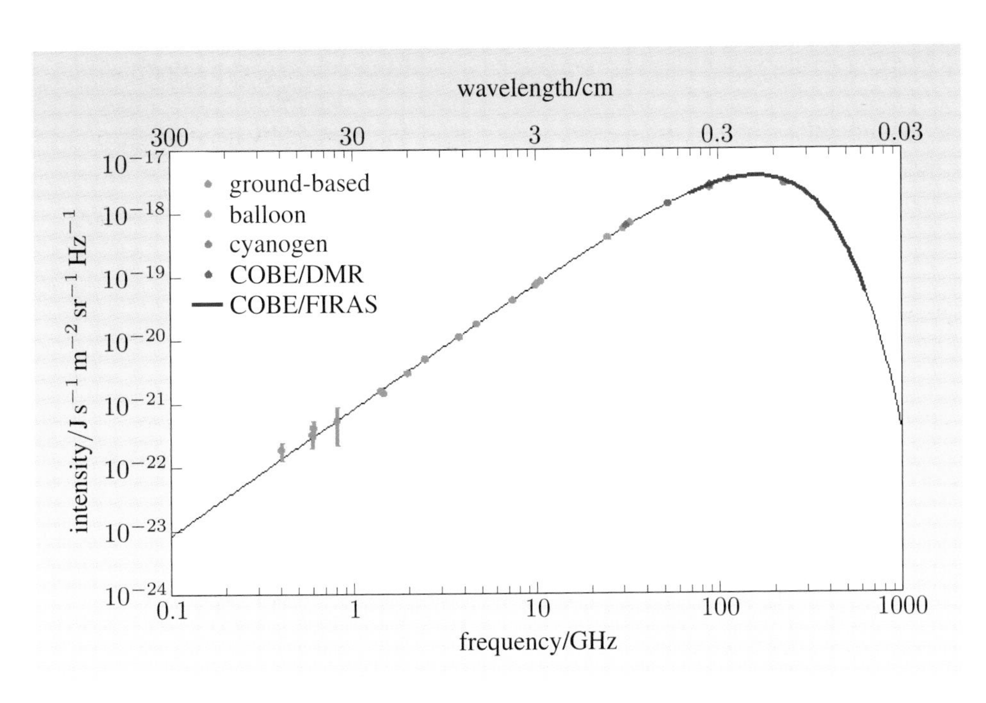

Figure 2.1 The intensity of the CMB radiation, measured by various techniques. Both the x-axis and y-axis are plotted logarithmically. The shallow slope at the left-hand side is known as the Rayleigh–Jeans regime, while the steeper slope at shorter wavelengths is known as the Wien regime. The uncertainties for the data from the FIRAS instrument on the COBE satellite are smaller than the plotting symbols, and the deviation from a perfect black body is less than 50 parts per million.

Universe. From a detailed statistical mechanical calculation[2] (we shall spare you the details), the fraction of ionized gas x near $z = 1000$ comes out as

$$x(z) \simeq 2.4 \times 10^{-3} \frac{\sqrt{\Omega_{\mathrm{m},0}\, h^2}}{\Omega_{\mathrm{b},0}\, h^2} \left(\frac{z}{1000}\right)^{12.75}, \tag{2.1}$$

i.e. it depends on the density of baryons relative to the critical density, $\Omega_{\mathrm{b},0}$, and on $\Omega_{\mathrm{m},0}$. We'll see later how the CMB gives estimates of $\Omega_{\mathrm{m},0}$ and $\Omega_{\mathrm{b},0}$. (The physical process of electrons binding with protons to make hydrogen is known as 'recombining' but this is a misnomer here, because they are in fact combining for the first time.)

Exercise 2.1 We can ionize a hydrogen atom by colliding it with a photon with an energy $E = h\nu = 13.6\,\mathrm{eV}$. Let's think of recombination as the inverse of ionization. Imagine that the electron binds to a proton and emits a photon with an energy $E = 13.6\,\mathrm{eV}$. Won't this photon go on to ionize another atom? How is it that the Universe is able to recombine at all? (It's *not* that the Universe expands and makes the number density of atoms sufficiently dilute that ionizing is rare.) ■

The CMB was discovered in 1964 by Arno Penzias and Robert Wilson at the Bell Laboratories in New Jersey. They were given the task of calibrating a microwave antenna for use in telecommunications and astronomy, but found a surprising and unexplained constant noise source distributed isotropically across the sky. The isotropy and constancy themselves ruled out several potential causes: Galactic sources should be preferentially found in the Galactic plane; Solar System sources should be preferentially in the ecliptic plane; the signal was the same in the direction of extraterrestrial radio sources (which we shall meet in Chapter 4); sources related to nuclear tests should decay with time; other sources related to human activity might be expected to be stronger in the direction of nearby

[2] Jones, B.J.T. and Wyse, R.F.G., 1985, *Astronomy and Astrophysics*, **149**, 144.

cities. They also found pigeons roosting in the antenna and needed to remove what Penzias later called a 'white dielectric material'. Although they trapped the pigeons and released them thirty miles away, the pigeons kept returning; reluctantly, the birds were eventually shot. The anomalous isotropic noise source remained, and neither Penzias nor Wilson could find its source.

Unknown to both, a rival group in Princeton, led by Robert Dicke, had just predicted an isotropic CMB from the Big Bang theory. Dicke's group were planning an experiment to detect it. Once Dicke heard of Penzias and Wilson's anomalous noise, Dicke spoke on the telephone to the Bell Laboratories group, after which he announced to his team: 'Boys, we've been scooped'.

The result was that two back-to-back papers in the *Astrophysical Journal* announced the discovery to the world: Penzias and Wilson published the discovery itself, while Dicke's group published the theoretical explanation. Penzias and Wilson co-won the 1978 Nobel Prize in Physics for their discovery (along with Pyotr Kapitsa for a different discovery).

The story has further twists. It turned out that the CMB prediction was implicit in calculations published in 1948 by George Gamow, Ralph Alpher and Robert Herman. Furthermore, in 1934 Andrew McKellar measured the 'effective temperature of interstellar space' in a careful experiment using spectroscopy to derive the typical energy level excitations of molecules in the interstellar medium (we shall see how in Section 2.2 and in later chapters). He found this to be about 2.3 K, but the work pre-dated the Big Bang predictions and neither he nor any reader appreciated its significance at the time. The currently accepted value of the CMB temperature is 2.725 ± 0.001 K.

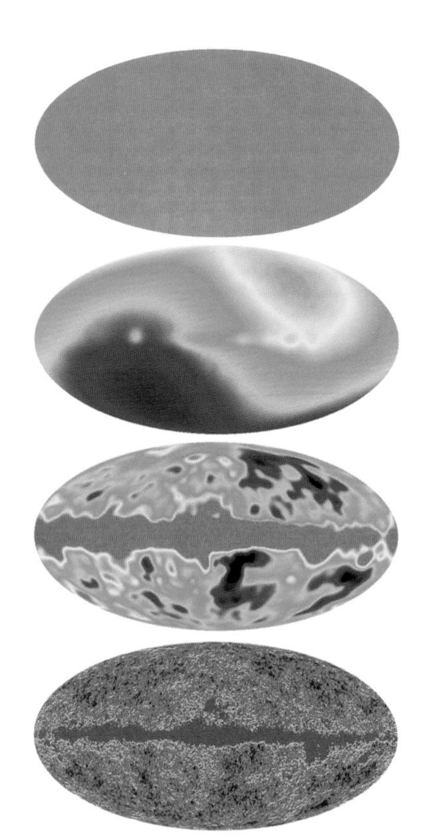

Figure 2.2 All-sky maps of the CMB temperature. The top image is scaled from 0 K to 4 K and looks very uniform. The next image has a much smaller scaling, with a range of just 3.353 mK. The pattern is the dipole that results from the Doppler shift due to the Earth's motion relative to the cosmic rest frame (Section 2.11). After accounting for this motion and further restricting the temperature range to just 18 μK, we see a large band due to our Galaxy and the background primordial fluctuations where the Galactic foreground does not outshine them. These three images were taken with the COBE satellite; the final image is higher-resolution data from the WMAP satellite.

Exercise 2.2 Stefan's law can be shown to imply that the energy density is $4\sigma T^4/c$. Use this to show that the present-day CMB radiation energy density of the Universe is $\Omega_{\mathrm{r},0}\, h^2 = 2.5 \times 10^{-5}$. (The value of Stefan's constant σ is $5.67 \times 10^{-8}\,\mathrm{W\,m^{-2}\,K^{-4}}$.) ■

There is also an additional contribution from light neutrinos to Ω_{r} of about 68%, so the total relativistic energy density is slightly larger (Equation 1.27 in Chapter 1) with a value $\Omega_{\mathrm{r},0}\, h^2 = 4.2 \times 10^{-5}$.

The CMB is also remarkably uniform (Figure 2.2); we'll show that this presents a very serious cosmological problem. If we increase the contrast level, we first find a characteristic hot and cold pattern (Figure 2.2). We'll show in Section 2.11 that this is caused by our motion relative to the cosmic rest frame. Correcting for this motion, we find a strong signal from our Galaxy (the horizontal band in Figure 2.2), and apart from this we find that the CMB has intrinsic fluctuations at the level of microkelvins. There is no single clear physical theory for the level of these fluctuations, though we'll see in Section 2.8 what we know about how they may have been generated.

2.2 The CMB temperature as a function of redshift

Why is the CMB such a perfect black body? Shouldn't the redshifting of the photons distort the spectrum? It turns out that black body radiation has the remarkable property that it keeps a black body shape, independently of cosmic expansion. The black body spectrum is

$$I(\nu, T)\,\mathrm{d}\nu \propto \frac{\nu^3}{\mathrm{e}^{h\nu/kT} - 1}\,\mathrm{d}\nu, \tag{2.2}$$

where $I(\nu, T)\,\mathrm{d}\nu$ is the energy per unit area in a frequency interval ν to $\nu + \mathrm{d}\nu$, T is the temperature, k is Boltzmann's constant, and h is Planck's constant. If we substitute in $\nu' = \nu/(1+z)$, we find that

$$I(\nu', T)\,\frac{\mathrm{d}\nu'}{1+z} \propto \frac{(\nu')^3(1+z)^3}{\mathrm{e}^{h\nu'(1+z)/kT} - 1}\,\frac{\mathrm{d}\nu'}{1+z}$$

thus

$$I(\nu', T)\,\mathrm{d}\nu' \propto \frac{(\nu')^3}{\mathrm{e}^{h\nu'(1+z)/kT} - 1}\,\mathrm{d}\nu',$$

omitting the constant $(1+z)$ factors in the proportionality, so

$$I(\nu', T')\,\mathrm{d}\nu' \propto \frac{(\nu')^3}{\mathrm{e}^{h\nu'/kT'} - 1}\,\mathrm{d}\nu',$$

where $T' = T/(1+z)$. This has exactly the same form as Equation 2.2. There are some underlying physical reasons for this that we shall explore in this section.

The wavelengths have increased by a factor of $(1+z)$ since the photons were emitted, and the volumes have increased by a factor $(1+z)^3$. Since photon energy is related to frequency by $E = h\nu$, the energy density (i.e. energy per unit volume) has decreased by a factor of $(1+z)^{-4}$ since the photons were emitted. But the Stefan–Boltzmann law states that the energy density is proportional to T^4, where T is the temperature. Therefore

$$\frac{T_{\mathrm{emitted}}}{T_{\mathrm{observed}}} = (1+z). \tag{2.3}$$

We therefore have a clear prediction that the CMB was warmer in the past: $T_{\mathrm{CMB}}(z) = (1+z)\,T_{\mathrm{CMB}}(0)$, where $T_{\mathrm{CMB}}(0)$ is the present-day observed temperature of the CMB. But how could we tell? There are no friendly aliens beaming us their measurements of T_{CMB} from earlier in the Universe (or at least, none that we know of). But the CMB sets a minimum temperature for astronomical objects, because anything cooler would be heated by the CMB, and we can detect the effects of this minimum temperature.

The temperature of a gas affects the energy levels of the gas molecules. Energy levels of the order kT tend to be populated by the molecules, while energy levels $\gg kT$ or $\ll kT$ are not. The relative numbers in states close to kT will depend on whether the gas is monatomic, diatomic (and so with extra rotational and vibrational modes) or more complex, but the temperature-dependence of these numbers is calculable and known for very many molecules. When a molecule changes state, it involves the emission or absorption of a photon, and these emission or absorption lines can be used to both identify the molecular species and derive the redshift (see, for example, Figure 1.10 in Chapter 1). The relative strengths of emission lines of any particular molecular species will depend on the temperature, because the amount of any particular emission will depend on the number of molecules in the emission process's initial state. We can therefore use the relative strengths of emission lines in high-redshift objects to constrain the gas temperatures, and so place upper limits on the CMB temperature.

Srianand, R. et al., 2008, *Astronomy & Astrophysics*, **482**, L39–42.

One recent measurement of this temperature at a redshift of $z = 2.418\,37$ yielded $T = 9.15 \pm 0.32$ K using carbon monoxide rotational modes, and the CMB has been argued to dominate the CO excitation in this system. This beautifully confirms the predicted CMB temperature of 9.315 ± 0.007 K at that redshift.

One subtle question often asked about the CMB temperature is: where did the photons' energy go? The emitted energy of any photon would have been $h\,\nu_{\text{emitted}}$, but the energy received is $h\,\nu_{\text{emitted}}/(1+z)$.

Could it be a gravitational redshift? A gravitational field can certainly redshift photons. A light signal sent from the surface of the Earth will be received in space with a slightly redder wavelength, because energy has been lost by climbing up the gravitational potential well. However, this cannot be the case in the expanding Universe, because it is homogeneous and isotropic.

An analogy is sometimes made to the adiabatic expansion of a photon gas. If you have a photon gas contained in a box, and expand the box by a factor $(1+z)$ (so the volume increases by $(1+z)^3$), the energy density will decrease by a factor $(1+z)^{-4}$ and the temperature by $(1+z)^{-1}$, as we have argued above. However, in this case the photon gas does $p\,\mathrm{d}V$ work against the sides of the box. In the cosmological case, though, the issue of the $p\,\mathrm{d}V$ work is much more subtle.

Unfortunately, this issue takes us into very deep waters. In general relativity, the separate concepts of energy conservation and momentum conservation are replaced by zero derivatives of the 'energy–momentum tensor'. The short answer is that 'energy per unit volume' becomes dependent on the reference frame, so 'energy' on its own cannot be said to be conserved, although a more general conservation law *does* apply (see the further reading section).

To see why 'energy per unit volume' is dependent on the reference frame, imagine that the Universe is filled with a pressureless gas of particles, and for simplicity assume just the flat Minkowski metric of special relativity. Suppose that the particles each have a rest mass m and that their number density is n particles per unit volume. The energy density will therefore be $mc^2 \times n$. If we now make a Lorentz transformation to a reference frame that's moving relative to the first, we can see that the mass will be increased by a factor of γ to γm, while Lorentz contraction of the volumes will result in an increase in n by the same factor, to γn. The moving observer will therefore see an energy density of $\gamma mc^2 \times \gamma n = \gamma^2 mc^2 n$. Therefore the energy density isn't an invariant scalar, because different observers see different energy densities. It also can't be a component of a four-vector, because when you Lorentz transform a four-vector you get only one γ factor. So 'energy density' has to be part of a different sort of mathematical object. This object is a **tensor**; we shall meet more examples of tensors later in this book.

Exercise 2.3 Show that the redshift of matter–radiation equality z_{eq}, when the energy densities of matter and radiation (including neutrinos) were comparable, is given by

$$1 + z_{\text{eq}} = 23\,800\,\Omega_{\text{m},0}\,h^2 (T_{\text{CMB},0}/2.725\,\text{K})^{-4}, \tag{2.4}$$

where $T_{\text{CMB},0}$ is the present-day CMB temperature. ■

At early enough times, the Universe was radiation-dominated, with $\Omega_r \simeq 1$ and negligible contributions from other density parameters. Figure 1.11 shows the evolution of density parameters.

Exercise 2.4 In Section 1.6, we showed that a universe with $\Omega_m = 1$ and all other density parameters zero obeys $a = R/R_0 \propto t^{2/3}$. Make a similar analysis for a universe in which $\Omega_r = 1$ and the other density parameters are negligible, and show that in this case $a = R/R_0 \propto t^{1/2}$. ■

2.3 Why is the CMB a black body?

The previous section asked why the CMB radiation should be such a perfect black body, but arguably failed to answer the question; we showed only that a black body should stay a black body, not why it should be black body radiation in the first place. Figure 2.1 shows the spectrum observed from the COBE satellite. (The COBE FIRAS error bars are smaller than the plot symbols.)

One reason why we might expect the spectrum *not* to be thermal is that the early Universe was very young — was there enough time for the particles to thermalize? A relatively simple argument shows that the young age of the Universe isn't a problem. The early Universe is expanding very fast and (on purely dimensional grounds) we can define a dynamical timescale of $\tau_{\text{dyn}} \propto (G\rho)^{-1/2}$, where ρ is the energy density. We won't worry about the constant of proportionality for now. The early Universe must have been radiation-dominated (Chapter 1), so $\rho \propto R^{-4}$, where R is the scale factor of the Universe. Therefore the dynamical timescale will scale as $\tau_{\text{dyn}} \propto R^{+2}$, and so the dynamical inverse timescale varies as $1/\tau_{\text{dyn}} \propto R^{-2}$. Meanwhile, thermalization of the particles will happen through interactions between the particles. Even photons can interact with each other, though the interaction probability is low except at the highest energies. The number of collisions that a single particle will experience per unit time will be proportional to the number density of particles that it's interacting with, n. To find the *total* number of collisions for all the particles, we multiply this again by the number density of particles. The collision rate must therefore be proportional to n^2, which scales as $n^2 \propto (R^{-3})^2 = R^{-6}$. So as R tends to zero, the collision rate ($\propto R^{-6}$) increases much faster than the dynamical rate ($\propto R^{-2}$). Regardless of the constants of proportionality, thermalization will be increasingly easy at earlier times, and we must reach a time in the early Universe when the particles are thermalized. The current Big Bang model is sometimes referred to as the **Hot Big Bang**, distinguishing it from a previous (but now disfavoured) model.

The closeness of the spectrum to a black body is also strong evidence against the early Steady State models of the Universe, which have no Big Bang but a continual creation of matter throughout all space. (The term 'Big Bang' was originally a pejorative coined by a Steady State proponent, Fred Hoyle.) The Steady State model supposed that the CMB was thermal radiation from dust clouds, which were themselves heated by stars (much like the Orion nebula). However, this would predict that the CMB is comprised of a range of temperatures from a variety of redshifts, so the spectrum would not resemble a single black body. Steady State models have fallen out of favour, partly because of the lack of a plausible mechanism for the steady state creation of matter throughout all space, and partly because they struggle to account for other observations (microwave background and its anisotropies, the evolution of large-scale structure, Big Bang nucleosynthesis, etc.) without appearing contrived to many. For these reasons we won't discuss these models further in this book.

2.4 Baryogenesis

One of the deep unsolved problems in fundamental physics is why our Universe has more matter than antimatter. In the standard model of particle physics, **baryon number** is strictly conserved. (A baryon such as a proton or a neutron has a baryon number of $B = +1$, while an antibaryon such as an antiproton or antineutron has a baryon number of $B = -1$.) Baryons are created and destroyed in baryon–antibaryon pairs, conserving baryon number, yet somehow we find ourselves in a Universe composed almost entirely of matter. We could adopt the position that it's somehow set in the initial conditions, as is also argued for 'explaining' the expansion of the Universe, but in both cases this arguably evades the question. One might suppose that there are distant galaxies made of antimatter rather than matter, but the intergalactic medium is not empty and there should be very clear observable signatures from ongoing annihilation along the boundary between the matter and antimatter regions.

While the Universe was hot enough that kT was above the proton rest mass energy, protons and antiprotons would have been being created and destroyed from particles colliding at their thermal velocities. As the Universe expanded and cooled, this baryon–antibaryon creation eventually ceased and the protons and antiprotons were free to annihilate, and it turns out that the collision rates were high enough for this to happen very efficiently. What we see now as a Universe with mainly matter (rather than antimatter) is in fact the relic of a subtle asymmetry in the early baryon versus antibaryon numbers, of the order of $1 + 10^{-9}$ protons for every antiproton. What generated this initial imbalance? Why is there any matter left in the Universe?

This on its own tells us that there must be new physics beyond the standard model of particle physics. What sort of theory could explain baryosynthesis? The answer may come from so-called grand unified theories (GUTs) that unify three of the four fundamental forces: the strong nuclear force, the weak nuclear force and electromagnetism. These forces appear distinct with different strengths, but their strengths are predicted in GUTs to converge at energies of $\sim 10^{15}$ GeV. GUT reactions above these energies could be the source of the present-day baryon asymmetry in the Universe. However, it's by no means certain that this is the correct answer. An epoch of inflation (which we shall meet in Section 2.8) would erase any pre-existing baryon asymmetry. Inflation is thought to be triggered by a GUT-scale phase transition (see Section 2.8), and once inflation has finished, it leaves the Universe at a lower temperature than the GUT scale. This would leave us with no baryon asymmetry, unless generated during the processes that end inflation.

There must also have been a primordial lepton asymmetry, otherwise the excess number of protons over antiprotons would have left the Universe with an overall electric charge. GUTs view baryons and leptons as different states of one common species of particle, and in many GUTs, $B - L$ is conserved (L being the lepton number, e.g. +1 for e and ν_e, −1 for e^+ and $\overline{\nu}_e$), so this lepton asymmetry would be a natural consequence of the baryon asymmetry. The present-day lepton asymmetry would now reside in the cosmic neutrino background, which we shall meet in Section 2.6.

2.5 The entropy per baryon

Once the baryon–antibaryon annihilation finished, leaving the residual baryon excess, the comoving baryon density was conserved. The next major annihilation stage in the Universe's history was $e^\pm$ annihilation through the reaction $e^- + e^+ \rightarrow \gamma + \gamma$, i.e. creating two photons. After this point and until the first starlight illuminated the Universe (Chapter 8), the comoving number density of photons is conserved. It's conventional to measure the baryon asymmetry in terms of the photon abundance: we define a new quantity η as

$$\eta = \frac{n_b - n_{\overline{b}}}{n_\gamma}, \tag{2.5}$$

where n_b is the comoving density of baryons, $n_{\overline{b}}$ is that of antibaryons, and n_γ is that of photons. (Sometimes the notation $\eta = 10^{-10}\eta_{10}$ is used, though we won't make use of this notation in this book.)

The quantity η is sometimes referred to as a measure of the **entropy per baryon**. A photon gas has an entropy that's proportional to the number of photons, so η is in fact proportional to the reciprocal of the entropy per baryon. Another way of measuring the baryon density of the Universe is as a fraction of the critical density, i.e. $\Omega_{b,0}$ (see Chapter 1). The η parameter is related to $\Omega_{b,0}$ via

$$\Omega_{b,0}\, h^2 = \frac{\eta}{2.74 \times 10^{-8}}. \tag{2.6}$$

If we take the entropy in the photon gas and divide it by the cosmological matter density, the entropy per unit mass comes out as $1.09 \times 10^{12}(\Omega_{m,0}\, h^2)^{-1}\,\mathrm{J\,K^{-1}\,kg^{-1}}$. To give you some sense of the scale of this number, taking 300 K water and raising its temperature by 1 K raises the entropy by only $14\,\mathrm{J\,K^{-1}\,kg^{-1}}$. The entropy content of the Universe is overwhelmingly dominated by the entropy of the CMB (neglecting black holes), which is why the expansion can be treated as adiabatic and reversible.

2.6 Primordial nucleosynthesis: a thousand seconds that shaped the Universe

2.6.1 The primordial fireball

If we mentally rewind the history of the Universe, we shall reach a point where the Universe becomes opaque like the photosphere of a star. As we've seen, the end of that opaque epoch is the CMB, i.e. the surface of last scattering. If we continue rewinding, the Universe will become hotter and denser, and should ultimately reach the temperature of the Sun's core. Should we therefore expect nuclear reactions? It turns out that most of the energy density then was in radiation rather than matter (Chapter 1) so baryons (e.g. protons and neutrons) would have been much rarer relative to photons, compared to the Sun's core. Nuclear reactions at that time were therefore slow. However, it turns out that the nuclear reaction rates were significant earlier on, at higher temperatures of around 10^9 K.

It's an extraordinary intellectual triumph of the Big Bang theory that it's possible to calculate the nuclear reaction rates in the early Universe and estimate the abundances of nuclei and particles from this primordial nucleosynthesis. These estimates are in fairly good agreement with observations, as we shall see in this section. The key concept is **freeze-out**, which we shall meet first in the context of protons and neutrons.

In thermal equilibrium, the relative numbers of protons and neutrons, n_p and n_n respectively, will be related through a Boltzmann distribution:

$$\frac{n_\mathrm{n}}{n_\mathrm{p}} = \exp\left(\frac{-\Delta m\, c^2}{kT}\right) \simeq \exp\left(\frac{-10^{10.176}\,\mathrm{K}}{T}\right), \tag{2.7}$$

where $\Delta m \simeq 1.29\,\mathrm{MeV}/c^2$ is the mass difference between a neutron and a proton. (Strictly speaking this is true only when the protons and neutrons are non-relativistic, but this is the case in the following discussion.) Protons and neutrons will be converting between each other through the reactions $\mathrm{p} + \mathrm{e}^- \rightleftharpoons \mathrm{n} + \nu_\mathrm{e}$ and $\mathrm{p} + \overline{\nu}_\mathrm{e} \rightleftharpoons \mathrm{n} + \mathrm{e}^+$. The rate v of either reaction can be calculated through the theory of the weak nuclear force, and in the high-temperature limit it turns out that the reaction rates are the same and have a very strong temperature-dependence:

$$v = \left(\frac{10^{10.135}\,\mathrm{K}}{T}\right)^{-5} \mathrm{s}^{-1}. \tag{2.8}$$

Meanwhile, the ambient temperature is changing as the Universe expands, varying as $(1+z)$ (see Section 2.2). The early Universe was approximately spatially flat (because $\Omega_k \simeq 0$ in the early Universe — see Chapter 1) and radiation-dominated (also Chapter 1), from which we find that the scale factor satisfies $R \propto t^{1/2}$ (Exercise 2.4). From this it follows that the temperature varies with time as $t \propto T^{-2}$, and putting in the constants of proportionality gives

$$t = \left(\frac{10^{10.125}\,\mathrm{K}}{T}\right)^{2} \mathrm{s}. \tag{2.9}$$

As a result, the reaction timescale $(1/v)$ for $\mathrm{p} + \mathrm{e}^- \rightleftharpoons \mathrm{n} + \nu_\mathrm{e}$ will quite suddenly become longer than the age of the Universe at a time $t = 1/v$. After this point the neutron–proton reactions cease and the neutron–proton ratio (Equation 2.7) is **frozen out** at the equilibrium value that it had at that time. The time $t = 1/v$ corresponds to a temperature of

$$\left(\frac{10^{10.125}\,\mathrm{K}}{T}\right)^{2} = \left(\frac{10^{10.135}\,\mathrm{K}}{T}\right)^{5}, \tag{2.10}$$

which we can rearrange to find $T = 10^{10.142}$ K. Plugging this into Equation 2.7, we find that the relic neutron–proton ratio must be

$$\frac{n_\mathrm{n}}{n_\mathrm{p}} \simeq \exp\left(\frac{-10^{10.176}\,\mathrm{K}}{10^{10.142}\,\mathrm{K}}\right) \simeq 0.34. \tag{2.11}$$

At this time, the Universe was only about one second old (Equation 2.9).

There are some slight corrections to this, and more detailed calculations give $n_\mathrm{n}/n_\mathrm{p} \simeq 1/7$ for the present-day relic abundance of neutrons and protons. For example, we've assumed an instantaneous transition. Also, the temperature-dependence of the rate will be slightly modified at lower temperatures, because Equation 2.8 is the high-T limiting case. Another potential complication is the fact that other reactions will be happening at the same time as, for example, $\mathrm{p} + \mathrm{e}^- \rightleftharpoons \mathrm{n} + \nu_\mathrm{e}$. For instance, atomic nuclei will form from the protons and neutrons. While the ambient thermal energy kT of particles is much larger than nuclear binding energies, these nuclei will quickly be destroyed again, so we can ignore these reactions for now.

We've gone carefully through this freeze-out process because it's one of the key physical principles in the nuclear reactions of the primordial fireball. For example, the electron–positron annihilation creates neutrinos through the weak interaction reaction $\mathrm{e}^- + \mathrm{e}^+ \rightleftharpoons \nu_\mathrm{e} + \overline{\nu}_\mathrm{e}$ (as well as annihilating through the electromagnetic interaction $\mathrm{e}^- + \mathrm{e}^+ \rightleftharpoons \gamma + \gamma$). Neutrino production freezes out at a temperature of approximately $10^{10.5}$ K, corresponding to a cosmic time of about 0.18 seconds, leaving most of the electrons and positrons to annihilate to make more photons, which happens at a temperature of $T \simeq m_\mathrm{e}c^2/k \simeq 10^{9.77}$ K, corresponding to a time (Equation 2.9) of about five seconds, i.e. shortly after the neutron–proton freeze-out. Neutrinos interact only very weakly with other matter, and their freeze-out happened before the epoch of the CMB; there should be a cosmic neutrino background from much earlier cosmic epochs than the CMB. Detailed calculations of the relic abundances of protons and neutrons take into account the ongoing changes in the neutrino population, though we won't discuss this in this book. There are formidable experimental challenges to the direct detection of the primordial neutrino background, though the presence of these neutrinos can be inferred indirectly from the structures in the CMB. (Nevertheless, neutrinos from astrophysical sources have been detected, most famously from the supernova SN 1987A.)

Neutrons aren't stable but instead decay with a half-life of $\tau = 885.7 \pm 0.8$ s. Why are there any left? Why isn't the Universe pure hydrogen? Luckily for us, the temperatures soon dropped enough to allow the formation of atomic nuclei. To see why, compare the binding energy of the first nuclide heavier than hydrogen (deuteron, 2.225 MeV) with the electron rest mass energy (0.511 MeV) and the neutron–proton mass difference (1.3 MeV). The Universe was still only a few seconds old.

A note on terminology

Children are taught at school that nuclear reactions are not combustion, so it's not correct to refer to nuclear reactions as 'burning'. This is quite right. However, at this level it's usually felt that there is no danger of confusing this with combustion, so the technical literature makes free use of the verb 'to burn' and related words. For example, the early Universe is sometimes referred to as the primordial fireball. I once heard a supernova described in a seminar as 'like a forest fire, but the trees can run away'.

The nuclear reactions in the next 1000 seconds or so shaped the baryonic content of the Universe. To calculate the final mix of elements left at the end of these nuclear reactions, you need to consider all the different reactions. We won't do this here (see the further reading section for more details), but the main processes are summarized in Table 2.1 and Figure 2.3.

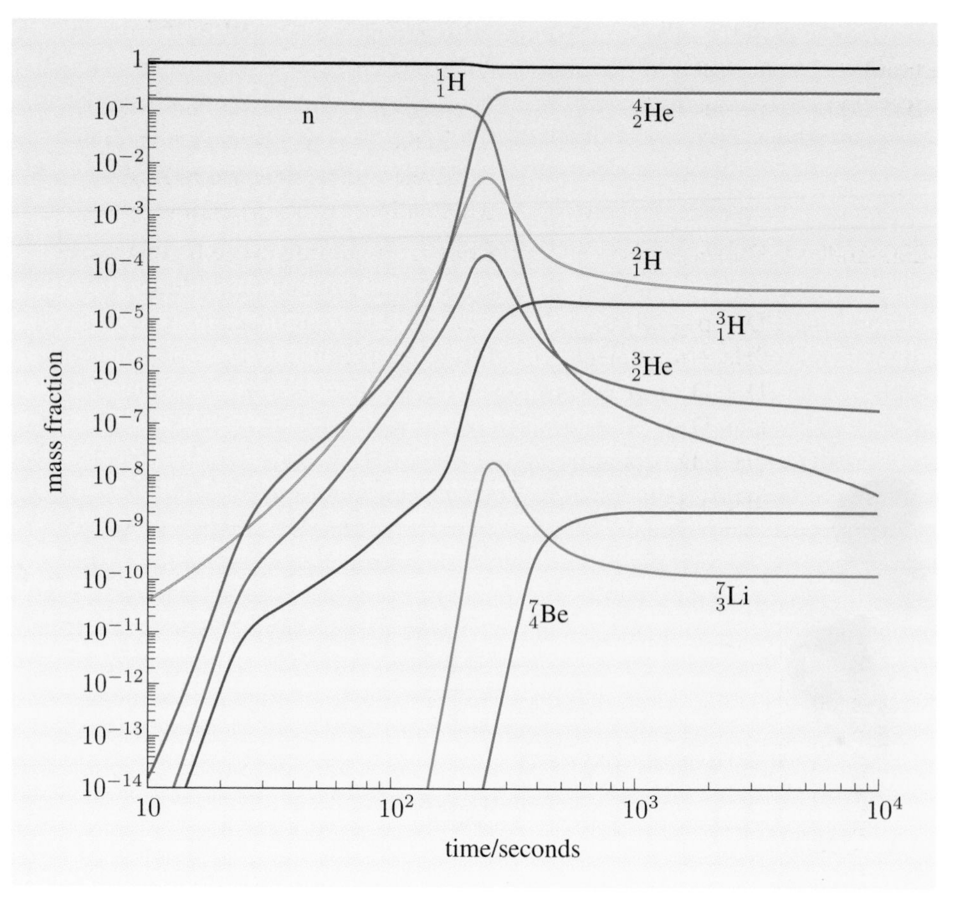

Figure 2.3 Predicted light element abundances during the primordial fireball. Many of the reactions involved in changing these abundances are listed in Table 2.1.

2.6.2 The primordial element abundances

So, after this fireball, how did the Universe end up? The element abundances depend on the number of baryons per photon, or (equivalently) on η or $\Omega_{b,0}\, h^2$.

The value of η has been very well determined by the WMAP satellite (of which more later) to be $\eta = (6.14 \pm 0.25) \times 10^{-10}$. Big Bang nucleosynthesis therefore makes very clear predictions for the primordial abundances of elements created in the first half hour of the Universe's existence. We can test these predictions, and the overall level of agreement with observations is one of the many successes of the Big Bang model (and challenges to rivals such as Steady State models). However, the tricky part of these experiments is to find baryonic matter that has remained in its primordial condition for the ~13.7 billion years since primordial nucleosynthesis. Figure 2.4 shows a compilation of constraints on η.

Table 2.1 Some of the important reactions in primordial nucleosynthesis.

Time since the Big Bang	Reactions	Description
$< 1\,\mathrm{s}$	$\mathrm{p} + \mathrm{e}^- \rightleftharpoons \mathrm{n} + \nu_\mathrm{e}$, $\mathrm{n} + \mathrm{e}^+ \rightleftharpoons \mathrm{p} + \overline{\nu}_\mathrm{e}$	Neutron–proton freeze-out sets the subsequent neutron–proton ratio.
$\sim$1–100 s	$\mathrm{n} \to \mathrm{p} + \mathrm{e}^- + \overline{\nu}_\mathrm{e}$	Neutrons then decay.
$\sim$100–200 s	$\mathrm{p} + \mathrm{n} \rightleftharpoons \mathrm{D} + \gamma$	The reason why there are any neutrons left in the Universe is nuclear reactions that create stable deuterium nuclei, which allow further reactions.
$\sim$200–1000 s	$\mathrm{D} + \mathrm{n} \to {}^3_1\mathrm{H} + \gamma$, ${}^3_1\mathrm{H} + \mathrm{p} \to {}^4_2\mathrm{He} + \gamma$, $\mathrm{D} + \mathrm{p} \to {}^3_2\mathrm{He} + \gamma$, ${}^3_2\mathrm{He} + \mathrm{n} \to {}^4_2\mathrm{He} + \gamma$, $\mathrm{D} + \mathrm{D} \to {}^3_2\mathrm{He} + \mathrm{n}$, $\mathrm{D} + \mathrm{D} \to {}^3_1\mathrm{H} + \mathrm{p}$, ${}^3_1\mathrm{H} + \mathrm{D} \to {}^4_2\mathrm{He} + \mathrm{n}$, ${}^3_2\mathrm{He} + \mathrm{D} \to {}^4_2\mathrm{He} + \mathrm{p}$	Deuterium burning exceeds deuterium creation. The net effect of these deuterium burning reactions is $\mathrm{D} + \mathrm{D} \to {}^4_2\mathrm{He} + \gamma$.

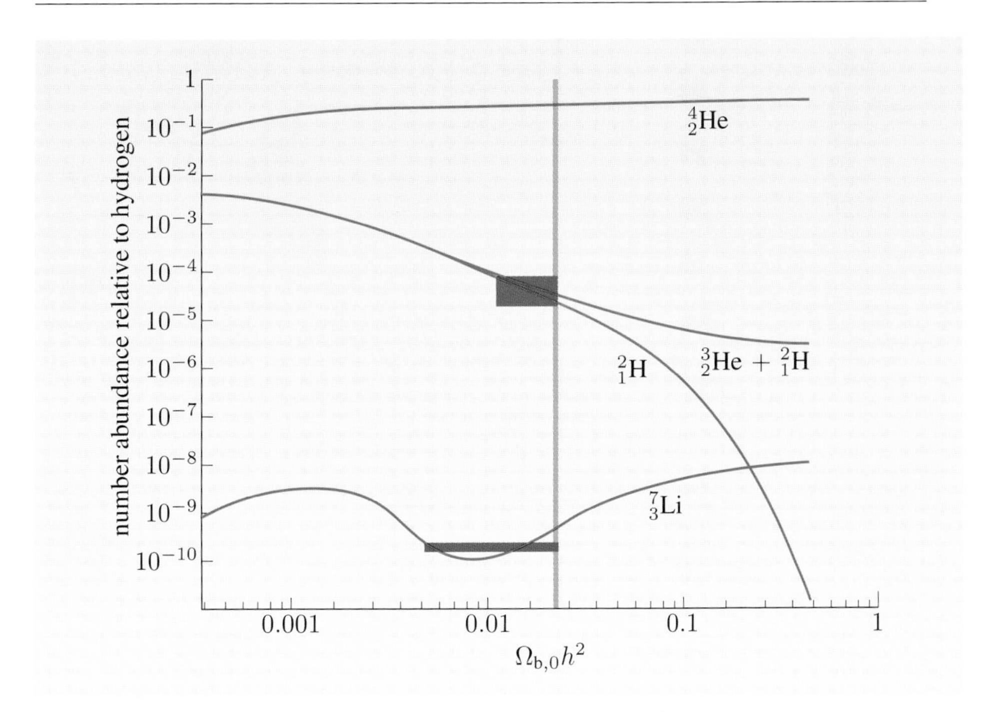

Figure 2.4 The light element abundances relative to ${}^1_1\mathrm{H}$, as a function of the present-day baryon density times h^2. The vertical area is the constraint on the baryon density from WMAP. The curves show the predictions from nucleosynthesis calculations, while the horizontal boxes show the observational constraints. There is a broadly consistent picture apart from the ${}^7_3\mathrm{Li}$ abundance, but this element can be destroyed in stars so is difficult to measure.

The best constraints on the deuterium abundance have come from the absorption lines in neutral hydrogen clouds. Here, the light from a background source (such as a quasar) passes through a foreground dense neutral hydrogen clump. The abundance of deuterium is low, but if that clump is sufficiently dense, characteristic absorption lines can be detected. We shall return to this in Chapter 8. Deuterium is destroyed in stellar nucleosynthesis, so the observed deuterium abundance is more accurately described as a lower bound to the primordial abundance.

The lithium abundance is very difficult to measure because lithium is relatively rare. Cosmic rays and stellar nucleosynthesis can both produce ^{7}Li long after primordial nucleosynthesis, and ^{7}Li can also be destroyed in stellar interiors. Attempts have been made to measure the ^{7}Li abundance in old, metal-poor stars for which one might hope that these reactions are minimized. However, as we see in Figure 2.4, there seems to be some discrepancy with Big Bang nucleosynthesis. There is a very active debate over whether this discrepancy is the signature of new physics, or whether there are unknown systematic errors in the abundance calculations, or whether we are seeing the destruction of lithium in the previous generation of stars, or whether there are uncertainties in the stellar temperature scale that is used to convert the lithium absorption line depths to lithium abundances.

See, for example, Steigman, G., 2007, *Annual Review of Nuclear and Particle Science*, **57**, 463.

The $^{3}_{2}$He abundance is also complicated and model-dependent. Again, it's both created and destroyed in stars. The strongest constraints on the $^{3}_{2}$He abundance currently come from the metal-poor H II region that is most distant from the Galactic Centre. $^{3}_{2}$He is less sensitive to changes in η than D (see Figure 2.4) so the constraints on η in Figure 2.4 are correspondingly weaker.

Perhaps the biggest surprise is the observed $^{4}_{2}$He abundance. The abundance depends only very weakly on η, and with the WMAP value of η, the abundance of $^{4}_{2}$He predicted by primordial nucleosynthesis is $Y_p = 0.2485 \pm 0.0008$. The $^{4}_{2}$He abundance monotonically increases with time in stellar nucleosynthesis, but as the oxygen abundance (also created in stellar nucleosynthesis) in stars tends to zero, the $^{4}_{2}$He abundance should tend towards its primordial value. Some measurements of the primordial $^{4}_{2}$He abundance (also written Y_p) differ from the nucleosynthesis prediction by more than 2σ, i.e. two standard deviations! However, there is considerable debate over the *systematic* errors present in these data, which appear to be significant. One approach is to derive a conservative upper limit to Y_p using only the best-studied systems for which systematics are best characterized. In this way a 2σ limit of $Y_p < 0.254$ has been found, consistent with Big Bang nucleosynthesis. Alternatively, the currently most defensible compilation of observations for which systematics can be well characterized gives $Y_p = 0.240 \pm 0.006$. This is only marginally discrepant ($\sim 1.4\sigma$) with the Big Bang nucleosynthesis value.

Again see, for example, Steigman, 2007.

If the $^{4}_{2}$He abundance is found ultimately to be discrepant with primordial nucleosynthesis, it may be a signature of unknown new physical laws, for example modifying the expansion rate at early post-inflationary epochs (but only by at most a few tens of per cent). This would change the amount of time available for primordial nucleosynthesis and so change the final abundances.

2.7 The need for new physics

The primordial abundances of light elements have long been felt by some to hint tantalizingly at new unknown physical laws, but there are many much stronger hints. For example, where did the matter–antimatter asymmetry of the Universe come from (Section 2.4)? Also, what caused the initial inhomogeneities in the Universe? If the Universe were perfectly homogeneous, it would have stayed homogeneous and no stars or galaxies would have formed. Something must have given the Universe its initial density perturbations. Also, what triggered the initial expansion? This is often written off as part of the initial conditions, but isn't that evading the question? We've also met the fine-tuning problems for the cosmological density parameters Ω_m and Ω_Λ in Chapter 1 (Section 1.7), known as the **flatness problem**.

There is another very fundamental problem posed by the uniformity of the microwave background itself, known as the **horizon problem**. Suppose that you are at some place in the early Universe, arbitrarily close to the time of the Big Bang. You send a photon out. Neglecting the opacity of the Universe, how far will that photon travel? Anything further could not have been in causal contact with you since the Big Bang, so the distance that the photon travels sets the size of the causally-connected region.

We start from the Robertson–Walker metric (Equation 1.6) with the approximation of a spatially-flat universe (Figure 1.11) so $k = 0$. We can set the origin to where the photon starts so the light ray is radial, so $\mathrm{d}\theta = \mathrm{d}\phi = 0$. Also, all light rays have $\mathrm{d}s = 0$, so we have that $R(t)\,\mathrm{d}r = c\,\mathrm{d}t$, where $R(t)$ is the scale factor of the Universe, as in Chapter 1. The proper distance travelled by the photon will therefore be

$$r = \int_0^t \frac{c\,\mathrm{d}t}{R(t)}. \tag{2.12}$$

In the early radiation-dominated Universe, $R \propto t^{1/2}$ (Exercise 2.4) so this integral converged to a finite value. In the later matter-dominated Universe (before Ω_Λ became significant), we had $R \propto t^{2/3}$, which again converges. The size of this causally-connected region is known as the **particle horizon**. Note that this is different to the *event* horizon (Section 1.11). To calculate the event horizon, you'd integrate from t to infinity, not from 0 to t.

We saw in Section 1.9 that Equation 1.44 implies that the size of the comoving distance to $z = \infty$ in an $\Omega_\mathrm{m} = 1$, $\Lambda = 0$ universe is $2c/H_0$. The proper distance to the particle horizon must therefore be $2c/H(z)$, where $H(z)$ is the Hubble parameter at redshift z. (If we had assumed a radiation-dominated universe, this would come out as $c/H(z)$, which is a factor of two smaller.) The value of $H(z)$ at the time of recombination comes out as $18\,200 \times H_0$ using Equation 1.33 and the values of the density parameters in Section 1.5. The particle horizon size then comes out as $2c/H = 2c/(H_0 \times 18\,200) = (2c/H_0) \times 5.5 \times 10^{-5}$, or about $0.46\,\mathrm{Mpc}$ for an H_0 of $72\,\mathrm{km\,s^{-1}\,Mpc^{-1}}$.

The horizon problem is that $0.46\,\mathrm{Mpc}$ on the CMB is very small: by numerically integrating Equation 1.44, we find that the comoving distance to $z = 1090$ comes out as $14\,189\,\mathrm{Mpc}$, so the angular diameter distance (Equation 1.47) is just $14\,189/1091 \simeq 13\,\mathrm{Mpc}$. The angular size of the particle horizon at the time of recombination is just $0.46/13$ radians, or about 2 degrees (or slightly less if we take into account the early radiation-dominated phase). We've just shown that objects further apart than this distance could not have been in causal contact, so how is it that parts of the CMB sky more distant than two degrees ever managed look so similar?

There is also a problem that arises from almost all grand unified theories (GUTs) that seek to unify three of the four fundamental forces (electromagnetism, strong nuclear force, weak nuclear force). As the Universe expanded and cooled, the GUT field (whatever it was) would settle into particular configurations. This is rather like paramagnetism, where below a critical temperature (the 'Curie temperature') the magnetic moments of molecules align with those of their neighbours into magnetic domains. In the cosmological case these domains can have various sorts of boundaries, including a monopole state where the local field points radially away from a particular point. Macroscopically this would look like a magnetic monopole. GUTs predict about one monopole per horizon size at the time when the Universe was at the critical GUT temperature, but as this was very early in the Universe, the horizon size was small. Therefore the present-day Universe should have many magnetic monopoles — so many, in fact, that they would dominate the energy density of the Universe. Why do we not see them in the Universe? This is known as the **monopole problem**.

Perhaps the solution to all these problems is at the Planck epoch. We currently have no consistent theory that unifies quantum mechanics and general relativity. Where should we expect such a theory to be needed? Presumably the theory would need to use $\hbar$, G and c, so we can use these to derive a characteristic length, mass and time:

$$m_{\rm Pl} = \sqrt{\frac{\hbar c}{G}} \simeq 10^{19}\,{\rm GeV}/c^2, \tag{2.13}$$

$$r_{\rm Pl} = \sqrt{\frac{\hbar G}{c^3}} \simeq 10^{-35}\,{\rm m}, \tag{2.14}$$

$$t_{\rm Pl} = \sqrt{\frac{\hbar G}{c^5}} \simeq 10^{-43}\,{\rm s}. \tag{2.15}$$

These are known as the **Planck mass**, **Planck length** and **Planck time**, respectively. When a mass, length or time interval under consideration is of the order of the Planck scales, we should expect an unknown theory of quantum gravity to be needed. Clearly the initial singularity at $t = 0$ in the Robertson–Walker metric is an example, as is the singularity at the centre of a black hole (Chapter 6). Note that the GUT energy scale of 10^{15} GeV is a factor of 10^4 from the Planck scale. While 10^4 might be considered a large factor, the current temperature of the CMB of 2.7 K is equivalent to about 2×10^{-4} eV, i.e. a factor of about 10^{28} from the GUT epoch.

Exercise 2.5 Show on dimensional grounds that the only characteristic timescale involving $\hbar$, G and c is proportional to $\sqrt{\hbar G/c^5}$. ■

It may be that in order to solve all these problems (monopole, flatness, horizon, baryon asymmetry, density perturbations, initial expansion) we need the unknown theory of quantum gravity at the earliest times in the Universe ($t \simeq t_{\rm Pl}$). However, there has been a proposal to solve these problems at the later GUT epoch in the Universe's history when the characteristic temperature was around the GUT scale of approximately 10^{15} GeV, in which GUT-scale physics triggers a very rapid phase of expansion known as **inflation**. Before describing what triggers this phase, we'll first look at how this solves some of these problems.

We've shown that the particle horizon size is $r = \int_0^t c\,{\rm d}t'/R(t')$. If we want this to diverge, we'll need an expansion rate $R(t)$ much faster than the $t^{1/2}$

in the radiation-dominated era or the $t^{2/3}$ in the matter-dominated era. If we suppose that $R(t) \propto t^\alpha$, then this integral gives the horizon size as $ct^{1-\alpha}/(1-\alpha)$ evaluated from $t = 0$ to t, i.e.

$$r = \frac{c}{1-\alpha}\left(t^{1-\alpha} - 0^{1-\alpha}\right). \tag{2.16}$$

If $\alpha > 1$, then the horizon size formally diverges, if the t^α expansion operated right back to the Big Bang at $t = 0$.

Physically, the t^α phase would be a period of very rapid expansion in the Universe. This would immediately solve the horizon problem, because the regions that appear to us to be causally disconnected were in fact once part of a much smaller, causally-connected region that was then inflated to a much bigger size. We can calculate the minimum amount of inflation needed to solve the horizon problem: we need that the comoving horizon size when the Universe had a temperature at the GUT scale ($E \simeq 10^{15}$ GeV, i.e. $T = E/k \simeq 10^{28}$ K) was inflated to at least the horizon size today, when the CMB temperature is 2.7 K. The redshift of the GUT epoch was therefore $1 + z \simeq 10^{28}/2.7 \simeq 10^{27.5}$. We can estimate the size of the proper causal horizon (in the absence of inflation) as ct_{GUT}, where t_{GUT} is the age of the Universe in the GUT epoch — about 10^{-35} s. The comoving size will be $cz\,t_{\text{GUT}}$ (using $1 + z \simeq z$), which comes out as about 10 metres. This is about a factor of e^{60} smaller than the current horizon size. Therefore at least about 60 e-foldings of inflation are needed; were it not for inflation, the present-day Universe would be strongly inhomogeneous everywhere on scales more than a few metres.

Inflation gives a mechanism for generating the initial density perturbations in the Universe. As we saw in Section 1.11, a universe dominated by a cosmological constant has an event horizon with a proper radius of c/H. This event horizon will generate **Hawking radiation**, which can be understood qualitatively as follows. Quantum mechanics predicts that virtual particle–antiparticle pairs will be created close to the event horizon, but sometimes one part of the pair will fall inside the event horizon while the other escapes. Event horizons should therefore have an associated energy radiation that has been predicted for black holes. Detailed quantum field theory calculations show that this radiation has a thermal spectrum. In the context of inflation, the resulting random quantum thermal fluctuations are ultimately the source of the initial cosmological density perturbations that eventually formed stars and galaxies. All forms of radiation should contribute to the Hawking radiation, including gravitational waves, so a prediction of inflation is a primordial gravitational wave background.

The monopole problem could also be solved by a period of inflation. There would have been many magnetic monopoles in the early Universe, but once made, the total number of cosmological monopoles is conserved. The period of inflation would then have greatly diluted their space density. If this inflation epoch is allowed to run for sufficiently long, the probability of finding even one monopole in the observable Universe could be vanishingly low.

We can infer more about the nature of the substance driving inflation if we use Equation 1.8, which we'll reproduce in a slightly modified form here:

$$\frac{1}{R}\frac{d^2R}{dt^2} = \frac{-4\pi G}{3}\left(\rho + \frac{3p}{c^2}\right) + \frac{\Lambda c^2}{3}. \tag{2.17}$$

As we found in Chapter 1, the Λ term makes a negligible contribution to the dynamics of the expansion in the early Universe, so we can neglect it. Let's suppose that R is proportional to t^α: i.e. $R(r) = nt^\alpha$, where n is some constant. The time derivative is $\mathrm{d}R/\mathrm{d}t = n\alpha t^{\alpha-1}$, and the second derivative is $\mathrm{d}^2R/\mathrm{d}t^2 = n\alpha(\alpha-1)t^{\alpha-2}$. Dividing this by R gives us $(1/R)\mathrm{d}^2R/\mathrm{d}t^2 = \alpha(\alpha-1)t^{-2}$, but this is also the left-hand side of Equation 2.17. In order to solve the horizon problem, we need that $\alpha > 1$, which means that $\alpha(\alpha-1)t^{-2}$ must be positive. This means that the left-hand side of Equation 2.17 must also be positive. In order for the right-hand side to be positive, we need that $\rho + (3p/c^2) < 0$.

What sort of substance would satisfy this? We characterize the equation of state of a gas as $p = w\rho c^2$, where p is the pressure and ρ is the density. The parameter w defines the equation of state. For example, $w = 0$ is pressureless matter (sometimes called 'dust'), while for a monatomic gas $w = 2/3$. A photon gas has $w = 1/3$. In this case we have $\rho + (3p/c^2) < 0$, which implies that $w < -1/3$. In other words, inflation needs a sort of negative pressure!

Exercise 2.6 Show that the inflation condition that $\alpha > 1$ is equivalent to the scale factor accelerating, i.e. $\mathrm{d}^2R/\mathrm{d}t^2 > 0$. ■

Inflation can also explain why the Universe is so close to being spatially flat (the flatness problem). From thermodynamics, an adiabatic expansion of a gas with an equation of state parameter w satisfies $p \propto V^{-(1+w)}$, where p is the pressure and V is the volume. If the rest mass density is negligible, then $\rho \propto V^{-(1+w)}$ too. In this case we can write $\rho \propto R^{-3(1+w)}$, where R is the scale factor. (As in the photon gas in Section 2.2, we're neglecting the issue of $p\,\mathrm{d}V$ work, but this equation turns out to be true in the fully general relativistic case.) The key to solving the flatness problem is in Equation 1.7 from Chapter 1, which we'll reproduce here:

$$\left(\frac{\mathrm{d}R}{\mathrm{d}t}\right)^2 = \frac{8\pi G\rho R^2}{3} - kc^2 + \frac{\Lambda c^2 R^2}{3}. \qquad \text{(Eqn 1.7)}$$

Again, we'll neglect the Λ term because it makes a negligible contribution to the dynamics in the early Universe. Suppose that the Universe had some arbitrary curvature constant k before inflation. As inflation expanded the Universe, the ρR^2 term in the equation above will vary as $\rho R^2 = R^{-3(1+w)}R^2 = R^{-3(1+w)+2} = R^{-(3w+1)}$. If $w < -1/3$, then the ρR^2 term increases with the scale factor. This means that the ρR^2 term will eventually dominate and the kc^2 term will be very small in comparison, and can be neglected or taken to be zero. Thus after inflation, the Universe is left in a state that is close to spatial flatness. Another way of thinking of this is that the process of inflation takes one tiny local patch that appears locally flat, and expands it enormously. Thus no matter how wrinkly the initial state of spacetime before inflation, a small enough local region will appear locally flat, so the result after inflation is a spacetime that's spatially flat. Spatial flatness is in fact a key prediction of inflation.

Finally, inflation also changes the estimates of the age of the Universe, because the epoch of inflation could be arbitrarily long. In principle this is one way of solving the singularity at $t = 0$ in the Robertson–Walker metric. Having said that, we can calculate the minimum time needed for inflation to solve the horizon problem. We found earlier that there need to be at least 60 e-foldings of expansion, and these will take a time $\Delta t \simeq 60H^{-1}$, where H is the Hubble parameter during inflation. This comes out as only 10^{-33} seconds.

But how is inflation triggered? We shall look at some possibilities in the next section.

2.8 The inflaton field

The idea behind inflation is to speculate that the Universe is filled with a **scalar field**. This is very different to the fields associated with the four fundamental forces that you have met so far. The electromagnetic force, weak nuclear force, strong nuclear force and (Newtonian) gravitational force are all **vector fields**, i.e. we can draw the force vector at every point in space, as illustrated in Figure 2.5. A scalar field, however, has no direction. It's just an intensity or strength at every point in space, also illustrated in Figure 2.5. We have never detected a *fundamental* scalar field in physics, though derived quantities such as temperature are scalar fields (but not fundamental ones). Should the Large Hadron Collider detect the Higgs boson, then this will be evidence for a fundamental scalar field known as the Higgs field, though it's known that the Higgs field can't be identical to the field responsible for inflation.

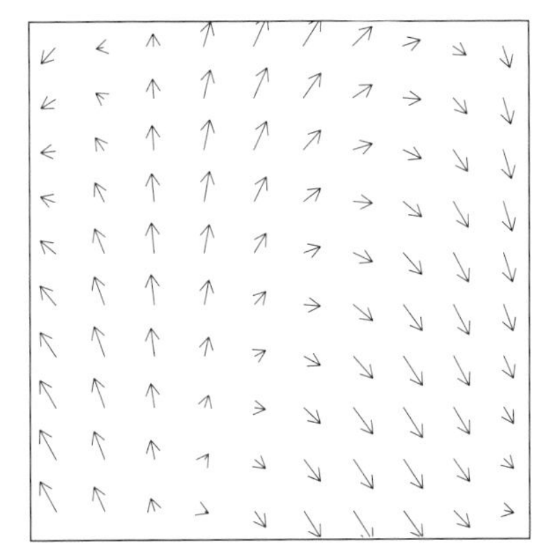

Figure 2.5 An example of a vector field and a scalar field. Vector fields have an amplitude (the length of the arrows) and a direction (where the arrows point) at every position. We've drawn arrows at only a few points so the figure isn't awash with overlapping arrows, but vector fields have position and direction everywhere, not just at a finite number of points. Scalar fields, on the other hand, have only an amplitude. We've represented this as a greyscale image.

The scalar field associated with inflation takes a value that we shall symbolize as ϕ. In general, ϕ can vary with time and space, though to a first approximation everywhere in the Universe will have the same value of ϕ at any one time. The time-variation of ϕ throughout the Universe will prove to be very important. The field ϕ has a particle associated with it, just as the electromagnetic field is associated with the photon. In this case the particle is known as the **inflaton** and the corresponding field ϕ is known as the **inflaton field**. Note that this is 'inflaton field', not 'inflation field'.

The value of the scalar field ϕ has an energy associated with it (in fact, it's an energy *density*), which we write as $V(\phi)$. We can imagine this as a valley with an object in a potential well, as shown in Figure 2.6. The height of the object in Figure 2.6 represents the energy $V(\phi)$, while the horizontal position represents the value of ϕ. If the object is offset from the minimum of $V(\phi)$, then it will slide down the well in this analogy, and indeed we imagine that there should be an energy term associated with $\mathrm{d}\phi/\mathrm{d}t$.

However, we don't know the shape of the potential. Taking into account the possibility of a temperature-dependent interaction with the other particles, it's expected that there should be a temperature-dependent 'effective potential' that could, for example, look like Figure 2.7. The Universe starts in the minimum configuration for ϕ, but as the temperature drops, the shape of the effective potential changes. The Universe may find itself in a secondary minimum or may find that the only minimum has shifted. In the former case, the Universe is in a **false vacuum** and could quantum tunnel through the barrier, at which point the value of ϕ could fall to the new state; in the latter case, the value of ϕ in the Universe will simply slide down to the new state.

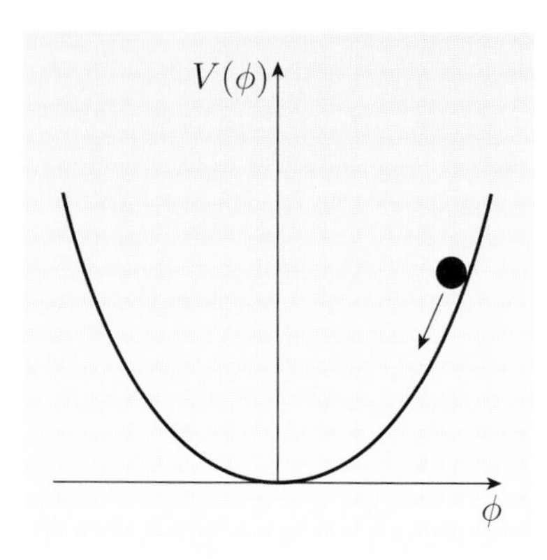

Figure 2.6 Schematic representation of the value of the inflaton field ϕ, versus the energy associated with the field $V(\phi)$.

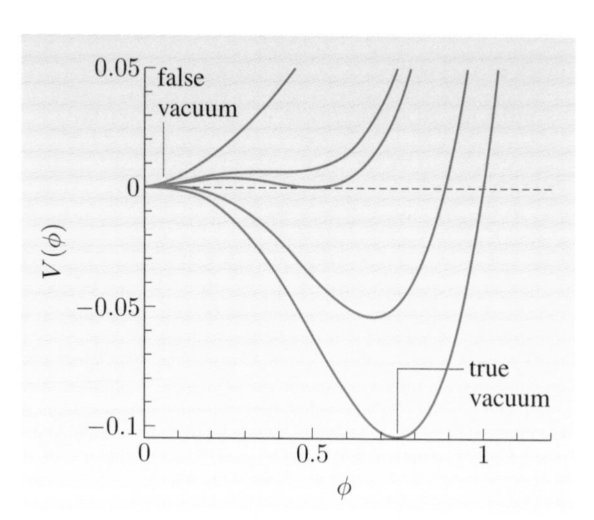

Figure 2.7 Illustration of how temperature-dependent effects can create a false vacuum. Early in the history of the Universe, the inflaton field is around the energy minimum at $\phi = 0$, but as the Universe cools, a second, deeper minimum appears elsewhere. The Universe slides or (if necessary) quantum tunnels to the new minimum.

In either case, there is an energy difference between the upper and lower vacuum states of ΔV. If we take $V = 0$ as the true vacuum, then the elevated state has an effective cosmological constant (though strictly speaking this is a misnomer as it would not be constant in this case). The order of magnitude for ΔV expected in GUTs is huge, giving *prima facie* plausibility to inflation: the characteristic energy density can be shown to be (on, for example, purely dimensional grounds) $\rho \simeq E_{\text{GUT}}^4/(\hbar^3 c^5) \simeq (10^{15}\,\text{GeV})^4/(\hbar^3 c^5) \simeq 10^{80}\,\text{kg}\,\text{m}^{-3}$.

Going back to our analogy of an object sliding inside a potential well (Figure 2.6), the full equation of motion turns out to be (in natural units of $c = \hbar = 1$, see box):

$$\ddot{\phi} = 3H\dot{\phi} - \nabla^2\phi + \frac{\text{d}V(\phi)}{\text{d}\phi} = 0. \tag{2.18}$$

The derivation of this formula is lengthy, but it follows ultimately from energy conservation considerations in the quantum scalar field. The gradient $\boldsymbol{\nabla}$ is with respect to proper spatial coordinates (not comoving ones), and the dots are time derivatives. Note that it involves the Hubble parameter H. In the analogy of an object sliding down the valley, the $H\dot{\phi}$ term is equivalent to a friction term, while $\text{d}V/\text{d}\phi$ is the force acting on the object.

Natural units

Particle physicists sometimes opt to use 'natural units' in which $\hbar = c = 1$ to keep the algebra simpler, avoiding fiddly factors of $\hbar$ and c that can be determined at the end from dimensional analysis. The thinking is to treat $\hbar$ and c as implying 'conversion factors' between different dimensions. For example, c could be thought of as the conversion factor between space measurements and time measurements. What defines these conversion factors? Well, for us it's about how we choose to measure lengths (e.g. metres) and times (e.g. seconds). The Universe doesn't care whether we use metres or seconds, or miles and years, so why not choose units in which c is set to one? For us c has dimensions LT^{-1} (e.g. metres per second), so $c = 1$ has the effect of treating space units in the same way as time units. In natural units, energies have the same dimensions as mass (because $E = mc^2$) and 1/time (because $E = h\nu$).

A similar consideration turns out to give the pressure and energy density, again in natural units ($c = \hbar = 1$):

$$p = \tfrac{1}{2}\dot{\phi}^2 - \tfrac{1}{6}\left(\boldsymbol{\nabla}\phi\right)^2 - V(\phi), \tag{2.19}$$

$$\rho = \tfrac{1}{2}\dot{\phi}^2 + \tfrac{1}{2}\left(\boldsymbol{\nabla}\phi\right)^2 + V(\phi). \tag{2.20}$$

In natural units the equation of state parameter $w = p/(\rho c^2)$ is written as $w = p/\rho$. Equations 2.19 and 2.20 could generate a negative equation of state parameter: for example, if $V \gg \dot{\phi}^2$ and spatial derivatives are negligible, then $w = -1$. If we define ϕ to have the units of energy, then Equations 2.19 and 2.20 come out in conventional units as

$$p = \frac{1}{2\hbar c^3}\dot{\phi}^2 - \frac{1}{6\hbar c}\left(\boldsymbol{\nabla}\phi\right)^2 - V(\phi), \tag{2.21}$$

$$\rho c^2 = \frac{1}{2\hbar c^3}\dot{\phi}^2 + \frac{1}{2\hbar c}\left(\boldsymbol{\nabla}\phi\right)^2 + V(\phi), \tag{2.22}$$

so each term has the dimensions of energy density.

It's usual in inflationary calculations to assume that the spatial derivatives are negligible, because we're inflating a small, locally-homogeneous region to a giant size, so any inhomogeneities will become negligible. This means that in most contexts, ϕ is the value of the field throughout the observable Universe. If we assume that the field ϕ is approximately the same everywhere, then $\boldsymbol{\nabla}\phi \simeq 0$ and $\nabla^2\phi \simeq 0$.

We also found in Section 2.7 that we need $w < -1/3$ for inflation to happen. In order to achieve this, we need that the potential V in Equations 2.19 and 2.20 starts off by dominating over the kinetic energy term involving $\dot{\phi}$. When this is no longer true, inflation will cease. At this point the 'object' in Figure 2.6 will oscillate around the minimum, with the oscillations damped by the $H\dot{\phi}$ term. In addition, it's expected that the inflaton field will then decay into conventional matter and radiation. This particle generation would appear as another friction-like term in the equation of motion. At this point in the history of the Universe, the temperature would have been very cold, because the energy densities of matter and radiation will have been reduced by factors of a^3 and a^4, respectively, where a is the dimensionless scale factor of the Universe. The subsequent particle generation process is known as **reheating**, but the exact mechanism is not known since the underlying physics of the inflaton field is not known. During this process, the matter–antimatter asymmetry of the Universe may have been generated. The end result of inflation is that the Universe is left with more or less the same energy density as when it started, but in the form of radiation and matter, and with an imbalance of matter over antimatter.

The requirement that V starts out much bigger than the kinetic energy term can also be shown to imply that we need $\ddot{\phi}$ to be small, and that ϕ is homogeneous. The proof of this is very involved, but we can sketch a demonstration. Suppose that ϕ has some intrinsic variation over a spatial scale δx. We'd expect there also to be intrinsic temporal variations over timescales of $\delta t = \delta x/c$. We could think of this as being equivalent to a kinetic energy term $\phi^2/(\delta t)^2$. In order for the potential to dominate, we need that $V(\phi)$ is much bigger than $\phi^2/(\delta t)^2$, i.e. $V(\phi) \gg \phi^2/(\delta t)^2$. If we differentiate this with respect to ϕ, we find that $\mathrm{d}V/\mathrm{d}\phi \gg 2\phi/(\delta t)^2$. But this will be of the order $\ddot{\phi}$. We should therefore expect to be able to neglect the $\ddot{\phi}$ term in Equation 2.18. This approximation is known as

The expression 'slow-roll' is perhaps misleading because it seems to suggest that the object in Figure 2.6 acquires some angular momentum. To avoid this, we've used the verb 'slide' in preference to 'roll' where we can, but be aware that most of the technical literature and textbooks use 'roll' in this context.

the **slow-roll approximation**. The result is that the slow-roll approximation leads to us approximating the equation of motion as

$$3H\dot{\phi} = -\frac{\mathrm{d}V}{\mathrm{d}\phi} = -V'. \tag{2.23}$$

The next step is to substitute this into the Friedmann equation (Equation 1.7) rewritten in natural units. (To do this, we replace the factor of G with one of the Planck scales in Equations 2.13–2.15 — conventionally, mass.) We then use $V \gg \dot{\phi}^2$ to show that

$$\begin{aligned} H^2 &= \frac{8\pi}{3m_{\mathrm{Pl}}^2}\frac{1}{\hbar c}\left(\frac{1}{2\hbar c^3}\dot{\phi}^2 + \frac{1}{2\hbar c}(\boldsymbol{\nabla}\phi)^2 + V(\phi)\right) \\ &\simeq \frac{8\pi}{3m_{\mathrm{Pl}}^2}V(\phi). \end{aligned} \tag{2.24}$$

Putting these together, one can show that the requirement that $V \gg \dot{\phi}^2$ can be expressed as constraints on two new dimensionless quantities:

$$\varepsilon = \frac{-\dot{H}}{H^2} = \frac{m_{\mathrm{Pl}}^2}{16\pi}\left(\frac{V'}{V}\right)^2 \ll 1, \tag{2.25}$$

$$\eta = \frac{\ddot{\phi}}{H\dot{\phi}} = \frac{m_{\mathrm{Pl}}^2}{8\pi}\left(\frac{V''}{V}\right) \ll 1, \tag{2.26}$$

where $V' = \mathrm{d}V/\mathrm{d}\phi$ and $V'' = \mathrm{d}^2V/\mathrm{d}\phi^2$.

These equations are a dimensionless way of expressing the constraint that the potential V must be shallow and flat enough to allow slow-rolling. These criteria are requirements for inflation to start, and inflation will end when, for example, $\varepsilon \simeq 1$.

We don't know the shape of the inflation potential. There are many varieties of inflation, each of which hypothesizes a differently shaped potential. However, the observational consequences of inflation all rely on the last stages of inflation when the 'object' in Figure 2.6 is close to the minimum, so they don't depend strongly on the shape of the potential. In a sense this is a pity, because it restricts our ability to constrain this new physics experimentally, but it also greatly simplifies the predictions of inflation and makes them more robust to changes in the underlying assumptions. We'll describe some of the observational consequences of inflation in the next section.

2.9 The primordial density power spectrum

One of the key observables that are predicted to result from inflation is the 'clumpiness' of the CMB. The statistics of the clumpiness of the CMB are a key cosmological constraint and the key to modern precision cosmology. To show you how this works, we'll need some mathematical way of describing this clumpiness. Clumps can be small or large, so one way of describing clumpiness could be with a Fourier series. It's conventional in this field to use complex Fourier series, so if you've not met these before, see the box below. This box will also briefly mention Fourier transforms, though we won't use these for the most part in this book.

Fourier series and transforms

Here's a quick reminder of what a Fourier series expansion looks like. We have a function $f(x)$ that's periodic over the interval $-L/2 < x < L/2$ (e.g. waves in a box of length L), and we find that it can be expressed as

$$f(x) = \frac{a_0}{2} + \sum_{n=1}^{\infty}\left(a_n \cos\left(\frac{2\pi n x}{L}\right) + b_n \sin\left(\frac{2\pi n x}{L}\right)\right), \tag{2.27}$$

where the coefficients a_n and b_n are given by

$$a_n = \frac{2}{L}\int_{-L/2}^{L/2} f(x)\cos\left(\frac{2\pi n x}{L}\right)\mathrm{d}x, \tag{2.28}$$

$$b_n = \frac{2}{L}\int_{-L/2}^{L/2} f(x)\sin\left(\frac{2\pi n x}{L}\right)\mathrm{d}x. \tag{2.29}$$

Now, we can simplify this slightly if we use complex numbers, i.e. using $\mathrm{i} = \sqrt{-1}$. It's known that $\mathrm{e}^{\mathrm{i}\theta} = \cos\theta + \mathrm{i}\sin\theta$, which we can use to write the Fourier series as

$$f(x) = \sum_{n=-\infty}^{+\infty} A_n \mathrm{e}^{2\pi \mathrm{i} n x/L}. \tag{2.30}$$

This corresponds to the previous Fourier series if $A_n = \frac{1}{2}(a_{|n|} + \mathrm{i}b_{|n|})$ for $n < 0$, $a_0/2$ for $n = 0$, and $\frac{1}{2}(a_n - \mathrm{i}b_n)$ for $n > 0$. This can be shown to lead to the following expression for A_n:

$$A_n = \frac{1}{L}\int_{-L/2}^{L/2} f(x)\,\mathrm{e}^{-2\pi \mathrm{i} n x/L}\,\mathrm{d}x. \tag{2.31}$$

Sometimes the complex Fourier series is written as

$$f(x) = \sum_{n=-\infty}^{+\infty} A_n \mathrm{e}^{\mathrm{i}x k_n}, \tag{2.32}$$

where $k_n = 2\pi n/L$ is known as the wave number. We *won't* ask you to manipulate complex Fourier series in this book, but we *do* want you to have met them.

Fourier series occur very often in physics, but what happens when the box that you're using becomes limitingly big? In this case, sums become integrals, and Fourier series become integrals. These integrals are known as **Fourier transforms**. In cosmology we deal with only finite volumes, so in practice we need only Fourier series, but just so you've met them, the Fourier transforms are

$$f(x) = \int_{-\infty}^{+\infty} F(k)\,\mathrm{e}^{2\pi \mathrm{i} k x}\,\mathrm{d}k, \tag{2.33}$$

$$F(k) = \int_{-\infty}^{+\infty} f(x)\,\mathrm{e}^{-2\pi \mathrm{i} k x}\,\mathrm{d}x. \tag{2.34}$$

Note how similar these equations are. $F(k)$ is known as the Fourier transform of $f(x)$. Transforming twice gets you almost back where you started: if you make a Fourier transform of $F(k)$, you get $f(-x)$ back. Fourier transforms occur throughout physics. For example, diffraction in optics involves Fourier transforms. The image of a star seen through a telescope is the Fourier transform of the telescope aperture — well, almost. The amplitudes of the waves hitting your detector *are* the Fourier transform of the telescope aperture, but what you measure is the energy of the light on your detector, which is proportional to the amplitude of the electromagnetic wave squared, so your image will be the Fourier transform of the telescope aperture, squared.

Let's write the average density of matter as $\overline{\rho}$ and the deviation from this average as $\delta\rho$. This deviation will vary with position. It's common to express the clumpiness in terms of the fractional overdensity or underdensity, $(\delta\rho)/\overline{\rho}$. Often this fractional overdensity is simply abbreviated as δ. By definition, the mean value of δ is zero. Since δ will vary as a function of position, we'll write this as $\delta(\boldsymbol{x})$.

Imagine that we are considering a large box in the Universe with side length L. We can write $\delta(\boldsymbol{r})$ as a Fourier series in three dimensions. For simplicity for now, however, let's just consider a one-dimensional universe so the 'box' is just a length L. The density of matter expanded as a Fourier series will be

$$\delta(x) = \frac{\delta\rho}{\rho} = \sum_{n=-\infty,\infty} C_{k_n} \mathrm{e}^{\mathrm{i}k_n x}, \tag{2.35}$$

where $k_n = 2\pi n/L$ (with n as an integer) is known as the **wave number**, and

$$C_{k_n} = \frac{1}{L}\int_{-L/2}^{L/2} \delta(x)\,\mathrm{e}^{-\mathrm{i}k_n x}\,\mathrm{d}x. \tag{2.36}$$

In some sense, the coefficients C_{k_n} characterize how much structure there is at any wavelength. We can carry this over into three dimensions:

$$\delta(\boldsymbol{r}) = \sum C(\boldsymbol{k})\,\mathrm{e}^{\mathrm{i}\boldsymbol{k}\cdot\boldsymbol{r}}, \tag{2.37}$$

where the sum is over all wave numbers $\boldsymbol{k} = (k_x, k_y, k_z)$ in the box, e.g. $k_x = 2\pi n/L$ and similarly for y and z. The conventional symbol used to represent the Fourier coefficients is δ_k:

$$\delta(\boldsymbol{r}) = \sum \delta_k \mathrm{e}^{\mathrm{i}\boldsymbol{k}\cdot\boldsymbol{r}}, \tag{2.38}$$

$$\delta_k(\boldsymbol{k}) = \frac{1}{L^3}\int_{\text{within } L^3} \delta(\boldsymbol{r})\,\mathrm{e}^{-\mathrm{i}\boldsymbol{k}\cdot\boldsymbol{r}}\,\mathrm{d}\mathbf{r}. \tag{2.39}$$

So far we've just written down Fourier series; how can we use these to characterize the clumpiness? One approach is to measure how much variation there is in the Fourier coefficients. Since root-mean-square (RMS) is a measure of the standard deviation of the random sample, we can estimate the variance by averaging the squares of the Fourier coefficients over different realizations of the density field for a fixed $\boldsymbol{k}$, i.e. $\langle|\delta_k|^2\rangle$, where $|\delta_k|^2 = \delta_k\delta_k^*$ (remember that δ_k is a complex number, so δ_k^* is its complex conjugate).

How might this work in practice? First, if the $\delta(\boldsymbol{x})$ distribution is isotropic on average, the Fourier coefficients won't *on average* depend on the direction of $\boldsymbol{k}$. Second, the amount of clumpiness could depend on how closely you look at the density field map. For example, the density distribution could be clumpy on medium-sized scales, but look smooth on larger scales and on smaller scales. For this reason it's useful to calculate the variance in the Fourier coefficients as a function of the length of the wave number vector, $k = |\boldsymbol{k}|$. This is known as the **power spectrum** and is written as

$$P(k) = \langle |\delta_k|^2 \rangle. \tag{2.40}$$

In the present-day Universe, an overdensity ($\delta(\boldsymbol{x}) > 0$) will attract surrounding matter through gravity and will tend to increase the value of δ. Similarly, underdensities ($\delta < 0$) will empty out of matter, causing the value of δ to become more negative. The density perturbations $\delta(\boldsymbol{r})$ will initially evolve from self-gravity in such a way that each Fourier mode evolves independently. This is also referred to as the 'linear regime' in the evolution of the density field. This is one reason why the power spectrum is used in cosmology, rather than other measures of clustering.

These effects of self-gravity can be neglected during inflation, but inflation makes very clear predictions for the initial density power spectrum. The key idea for inflation is that the *gravitational potential* laid down by the inflating Universe was invariant under time translation, i.e. the Universe should look the same on average if you make the transformation $t \longrightarrow t + \Delta t$, regardless of your choice of Δt (as long as it's shorter than the duration of inflation). Therefore there must be a constant level of fluctuations on (say) the scale of the horizon. In other words, there must be a continuous time-invariant process in which quantum fluctuations are being created within the Hubble volume then inflated out of it. Also, these fluctuations cannot have any characteristic length scale (or the Universe would not look the same regardless of the choice of Δt). This is a fractal universe.

To see what this means in terms of the power spectrum, we need to express the fluctuations in a scale-invariant way, then state that the fluctuations are constant. For example, it's no good measuring the power spectrum on scales of $k = 1\,\mathrm{m}^{-1}$ to $k = 2\,\mathrm{m}^{-1}$, because this invokes the characteristic scale length of the metre. What we *can* do, however, is measure the power spectrum in an interval between any wave number k and double that wave number, $2k$. We'd then require that the value of the power spectrum shouldn't depend on the choice of k. In other words, in any factor-of-two interval in k, we should measure the same power spectrum. Another way of expressing this is to use the natural log of the wave number, $\ln k$, and require that the power spectrum is the same in any logarithmic interval $\Delta \ln k = \ln 2$. Of course, there's nothing special about the choice of 2. The general way of expressing this scale-invariance is to say that the variance in the density field per logarithmic k interval is constant:

Note that k here is the wave number *not* the curvature parameter!

$$\frac{\Delta\sigma^2}{\Delta \ln k} \simeq \frac{\mathrm{d}\sigma^2}{\mathrm{d}\ln k} = \text{constant}. \tag{2.41}$$

This is known as the **scale-invariant spectrum** or the **Harrison–Zel'dovich spectrum**.

The quantity $\mathrm{d}\sigma^2/\mathrm{d}\ln k$ is known as the **dimensionless power spectrum** of the gravitational potential Φ and is conventionally (but perhaps confusingly) given the symbol Δ_Φ^2:

$$\Delta_\Phi^2(k) \equiv \frac{\mathrm{d}\sigma^2}{\mathrm{d}\ln k}. \tag{2.42}$$

To relate this to the matter power spectrum, we need to use Poisson's equation, $\nabla^2\,\delta\Phi = 4\pi G\rho_0\delta$, where as before the quantity δ is used as a shorthand for the fractional overdensity $\delta\rho/\rho_0$. From a Fourier transformation of this equation, it turns out that the Fourier transform of the potential fluctuations $\delta\Phi_k$ satisfies $\delta\Phi_k = -4\pi G\rho_0\delta_k/k^2$, where δ_k is the Fourier transform of δ. The scale-invariant potential fluctuations thus give a dimensionless matter power spectrum: $\Delta^2 \propto k^4$.

This turns out to be similar to the power spectrum that we defined in Equation 2.40, but not quite identical. The difference rests on how the averaging of the Fourier modes is done, because some ranges of k have more Fourier modes than others.

To see why this is, imagine that we have a cubical box of the Universe with a volume $V = L \times L \times L$, which we're describing with a Fourier series. The allowed wavelengths along the x-axis will be $\lambda = L/n$, where n is an integer, so the wave numbers along the x-axis are $k_x = 2\pi n/L$. Therefore the number of modes from k_x to $k_x + \mathrm{d}k_x$ in the x-direction will be $(L/2\pi)\,\mathrm{d}k_x$. Now, instead of just considering the modes along the x-axis, consider all three axes. How many modes are there in a radial shell of thickness $\mathrm{d}k$? (Note that we're writing $k = |\boldsymbol{k}| = \sqrt{k_x^2 + k_y^2 + k_z^2}$.) This will be the density of modes times the volume of the shell, which is $(L/2\pi)^3 \times$ the volume of the shell. This volume is $\mathrm{d}^3k = 4\pi k^2\,\mathrm{d}k$. The number of modes in our shell is therefore

$$\left(\frac{L}{2\pi}\right)^3 \mathrm{d}^3k = \frac{V}{(2\pi)^3}\,\mathrm{d}^3k = \frac{V}{(2\pi)^3}4\pi k^2\,\mathrm{d}k = \frac{V}{2\pi^2}k^2\,\mathrm{d}k.$$

When we're calculating the dimensionless power spectrum, we're asking what the variance is per *logarithmic* interval of k, and $\mathrm{d}\ln k = (1/k)\,\mathrm{d}k$, so $\mathrm{d}k = k\,\mathrm{d}\ln k$. It perhaps shouldn't be a huge surprise, therefore, that the dimensionless power spectrum comes out as

$$\Delta^2(k) = \frac{V}{2\pi^2}k^3\langle|\delta_k|^2\rangle = \frac{V}{2\pi^2}k^3P(k). \tag{2.43}$$

We'll see in Chapter 4 how this is also used in measuring the clustering of galaxies.

Often the power spectrum is written as

$$P(k) \propto k^{n_\mathrm{s}} \tag{2.44}$$

or equivalently

$$\Delta^2(k) \propto k^{n_\mathrm{s}+3}, \tag{2.45}$$

where n_s is known as the spectral index of scalar perturbations, which satisfies $n_\mathrm{s} = 1$ for a scale-invariant spectrum. White noise (e.g. putting down atoms at random with the same probability everywhere) would have $n_\mathrm{s} = 0$.

But how big are these fluctuations? The detailed quantum field theoretical calculation is complicated, but we can get some idea through the following argument. Quantum fluctuations in the field ϕ will result in regions of the Universe finishing inflation at slightly different times. Figure 2.8 illustrates this as two objects sliding down the potential well $V(\phi)$, slightly offset. The field at these two positions will have the same slow-rolling behaviour, but will finish at times that are offset by δt, where

$$\delta t = \frac{\delta\phi}{\dot{\phi}}. \tag{2.46}$$

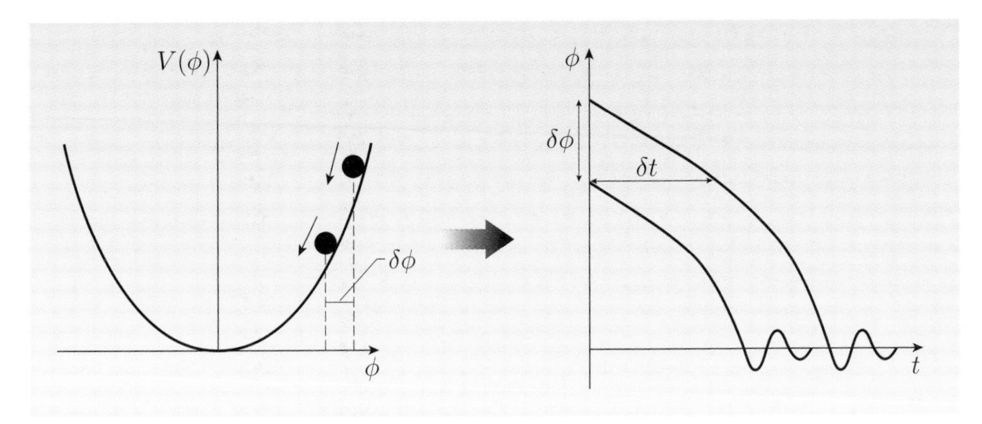

Figure 2.8 A slight difference in the scalar field value $\delta\phi$ means that inflation finishes with a time separation of δt, which creates a density fluctuation of $\delta = H\,\delta t$ (in natural units).

The difference in density between these two regions at the scale of the horizon $\delta_{\rm H}$ will be roughly $H\,\delta t$, where H is the Hubble parameter during inflation. Quantum field theory (and in fact dimensional analysis) predicts that the RMS of $\delta\phi$ on the scale of the horizon is equal to $H/(2\pi)$ in natural units, so the horizon-scale fluctuations will be of the order $\delta_{\rm H} = H^2/(2\pi\dot{\phi})$. This will depend on the shape of the inflation potential and is one of the free parameters in fitting inflationary models to data on the large-scale structure of the Universe.

Scale-invariance breaks down once the expansion ceases to be exponential, so we expect a slight deviation from scale-invariance to be imprinted on the fluctuations as inflation ends. This will depend on the shape of the inflation potential near the end of inflation, i.e. on the parameters ε and η (defined in Equations 2.25 and 2.26). The result is that $n_{\rm s} \simeq 1 + 2\eta - 6\varepsilon$ for the values of ε and η near the end of inflation. There is also a prediction of a clustered background of gravitational waves (which we shall meet in Section 2.16), which also has a dependence on ε.

Two other key predictions of inflation are worth mentioning. First, the perturbations of the matter and radiation number densities should be equal; these are known as **adiabatic** perturbations because adiabatic expansion conserves the ratio of matter and radiation number densities. Second, the phases of the Fourier decomposition should be random and uncorrelated with each other. Intuitively this seems reasonable since the quantum fluctuations at one time should be uncorrelated with the quantum fluctuations at a later time; the earlier quantum fluctuations give rise to the Fourier components on larger spatial scales, while the later quantum fluctuations are responsible for the Fourier modes on smaller spatial scales. It can be shown that random phases imply that the fluctuations are a **Gaussian random field**, which means that the joint probability distribution for the density at any number of points must be a multivariate Gaussian distribution. Because a Gaussian random field has no information contained in the phases (i.e. they are all uniformly randomly distributed), all the statistical information about the density field is contained in the amplitudes, so the power spectrum completely characterizes the density fluctuations.

In summary, inflation predicts a nearly scale-invariant spectrum of primordial Gaussian density fluctuations, spatial flatness, a gravitational wave background, and adiabatic fluctuations.

The overall amplitude of the initial perturbations depends on the shape of the inflation potential, as does the deviation from scale-invariance. We'll see in Section 2.16 that the gravitational wave background does too. The CMB fluctuations have been shown to be consistent with adiabatic perturbations, so we won't discuss alternative sources of perturbations here (e.g. 'isocurvature' perturbations); if there is a non-adiabatic contribution, it must be small. Many tests have been made of the CMB clustering to search for non-Gaussian character, though no unequivocal signal has yet been found. The expectation is that the reheating at the end of inflation was the time of baryogenesis, which set the subsequent entropy per baryon, but the GUT-scale physics that determined these processes (and inflation itself) is still uncertain.

Trotta, R., 2007, *Monthly Notices of the Royal Astronomical Society*, **375**, L26.

One implication of inflation is that there may be regions far off the minimum $V(\phi)$ that inflate eternally. If ϕ is very large, the quantum fluctuations in ϕ would make ϕ perform a random walk that overwhelms the drift towards the minimum $V(\phi)$. Our observable portion of Universe could be just an infinitesimal part of a much, much larger complex. One of the enduring surprises of observational cosmology is that it is possible at all — that is, we can build telescopes that are big enough to detect light from most of the way back to the Big Bang, and observe galaxies throughout most of the Hubble volume (Section 1.9). However, if one of these variants of inflation is correct, the *observable* part of the Universe is a very tiny part of it indeed. This boggles the mind.

Finally, it's worth remembering that one of the motivations of inflation is to solve the horizon problem and many others without invoking Planck scale physics such as quantum gravity. We're describing a general relativistic Universe, which inevitably involves the gravitational constant G, and quantum mechanics, which inevitably involves Planck's constant $\hbar = h/(2\pi)$. It's therefore perhaps inevitable but a little disappointing that the Planck scale should occur in various forms in the inflation equations. The following exercise will demonstrate the inevitability of the Planck scale in inflation.

Exercise 2.7 The number of e-foldings of inflation is roughly $N = \int H\,\mathrm{d}t$. Use the slow-roll approximation to show that

$$N = \frac{-8\pi}{m_{\mathrm{Pl}}^2} \int_{\phi_2}^{\phi_1} \frac{V}{V'}\,\mathrm{d}\phi, \tag{2.47}$$

where ϕ_2 and ϕ_1 are values of the inflaton field at the start and end points of inflation, respectively. We can choose $\phi_1 = 0$ without loss of generality.

Next, make the assumption that V' is roughly of the order of V/ϕ (which should be true if the potential is reasonably smooth and slowly-varying) to show that $N \sim (\phi_2/m_{\mathrm{Pl}})^2$ and hence that we need ϕ_2 significantly larger than m_{Pl}. ■

Similarly, the same criteria can be used to show that the parameters ε and η are both $\ll 1$ (Equations 2.25 and 2.26). Inflation ends when $\phi \sim m_{\mathrm{Pl}}$, so we have not escaped consideration of the Planck scale.

2.10 The real music of the spheres

We've already seen that the microwave background is strikingly uniform, in marked contrast to the present-day matter density of the Universe.

- Why is it that the photons in the Universe are so uniformly distributed, while the distribution of matter is so varied?

- The photon uniformity reflects the distribution at the time of recombination, i.e. the time when the Universe was last opaque. This was the last time that the photon distribution was strongly coupled to the matter distribution. Since then, the matter distribution has evolved while the photons have travelled more or less unimpeded through the Universe.

However, as we saw in Figure 2.2, there are slight inhomogeneities. Inflation is one potential mechanism for generating these irregularities, as we saw in Section 2.8, though we don't yet know the shape of the inflation potential. These inhomogeneities are the fluctuations that grew through gravity into the present-day matter distribution of stars, galaxies and clusters of galaxies.

Again we'll be characterizing these with the power spectrum, which is essentially the RMS as a function of angular scale, using Fourier transforms. However, our Fourier analysis implicitly used a flat space. The sky isn't flat, so we need some equivalent that works on the spherical surface of the sky. The idea is to replace the sin and cos functions with some other functions that are appropriate for a sphere.

The functions usually chosen are the **spherical harmonics** Y_{lm}, defined as

$$Y_{lm}(\theta, \phi) \propto \mathrm{e}^{\mathrm{i}m\phi} P_l^m(\cos\theta), \tag{2.48}$$

A simpler way to express the definition of spherical harmonics is as the eigenfunctions of the angular part of the ∇^2 operator, but this takes us outside the scope of this course.

where the P_l^m are the **associated Legendre polynomials**. The θ and ϕ coordinates are the ones from spherical coordinates. We won't derive the Legendre polynomial functions, but just for completeness they are defined as

$$P_l^m(x) = \frac{(-1)^m}{2^l l!}\left(1 - x^2\right)^{m/2} \frac{\mathrm{d}^{l+m}}{\mathrm{d}x^{l+m}}\left(x^2 - 1\right)^l, \tag{2.49}$$

where $\mathrm{d}^{l+m}/\mathrm{d}x^{l+m}$ is the $(l+m)$th-order derivative. When the CMB structure is expanded (like a Fourier series) in terms of spherical harmonics, the coefficients used are named the monopole, dipole, quadrupole, octopole, and so on. The $l = 1$ Legendre polynomials have just one trigonometric function of θ (e.g. $\sin\theta$), while the $l = 2$ polynomials have two (e.g. $\sin^2\theta$), and so on. Spherical harmonics are also used in quantum mechanics, especially in describing electron orbits in atoms, and in helioseismology.

We calculate $(\delta T)/\overline{T}_{\mathrm{CMB}}$, where $\overline{T}_{\mathrm{CMB}}$ is the average CMB temperature and $\delta T = T - \overline{T}_{\mathrm{CMB}}$. As with the power spectrum above, we shall write this as δ, though in this case δ will depend on the angular position $\boldsymbol{q} = (\theta, \phi)$ on the sky rather than the spatial position $\boldsymbol{x}$.

The spherical equivalent of the Fourier transform is

$$\delta(\theta, \phi) = \sum_{l=0}^{+\infty} \sum_{m=-l}^{+l} a_{lm} Y_{lm}(\theta, \phi). \tag{2.50}$$

The a_{lm} are the equivalent of Fourier coefficients:

$$a_{lm} = \int \delta(\boldsymbol{q}) Y_{lm}^* \, \mathrm{d}^2 q, \tag{2.51}$$

where the integral is done over the whole sky.

One important quality in defining this spherical alternative to Fourier transforms is **orthogonality**. In trigonometry, $\int_{-\pi}^{\pi} \sin(nx) \sin(mx) \, \mathrm{d}x$ for integers m and n is always zero unless $m = n$. A similar result holds in two dimensions. Here $\int Y_{lm} Y_{l'm'}^* \, \mathrm{d}^2 q = 0$ unless $l = l'$ *and* $m = m'$. It is this orthogonality that is important in the Fourier components of density perturbations evolving independently in the linear regime (Section 2.9 and Chapter 4). We'll return to this in Chapter 8 on the Lyman α forest clustering, and in Chapters 3 and 4 on the large-scale structure of galaxies.

Because of isotropy, the coefficients a_{lm} are a function of l only, not m, so the sum in Equation 2.50 can be expressed as $(2l + 1)$ times the sum over l only. The quantities l and m are like wave numbers on the sky, so the smaller the angle, the *larger* the value of l. As a rough rule of thumb, the scales l are related to angular sizes θ on the sky through $l \simeq 180^\circ/\theta$.

The power spectrum of the CMB is usually written as

$$C_l = \langle a_{lm}^2 \rangle. \tag{2.52}$$

Conventionally, the C_l power spectrum tends to be plotted as $(\Delta T)^2 = l(l + 1) C_l \overline{T}_{\mathrm{CMB}}^2$, or sometimes with an additional divisor of 2π. This measures the power per *logarithmic* interval in l, so a scale-invariant spectrum looks horizontal in such a plot.

See Chapter 18 of Peacock, J.A., 1999, *Cosmological Physics*, Cambridge University Press.

How precisely can the C_l power spectrum be measured? There are only $2l + 1$ m-samples of power at any fixed l, which limits the precision of the measurements of C_l. This limit is known as **cosmic variance**. Formally, the precision limit from cosmic variance comes out as

$$\Delta C_l = \sqrt{\frac{2}{2l + 1}} \, C_l. \tag{2.53}$$

This is the *best possible* measurement, in the absence of any instrumental noise or astrophysical foreground systematic effects. In order to measure the C_l modes better than this, we'd need a bigger sky!

There have been many attempts to measure the CMB power spectrum. The first great breakthrough was with the COBE (Cosmic Background Explorer) satellite, which measured the anisotropies on scales larger than about 0.1°. This confirmed that the power spectrum is approximately scale-invariant, exactly as predicted by inflation. COBE also showed that the CMB spectrum was an excellent black body (Figure 2.1), exactly as predicted by the Hot Big Bang model. The COBE results won John Mather and George Smoot the 2006 Nobel Prize in Physics. The Wilkinson Microwave Anisotropy Probe (WMAP) has now constrained much more of the C_l spectrum, shown in Figure 2.9. We shall see why the power spectrum has peaks in Section 2.12. Many ground-based and balloon-borne experiments have made constraints on the highest-l region, though the maps at this resolution are not yet all-sky. This will change shortly with the European Space Agency Planck mission, which launched on 14 May 2009. Planck is also expected to make tremendous advances in measuring the clustering of the *polarized* CMB, about which we shall hear more later.

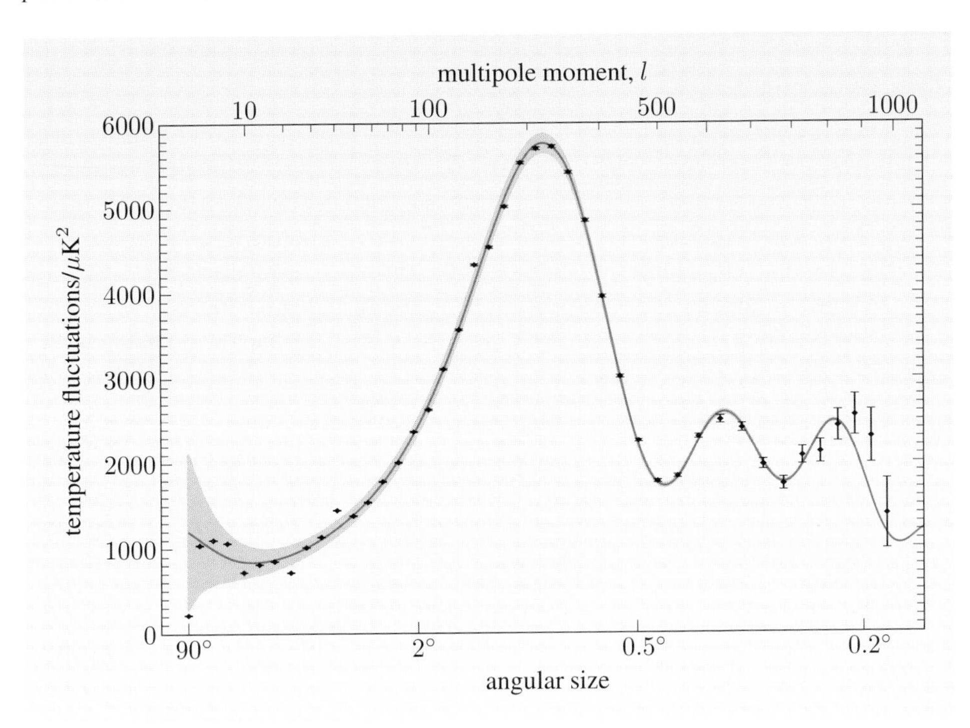

Figure 2.9 The CMB power spectrum measured in the first five years of the WMAP satellite. The curve shows the best fit to the data, in which the cosmological parameters are inferred. The grey region shows the scatter in the data that one would expect from cosmic variance, i.e. the fact that you're sampling only a finite region of the Universe. The fluctuations plotted in the y-axis are $l(l+1)C_l\overline{T}^2_{\text{CMB}}/2\pi$.

We've seen how the theory of inflation predicts a roughly scale-invariant spectrum of density perturbations, and that the real horizon size at recombination was in fact much larger than one would predict without inflation. Nevertheless, the apparent (i.e. non-inflationary) horizon size is still a useful scale length. On sizes much smaller than this scale, regions will have had time since inflation to affect each other. On sizes much larger than this scale, the only causal contact could have been during or prior to inflation. We'd therefore expect that the power spectrum on large scales should have the roughly scale-invariant behaviour predicted by inflation. This is known as the **Sachs–Wolfe plateau** and is indeed what's seen in observations. (The clustering amplitude on the Sachs–Wolfe plateau also agrees with the amplitude of matter fluctuations in the local Universe on 8 Mpc scales, known as σ_8, of which more later.) However, we'll see in Section 2.13 that the passage of photons through the Universe over the past 13 or so billion years can cause some additional distortions on the largest scales.

Another effect that might leave its imprint on the CMB is the topology of the Universe. If we travel for long enough in one direction, might we go right round the Universe and back to where we started? It's possible to show that in any $k > 0$ (spatially spherical) universe, the expansion is too fast to permit this motion. However, there's another way to make this happen. Imagine a sheet of paper. We can easily draw geodesics on this surface: they are just straight lines that you would draw with a ruler. Now curl the paper into a tube. The lines previously drawn are still geodesics. However, if you travel in one direction for long enough, you get back where you started, despite the fact that *geodesically* the surface is spatially flat. This is rather like the 1970s arcade game Asteroids in which you can disappear off the edge of the screen in one direction and reappear at the opposite edge. Curvature that changes the geodesics is called **intrinsic curvature**, while curvature that doesn't is called **extrinsic curvature**. Einstein's theory of general relativity makes predictions only for intrinsic curvature; we have no theory making any prediction for extrinsic curvature. Going back to our tube made of a piece of paper, we can't link the two ends of the tube in three dimensions without bending the tube and so generating intrinsic curvature, but if we had four spatial dimensions, we could link the two ends and still have zero intrinsic curvature. The paper would then be arranged into a torus shape, i.e. it has assumed a different **topology** to a single sheet. What is the topology of our Universe? A complex topology would leave characteristic imprints on the CMB if the wrap-around scales were small enough. No such features have been found, implying that any wrap-around topology in our Universe has to be at least around the size of the Hubble volume.

We'll see below how some of the fluctuations that *are* seen are due to acoustic oscillations in the early Universe. Some audio representations of the acoustic oscillations after the Big Bang can be found in the further reading section. The cosmologist Peter Coles estimated the amplitude of these acoustic oscillations in decibels (setting aside the obvious objections that there were no people to hear them and the conditions were too hot and dense for terrestrial life anyway) and found that the Big Bang was no louder than a rock band.

2.11 The CMB dipole

The CMB is extraordinarily uniform across the whole sky (Figure 2.2), as we've seen. If we increase the contrast ratio of the image (Figure 2.2), we see that the CMB is dominated by a characteristic pattern of hot and cold regions. This is the **dipole** caused by the Doppler effect of our motion relative to the CMB rest frame.

We can derive this quite quickly in special relativity using the wave four-vector

$$\boldsymbol{k} = \left(\frac{\omega}{c}, k_x, k_y, k_z\right), \tag{2.54}$$

where ω is the angular frequency, related to the frequency ν and period T by $T = 1/\nu = 2\pi/\omega$, and the k values are wave numbers, related to wavelengths λ by $\lambda = 2\pi/k$. This four-vector describes the light waves and has zero invariant length, i.e.

$$\left(\frac{\omega}{c}\right)^2 - k_x^2 - k_y^2 - k_z^2 = 0. \tag{2.55}$$

Like any four-vector, it transforms with the Lorentz transformation (see Appendix B). For simplicity (but without loss of generality) we'll assume that our motion is along the x-axis and we'll consider a light ray in the xy-plane. There is no z-axis component of the light ray's motion, so the z-axis component of the wave vector is zero, which is also true for all observers. We'll also assume that there is a CMB rest frame in which it appears uniform.

First, imagine a stationary observer on the Earth. He or she receives a CMB photon in the xy-plane. Now we imagine making a Lorentz transformation to the CMB rest frame, which we've chosen to be a velocity boost along the x-axis. We'll give the CMB rest frame primed coordinates. Applying the Lorentz transformation (Appendix B, Section B.4), we find that an observer moving relative to the Earth along the x-axis with velocity v will see a wave vector

$$\mathbf{k}' = \left(\frac{\omega'}{c}, k'_x, k'_y, k'_z\right) = \left(\gamma\frac{\omega}{c} - \frac{v}{c}\gamma k_x, \gamma k_x - \frac{v}{c}\gamma\frac{\omega}{c}, k_y, 0\right). \tag{2.56}$$

Focusing on the time-like (zeroth) component, we find that the observer in the CMB rest frame will see the light at a different frequency:

$$\frac{\omega'}{c} = \gamma\frac{\omega}{c} - \frac{v}{c}\gamma k_x. \tag{2.57}$$

We can relate k_x to ω using the null length of the wave vector (Equation 2.55) and $k_z = 0$:

$$\left(\frac{\omega}{c}\right)^2 = k_x^2 + k_y^2. \tag{2.58}$$

This is Pythagoras's theorem, with the hypotenuse of the triangle as ω/c. The angle that the light ray makes with the x-axis, θ, can be found from trigonometry: $\cos\theta$ = adjacent divided by hypotenuse, or $k_x/(\omega/c)$. Therefore $k_x = (\omega/c)\cos\theta$. Plugging this into Equation 2.57 and rearranging, we find that

$$\omega' = \gamma\omega\left(1 - \frac{v}{c}\cos\theta\right), \tag{2.59}$$

i.e. there is a θ-dependent blueshifting or redshifting. We've already found that a redshifted or blueshifted black body spectrum is still a black body spectrum, though with a different temperature. Therefore we can write

$$T' = \gamma T\left(1 - \frac{v}{c}\cos\theta\right), \tag{2.60}$$

where T' is the temperature in the CMB rest frame, while T is the temperature as seen from Earth. But we've assumed that the CMB has a uniform temperature in the CMB rest frame, i.e. T' = constant, so we must see a fractional temperature variation

$$\begin{aligned}
\frac{T}{T'} &= \frac{1}{\gamma\left(1 - \frac{v}{c}\cos\theta\right)} \\
&= \left(1 - \left(\frac{v}{c}\right)^2\right)^{1/2}\left(1 - \frac{v}{c}\cos\theta\right)^{-1} \\
&= \left(1 - \frac{1}{2}\left(\frac{v}{c}\right)^2 + \cdots\right)\left(1 + \frac{v}{c}\cos\theta + \frac{v^2}{c^2}\cos^2\theta + \cdots\right) \\
&= 1 + \frac{v}{c}\cos\theta + \frac{v^2}{c^2}\left(\cos^2\theta - \frac{1}{2}\right) + \cdots.
\end{aligned}$$

Comparing this to Section 2.10, we see that our motion relative to the CMB induces a dipole as well as having smaller effects on higher-order multipoles.

- Can we use the CMB to measure the primordial dipole?
- No, unless we can find an alternative way to measure our motion relative to the cosmic rest frame.

2.12 The acoustic peaks in the CMB

For most of the time that cosmology has existed as a subject for study, it's been extremely difficult to measure most fundamental parameters to anything much better than a factor of two. This embarrassing situation changed dramatically in the last decade, and **precision cosmology** is now possible. Much of this new precision has come from measurements of microwave background fluctuations, and we'll see in this section why they are so uniquely powerful in cosmology.

The density fluctuations following inflation were imprinted jointly on the dark matter density *and* the photon density *and* the baryon density. Before recombination (Section 2.1), the motion of the baryons was strongly coupled to the photons, because the photons were scattered against the free electrons (Thomson and Coulomb scattering), while the electrons were themselves strongly coupled electrostatically to the baryons (e.g. protons). Therefore we can think of the photons and baryons as a joint photon–baryon gas or photon–baryon fluid. The distribution of dark matter dominates the gravitational potential.

Gravitational attraction caused the photon–baryon gas to fall out of underdense regions and into overdense regions. As the gas fell in and compressed towards the centre of the overdense region, photon pressure outwards resisted the inward flow. This sets up oscillations. The frequency of the oscillations is $\nu_{\rm osc} = c_{\rm s}/\lambda$, where $c_{\rm s}$ is the sound speed in the early Universe, which comes out as $c_{\rm s} = c/\sqrt{3 + 2.25\,\Omega_{\rm b}/\Omega_{\rm r}}$. The value of λ depends on the size of the inhomogeneity. The density perturbations were roughly scale-invariant (Section 2.8), but some oscillations were still particularly favoured over others: nearly the *entire observable Universe* acted as a resonating cavity!

The size of this cavity is the size of the **sound horizon** after inflation, i.e. the distance that a sound wave could have travelled since the end of inflation. As with a musical instrument, this resonating cavity has a fundamental note and has overtones. The fundamental note has a wavelength that's twice the size of the sound horizon, which is the first and biggest peak in Figure 2.9. Overtones have wavelengths that are integer multiples of the sound horizon size, which are the subsequent peaks in Figure 2.9. Sometimes these peaks are called 'Doppler peaks', though Doppler motions are only a small part of the physics of the generation of these peaks. A more accurate terminology is 'acoustic peaks' since they arise mainly from the effects of acoustic waves. Each of these acoustic peaks gives us a precision measurement of some cosmological parameters.

Exercise 2.8 Imagine that you are in the early Universe shortly after recombination, watching the surface of last scattering recede from you. Would the CMB at this time have the same acoustic peaks that we see today? ■

The first acoustic peak is determined mainly by the sound horizon size. This in turn is mainly dependent on the Hubble parameter at that time, H, and therefore on H_0. The angular size of this structure is found by calculating the *angular diameter* distance to the surface of last scattering, which in turn has an H_0-dependence. The ratio of the two is therefore H_0-independent. The apparent size of the first acoustic peak is therefore determined almost entirely by the geometry of the Universe, i.e. open, flat, or closed (see Figure 1.7). Detailed calculations show that the l value of the first acoustic peak is predicted as $l \simeq 200$ in a flat Universe.

Exercise 2.9 The Universe at the time of recombination (redshift $z_{\text{recomb}} \simeq 1000$) was matter-dominated, because the epoch of matter–radiation equality was much earlier ($z \simeq 24\,000$; see Exercise 2.3). Show that the size of the sound horizon at recombination was roughly $0.27\,\text{Mpc}$, assuming $c_{\text{s}} \simeq c/\sqrt{3}$. ■

It's sometimes said that the sound horizon is a **standard rod**, i.e. an object with a known fixed length in metres. If we measure the angular size of a standard rod, we can calculate the exact angular diameter distance to it. However, the sound horizon is not quite so simple since it depends on Ω_{b} and Ω_{r}. The good news is that we can determine $\Omega_{\text{b}}/\Omega_{\text{r}}$ from the other acoustic peaks. The sound horizon has sometimes been called a 'standardizable rod'. We'll meet another sort of standard measure in Chapter 3, the 'standard candle', but this will also turn out to be standardizable in practice rather than standard.

Now, Ω_{r} can be found ultimately from the normalization of the present-day black body radiation spectrum. Meanwhile, the second acoustic peak in the CMB gives us Ω_{b}, as follows. The more baryons are swept along with the photons, the deeper into the potential wells the photon–baryon flux goes (a process called 'baryon drag'). The effects are different on the odd-numbered and even-numbered acoustic peaks, because only the odd-numbered peaks contain a half-wavelength, so the odd-numbered peaks are particularly sensitive to the amplitude of this oscillation. The strength of the odd-numbered peaks is mainly about how far into the potential well the baryons move, so increasing the baryon density will tend to enhance the odd-numbered peaks. The measurement of the second acoustic peak by WMAP is now the best experimental constraint on $\Omega_{\text{b},0}$, against which one can test the predictions of primordial nucleosynthesis (Section 2.6).

The peaks on smaller angular scales depend on oscillations that started earlier, when the sound horizon was smaller, probing times even earlier than matter–radiation equality (i.e. when $\Omega_{\text{m}} = \Omega_{\text{r}}$). It turns out that this extra information is enough to unpick the dark matter density (which dominates Ω_{m}).

Figure 2.9 also shows that the higher acoustic peaks are suppressed. The reason for this is that the transition to a transparent Universe wasn't instantaneous. As the opacity of the Universe dropped, the photons started diffusing away from the positions that they had while the Universe was opaque. Once the Universe became transparent, the photons travelled in a straight line ('free-streamed'), but just before that point they were undergoing a random walk. This diffusion smoothed out the structure on the smallest scales. The strength of this effect depends on how long it took the Universe to make the transition from opaque to transparent. Alternatively, another way of looking at this is that structures are smoothed out if they are smaller than the thickness of the last scattering surface. This effect is known as **diffusion damping** or **Silk damping**, after the cosmologist Joe Silk who first characterized this effect. It's also dependent on Ω_{b} because increasing the numbers of baryons also increases the rate of collisions that the photons experience. The shape of this damping tail is an important consistency check for the cosmological parameters derived from the acoustic peaks.

One final surprise is that this frozen cosmological sound wave structure has never gone away — we'll see in Chapter 3 that it's still visible in the galaxy distribution!

2.13 The Sachs–Wolfe effect

After recombination, the density perturbations in the Universe slowly evolve. Overdensities of dark matter attract the surrounding matter, so matter flows from the underdense regions into the overdense regions. The Universe is now transparent, so there is no photon pressure to prevent matter falling in, so instead of forming oscillations the overdensities grow as more and more matter is drawn in. Similarly, the voids empty out of matter, becoming more and more significant underdensities.

The time evolution of matter density now has a measurable effect on the photons. For example, a photon could pass into a density enhancement, but by the time the photon crosses the overdensity and emerges out the other side, the overdensity has grown. The photon finds that it has to climb out of a bigger potential well than it originally fell into, so the photon will end up with a slight redshift. Similarly, photons passing through a void will find themselves with a net blueshift as the voids become more significant underdensities. This is known as the **Sachs–Wolfe effect** after its discoverers. This effect is detectable in the CMB, known as the **early Integrated Sachs–Wolfe effect**. Much later, large-scale density inhomogeneities such as galaxy clusters affect the passage of photons in exactly the same way. The early inhomogeneities had no luminous matter to mark them, but the later, larger clusters and voids can be traced by optical galaxies. This is sometimes known simply as the **Integrated Sachs–Wolfe effect** . Finally, as the Universe began to enter the $\Omega_\Lambda > 0$ phase, the acceleration of the expansion rate caused the depths of potential wells to decrease, leading to a further **late-time Integrated Sachs–Wolfe effect**. In other words, photons gain energy falling into galaxy clusters, but do not lose all this energy on leaving them again. Meanwhile, photons expend energy to climb up the potential well into a void, but do not gain all this energy back on leaving it. In both cases the accelerating expansion of the Universe has reduced the strength of the overdensity or underdensity. The effect is sometimes abbreviated as ISW.

See, for example, Sachs, R.K. and Wolfe, A.M., 1967, *Astrophysical Journal*, **147**, 73.

It is sometimes also known as the **Rees–Sciama effect**, though beware: sometimes the latter is used for the non-linear terms in the derivation of this effect, while 'Integrated Sachs–Wolfe' is used for the linear terms.

The key prediction of the ISW effect is that there should be a correlation between CMB hot spots and foreground galaxy clusters. Similarly, one can predict a correlation between CMB cold spots and foreground voids.

This is exactly what is seen in correlating the galaxy clusters and voids in the Sloan Digital Sky Survey with the CMB. Figure 2.10 shows part of the CMB map from WMAP, with the positions of foreground galaxy clusters marked in red and foreground voids marked in blue. The team who did this claimed a trend for foreground clusters to be associated with warm (red) regions, while voids tended to be cooler (blue). Since this trend does not appear compelling from this image alone, they took small segments of the WMAP sky map around each of their galaxy clusters and *averaged* these small images together. They did the same on the voids. The resulting averaged images are shown in Figure 2.11. The resulting detection is significant at over 4 standard deviations.

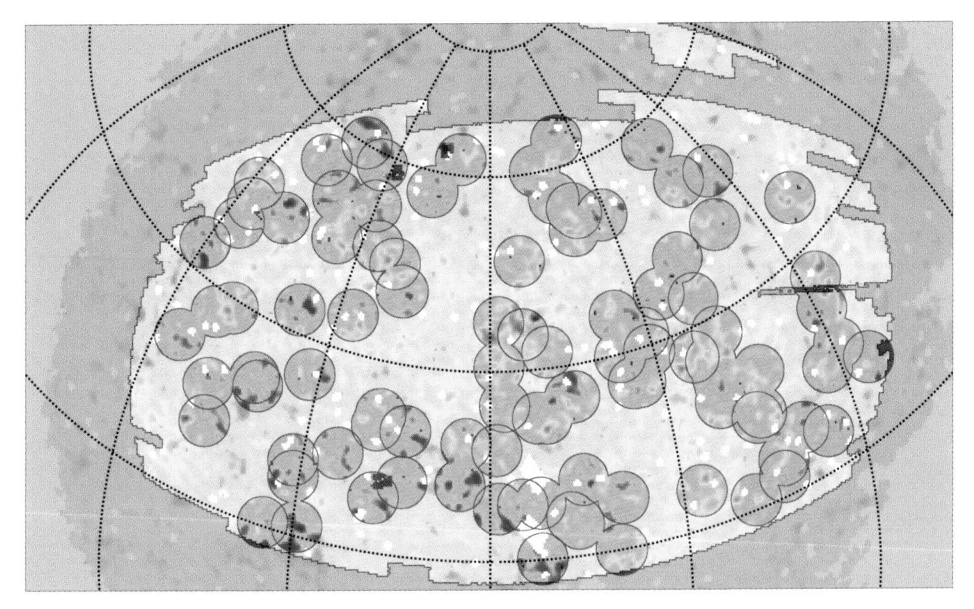

Figure 2.10 The CMB, with the positions of known foreground galaxy clusters (red) and voids (blue) marked.

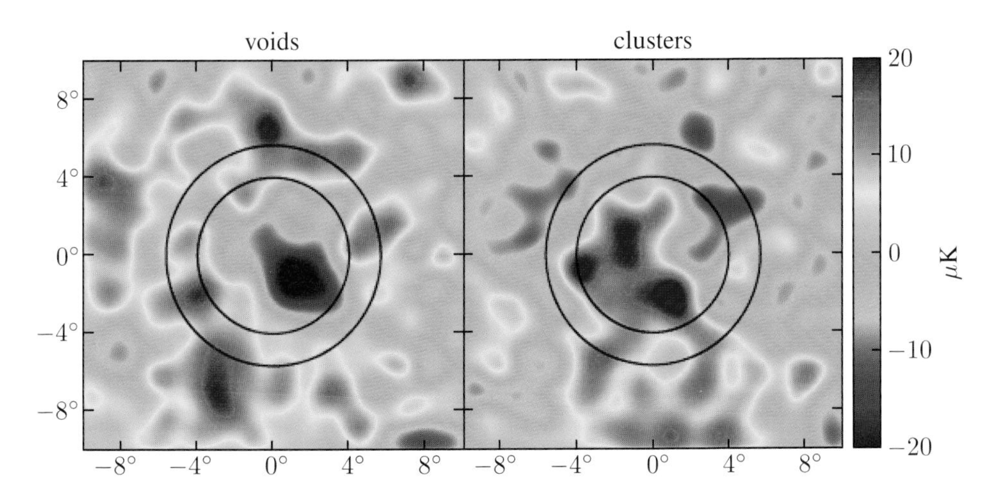

Figure 2.11 Average CMB images at the positions of voids and clusters.

We know from the acoustic peaks in the CMB that we live in a spatially flat universe to a good approximation (Section 2.12); the detection of a late-time ISW effect can be reconciled with spatial flatness only if there is a cosmological constant (or, more generally, dark energy — see later). We'll see below that it is very difficult to measure Ω_Λ from the CMB alone, so cross-correlating foreground populations with the CMB to search for the ISW is a very valuable additional use of the CMB maps.

This technique (averaging images of separate objects in order to detect the average signal from the population) is an example of a **stacking analysis**. This is used very widely in observational cosmology.

2.14 Reionization

After the Big Bang, what generated the first light in the Universe? Were stars the first luminous objects in the Universe that illuminated the darkness, or accreting black holes? In Chapter 8 we'll discuss what observational constraints we have on the first light in the Universe, and also another way of finding Ω_b. However, the CMB gives its own unique constraint on the first light in the Universe.

The effect is similar to Silk damping (Section 2.12). Once the first luminous objects have reionized the Universe, the free electrons liberated by ionizing the atoms can once again scatter CMB photons through Thomson scattering. As with Silk damping, the effect is to suppress the acoustic peaks in a characteristic manner. This time, however, the Universe is transparent and the mean free path of the photons is much larger, of the order of the horizon size. (This also implies that new acoustic oscillations won't form.) The suppression therefore acts on both large and small scales. The overall effect of reionization resembles a change in the overall normalization of the fluctuations, except on the largest scales.

'Reionization' is perhaps a misleading term. We speak of 'recombination' at $z \simeq 1000$ to describe the formation of the first neutral atoms. These are the first atoms, so we can hardly speak of their 'recombining', since they are combining for the first time! Nevertheless, the term 'recombination' is the one used. Similarly, 'reionization' is open to criticism, since these atoms are being ionized for the first time. They are re-making an ionized plasma that existed at $z > 1000$, but this is the first example of the ionization process, so the term 'reionization' at $z > 6$ could be considered as inappropriate as 'recombination' at $z \simeq 1000$. Nevertheless, these are the terms in use. One can only apologize.

Another effect of reionizing the Universe is to change the *polarized* components of the CMB, because scattered light is polarized. We'll discuss the polarization of the CMB in Section 2.16. The current CMB constraints on the epoch of reionization from the five-year WMAP results are shown in Figure 2.12. While the redshift of reionization is not very well determined, the **optical depth** τ to Thomson scattering is better measured: the probability of a photon undergoing Thomson scattering is defined as $e^{-\tau}$, where the five-year WMAP CMB data set the constraint $\tau = 0.087 \pm 0.017$.

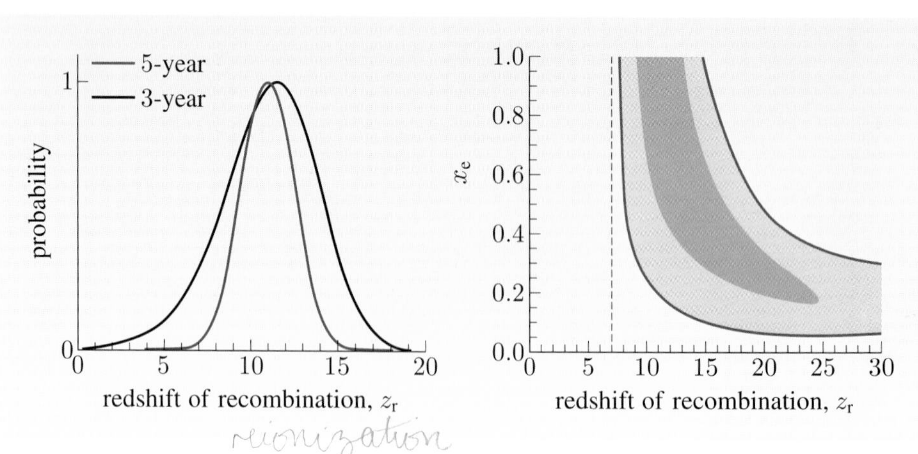

Figure 2.12 The constraints on reionization from the WMAP CMB maps. The left-hand panel is the likelihood constraints from the WMAP 3-year and 5-year data sets (note the improved constraint from the extra two years of data), assuming an instantaneous reionization at a redshift z_r. But the CMB data don't in themselves require the reionization to be instantaneous. The right-hand panel shows the constraints if we assume a two-step model: the reionization was instantaneously set to a level of x_e at redshift z_r, then instantaneously completely ionized at redshift 7. The dark shaded region is the 1σ contour, i.e. there is an approximately 68% chance that the underlying value is in that region, while the light shaded region is the 2σ contour ($\simeq$ 95%).

2.15 Cosmological parameter constraints

Figure 2.13 shows the effect that varying some cosmological parameters has on the acoustic peaks. As we saw in Section 2.12, these effects are quite strong and are our best experimental constraints on many cosmological parameters at the time of writing. Unfortunately, the CMB on its own isn't quite enough to constrain all the cosmological parameters. To see why, see panel (b) of Figure 2.13.

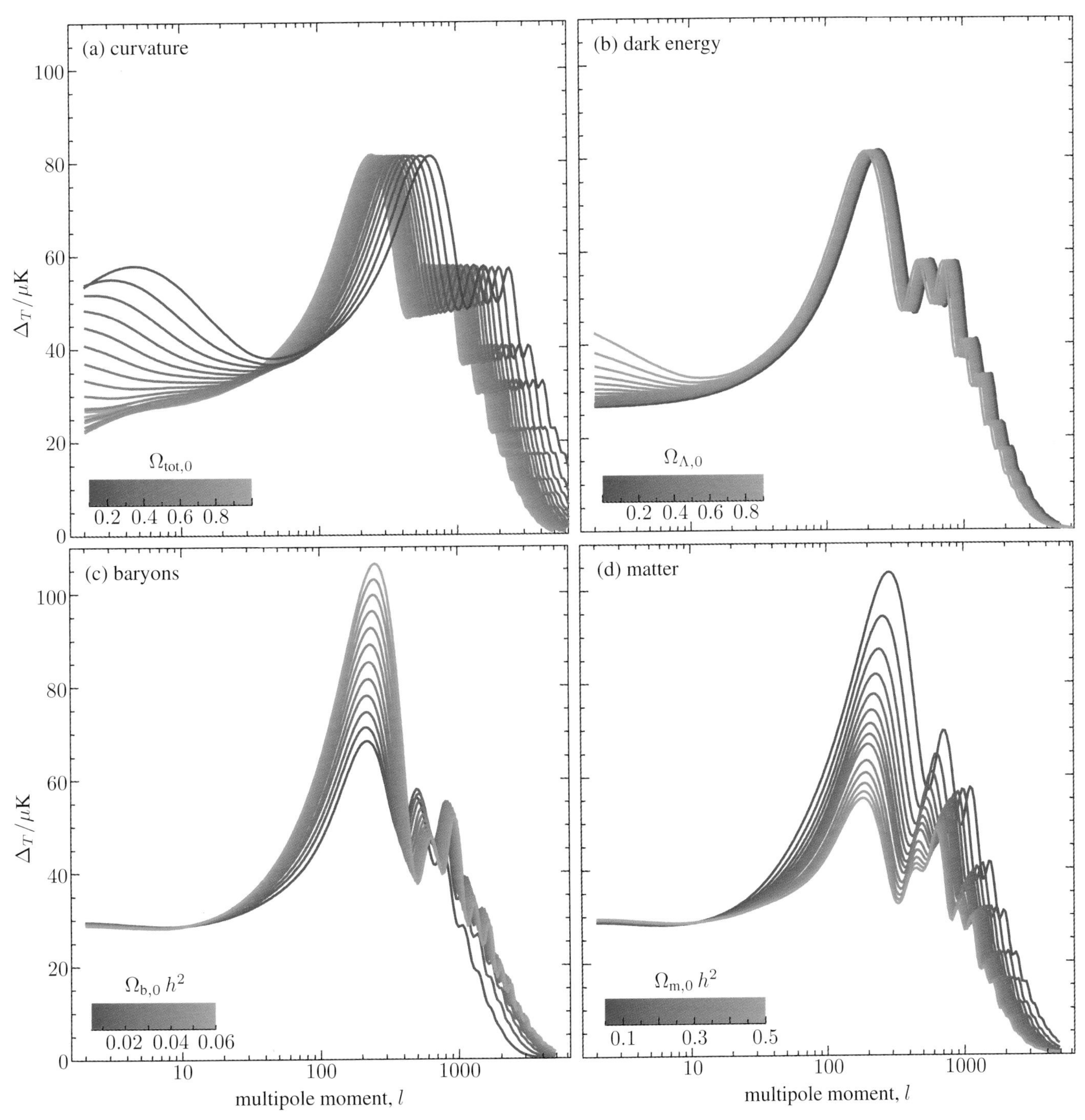

Figure 2.13 How the shapes of the acoustic peaks depend on present-day cosmological parameters. All the models are varied around a common starting point of $\Omega_{total,0} = 1$, $\Omega_{\Lambda,0} = 0.65$, $\Omega_{b,0}\,h^2 = 0.02$, $\Omega_{m,0}\,h^2 = 0.147$, a scale-invariant power spectrum and no reionization. The temperature fluctuation Δ_T plotted is defined as $\Delta_T = \sqrt{l(l+1)C_l/2\pi}\ T_{CMB}$.

The positions of the acoustic peaks are only very weakly dependent on $\Omega_{\Lambda,0}$, for a fixed $\Omega_{k,0}$. If $\Omega_{b,0}$ is fixed and $\Omega_{k,0} = 0$, then we can constrain $\Omega_{m,0}$ (panel (d) of Figure 2.13), but there is not enough information in the CMB to constrain $\Omega_{k,0}$ *and* $\Omega_{m,0}$ *and* $\Omega_{b,0}$ *and* $\Omega_{\Lambda,0}$ all simultaneously.

The fact that constraints on one parameter can correlate with constraints on another is known as **parameter degeneracies**. These degeneracies are intrinsic to CMB experiments, but the degeneracies can be broken by including a comparison with other, non-CMB experiments. For example, Figure 2.14 shows the constraints on $\Omega_{m,0}$ and $\Omega_{\Lambda,0}$ from the CMB and from high-redshift supernovae (which we shall meet in Chapter 3). The supernovae and CMB constraints both have degeneracies, but they are in opposite directions, so combining the constraints makes it possible to measure $\Omega_{m,0}$ and $\Omega_{\Lambda,0}$ separately. Another example that we've already met is the late-time ISW (Section 2.13), from which one can infer $\Omega_{\Lambda,0}$ if $\Omega_{k,0}$ is known.

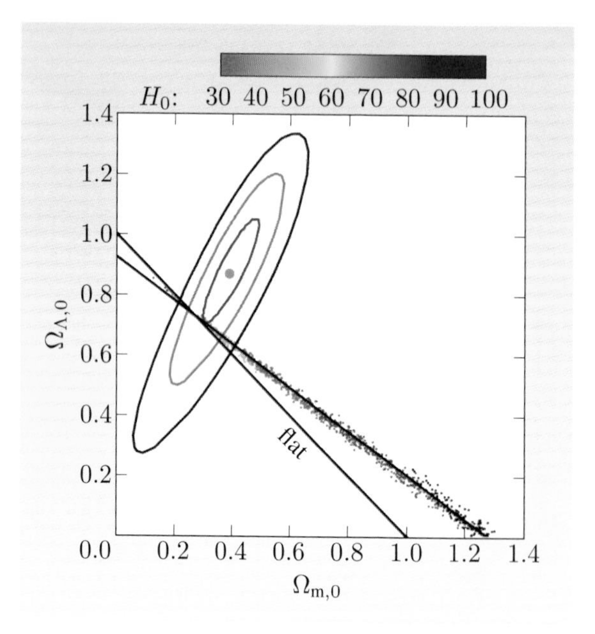

Figure 2.14 The constraints on the cosmological density parameters from the WMAP CMB measurements and from high-redshift supernovae. The contours show the 1σ (inner ring), 2σ (middle ring) and 3σ (outer ring) allowed range for the supernova data of Kowalski et al (2008). The dots show some Monte Carlo (i.e. random) realizations of the WMAP data, in which model universes are selected in proportion to their likelihood of fitting the WMAP data of Dunkley et al. The value of the Hubble parameter H_0 is colour-coded for these model universes. The combination of the WMAP data and the supernova data leaves only a very small region of this plane mutually allowed, thus breaking the parameter degeneracy in the WMAP data alone.

Together, the parameter constraints from the CMB and other sources have converged remarkably on the parameters. Overall, the level of agreement between the CMB and other constraints has led to the resulting cosmological model being called the **concordance cosmology**.

A completely different (and controversial) type of constraint is also worth mentioning. The Universe contains intelligent life. Can we use this fact on its own to constrain the cosmological parameters? This line of argument is known as the **anthropic principle**. This might explain why, for example, we find ourselves in a Universe after recombination and at a time when $\Omega_m > \Omega_r$ so gravitational collapse is possible and stars can form. The Universe also cannot be so old that stars no longer form, as might be the case if the age were measured in hundreds of billions of years, or when $\Omega_m \simeq 0$. Using anthropic arguments to constrain our position in time and space within a given cosmological model is known as the **weak anthropic principle**. This is quite a departure from the Copernican principle! It is also an example of a 'selection effect', about which we shall say more in Chapter 4. Anthropic arguments attracted some interest prior to precision cosmology, but the experimental constraints on the concordance cosmological model are now much stronger than the anthropic constraints.

A more radical variant is to suppose that an ensemble of universes exists (a so-called 'multiverse'), each universe having different fundamental physical constants. Anthropic arguments can then be used to constrain which parts of this ensemble intelligent observers could inhabit. The underlying assertion that our own Universe *must* be suitable for the formation of intelligent life (from which one might constrain the fundamental physical constants) is sometimes known as the **strong anthropic principle**. While this could explain various apparent fine-tunings in physical constants, the disadvantage of these arguments is that they give no physical mechanism for explaining parameter values. There is also predictably some disagreement over how to best calculate likelihood distributions for fundamental physical constants on anthropic grounds. Also, is this part of testable science? It may be or may become so if a testable theoretical framework could be found for explaining this ensemble. In any case, is 'science' exclusively concerned with things that are testable in practice, now? We won't rehearse the debates in this book but you will no doubt sense that this is an area that can generate a great deal of controversy.

2.16 The polarization of the CMB

A small percentage of the CMB is polarized, and like the unpolarized CMB this polarized component also has structure. This polarized signal is an independent measure of the cosmological parameters and could be the key to uncovering the physics of inflation. At the time of writing only part of this polarized clustered signal has been reliably detected, but CMB polarization will be a major focus of observational cosmology in the coming decades.

The primordial CMB polarization is generated only by the scattering of CMB photons, so it is therefore sensitive to both recombination and reionization. The CMB has no circular polarization, because this can't be generated by scattering. The scattering is also wavelength-independent, so the scattered CMB spectrum is the same as the unpolarized CMB spectrum. For that reason it's usual to use the fractional polarization, in particular to measure the temperature difference that the polarized light has relative to the mean CMB temperature, e.g. $(T_{\text{pol}} - T_{\text{CMB}})/\langle T_{\text{CMB}} \rangle$, where the angle brackets indicate an average. There is also more scope for information in polarization on the sky because polarizations have directions.

There are many ways of representing polarization mathematically. CMB science tends to break the linear polarization into two components known as **E-mode** and **B-mode**. (The origins of these names are an analogy to electromagnetism — these are *not* the electric and magnetic field components of the CMB electromagnetic wave! It's not essential to our story, but if you want details on this analogy, see the box below.) There's a physical reason for doing this: it turns out that the primordial B-mode is due entirely to primordial gravitational waves!

B-modes are sometimes also called tensor modes, though we don't usually use that expression in this book. The term is related to how they can be expressed as perturbations of the metric.

The Helmholtz–Hodge theorem

The electromagnetic analogy is as follows. There's a general mathematical theorem, known as the *Helmholtz–Hodge theorem*, that any vector field $\boldsymbol{v}$ can be expressed in two parts: $\boldsymbol{v} = \boldsymbol{B} + \boldsymbol{\nabla}\phi$, where ϕ is a scalar field and $\boldsymbol{B}$ has no divergence, i.e. $\boldsymbol{\nabla} \cdot \boldsymbol{B} = 0$. This is like electromagnetism, where the electric field is $\boldsymbol{E} = \boldsymbol{\nabla}\phi$ and ϕ is the scalar potential in electromagnetism; also, the lack of magnetic monopoles in electromagnetism implies that $\boldsymbol{\nabla} \cdot \boldsymbol{B} = 0$, where $\boldsymbol{B}$ is the magnetic field. It's also generally true that a curl of a gradient is zero, i.e. $\boldsymbol{\nabla} \times \boldsymbol{\nabla}\phi = \boldsymbol{0}$ for any scalar field ϕ, implying $\boldsymbol{\nabla} \times \boldsymbol{E} = \boldsymbol{0}$. So any vector field can be broken into a 'magnetic' (i.e. divergence-free) component and an 'electric' (i.e. curl-free) component. Now, in the case of our CMB linear polarizations we are dealing not with a vector field but with something subtly different, because if we rotated the polarization by 180° we'd get the same polarization, which isn't true of a vector. The mathematical expressions for the 'electric' and 'magnetic' components of CMB polarization are therefore slightly different. We won't go into these differences here, but there is more information in the further reading section.

How do you measure the clustering of the polarized CMB? Using a mathematical formalism similar to the unpolarized CMB structure, it's possible to quantify the structure in the polarized background as a function of angular size. Instead of measuring the difference in unpolarized temperatures between two locations (known as the TT power spectrum, with T standing for temperature), we could measure the differences in (say) the temperatures of the E-component of the polarization between two locations. This is sometimes referred to as the EE power spectrum. Similarly, we could measure the clustering of the B-mode, which would be called the BB power spectrum. We could also compare, say, the E-mode polarization temperature in one place with the unpolarized temperature in another. This cross-correlation would go by the name TE. In all there are six possible permutations: TT, EE, BB, TE, TB, EB. Of these, it can be shown from parity arguments that TB and BE should be zero, so there are four astrophysically useful cross-correlations. The predicted levels of these clustering strengths are shown in Figure 2.15. Note that the EE oscillations are out of phase with the TT oscillations, for reasons related to how the light is scattered at the time of recombination (see the further reading section for more details).

The detection of B-mode polarized clustering would be tremendously exciting, because the primordial gravitational wave background constrains the shape of the inflation potential (Section 2.8) and would be the first genuine consistency test of inflation. If we describe the scalar clustering power spectrum as $C_l^{\mathrm{S}} \propto l^{n_{\mathrm{s}}-3}$ (where n_{s} is as defined in Equations 2.44 and 2.45) and the tensor clustering as $C_l^{\mathrm{T}} \propto l^{n_{\mathrm{T}}-3}$, then inflation's predictions for powers are $n_{\mathrm{s}} \simeq 1 + 2\eta - 6\varepsilon$ and $n_{\mathrm{T}} \simeq 1 - 2\varepsilon$, respectively, where ε and η are defined in Section 2.8 and are related

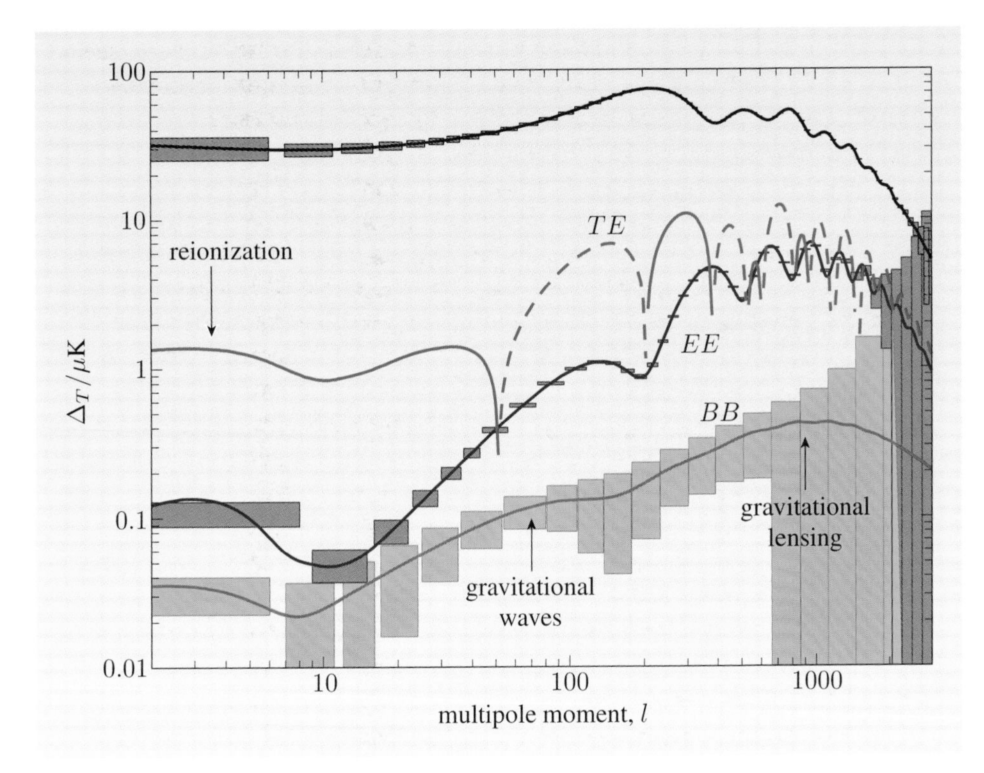

Figure 2.15 The predicted unpolarized CMB power spectrum (upper line), compared to the TE power spectrum, the EE power spectrum and the BB power spectrum. Negative values are dashed, and the predicted 1σ (i.e. 68% confidence) uncertainties from the European Space Agency Planck satellite are shown as bars. Note that the polarized signals are much weaker than the unpolarized signal.

to the shape of the inflation potential. These powers are sometimes referred to as **spectral indices** of the density perturbations. There is also a prediction for the amplitudes of the scalar and tensor clustering to be related by $C_l^{\mathrm{T}}/C_l^{\mathrm{S}} = 12.4\varepsilon$. The consistency test of inflation is whether the tensor spectral index agrees with the relative strengths of the scalar and tensor modes. Unfortunately, the closer the scalar power spectrum is to scale-invariance, the harder it is to test inflation; the current experimental constraint on the scalar spectral index is $0.0081 < 1 - n_{\mathrm{s}} < 0.0647$ (WMAP five-year constraint).

Experimentally, this is a very challenging experiment for the current generation of detectors, as can be seen from Figure 2.15. Currently, the only constraints on the polarized signals are from the TE cross-correlation power spectrum from the WMAP satellite, shown in Figure 2.16. The detection of primordial gravitational waves is the next great challenge for CMB experiments and could show us the path to new physics. One of the problems is the existence of astronomical foregrounds that could contribute to the BB power spectrum. Gravitational lensing of the CMB by the intervening large-scale structure of the Universe deflects the CMB photons so changes the C_l clustering signal (see Figure 2.15). It can produce a BB correlation, particularly at smaller scales of $l > 500$ or so. We discuss gravitational lensing in more depth in Chapter 7. It's by no means certain that there *is* a gravitational wave background to detect: a rival theory to inflation known as the 'ekpyrotic Universe' predicts no detectable CMB gravitational wave background. This model is based on an extension of superstring theory known as

M-theory. These theories predict more dimensions than our usual three space and one time. Our Universe is imagined to be a 'sheet' in a higher-dimensional spacetime, and the trigger for the Big Bang in this model was the collision between two such sheets.

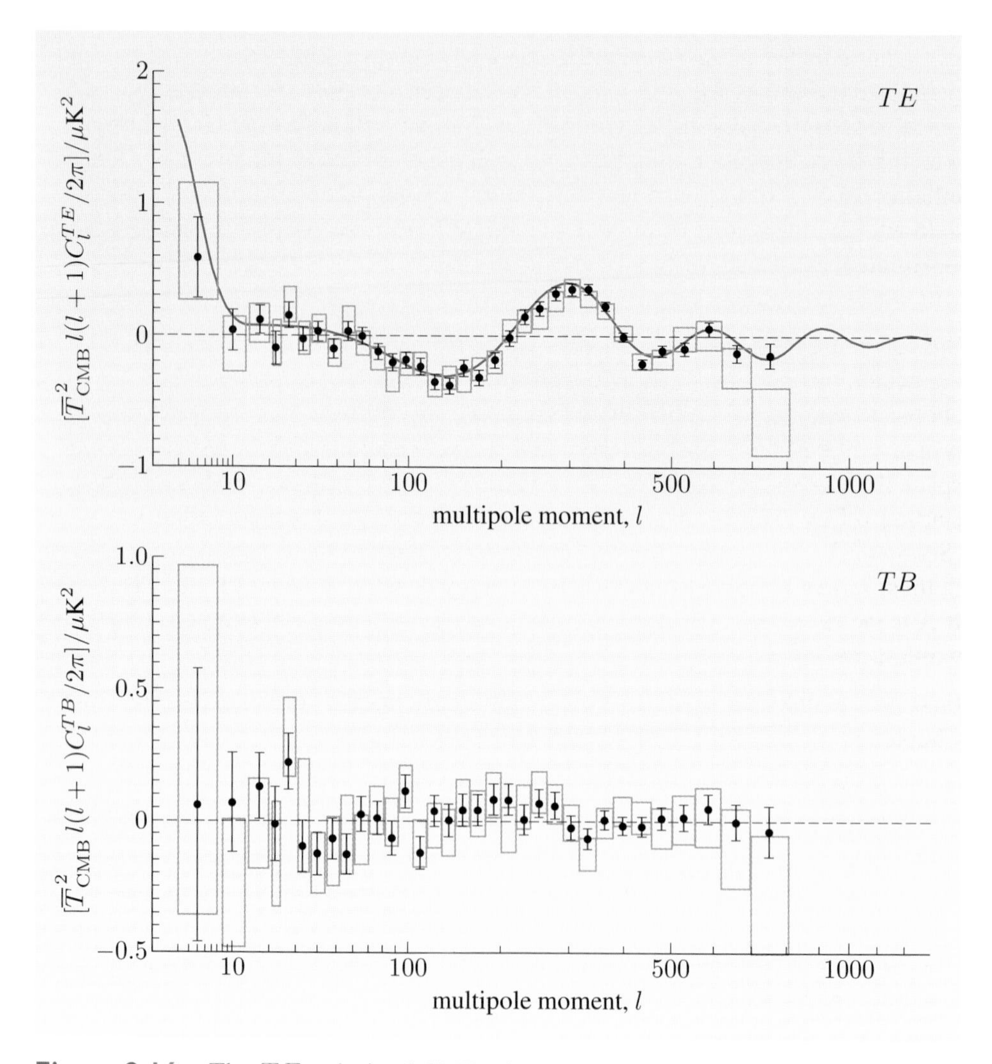

Figure 2.16 The TE polarized CMB clustering detected by the first seven years of WMAP, and WMAP's non-detection (as expected) of the TB clustering signal. The boxes show WMAP's five-year constraint, showing the improvement from adding a couple of years' more data.

We shall meet gravitational wave observatories briefly in Chapter 6. The direct detection of primordial gravitational waves in the new generation of gravitational wave observatories is also very challenging, partly because the energy density in gravitational radiation redshifts in the same way as the photon energy density.

Another source of CMB photon scattering is the electrons that are liberated at the epoch of reionization. As with the surface of last scattering at recombination, the scattered CMB light will be partly polarized. The polarized structure of the CMB is therefore simultaneously sensitive to physical processes at both recombination and reionization. This shows up as a bump in the E-mode polarization on large angular scales, around l of a few.

2.17 Dark energy and the fate of the Universe

Einstein's field equations of general relativity equate a measure of the spacetime curvature (the Einstein tensor) with a measure of the mass-energy and momentum (the energy–momentum tensor). What does this mean? In what sense is a curvature equal to a mass-energy? Is the mass-energy causing the curvature? Or perhaps the reverse? The origin of mass is a deep and obscure problem in particle physics. It may be that the Large Hadron Collider will detect the Higgs boson, which is a prediction of one candidate explanation of the origin of mass. (Like inflation, this also invokes a new scalar field, though it's known that the Higgs field and the inflaton field cannot be identical.) However, even if the Higgs boson is detected, the causes of the link between spacetime curvature and mass-energy will still be a mystery.

Spacetime also has an in-built capacity or tendency to expand, characterized by the cosmological constant Λ as we noted in Chapter 1. According to how we choose to interpret Einstein's field equations, we could choose to regard Λ as a property of spacetime, or we could choose to regard it as some substance within space that drives an expansion. We've already treated inflation in this way, by starting off by asking what sort of content in space could explain inflation (Section 2.7). We've also already implicitly considered the cosmological constant in these terms, by giving it an effective energy density parameter Ω_Λ in Chapter 1. If Λ is the result of spacetime containing some substance, what sort of substance would it be?

First, let's remind ourselves of the fundamental equations of the expansion from Chapter 1:

$$\left(\frac{\mathrm{d}R}{\mathrm{d}t}\right)^2 = \dot{R}^2 = \frac{8\pi G\rho R^2}{3} - kc^2 + \frac{\Lambda c^2 R^2}{3}, \qquad \text{(Eqn 1.7)}$$

$$\frac{\mathrm{d}^2 R}{\mathrm{d}t^2} = \frac{\mathrm{d}}{\mathrm{d}t}\left(\frac{\mathrm{d}R}{\mathrm{d}t}\right) = \ddot{R} = -4\pi G\left(\rho + \frac{3p}{c^2}\right)\frac{R}{3} + \frac{\Lambda c^2 R}{3}. \qquad \text{(Eqn 1.8)}$$

If we want to regard the cosmological constant as having an effective energy density ρ_Λ and pressure p_Λ, we could write these as

$$H^2 = \left(\frac{\dot{R}}{R}\right)^2 = \frac{-kc^2}{R^2} + \frac{8\pi G}{3}\sum \rho_i, \qquad (2.61)$$

$$\frac{\ddot{R}}{R} = \frac{-4\pi G}{3}\sum\left(\rho_i + \frac{3p_i}{c^2}\right), \qquad (2.62)$$

where the sum is carried over all components, e.g. matter, radiation and in this case the cosmological constant. In order to reconcile Equations 1.7 and 2.61, one needs that

$$\frac{8\pi G\rho_\Lambda}{3} = \frac{\Lambda c^2}{3},$$

i.e. $\rho_\Lambda = \Lambda c^2/(8\pi G)$.

In Section 2.7 we wrote the equation of state of a gas as $p = w\rho c^2$, where p is the pressure and ρ is the density. The parameter w defines the equation of state. We can regard all the contents of the Universe as 'gases' in this sense. For example, pressureless matter has $w = 0$. We can re-cast Equation 2.62 in this way as

$$\frac{\ddot{R}}{R} = \frac{-4\pi G}{3} \sum \rho_i (1 + 3w_i), \tag{2.63}$$

where w_i is the equation of state parameter for the ith component. In order to reconcile this with Equation 1.8, we need that

$$\frac{-4\pi G}{3} \rho_\Lambda (1 + 3w_\Lambda) = \frac{\Lambda c^2}{3}$$

or

$$\frac{-4\pi G}{3} \frac{\Lambda c^2}{8\pi G} (1 + 3w_\Lambda) = \frac{\Lambda c^2}{3},$$

so

$$\frac{-1}{2} (1 + 3w_\Lambda) = 1$$

thus

$$w_\Lambda = -1,$$

i.e. the cosmological constant 'substance' has an effective negative pressure! The same is true of the inflaton field. However, inflation occurs at an energy scale of around 10^{24} eV, while the energy scale of the cosmological constant is more like 10^{-3} eV. The physical processes behind inflation and the cosmological constant would appear to be quite different.

We have no idea of the underlying physical causes of Λ. This has led some cosmologists to speculate that w_Λ might be different from -1, or even that w_Λ might be time-varying. In the latter case, the 'cosmological constant' would not be a constant. These modifications to the cosmological constant are generically known as **dark energy**. The cosmological constant Λ is therefore a special case of dark energy, in which $w = -1$ at all times. Note that dark energy appears to be very different to dark *matter*. Sometimes dark matter and dark energy are collectively called the 'dark sector'. In total, the dark sector comprises about 90% of the present-day energy density of the Universe, yet we know next to nothing about its physics.

Exercise 2.10 List at least four major differences between dark matter and dark energy. ■

Figure 2.17 shows the observational constraints on the dark energy equation of state. Note that there is a strong parameter degeneracy between the present-day dark energy density Ω_Λ and the equation of state. However, combining this with other observational constraints narrows the field considerably, such as the Hubble Space Telescope determination of the Hubble parameter, baryonic acoustic oscillations and high-redshift supernovae (which we shall meet in Chapter 3).

If we admit the possibility of a time-varying equation of state, the constraints worsen considerably. If we write $w = w_0 + w'z/(1 + z)$ for the dark energy equation of state, the corresponding constraints are shown in Figure 2.18. However, there are good reasons for believing that the constraints will improve. For example, the late-time ISW is sensitive to a late phase of accelerated expansion in the Universe, and changes in the dark energy equation of state produce changes in the ISW signal.

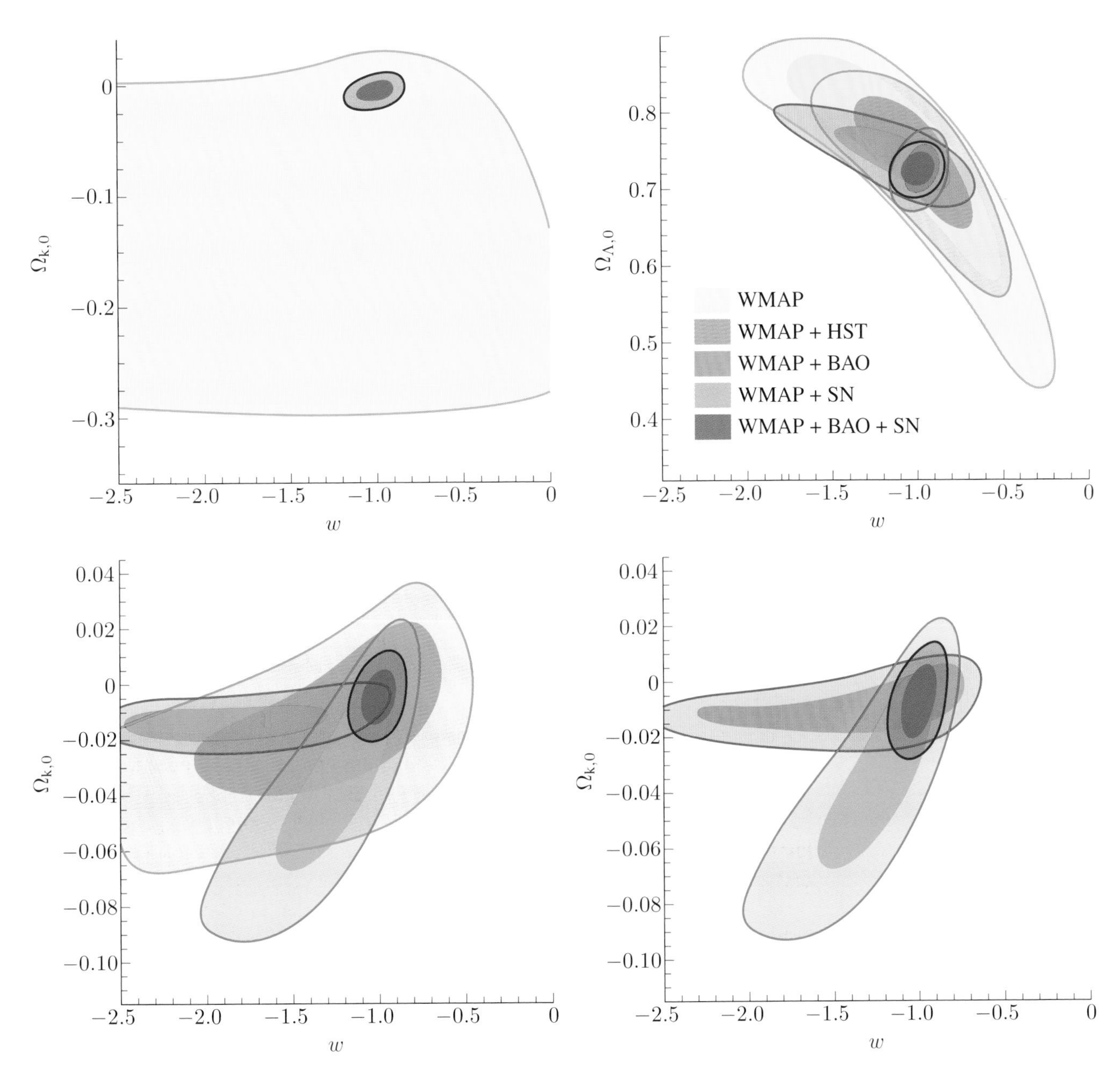

Figure 2.17 The constraints on the dark energy equation of state w from the WMAP CMB maps and other surveys, compared to the constraints on other parameters. The dark shaded regions are the 1σ contours, i.e. an approximately 68% likelihood that the underlying value is in that region, while the lighter shaded regions are 2σ ($\simeq$ 95% likelihood). The bottom-right panel uses BAOs (Chapter 3) from SDSS luminous red galaxies, while the bottom-left panel uses a wider compilation.

Are there any theoretical reasons for expecting w to depend on time? We could follow a similar line of reasoning to inflation and imagine that spacetime is filled with (another!) scalar field, which we'll call ϕ_Λ. (We'll use the Λ subscript to distinguish this field from the inflaton field.) If it has a potential of $V_\Lambda(\phi_\Lambda)$, then following Section 2.8, the field will satisfy $\ddot{\phi}_\Lambda + 3H\dot{\phi}_\Lambda = V'_\Lambda(\phi_\Lambda)$, where dots are time derivatives and the prime is a derivative with respect to ϕ_Λ. (Compare Equation 2.18 — again we're assuming that the field is the same everywhere in

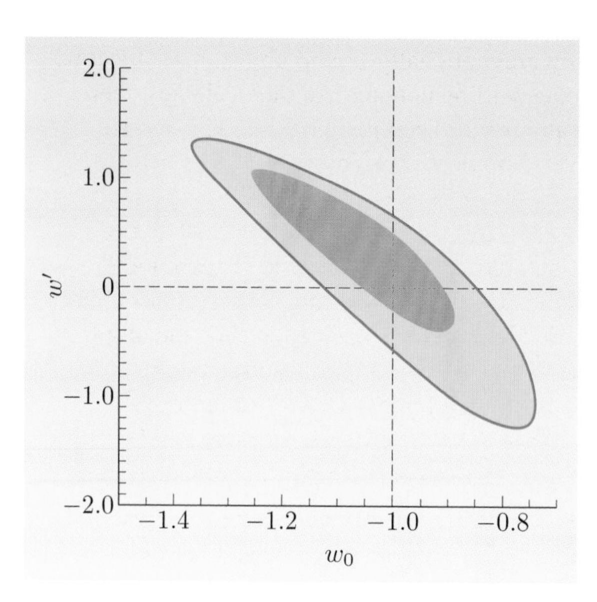

Figure 2.18 The constraints on the time-variation of the dark energy equation of state from the WMAP CMB data combined with supernovae, BAOs (Chapter 3) and nucleosynthesis constraints. The dark shaded regions are the 1σ contours, i.e. there is an approximately 68% likelihood inferred from the CMB data that the underlying value is in that region, while the lighter shaded regions are 2σ ($\simeq 95\%$ likelihood).

space.) Pursuing the analogy of a ball rolling or sliding down a hill, we could regard V_Λ as a potential energy, and $\frac{1}{2}\dot{\phi}_\Lambda^2$ as a kinetic energy that we'll call K_Λ. As with inflation, it turns out the energy density comes out as $\rho_\Lambda \propto K_\Lambda + V_\Lambda$, while the pressure is $p_\Lambda \propto K_\Lambda - V_\Lambda$ (compare Equations 2.19 and 2.20), so the equation of state is $w = (K_\Lambda - V_\Lambda)/(K_\Lambda + V_\Lambda)$. This would naturally vary with time. If the field is slowly-rolling, then $K_\Lambda \ll V_\Lambda$ so $w \simeq -1$, i.e. it would look like a cosmological constant. Before that point it would have w in the range $-1/3$ to -1 (Section 2.7). These models are sometimes called **quintessence** since they postulate a fifth fundamental field in addition to the four fundamental forces of nature. The mass of the particle associated with this field turns out to be very small indeed in particle physics terms, namely 10^{-33} eV, which leads to a whole new set of fine-tuning problems.

One curious regime that is not prohibited by experimental constraints is $w < -1$, sometimes called 'phantom energy'. These models radically change the projected fate of the Universe. To see why, we'll use the variation of energy density for an adiabatic expansion: $\rho \propto (R/R_0)^{-3(1+w)}$, which we also met in Section 2.7. We won't prove this thermodynamic formula here, but note that for a photon gas ($w = 1/3$) this gives $\rho \propto (R/R_0)^{-4}$, while for pressureless matter ($w = 0$) this gives $\rho \propto (R/R_0)^{-3}$. Adapting Equation 2.61, we find that

$$\frac{H^2}{H_0^2} = \sum_i \Omega_{i,0} \left(\frac{R}{R_0}\right)^{-3(1+w_i)}, \tag{2.64}$$

where the w_i are the equations of state of each of the contents of the Universe. Once we get to an epoch when dark energy dominates, we'll have that

$$H^2 \propto \left(\frac{R}{R_0}\right)^{-3(1+w_\Lambda)},$$

where we're now writing w_Λ for the equation of state of the only remaining component, dark energy. If $w_\Lambda = -1$ (i.e. a cosmological constant), then the right-hand side becomes a constant, so H tends to a constant value, which is what we found in Section 1.11. However, if $w_\Lambda < -1$, then H is perpetually increasing. This means that the radius of the cosmological event horizon will shrink. When it becomes smaller than the size of any gravitationally-bound object, that object will become unbound. Clusters of galaxies will be unbounded, then galaxies, then stars and planets. Eventually even atoms will become unbound. This model universe has been called the 'big rip'. The event horizon size reaches zero in a finite time, at which point the scale factor of the Universe becomes infinite. This singularity represents an effective end of the Universe. For $w_\Lambda = -3/2$, the Universe ends in about 21 Gyr. Galaxy clusters would be ripped apart about 1 Gyr before the end, galaxies about 60 Myr before the end, solar systems about 3 months before the end, planets about half an hour before the end, and atoms about 10^{-19} seconds before the end. The Universe would also spend most of its history in a dark-energy-dominated phase, avoiding the need for anthropic arguments. There are, however, ongoing debates as to whether any viable phantom energy model could be generated from particle physics considerations.

Finally, it's worth linking the discussion of dark energy with our earlier discussion of inflation. When we calculated the effective pressure and density of the inflaton field in Section 2.8, we found that $V \gg \dot{\phi}$ led to an effective equation of state $w = -1$ (Equations 2.19 and 2.20). As we've seen in this section, this is equivalent to a cosmological constant, so the slow-roll approximation leads to a Universe very much like a cosmological-constant-dominated de Sitter universe. This inflation was driven by the difference between the initial value of $V(\phi)$ and the minimum. We implicitly assumed that when the Universe reaches the minimum in $V(\phi)$, inflation stops, implying $V = 0$. However, it's by no means clear that $V = 0$ is the natural minimum value. The framework of inflation only deals with potential *differences*, but in general relativity the *absolute* value of the energy (in the form of the energy–momentum tensor) determines the curvature and dynamics of spacetime.

One might, for example, expect the zero-point energy to be set by Planck scale physics. Since Λ has dimensions of one over length squared, we might expect $\Lambda \sim r_{\rm Pl}^{-2} \simeq 10^{70}\,{\rm m^{-2}}$. However, as the following exercise shows, this is wildly out of kilter with the observations.

Exercise 2.11 Using the experimental values $\Omega_{\Lambda,0} = 0.742$ and $H_0 = 72\,{\rm km\,s^{-1}\,Mpc^{-1}}$, show that the observed value of Λ is $1.3 \times 10^{-52}\,{\rm m^{-2}}$. ■

This dimensional analysis gets the answer wrong by 122 orders of magnitude! We might make a slightly more physically-motivated estimate by imagining that the quantum vacuum is made up of quantum mechanical simple harmonic oscillators. It's a standard result in quantum mechanics that the wave function of a particle of mass m has a zero-point energy of $E_0 = h\omega/(4\pi)$, where ω is the angular frequency of the oscillator. We could consider a box in the Universe with a side L and add up all the zero-point energies of these oscillators: $E = (1/4\pi) \times \sum_j h\omega_j$, where $\omega^2 = k^2c^2 + m^2c^4/\hbar^2$ and $k = 2\pi/\lambda$, where λ is the de Broglie wavelength. We can let the dimension of the box L tend to infinity. The periodic boundary conditions of the box imply that the only wavelengths allowed are $\lambda_x = L/n_x$ in the x-direction (where n_x is an integer). Therefore in the interval k_x to $k_x + {\rm d}k_x$, there should be $(L/2\pi)\,{\rm d}k$ separate values of k_x. The same applies in y and z, in which case the energy per unit volume becomes

$$\frac{E}{L^3} = \frac{h}{4\pi}\int \frac{\omega(k)}{(2\pi)^3}\,{\rm d}^3k.$$

Unfortunately, this integral diverges. Perhaps this is OK, since we should expect our low-energy quantum mechanical calculations to break down at some length scale, which should be $1/\lambda = k \gg mc/\hbar$. If we set this minimum length scale to the Planck length, the vacuum energy density comes out again as 120 orders of magnitude less than the observed Λ. As the cosmologist John Peacock said: 'We are left with the strong impression that some vital physical principle is missing. This should make us wary of thinking that inflation, which exploits *changes* in the level of vacuum energy, can be the final answer.'

So this is the awkward position that we find ourselves in. We've tried to solve the horizon problem and other problems avoiding Planck scale physics with the GUT-scale inflation model, but we found that inflation doesn't quite escape all considerations of the Planck scale. We have no way of predicting even the order of magnitude of the cosmological constant from fundamental principles. We've filled space with at least three fundamental scalar fields (Higgs, inflaton, dark energy), while being unable to reconcile the fundamental conceptual bases of general relativity and quantum field theory at the Planck scale. Perhaps there is a tremendous conceptual breakthrough coming soon, but it's overdue, as the problems have been around for some decades. Perhaps there's some vital piece of experimental evidence that will provide the trigger, or perhaps an insight will provide a thrilling breakthrough. There are some contenders, but with experimental constraints and/or quantitative testable predictions hard to come by, they're still a little speculative to cover in depth in a book such as this. As a professional scientist one has a choice: do you take a punt on chasing these fundamental problems, or perhaps do you try something fundamental but more tractable? We have the astonishing capability to build telescopes that can detect galaxies throughout most of the Hubble volume, i.e. almost the entire observable Universe. In the next chapters we shall cover some of what's been discovered about the evolution of the Universe since the time of the CMB.

Summary of Chapter 2

1. The cosmic microwave background (CMB) is the light from the surface of last scattering from when the Universe was last opaque. At the time of last scattering the motion of the photon gas became decoupled from the baryonic matter.
2. A redshifted black body spectrum is also a black body spectrum.
3. The CMB is expected to be a black body because the photon–baryon collision rate scales as a^{-6} (where a is the dimensionless scale factor of the Universe), which increases faster as a tends to zero than the dynamical inverse timescale, so thermalization was increasingly easy at early epochs. The CMB spectrum is observed to be an excellent black body.
4. Baryon number conservation predicts perfect matter–antimatter symmetry (unless an asymmetry is incorporated into the initial conditions of the Universe), so baryon number non-conservation is expected in grand unified theories.
5. The η parameter is the ratio of the number density of baryons to that of photons. Since CMB photons dominate the entropy density of the Universe,

η is a measure of the number of baryons per unit entropy and is hence a measure of the (reciprocal of the) entropy per baryon.

6. Various particle creation and annihilation processes existed in equilibrium in the early Universe, with relative numbers determined by the Boltzmann distribution. When the ambient particle energies could no longer support the reactions, the number densities were frozen out.
7. The neutron–proton ratio predicted in freeze-out is the first step in Big Bang nucleosynthesis in calculating the abundances of nuclei. The observed abundances agree with observations, though there are cases where systematic uncertainties are under discussion.
8. Several lines of evidence point at the need for new physics: the horizon problem, the flatness problem, the baryon asymmetry problem, the monopole problem, and the origin of the primordial density perturbations. The theory of inflation gives a broad framework for new physics that could resolve these problems.
9. Inflation hypothesizes a scalar field ϕ (assumed roughly the same everywhere in the observable Universe) with associated energy density $V(\phi)$. The field ϕ is offset from the minimum of $V(\phi)$ and evolves sufficiently slowly to the minimum that the $\ddot{\phi}$ terms can be neglected. This is known as the slow-roll approximation. However, the shape of the inflation potential is not yet known. Many varieties of inflation theories exist, each with different assumptions about the shape of the $V(\phi)$ function.
10. Inflation implies an exponential phase of expansion in the early Universe, with an effective equation of state $w = -1/3$ to -1 (where a cosmological constant has $w = -1$, and $w < -1/3$ is necessary to solve the horizon problem). This cooled the Universe but the decay of the energy in the field into particles reheated the Universe; this is also a likely time for the creation of the matter–antimatter asymmetry in the Universe (baryogenesis).
11. Inflation predicts a roughly scale-invariant power spectrum of initial density perturbations, $P(k) \propto k$. This is in good agreement with observations. The slight departures from scale-invariance are imprinted at the end of inflation when the slow-roll approximation breaks down. These departures depend on the shape of the potential near the minimum.
12. Inflation also predicts adiabatic perturbations that are a Gaussian random field, i.e. with random phases.
13. The power spectrum of the CMB is measured using the C_l spectrum.
14. The CMB shows a dipole due to our motion relative to the cosmic rest frame.
15. The higher-order multipoles of the CMB show approximate scale-invariance; on smaller scales, the acoustic peaks in the C_l spectrum are due to resonances within the sonic particle horizon at the time of recombination.
16. The position of the first peak is determined by the sonic horizon particle size.

    The relative strength of the second peak is sensitive to the entropy per baryon.
17. Cosmological parameters derived from the CMB nevertheless have several degeneracies that can be resolved by incorporating other experimental constraints.

18. On the smallest scales, the Silk damping effect damps down the acoustic peaks. This effect is due to photon diffusion during the process of decoupling, between the times when photons were tightly coupled to matter and when the Universe became transparent.
19. The Sachs–Wolfe effect applies during the passage of photons through the Universe since the time of last scattering. The change in gravitational potentials as photons pass through the Universe leaves imprints that are detectable through stacking analyses of galaxy clusters in CMB maps. The strength of the late-time Integrated Sachs–Wolfe effect is sensitive to the evolution of dark energy.
20. The reionization of the Universe by the first luminous objects in the Universe (stars or accreting black holes) also increased the optical depth to Thomson scattering experienced by CMB photons. This optical depth is a free parameter in fitting the acoustic peaks of the CMB.
21. Primordial gravitational waves (also known as tensor modes) generate B-mode polarization. The spectral index and intensity of the gravitational wave background is determined by the same inflation parameters that predict the departure from scale invariance in the scalar perturbations, so detecting these primordial gravitational waves (either in B-mode CMB maps or in gravitational wave observatories) would be a direct test of the inflationary framework.
22. Dark energy models generalize the cosmological constant, whose effective equation of state parameter is $w = -1$, and consider the possibility of w varying with time. This could occur, for example, if Λ is associated with a scalar field, analogous (but not identical) to the inflaton.

Further reading

- Audio representations of CMB acoustic peaks can currently be found on Mark Whittle's web pages (http://www.astro.virginia.edu/~dmw8f) and John Cramer's web pages (http://faculty.washington.edu/jcramer).
- For more details on energy and momentum conservation in general relativity, see Lambourne, R., 2010, *Relativity, Gravitation and Cosmology*, Cambridge University Press.
- It's also possible for advanced undergraduate-level students to achieve a deeper knowledge of general relativity, though this is usually outside the scope of most undergraduate degrees. For readers who would like to try this immensely rewarding intellectual adventure, we would recommend an accessible text on general relativity such as Hobson, M.P., Efstathiou, G.P. and Lasenby, A.N., *General Relativity: An Introduction for Physicists*.
- The discovery papers of the CMB are: Penzias, A.A. and Wilson, R.W., 1965, 'A measurement of excess antenna temperature at 4080 Mc/s', *Astrophysical Journal*, **142**, 419; Dicke, R.H., Peebles, P.J.E., Roll, P.G. and Wilkinson, D.T., 1965, 'Cosmic black-body radiation', *Astrophysical Journal*, **142**, 414.
- Steigman, G., 2007, 'Primordial nucleosynthesis in the precision cosmology era', *Annual Review of Nuclear and Particle Systems*, **57**, 1 (available at arXiv:0712.1100).

- For more on the statistical physics that connects the microscopic world of molecules and collisions with the macroscopic world of densities and pressures, see, for example, Mandl, F., 1988, *Statistical Physics*, Wiley.
- For more about neutrino astronomy, see Spiering, C., 2008, 'High energy neutrino astronomy: status and perspectives', *Proceedings of the 4th International Meeting on High Energy Gamma-Ray Astronomy*, AIP Conference Proceedings Vol. 1085, pp. 18–29 (also available at arXIv:0811:4747), or Learned, J.G. and Mannheim, K., 2000, 'High-energy neutrino astrophysics', *Annual Review of Nuclear and Particle Science*, **50**, 679.
- For more on Fourier series and transforms, see, for example, Gillett, P., 1984, *Calculus and Analytic Geometry*, Houghton Mifflin Harcourt.
- For more about the physics of the CMB, try Hu, W. and Dodelson, S., 2002, 'Cosmic microwave background anisotropies', *Annual Review of Astronomy and Astrophysics*, **40**, 171, or Hu, W. and White, M., 1997, 'A CMB polarization primer', *New Astronomy Reviews*, **2**, 323.
- A more substantial introduction to many themes in this chapter can be found in the graduate-level text Peacock, J.A., 1999, *Cosmological Physics*, Cambridge University Press.

Chapter 3 The local Universe

'When I use a word,' said Humpty Dumpty, 'it means exactly what I intend it to mean.' 'The question is,' said Alice, 'can you use words this way?' 'The question is,' said Humpty, 'who is to be the Master?'

Lewis Carroll, *Alice's Adventures in Wonderland*

Introduction

You may have been surprised at the number of fundamental unknowns in the basic processes in the early Universe. In this chapter we'll cover our cosmic neighbourhood and find out some of what is, and what isn't, understood even here. We'll cover some of the phenomenology and terminology (a necessary evil) and some more tools of precision cosmology, and also start assembling the evidence for how the birth of galaxies like our own Milky Way happened.

3.1 Evidence for dark matter

Some of the best direct evidence for dark matter is in the local Universe. The rotation rate of a spiral galaxy can be used to infer the mass enclosed within a radius r measured from the centre of that galaxy, since $v^2/r = G\,M(r)/r^2$, where $M(r)$ is the mass at radii less than r. The velocity of stars and gas as a function of radius, $v(r)$, is known as the **rotation curve** of the galaxy. If there were no dark matter, the rotation curve should agree with predictions made from the sum of stars (detectable in visible light) and gas (detectable from radio fine-structure transitions that we shall meet in Chapter 8). Figure 3.1 shows the rotation curve of the galaxy NGC 1560. There is a clear and highly statistically significant deviation from the prediction using only stars plus gas, implying a dominant dark matter component at large radii. In due course there will be similar compelling evidence from our own Galaxy: the planned launch of the European Space Agency Gaia mission in the coming decade will map the kinematics of stars in our own Galaxy and infer any substructure (i.e. clumpiness) in its dark matter distribution.

One way of avoiding the conclusion of a dominant dark matter component is to suppose that the gravitational force law is different at large separations. We shall briefly discuss this possibility in Chapter 7. More evidence for dark matter comes from galaxy motions in clusters and from gravitational lensing which we shall also meet in Chapter 7. The latter provides significant challenges to modified gravitational force laws. As we've seen, the existence of dark matter is also supported by the relative strengths of the acoustic peaks in the CMB.

The current space density of stars and gas is lower than the baryon density $\Omega_{\rm b,0}$ inferred from primordial nucleosynthesis and the CMB acoustic peaks, so some 'dark' matter must be baryonic, perhaps in the form of neutral gas, brown dwarfs or MACHOs (see Chapters 7 and 8). However $\Omega_{\rm m,0}$ cannot be mainly baryonic, or it would violate the primordial nucleosynthesis constraints, so what is it? Despite the fact that neutrinos are now known to have mass, the estimated mass density of

$$\Omega_\nu h^2 \simeq \frac{\sum m_i}{93.5\,\mathrm{eV}} < 0.0076, \qquad (3.1)$$

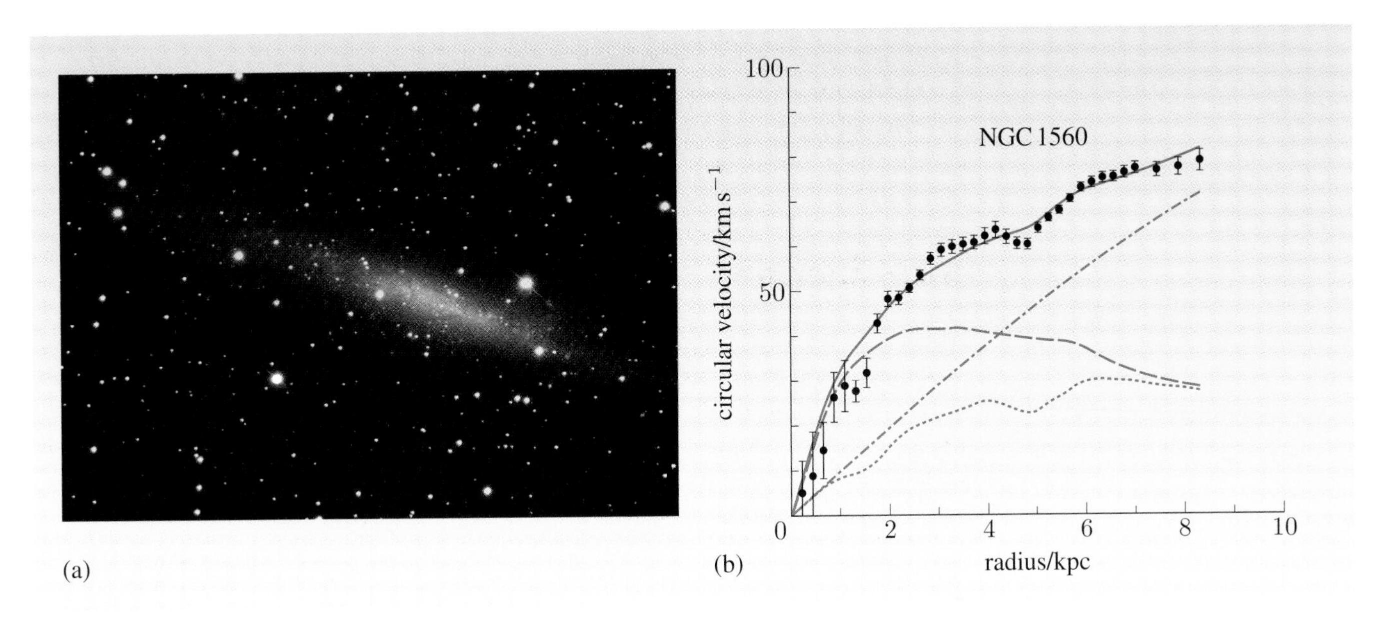

Figure 3.1 (a) An optical image of the approximately edge-on spiral galaxy NGC 1560 and (b) the galaxy's rotation curve. The data points show the observed velocities. The dashed curve shows the predicted contribution to the velocities from the inferred mass of stars, while the dotted line shows the contribution from the inferred mass of the gas. These are not enough on their own to explain the observed velocities. The dash–dotted line is the dark matter halo required to make up the remaining mass.

(where the sum is over the neutrino species) is too low to account for all the dark matter.

Two leading dark matter particle possibilities are the **neutralino** and the **axion**. The neutralino is a partner particle to the neutrino predicted in supersymmetry, which is an extension of the standard model of particle physics. The neutralino is an example of a weakly interacting massive particle (WIMP). If neutralinos exist in cosmologically-significant numbers, they may be found by direct detection experiments such as the one at Boulby Mine in the UK, through detecting the recoils of nuclei colliding with WIMPs. The interaction probability between a WIMP dark matter particle and normal matter is predicted to be extremely low, but with a sufficient flux of particles through the Earth (as the Solar System traverses the Galaxy) very rare collisions may be detected. As yet there are no uncontested claims of WIMP detection from these experiments. Axions, meanwhile, are particles proposed in quantum chromodynamics with a very low mass (10^{-3}–10^{-4} eV) that couple only very weakly to electromagnetism. They are predicted to decay into two photons in the presence of a strong magnetic field. Some recent laboratory claims of axion direct detection exploiting this decay (or its time reverse) were unfortunately later withdrawn or proved unrepeatable in other experiments.

In 2008 there was a flurry of excitement over the cosmic ray positrons detected by the PAMELA satellite (Payload for Antimatter Matter Exploration and Light-nuclei Astrophysics), as well as the ATIC (Advanced Thin Ionization Calorimeter) balloon-borne measurements of electrons and positrons (which ATIC couldn't distinguish). Both experiments inferred a peak in the spectrum of cosmic ray electrons at about 500 GeV. Could this be the signature of dark matter annihilation? The Fermi gamma-ray telescope is also sensitive to cosmic ray electrons, and though cosmic rays are usually a nuisance that needs to be carefully identified and removed from data, in this case Fermi could act as a powerful cosmic ray detector in its own right. Unfortunately, Fermi failed to confirm the PAMELA/ATIC signal. Although there is an excess relative to the predictions of cosmic ray propagation through the Galaxy, astrophysical processes could mimic this excess. It's not impossible, however, that future cosmic ray observations could detect dark matter annihilation processes.

It's usually assumed that dark matter interacts almost exclusively through gravity, but one can't rule out a set of forces only experienced by dark matter. For example, a 'dark electromagnetism' has been proposed. In this model there would be equal numbers of dark-negative and dark-positive particles, but their annihilations would be suppressed if the dark equivalent of the fine structure constant is sufficiently small. Dark matter galaxy haloes would be a sort of dark plasma in this model. Occam's razor has perhaps precluded much discussion of this imaginative proposal, but it still highlights how little we know of the dark sector in general.

Ackerman, L., Buckley, M.R., Carroll, S.M. and Kamionkowski, M., 2009, *Physical Review D*, **79**, 023519.

3.2 The Hubble tuning fork

Spiral galaxies are not the only cosmologically-relevant population. Edwin Hubble, the discoverer of the expansion of the Universe, is also remembered for making the first attempt at classifying galaxy morphologies. His division of spiral and elliptical galaxies into a hypothetical sequence is shown in Figure 3.2, with ellipticals called **early-type** and spirals called **late-type**. However, be warned that *there is no observed evolution from early to late*. This misleading terminology is still in use for historical reasons only. In fact, if anything there is evolution in the opposite direction, since spiral–spiral galaxy mergers are likely to generate an eventual merger product with an elliptical morphology. We'll see later that spirals and ellipticals are found in different places, and have different colours and different contents.

Astronomers could just as well use the word 'structure', but terms like 'morphology–density relation' are in such widespread use in astronomy that we need to use the standard convention in this book.

Galaxy classification is usually done by specialists, but a curious development has been the adaption of artificial neural networks to classify galaxies. This is a way of specifying a computer algorithm for classifying galaxies that can be compared to a human's classifications, and iteratively altered until its outputs match those of a human. Neural nets can reproduce galaxy classifications with the same level of disagreement with human classifiers as there is disagreement between human classifiers themselves. Indeed, neural nets may be better than humans, but we would never know. The algorithms found and used by neural nets are unfortunately not often easily interpretable, but other approaches to 'machine learning' can give interpretable classification schemes. Nevertheless, automatic classification is still not completely reliable, especially when there are irregular or disturbed galaxies. The **Galaxy Zoo** project has enrolled the general public to make morphological classifications of all the galaxies in Sloan Digital Sky Survey (SDSS) data release 6 — around a million galaxies. Astrophysics is unusual in that the amateur community makes a real scientific impact, from supernova and comet searches, to SETI@Home and Galaxy Zoo.

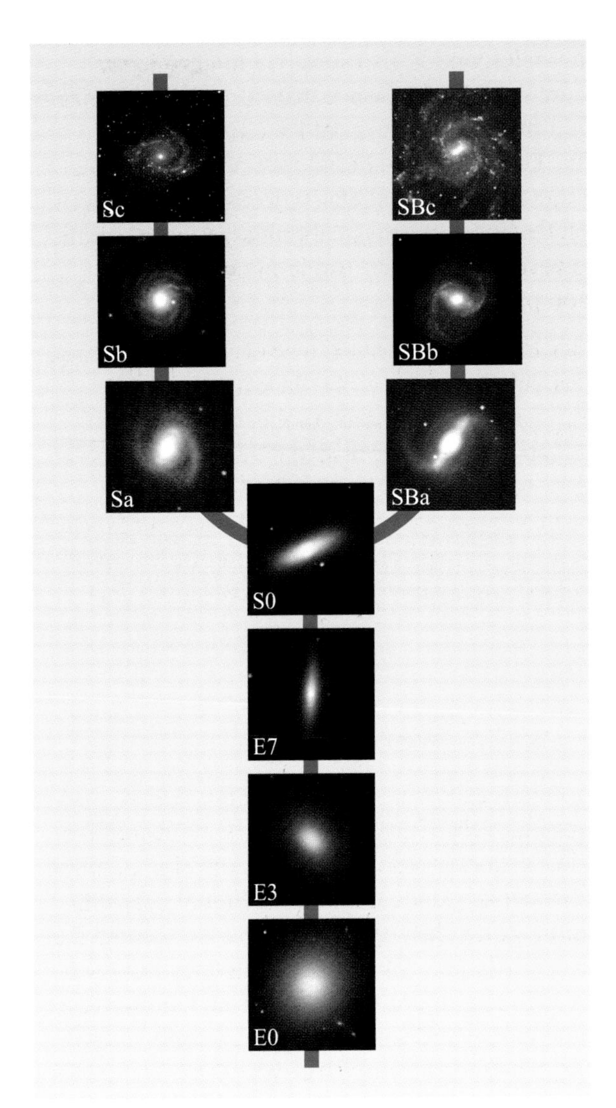

Figure 3.2 Digitized Sky Survey images of local galaxies illustrating points on the Hubble tuning fork for galaxy classification. Irregular galaxies lie outside this scheme.

In Exercise 1.6 in Chapter 1 we saw that the surface brightness of an object (i.e. its flux per square degree) declines as $(1 + z)^4$. If redshift is cosmological, we should expect this surface brightness variation in galaxies, but if for example 'tired light' models are responsible for redshift (Chapter 1), we should not. This is known as the **Tolman surface brightness test**. The attraction of this test is that the $(1 + z)^4$ prediction is independent of the cosmological parameters $\Omega_{\mathrm{m},0}$, $\Omega_{\Lambda,0}$, H_0 and so on. The result is that the surface brightness of elliptical galaxies between $z = 0$ and $z = 0.85$ declines as $(1 + z)^{2.80\pm0.25}$ in the **R-band filter** (a broad optical filter used in optical astronomy at about 550–750 nm), and $(1 + z)^{3.48\pm0.14}$ in the **I-band filter** (about 700–900 nm). However, this span of cosmic time is enough to expect evolution in the stellar populations of galaxies, as we shall see in Chapter 4. The colour-dependent surface brightness evolution is exactly as expected from the combination of $(1 + z)^4$ dimming combined with stellar population evolution models.

See, for example, Sandage, A., 2009, *Astronomical Journal*, **139**, 728.

The division between 'optical' and 'near-infrared' is often taken to be at around $1\,\mu$m, despite the fact that human eyesight loses sensitivity quickly above $0.75\,\mu$m. Perhaps this is because in observational astronomy, wavelengths $> 1\,\mu$m require different detector technology and observing techniques to wavelengths $< 1\,\mu$m.

A population of galaxies entirely overlooked by the Hubble scheme, indeed unknown to Edwin Hubble, is low surface brightness (LSB) galaxies. These are defined as having surface brightnesses fainter than 22.5 magnitudes per square arcsecond (recall that astronomical magnitudes m are related to fluxes S_ν by

$m = -2.5 \log_{10} S_\nu + \text{constant}$). Their rotation curves imply a much higher **mass-to-light ratio** (mass divided by luminosity) than is typical in the rest of the galaxy population, and are often dominated by dark matter at all radii. This makes them very different to, say, globular clusters of stars. LSB galaxies are often also very gas-rich and have low **metallicity** (i.e. low abundance of elements heavier than hydrogen and helium — recall the convention in astronomy to call all of these 'metals'), as well as sometimes having low star formation rates for their gas contents, implying that either they have yet to start the bulk of their star formation, or this star formation is suppressed.

3.3 Spiral galaxies and the Tully–Fisher relation

The rotation of spiral discs turns out to be closely related to other properties of spiral galaxies. In 1977, R. Brent Tully and J. Richard Fisher discovered that the velocity widths of spiral galaxies, Δv, correlate very strongly with their luminosities in the **B-band filter** (about 400–500 nm). Velocity widths can be measured from optical spectroscopy (via Doppler shifts of absorption lines) or at radio wavelengths (again via Doppler shifts of the 21cm-wave emission line of neutral hydrogen). These correlations have been found at other wavelengths too, and the correlations are typically stronger in the near-infrared, such as in the **K-band filter** (2–2.5 μm). The luminosity scales as $L \propto (\Delta v)^\alpha$, where the parameter α is typically between 3 and 4, depending on the wavelength at which the luminosity is measured.

We have now mentioned so many astronomical filters that we had better give you a full set, shown in Figure 3.3. There are also filters at longer wavelengths in what is sometimes known as the 'mid-infrared'. Ground-based astronomy tended to use letter names for these too, but with the advent of space-based mid-infrared astronomy this has been more or less superseded by simply quoting the wavelength in microns. There is no clear convention for which wavelengths are termed 'near'- or 'mid'-infrared (usage varies around 3–5 μm for the division), but 60 μm is often referred to as 'far-infrared'.

What physical processes give rise to the Tully–Fisher relation? Suppose that the rotation of the galaxy is Keplerian, i.e. the period P at the edge of the galaxy is related to the radius R by $P^2 \propto R^3/M$, where M is the galaxy mass. The rotation velocity is $v = 2\pi R/P$, so we can write the rotation width $\Delta v = 2v$ as $M \propto R(\Delta v)^2$. It's observed that the mass-to-light ratio of most spiral galaxies is roughly constant, i.e. $M/L \simeq$ constant, so $L \propto R(\Delta v)^2$. If we assume that galaxies have the same surface brightness, then L/R^2 must be constant, or $L \propto R^2$, but since $L \propto R(\Delta v)^2$ we must have $R \propto (\Delta v)^2$ so $L \propto (\Delta v)^4$. In practice, the surface brightness has a weak dependence on luminosity. For normal spiral galaxies on the Hubble sequence, the luminosity empirically appears to scale with the size of the galaxy as $L \propto R^{2.8}$ in the B-band, implying $L \propto (\Delta v)^{3.1}$.

The trouble is, the assumptions built into this quick calculation are clearly wrong. The mass-to-light ratio should be roughly constant if the mass is dominated by stars, but as we've seen, the mass is in fact dominated by dark matter. Also, we've already met galaxy populations that have much lower surface brightnesses —

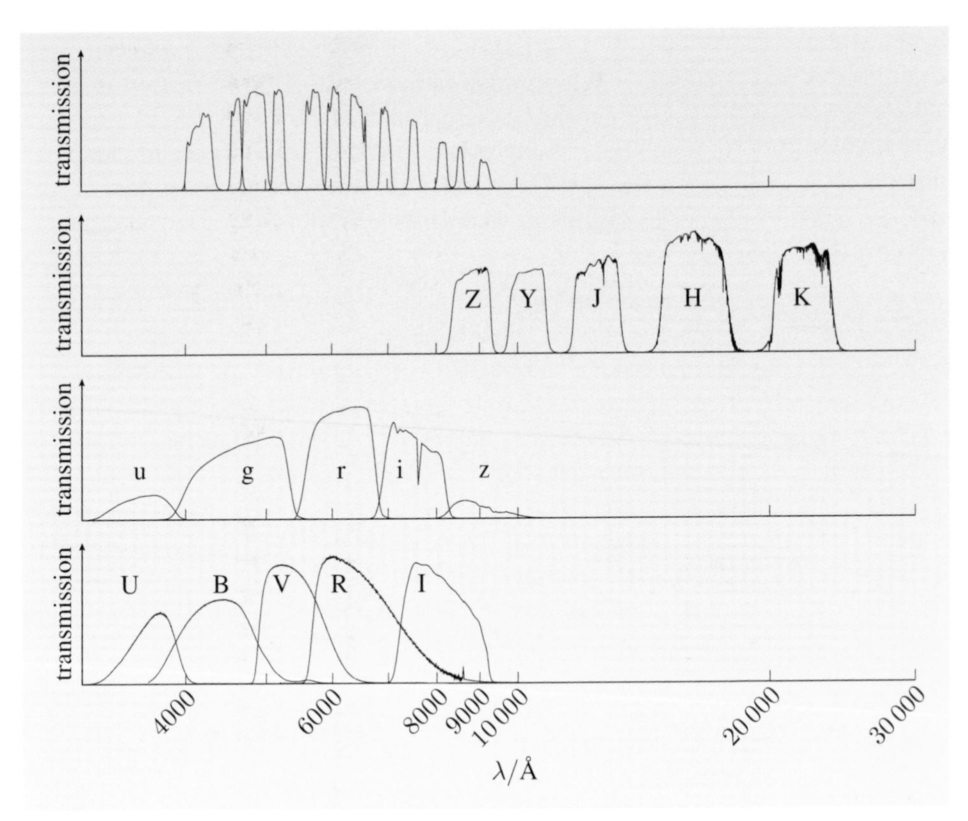

Figure 3.3 A selection of astronomical filters. The bottom row shows the UBVRI system at use at the Kitt Peak National Observatory. The next row up shows the Sloan Digital Sky Survey filters. The second from top shows the ZYJHK infrared filters in use at the United Kingdom Infrared Telescope, while the top row shows the filter set of the COMBO-17 survey (Chapter 4). The y-axis gives the relative transmission from the top of the atmosphere to the detector in the telescope, in arbitrary units.

the 'low surface brightness' galaxies. These also appear to follow the same Tully–Fisher relationship, but with a bigger scatter. The correlation is therefore somewhat surprising, and the low scatter about the best fit for high surface brightness galaxies is doubly surprising, so reproducing the observed Tully–Fisher relation has been a key constraint of models of spiral galaxy discs. Empirically, the star formation rate in galactic discs is proportional to the gas density ρ to the power n, where $n \simeq 1$–2, but star formation is a very complex process and the theoretical underpinning of this relation (known as the **Schmidt law**) is still sketchy.

Having said that, we'll see below that the Tully–Fisher relation is also empirically useful because it can be used to determine the luminosity distance to a galaxy, from which the Hubble parameter can be found.

The profiles of spiral discs have a surface brightness I that varies with radius typically as $I \propto \exp(-r/r_{\text{scale}})$, where r_{scale} is in this case known as the **scale length**. Similarly, the disc of our Galaxy (and indeed all spiral galaxies) has a density of stars that varies exponentially with the vertical height above (or below) the disc: $I \propto \exp(-h/h_{\text{scale}})$, where h_{scale} is known as the **scale height** of the disc. This scale height appears to vary with the type of stars: the youngest stars have $h_{\text{scale}} \sim 100\,\text{pc}$, while the oldest stars have $h_{\text{scale}} \sim 1.5\,\text{kpc}$. This latter population is also known as the **thick disc**. It's not known what generated the thick disc in our Galaxy, though these stars are observed to have a lower metallicity so may represent a relic from an earlier phase of the formation of our Galaxy, or perhaps they are a fossil of a disruption that our Galaxy underwent early in its history from another passing galaxy. Further out, stars are found throughout the halo of the Galaxy. These halo stars are typically much less enriched in heavy elements, suggesting that they are from a very early phase in the formation of our Galaxy.

The discs of spiral galaxies are far more abundant in gas than elliptical galaxies, which we'll meet in the next section. Hydrogen (mainly in the form of molecular H_2, but also neutral H I and ionized H II) dominates the gas mass of galaxies, but very often the easy-to-measure CO emission at radio wavelengths is used as a substitute measure for the gas content. CO molecules have a $J = 1 \rightarrow 0$ transition at 115 GHz, and other transitions at integer multiples of this frequency. The CO-to-H_2 abundance is sometimes taken as a fixed quantity in extragalactic astronomy, but the CO-to-H_2 conversion factor (known as X_{CO}) is only accurate to $\pm 50\%$ at 1σ (see the further reading section for more information). We'll see in later chapters that many galaxies are strongly luminous at far-infrared wavelengths, caused by thermal radiation from dust.

3.4 The fundamental plane of elliptical galaxies

Curiously, elliptical galaxies have a similar relation between velocity dispersion σ_v and luminosity L, known as the **Faber–Jackson relation**: $L \propto \sigma_v^\alpha$ with $\alpha \simeq 3$–4. However, in this case the velocities are not regular and confined to a disc, but are rather more like a gas of gravitationally-bound particles.

In fact, this is a special case of a more general relation in elliptical galaxies, known as the **fundamental plane**: $L \propto I_0^x \sigma_v^y$, where I_0 is the surface brightness within a given radius (conventionally the radius within which half the total light is generated, known as the **half-light radius**), and $x \simeq -0.7$ and $y \simeq 3$–4.

Empirically, elliptical galaxies appear to follow the **de Vaucouleurs law** or de Vaucouleurs profile:

$$I(r) = I_0 \exp\left(-(r/r_0)^{1/4}\right), \tag{3.2}$$

where $I(r)$ is the surface brightness, r is the radius measured from the centre, I_0 is a normalization that may vary from galaxy to galaxy, and r_0 is a characteristic scale length for that particular galaxy. Integrating this, the total luminosity comes out as $L \propto I_0 r_0^2$. The half-light radius is also proportional to r_0. Note that this profile is sometimes defined with $-(r/r_0)^{1/4} - 1$ in the exponent instead of just $-(r/r_0)^{1/4}$, though this is just equivalent to changing the definition of I_0. Sometimes a slightly more general law is used in which the $1/4$ index is changed to $1/n$, where n is a free parameter; this is known as a **Sérsic profile**. Nevertheless, $L \propto I_0 r_0^2$ still holds in all such profiles, though with different constants of proportionality.

Exercise 3.1 Show that any surface brightness law of the form $I(r) = I_0\, f(r/r_0)$ has a total luminosity that is proportional to $I_0 r_0^2$. ■

Again, a simple plausibility argument can be given for the existence of the fundamental plane if we accept the two-parameter characterization of elliptical galaxies in Equation 3.2. If we assume that the galaxies are gravitationally bound, then a dimensional analysis (or the virial theorem — see below and the further reading section) implies $\sigma_{\mathrm{v}}^2 \propto M/r_0$. Now suppose that there is a mass-to-light ratio that is a weak function of mass, so $M/L \propto M^a$. Together with $L \propto I_0 r_0^2$ this can be shown to imply that

$$L^{1+a} \propto \sigma_{\mathrm{v}}^{4-4a} I_0^{a-1}, \tag{3.3}$$

which fits the observations provided that $a \simeq 1/4$.

The fundamental plane is also sometimes known as the $\boldsymbol{D_n}$**–**$\boldsymbol{\sigma}$ **relation**. The parameter D_n is the diameter in which a galaxy's surface brightness is larger than a given constant. This diameter will depend approximately on the fundamental plane parameters as power laws: $D_n \propto r_0^{\alpha} I_0^{\beta}$ for some constants α and β. If the reference surface brightness is chosen carefully, one can find a definition for D_n in which it is also proportional to some power of σ_{v}, and so approximates the fundamental plane. In practice the D_n–σ relation is used as a distance indicator: knowledge of σ_{v} and a measurement of the angular size gives an angular diameter distance to the galaxy.

3.5 Clusters of galaxies

The largest gravitationally bound structures in the Universe are galaxy clusters. These assemblages are **virialized**, i.e. they have settled into a stable gravitationally-bound system for which the **virial theorem** applies: the average kinetic energy is equal to $-1/2$ times the average potential energy. This theorem is widely used in astrophysics, and a source for a proof of this useful relation is given in the further reading section. In this context, the virial theorem gives us direct evidence for dark matter, since the average kinetic energy of the constituent galaxies is too high to be accounted for by the gravitational potential of the visible matter.

You may be surprised to hear that the dominant baryonic component of galaxy clusters (by a factor of $\sim \times 10$ to $\times 15$) is not the galaxies themselves but a gas plasma. The potential well of the cluster is so deep that the gas that has fallen in and been entrained is heated to hundreds of millions of kelvins ($kT \sim 10\,\mathrm{keV}$), and emits X-rays through bremsstrahlung with luminosities $> 10^{35}$ W at 0.5–2 keV. (Accelerating or decelerating charges radiate electromagnetic waves, and in this case the acceleration/deceleration is during collisions between electrons and ions at thermal velocities.) Since thermal emission depends on collisions between particles, the X-ray luminosity scales as the number density n squared. The luminosity will also be proportional to the volume of gas. If we define a characteristic size r_0 for the cluster (a suitable definition will be given in Section 7.7 of Chapter 7), the X-ray luminosity is $L_{\mathrm{X}} \propto n^2 r_0^3$. Meanwhile, though, the mass of the baryonic gas will be given by the density times the volume, so $M_{\mathrm{gas}} \propto n r_0^3$. Putting these together, we have that $M_{\mathrm{gas}} \propto r_0^{3/2} L_{\mathrm{X}}^{1/2}$.

The *total* mass of the cluster (typically 10^{14}–$10^{15}\,\mathrm{M_\odot}$) will be governed by the equation of hydrostatic equilibrium. We shall cover this in more detail in Section 7.4, but for now we shall just quote the result that isothermal pressure balances lead to $M_{\mathrm{total}} \propto r_0$. Therefore the fraction of baryonic gas must be

$M_{gas}/M_{total} \propto r_0^{1/2} L_X^{1/2}$. So, if we know the distance to a cluster, we can calculate its size and X-ray luminosity, and find the baryonic mass fraction. This is further evidence for dark matter, i.e. the gas is not massive enough on its own to entrain itself in hydrostatic equilibrium. It's often assumed that the baryonic mass ratio shouldn't depend on redshift, since the mass is accreted into the cluster purely gravitationally, regardless of whether it's baryonic or not. This leads to another constraint on the cosmological parameters: if the angular size is θ_0 and the X-ray flux is F_X, we have that $\theta_0 = r_0/d_A$ (where d_A is the angular diameter distance), while $L_X \propto F_X d_L^2$ (where d_L is the luminosity distance). If M_{gas}/M_{total} is constant, then $\sqrt{r_0 L_X}$ must be constant, which equals $\sqrt{\theta_0 F_X}(1+z)^{1/2} r^{3/2}$, where r is the proper motion distance (using Equations 1.50) or the comoving distance if we're assuming a spatially-flat universe. The expression for r depends on the cosmology (Equation 1.44), so the assumed constancy of M_{gas}/M_{total} can be used to derive cosmological parameters.

Abell, G.O., 1958, *Astrophysical Journal Supplement*, **3**, 211.

Galaxy clusters were first discovered in optical photographic imaging surveys. The Abell cluster catalogue was compiled by an exhaustive visual inspection of photographic plates. Clusters were classified according to their 'richness', i.e. their density of galaxies, with Abell richness class 0 being the sparsest and 5 the densest. Abell found galaxy clusters out to around $z = 0.2$, though with some incompleteness in the catalogue at the higher redshifts and lower richness classes. X-ray sky surveys and modern digital optical sky surveys have both been used to find galaxy clusters to higher redshifts, and in the future many more are expected to be found in the next generation of CMB maps using the Sunyaev–Zel'dovich effect (Section 3.6).

The number counts of clusters are also a cosmological constraint. The formation of virialized clumps of dark matter can be calculated using a knowledge of the matter power spectrum and the cosmological model. If the Universe underwent an accelerated expansion in recent history (e.g. $z < 0.5$), this will have a strong effect on the gravitational collapse of clumps. The mass function of clusters is therefore a strong function of the dark energy parameter $w(z)$.

At the centre of a galaxy cluster is usually a large galaxy, usually the brightest in the cluster (known as the **brightest cluster galaxy** or BCG). Often this galaxy is a giant elliptical; otherwise it is a very large S0 galaxy known as a D galaxy (the D stands for 'diffuse', since they have unusually extended envelopes) or cD galaxy (a larger variant of D galaxies). These galaxies sometimes have multiple cores, suggesting that they are built up by ongoing galaxy mergers.

3.6 The Sunyaev–Zel'dovich effect

The CMB photons passing through a foreground galaxy cluster can be scattered by the electrons in the gas through the Compton scattering process. This will cause a change in the temperature of the CMB photons in the Rayleigh–Jeans tail of the CMB spectrum: $\Delta T/T \propto \int n_e \, d\ell \sim n_e \, \Delta\ell$ for a fixed cluster gas temperature T, where n_e is the electron number density and $\Delta\ell$ is the path length through the cluster. Meanwhile, the X-ray luminosity will scale as $L_X \propto \int n_e^2 \, d\ell \sim n_e^2 \, \Delta\ell$ (see above). So from comparing the X-ray luminosity with the temperature decrement, we can eliminate n_e to find $\Delta\ell$. If the cluster is symmetrical (which will be true on average if not necessarily in an individual case), the length $\Delta\ell$ should be the same as the angular size, so we can derive an angular diameter distance to the cluster. The Hubble parameter can be derived from cluster Sunyaev–Zel'dovich measurements, and though the constraints are not currently as tight as other determinations (61 ± 3 (random) ± 18 (systematic) $\mathrm{km\,s^{-1}\,Mpc^{-1}}$, where the systematics are from, for example, uncertainties in the radial temperature profiles in the cluster), it is important to have independent checks.

The Compton scattering conserves the number of CMB photons, but redistributes their energies. The CMB photons in the Rayleigh–Jeans regime are suppressed, but on the other side of the black body peak the number of photons is enhanced. This characteristic wavelength-dependent suppression and enhancement, known as the **Sunyaev–Zel'dovich effect** (or S–Z effect), will be searched for in the next generation of CMB maps to make comprehensive all-sky galaxy cluster surveys. Figure 3.4 shows the predicted absorption and emission from the Sunyaev–Zel'dovich effect.

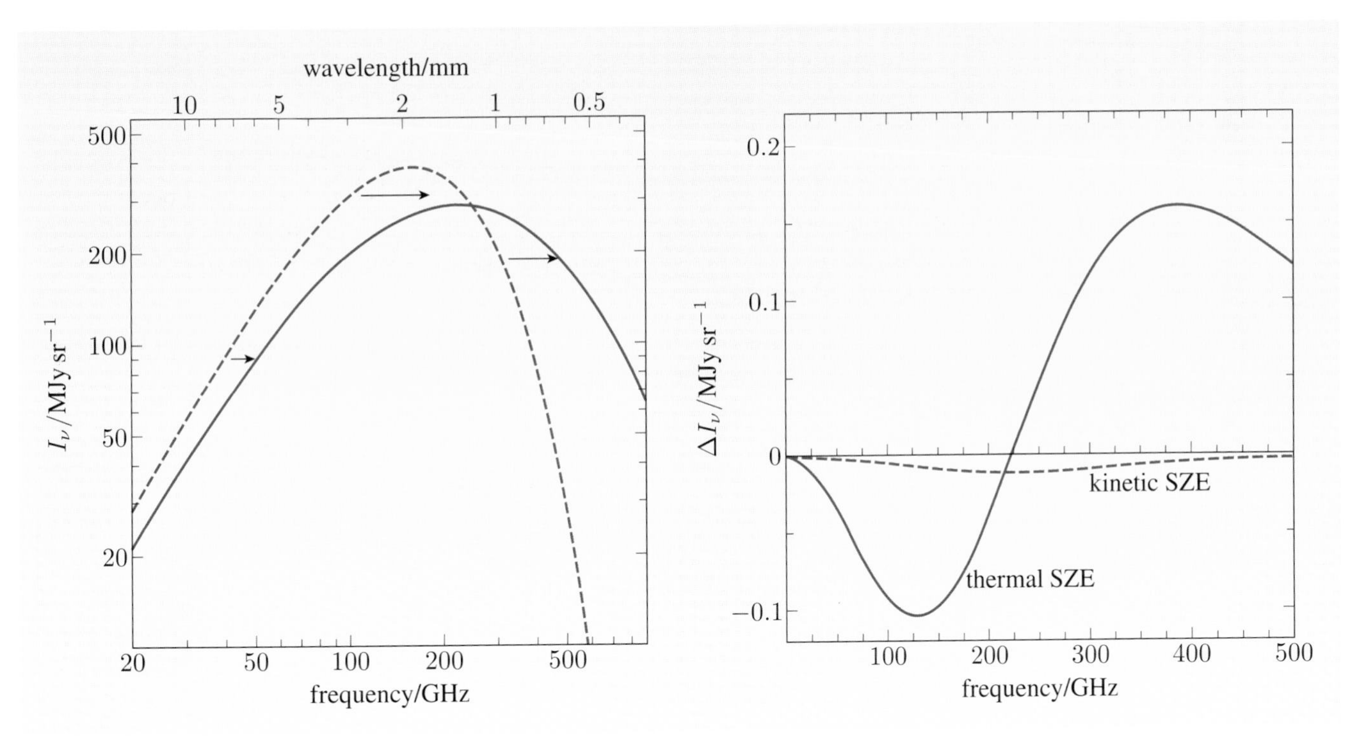

Figure 3.4 The Sunyaev–Zel'dovich effect on the intensity I_ν of the CMB. The left panel shows an exaggerated demonstration of the effect, while the right panel shows the change in the CMB intensity from the thermal and kinetic S-Z effect for an electron temperature of 10 keV, a Compton y-parameter of 10^{-4} and $500\,\mathrm{km\,s^{-1}}$ peculiar velocity.

An additional S–Z effect can occur if the electrons have additional kinetic energies from bulk motion of the cluster relative to the cosmic rest frame, though this effect (termed 'kinetic S–Z' to distinguish it from the 'thermal S–Z' discussed in this section) is much smaller, as shown in Figure 3.4. A more lengthy calculation of the S–Z distortion (see the further reading section) gives

$$\frac{\Delta T}{T_{\mathrm{CMB}}} = f(x)\,y = f(x) \int n_{\mathrm{e}} \frac{kT}{m_{\mathrm{e}}c^2} \sigma_{\mathrm{T}}\,\mathrm{d}\ell, \tag{3.4}$$

where $x = h\nu/(kT_{\text{CMB}})$ is known as the **dimensionless frequency**, and the parameter y defined in the above equation is the **Compton y parameter**. The parameter σ_{T} is the Thomson scattering cross section (if cross sections are unfamiliar, see the box below), and the function $f(x)$ is

$$f(x) = \left(x\frac{e^x + 1}{e^x - 1} - 4\right)\left(\frac{xe^x}{e^x - 1}\right), \tag{3.5}$$

provided the gas temperature is $\gg \overline{T}_{\text{CMB}}$. The important thing to note is that the distortion is *redshift-independent.*

Cross sections

Cross sections express the probability in quantum mechanics that objects interact, such as a photon being absorbed by a hydrogen atom. If atoms were classical spheres, the probability that the photon is absorbed would be 1 if the photon passed into the sphere, and 0 if it did not. The cross section σ in this case would be $\sigma = \pi r^2$, where r is the radius of the sphere. This is the size of the target that the photon must hit if it is to be absorbed. However, the atom does not have a sharp boundary. Instead, we can define the total cross section of N atoms to be R/S, where S is the number of incident photons per unit area per unit time, and R is the absorption rate, i.e. the number of absorptions per unit time. The cross section for one atom would then be $\sigma = R/(NS)$.

3.7 The morphology–density relation

Galaxy morphologies turn out to be very strongly dependent on their environments: elliptical galaxies are often (but not always) found in galaxy clusters, while spiral galaxies are often (but not always) in the 'field' (i.e. not in clusters). This has been quantified in the **morphology–density relation** shown in Figure 3.5.

What could cause this strikingly strong effect? Clearly, environment must have had a strong effect on the evolution of galaxies. As yet it's still not clear what causes dominate, though plausible contenders are ram pressure stripping of the gas from galaxies (which removes the fuel for star formation and so shuts it down), or perhaps the removal of gas from galaxies through tidal forces from the cluster (when the galaxy falls into the cluster), or from the tidal disruption caused by frequent, close, fast passes of other galaxies in the rich environment (termed **galaxy harassment**). In all cases, removal and/or disruption of gas-rich spiral discs will tend to transform morphologies. Indeed, elliptical galaxies tend to have very little cold gas from which stars could form (though they do contain hot gas) and to contain very little dust. We'll see in Chapter 4 that their red colours are related to their older ages relative to star-forming discs in spirals. Numerical simulations of similarly-sized spiral–spiral mergers show that the randomization of the stars' velocities eventually results in elliptical-like morphologies.

The hot gas, spheroidal morphology, redder stellar populations and cluster environments have led elliptical galaxies occasionally to be referred to in technical conference proceedings as 'warm, round, pink and friendly'.

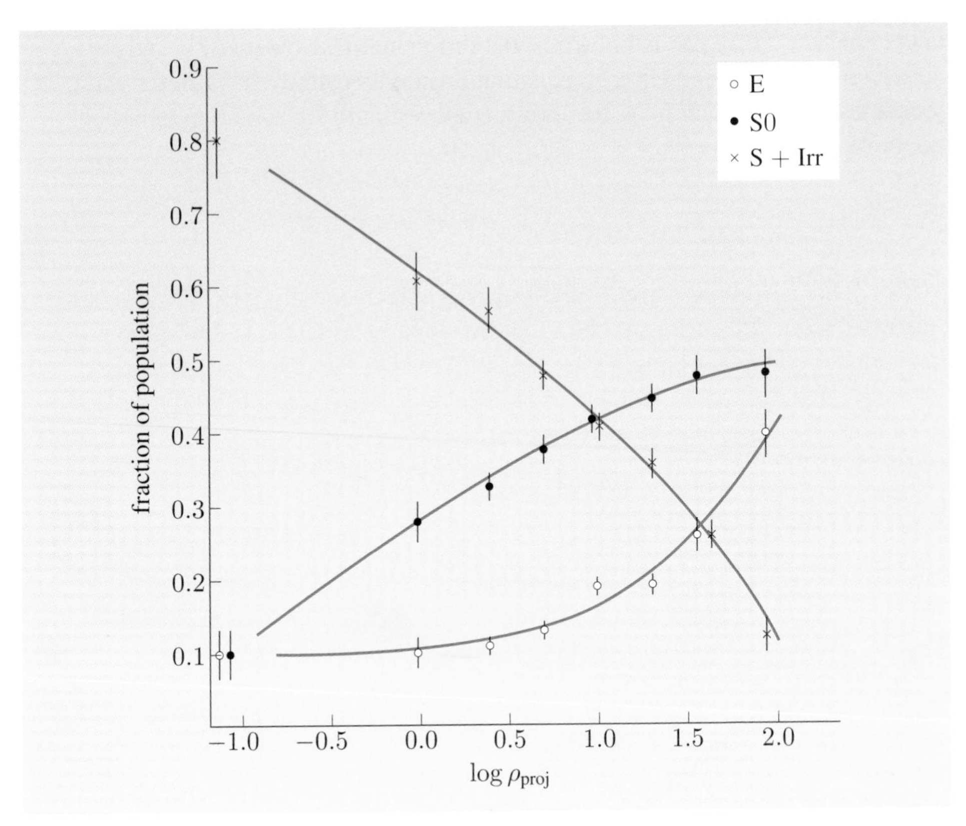

Figure 3.5 The morphology–density relation. The fractions of elliptical, S0 and spiral + irregular galaxies are plotted as a function of the logarithm of the projected galaxy density.

However, this doesn't appear to be the whole story as to how elliptical galaxies themselves form. The tightness of the fundamental plane relations and their old, red colours (consistent with highly evolved stellar populations) suggest instead a formation early in the history of the Universe (e.g. redshifts $z > 2$) in a large **starburst** (a strong burst of star formation), followed by **passive stellar evolution** (i.e. no subsequent star formation, so the galaxy colours change purely through the passage of their stars through the Hertzsprung–Russell diagram). This became known as the **monolithic collapse model** and reproduced the fundamental plane and the colour–magnitude diagrams of local elliptical galaxies. This model also explained why the metal-poor stars in the Galactic halo have highly elliptical orbits: they were formed when the initial gas cloud was in the process of collapse. Metal-rich stars in the disc, meanwhile, are formed later in this model. The alternative is the **hierarchical structure formation model** in which, as we'll see, galaxies are formed piecewise by the accretion and influence of neighbours. In particular, numerical simulations predict that the merger of spiral discs will result in an elliptical galaxy morphology. The relative dominance of these two mechanisms is currently the subject of some debate, and we shall return to this nature versus nurture issue in later chapters. The fundamental difficulty is that while the evolving distribution of dark matter can be (reasonably) easily calculated, because it only interacts through gravity, the evolving distribution of baryons such as stars and gas is much more complicated.

The classic reference for this model is Eggen, O.J., Lynden-Bell, D. and Sandage, A.R., 1962, *Astrophysical Journal*, **136**, 748.

The following exercises will derive the **Jeans mass**: the mass above which an object is unstable to gravitational collapse.

Exercise 3.2 Imagine a uniform sphere of radius R and density ρ. Write down the gravitational potential energy of a shell inside this sphere of radius r and thickness $\mathrm{d}r$, then integrate these shells to show that the gravitational binding energy of the sphere is

$$E_{\mathrm{GR}} = \frac{-3}{5}\frac{GM^2}{R}, \tag{3.6}$$

where M is the total mass of the sphere.

Exercise 3.3 According to the virial theorem, an object is stable against gravitational collapse when $2E_{\mathrm{K}} + E_{\mathrm{GR}} = 0$, where E_{K} is the kinetic energy. Assuming that the sphere is made of an ideal gas, show that the condition for gravitational collapse can be stated as

$$M > \left(\frac{5kT}{Gm_{\mathrm{p}}}\right)^{3/2} \left(\frac{3}{4\pi\rho}\right)^{1/2}. \tag{3.7}$$

This mass threshold is known as the Jeans mass. Put in the numbers from previous chapters to show that a baryonic overdensity of about a million solar masses was unstable to collapse at the epoch of recombination. ■

This mass is tantalizingly similar to the masses of present-day globular clusters, which are some of the oldest bound objects known in the Universe. However, it's unlikely that globular clusters formed as early as this, because of observed relationships between globular clusters and the galaxies that they inhabit. Nevertheless, you'll see in Exercise 6.6 of Chapter 6 that seeds of around this mass at redshifts higher than all known quasars are needed to explain the existence of supermassive black holes in quasars. The main difficulty, however, in making predictions is that baryons can do much more than just collapse under gravity, as we'll see in later chapters.

3.8 The Butcher–Oemler effect

Another hint at the formation histories of galaxies comes from the evolution in the colours of cluster galaxies: at higher redshifts (e.g. $z \sim 0.4$), there appears to be a larger fraction of blue galaxies in the cluster cores than are seen in the cores of clusters in the local Universe. This is known as the Butcher–Oemler effect, after its discoverers. Since young massive (short-lived) stars have higher temperatures and bluer colours, this suggests more star formation at higher redshifts in cluster core galaxies. This seemed to be a tantalizing clue on how environments affect the evolution of galaxies — was star formation shut down in rich environments in the recent past?

There's been some debate about the reality of this effect regarding underlying biases in the samples of objects studied, and the underlying causes of those effects that are present. We shall see in Chapters 4 and 5 how star formation in the Universe has evolved; the jury is still out on exactly how this evolution depended on galaxy environment.

See, for example, the discussion and review in Haines, C.P. et al., 2009, *Astrophysical Journal*, **704**, 126.

3.9 The cooling flow problem

We've already described how galaxies grow by mergers. Why don't the galaxies in galaxy clusters all merge to make one giant galaxy? One reason is that the surrounding galaxies are moving too fast, but what about the intra-cluster gas mentioned in Section 3.5? It needs to cool in order to be accreted, but the cooling time for the gas is longer than the current age of the Universe except in their cores. However, many clusters are observed to have dense cores, and the cooling time in the densest regions is shorter: we've already argued that the X-ray luminosity is proportional to the density n squared, so the energy loss per unit mass must be proportional to n. In these dense regions, the cooling timescale is still expected to be larger than the gravitational free-fall timescale (i.e. the time for an object to fall from rest to the centre of the cluster), so the gas would evolve as a quasi-hydrostatic equilibrium. If we imagine the gas in the core to have cooled, it will not have enough pressure to support the weight of the gas at larger radii, so the gas will be compressed, which further shortens its cooling time. This runaway cooling is known as a cooling flow.

However, if cooling flows exist, then the inferred amount of gas deposition onto the central cD galaxy turns out to be very large (hundreds of solar masses per year), so there should be evidence for large amounts of cool X-ray gas in the cluster cores. Cool gas is sometimes observed, but much less than expected, and the inferred infall rate should lead to many observational signatures, such as vigorous star formation, that are very rarely seen. The question then is why do cooling flows *not* exist? What are the hidden distributed heating mechanisms for the dense cluster cores? This is known as the **cooling flow problem**. We won't answer this immediately but will return to this question in Chapters 5 and 6.

3.10 The cosmological distance ladder

So far we have mentioned several ways of determining cosmological distances: the S–Z effect (via an angular diameter distance), Tully–Fisher (a luminosity distance), D_n–σ (an angular diameter distance), Faber–Jackson (a luminosity distance), and the position of the first acoustic peak in the CMB (an angular diameter distance). Many of these are based on the idea of a **standard candle** (an object with a known or derivable luminosity) or a **standard rod** (an object with a known or derivable size), from which one can calculate an angular diameter or luminosity distance, respectively. There are several other ways of determining cosmological distances that we have not yet covered, such as the following.

- The **tip of the red giant branch**: the brightest red giant stars turn out to have I-band magnitudes that are relatively insensitive to metallicity and age (a standard candle).
- **RR Lyrae stars**: these variable stars on the horizontal giant branch of the Hertzsprung–Russell diagram pulsate with a period shorter than one day that is closely correlated with their mean absolute magnitude (a standard candle).
- **Cepheid variables**: these are also variable stars with a well-defined relationship between period and luminosity. There are two types: type I are massive young stars while type II are low-mass stars. The period–luminosity

relationships for the two types are different; only type I Cepheid variables are used as primary distance indicators.

- **Planetary nebula luminosity function**: this is the number of planetary nebulae per unit volume per unit luminosity. The luminosity is usually measured in the [O III] 500.7 nm emission line of oxygen. It has a characteristic shape with a sharp cut-off above a fixed luminosity, which can be used as a standard candle.
- **Globular cluster luminosity function**: this is found to be approximately Gaussian. The mean of this Gaussian can be used as a secondary distance indicator (a standard candle).
- The **light echo** from the supernova SN 1987A is a standard rod.
- The rotation velocity of the **water maser** (a naturally-occurring microwave laser) in the galaxy NGC 4258 can be compared to its proper motion, finding a distance of 7.2 ± 0.5 Mpc.
- **Surface brightness fluctuations**: if there are on average N stars per square arcminute on the sky in a galaxy, this number will be subject to Poisson statistics and will fluctuate with standard deviation $\sqrt{N}$. The fluctuations in surface brightness can therefore give an estimate of the number of stars per unit area on the sky in that galaxy, from which an angular diameter distance to that galaxy can be derived.
- **Type Ia supernovae** are standard candles that will be discussed in more detail below.

Figure 3.6 summarizes how larger-scale distance indicators depend on earlier stages. The techniques have been labelled as standard rods, candles etc., but note that for the nearest objects the various types of distances are indistinguishable.

From a comparison with the redshifts, the Hubble parameter can be found through Equation 1.44 and its variants. In the low-redshift limit, the various distances (angular diameter, luminosity, comoving) are indistinguishable and the expression for H_0 reduces to $H_0 = cz/d$, where d is the distance. One of the primary goals of the Hubble Space Telescope (HST) was the determination of the Hubble parameter to 10% accuracy. This was achieved by measuring the periods and magnitudes of Cepheid variables in 18 spiral galaxies at distances < 20 Mpc. The high angular resolution of the HST was needed to identify the stars in these crowded fields. This provided the fundamental calibration of the Tully–Fisher relation, the fundamental plane, type Ia supernovae and surface brightness fluctuations; the final result was $H_0 = 72 \pm 8\,\text{km}\,\text{s}^{-1}\,\text{Mpc}^{-1}$.

Perhaps the most striking recent progress with the distance scale has been with type Ia supernovae. These supernovae are caused when a white dwarf is accreting matter from a companion star, which eventually sends the star above the threshold for ignition of carbon. The star explodes and briefly has a luminosity of about ten billion times that of our Sun. Type Ia supernovae occur only about once every two hundred years in our Galaxy (the last one in our Galaxy was seen by Tycho Brahe in 1572), but may occur at the rate of one per second in the observable Universe.

Since white dwarfs have a small range of masses, type Ia supernovae have a very restricted range of luminosities, making them ideal for consideration as standard candles. Empirically they have a peak luminosity that appears to be related to the

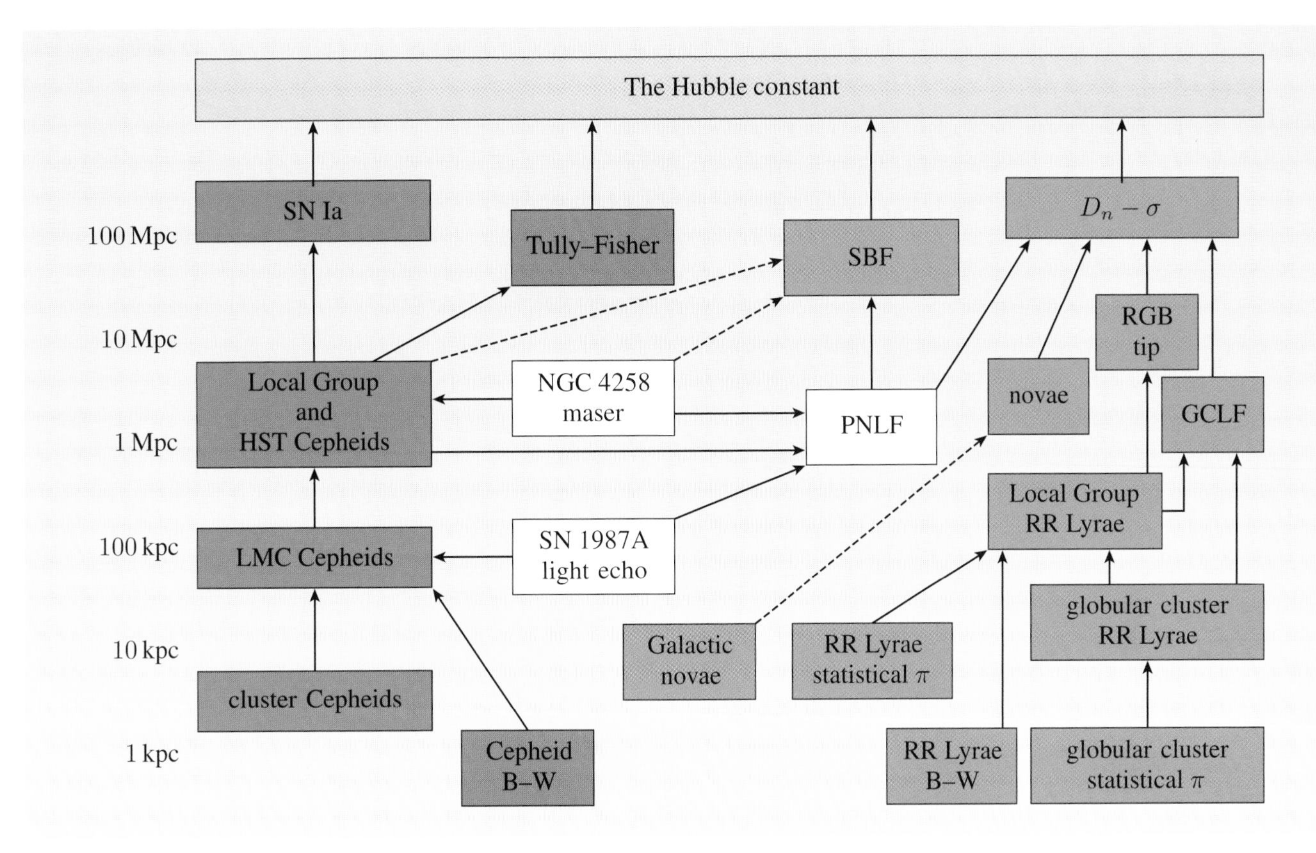

Figure 3.6 The cosmological distance ladder, showing how each method depends on the calibration for more nearby methods (arrows). Insecure calibration steps are shown as dashed lines. The pink boxes refer to methods that are useful in star-forming galaxies, while the blue boxes are useful in early-type galaxies. Open boxes are geometric distance determinations. The PNLF box (planetary nebulae luminosity function) works for all galaxy populations in the local supercluster.

subsequent rate of decay of brightness; although this relationship is not well understood, applying a correction for this makes type Ia supernovae excellent candidates for standard candles. The advent of large-scale digital CCD sky surveys made it possible to search for these supernovae at much higher redshifts. The Supernova Cosmology Project and the High-Redshift Supernova Search Team found supernovae out to $z \sim 1$. To many people's great surprise, the luminosity distances implied by the supernovae could only be fit with models in which $\Omega_\Lambda > 0$. Figure 2.14 in Chapter 2 showed how the simultaneous constraints from supernovae and the CMB were needed to constrain both $\Omega_{\rm m}$ and Ω_Λ. This conclusion was not immediately accepted by some in the astronomical community, who argued that (for example) dust extinction preferentially in the higher-redshift supernovae could mimic the effect of Ω_Λ by making the high-redshift supernovae preferentially fainter. Indeed, the opacity of spiral discs is expected to evolve. However, this would also be expected to increase the dispersion, which it did not appear to, and the supernovae showed no evidence of having redder optical spectra. Figure 3.7 shows the current magnitude–redshift diagram for type Ia supernovae. Plots of apparent magnitude against redshift are known as **Hubble diagrams** owing to their use in determining the Hubble parameter, and dating from Hubble's original 1929 discovery of the expansion of the Universe. A recent Cepheid-based calibration of the distances to the galaxies that hosted nearby supernovae, plus the galaxy NGC 4258 (with a water-maser-calibrated distance) has found $74.2 \pm 3.6\,\mathrm{km\,s^{-1}\,Mpc^{-1}}$.

Reiss, A. G. et al., 2009, *Astrophysical Journal*, **699**, 539.

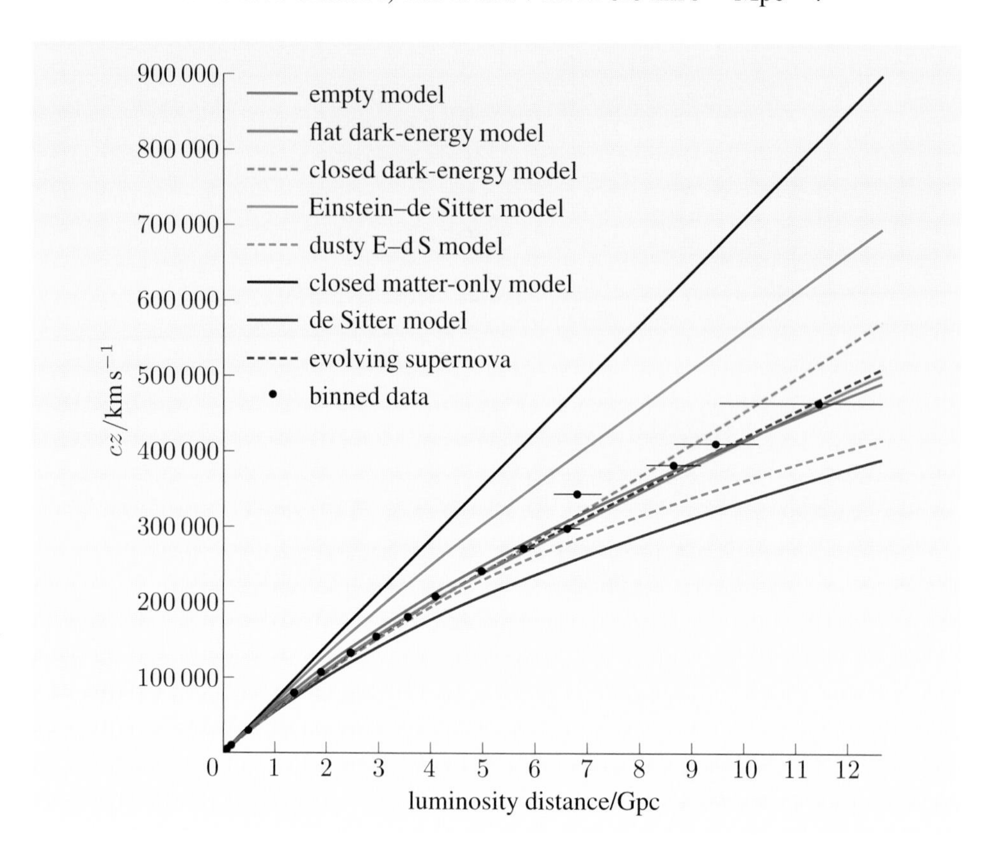

Figure 3.7 Constraints on the Hubble diagram of type Ia supernovae. The flat dark energy model has $\Omega_\mathrm{m} = 0.27$ and $\Omega_\Lambda = 0.73$. It is, of course, also possible to fit the data with a model in which supernovae have a specially-tailored evolution.

3.11 The large-scale structure of the Universe

Next we'll take you on a tour of the large-scale structure of the Universe. Our Galaxy, the Milky Way, is in the process of accreting two satellite galaxies, the Large Magellanic Cloud (LMC) and the Small Magellanic Cloud (SMC, Figure 3.8), which can be seen by the naked eye on a dark enough night in the southern hemisphere. These are our most famous satellite galaxies, but there are many more, many of which have only recently been discovered. Figure 3.9 shows our Galaxy's immediate cosmic neighbourhood. One close neighbour, the Sagittarius dwarf elliptical galaxy, was discovered only in 1994, despite being only half the distance to the LMC; its location on the sky is on the far side of the central bulge of our Galaxy. The Canis Major dwarf galaxy is even closer at only 7.6 kpc (closer to us than the centre of the Milky Way), yet was discovered only in 2003 using the Two Micron All-Sky Survey (2MASS), which highlighted the distinctive colours of the galaxy's stellar population. The Milky Way is accreting all these galaxies. The tidal shears from our Galaxy's gravitation field are distorting the satellite galaxies and in some cases pulling them into streams of stars. In Chapter 4 we'll see how this process has continued throughout the history of the Universe.

Figure 3.8 The Magellanic Clouds as seen from Queensland, Australia.

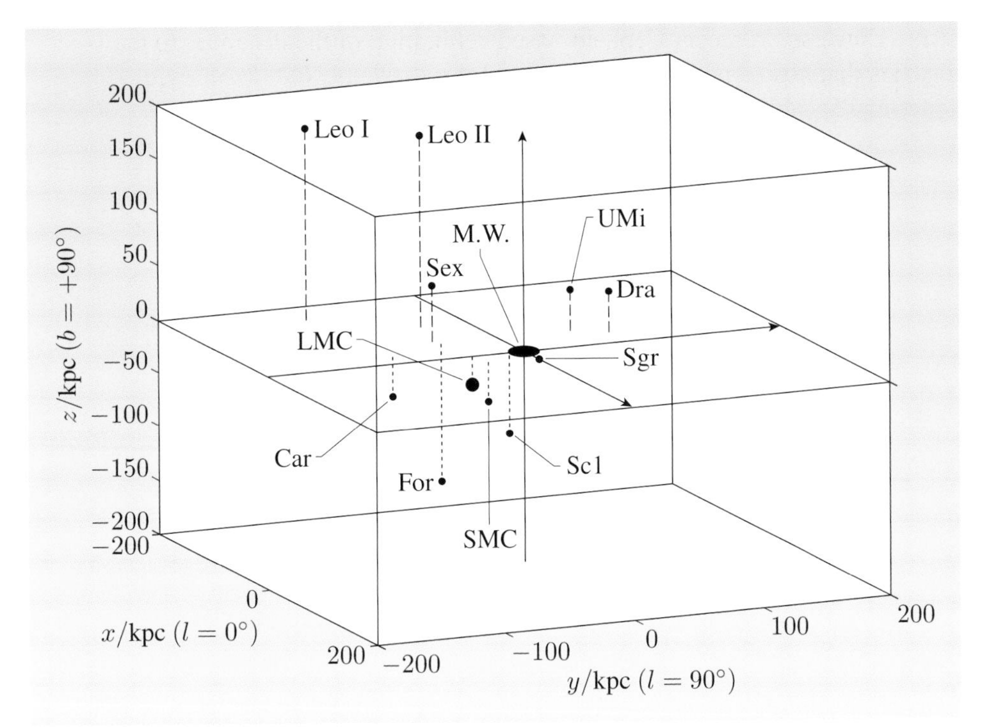

Figure 3.9 The neighbourhood of our galaxy. MW marks the Milky Way; l and b are the galactic longitude and latitude coordinates. The other annotations are abbreviated names of the nearby galaxies.

Figure 3.10 The Andromeda galaxy, M31, as seen in the optical.

Our Galaxy is falling towards and will eventually itself be accreted by its more massive neighbour, the Andromeda galaxy (Figure 3.10), also known as M31. In about a billion years' time, Andromeda will appear about the size and brightness of the Magellanic Clouds today. Even today, though, it is visible by the naked eye on a dark enough night in the northern hemisphere. M31 also has its own satellite galaxies currently being accreted.

Moving slightly out, both our Galaxy and M31 are members of the Local Group, known to contain over 40 galaxies. Edwin Hubble was the first to recognize that these neighbours represent an overdensity relative to the average galaxy density. Moving slightly out again, the Local Group is itself interacting with other galaxy groups, such as the Maffei 1 Group, the Sculptor Group, the M81 Group and the M83 Group (each named after one of their members). Galaxy groups are gravitationally bound but are not as massive as galaxy clusters; there is no established dividing line between the two, but $> 10^{14}\,M_\odot$ systems are usually regarded as clusters. Our nearest galaxy cluster is the Virgo cluster (Figures 3.11 and 3.12). The Virgo cluster currently has a positive redshift, but the Local Group is still likely eventually to be accreted by it. The Virgo cluster and the next nearest cluster, Coma, are themselves part of a structure known as the **local supercluster**. On these progressively larger scales (Figure 3.13), the Universe looks filled with sheets and filaments — indeed, the local supercluster galaxies are preferentially (though not exclusively) in a particular plane.

Figure 3.11 The centre of the Virgo cluster of galaxies, as seen in optical light. The cluster contains over 2000 galaxies and has a diameter ten times bigger than the full Moon on the sky, but is too faint to be seen by unaided eyesight.

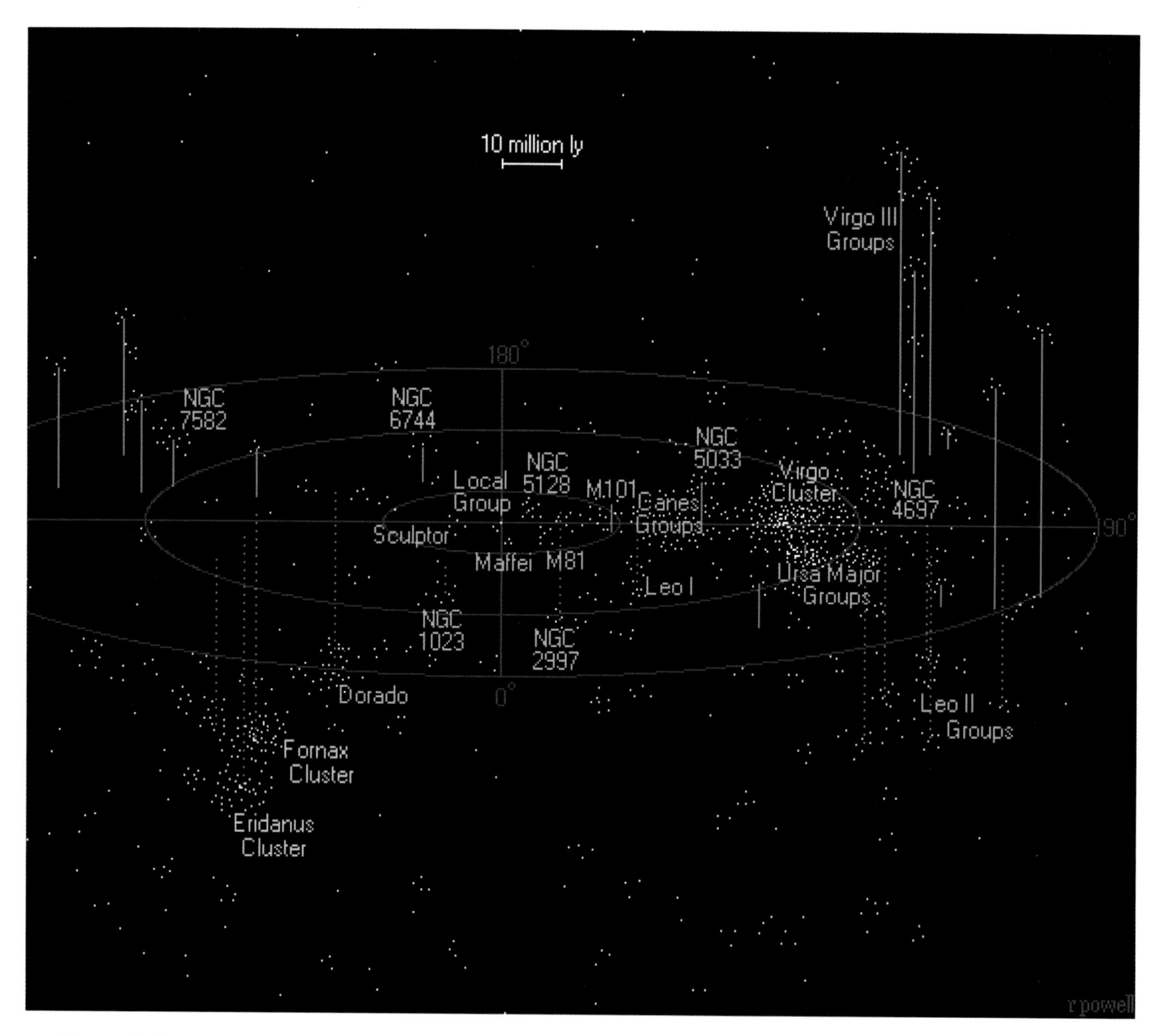

Figure 3.12 The location of our Local Group of galaxies and the Virgo cluster.

As we zoom out further, this filamentary structure is increasingly apparent. Figure 3.14 is the famous 'stick man' from the early CfA (Centre for Astrophysics) galaxy redshift surveys. The horizontal 'arms' featured in later surveys and became known as the Great Wall. This early survey is now dwarfed by the latest generation of redshift surveys, such as the 2dF galaxy redshift survey (2dFGRS, named after the fibre-optic spectroscope at the Anglo-Australian Telescope, with a two degree field on the sky) and the Sloan Digital Sky Survey (SDSS) galaxy redshift survey (a CCD imaging and spectroscopy survey of about π steradians on the sky). Figure 3.15 shows the cone diagram for the 2dF survey.

Sometimes referred to as π in the sky.

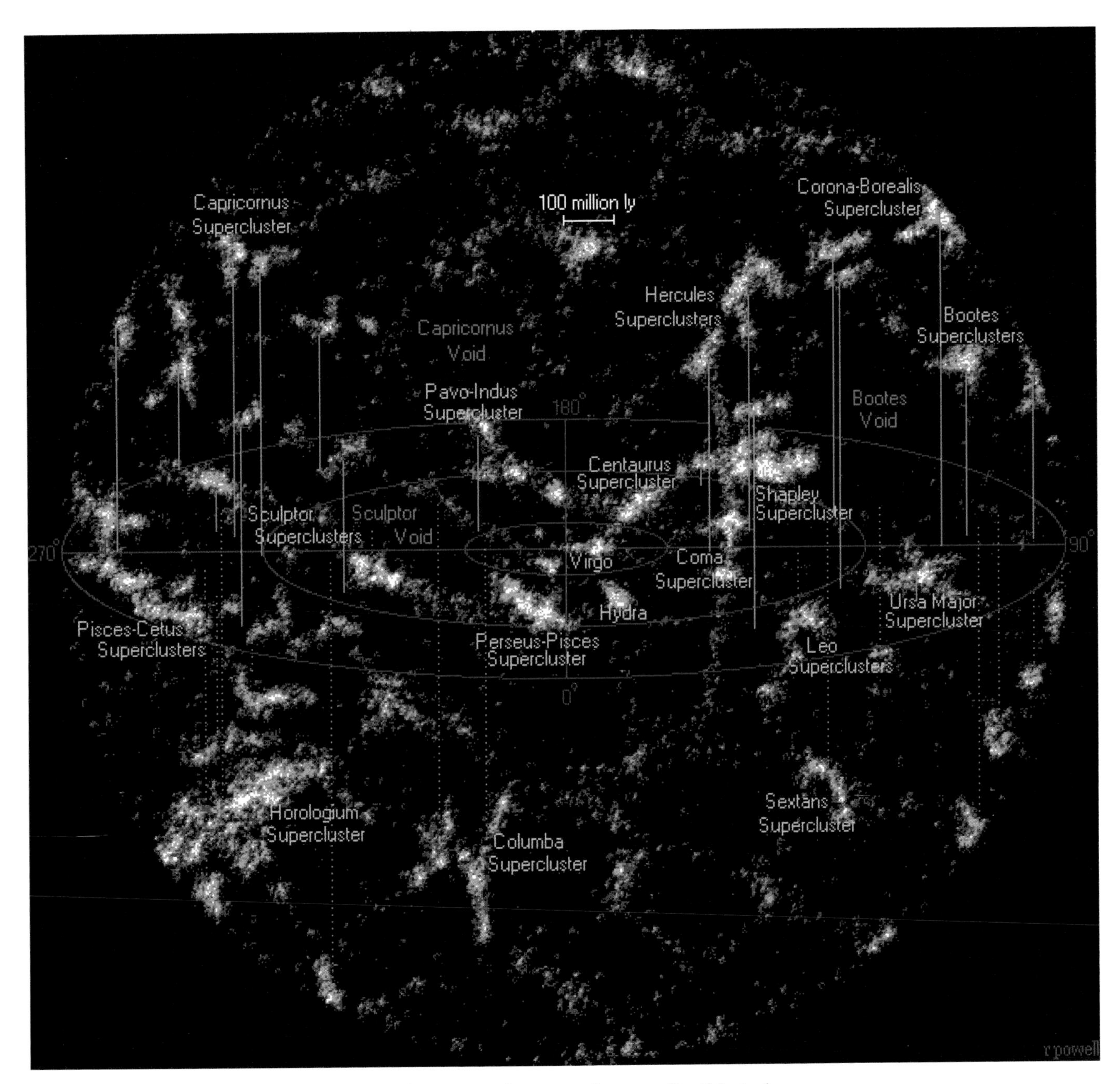

Figure 3.13 The Virgo cluster in relation to nearby superclusters. On this truly gigantic scale, the Local Group is barely discernible on the page.

Compare the scale to Figure 3.14. The large-scale structure of the Universe looks cobwebby, with superclusters, walls and giant voids. Note that the sensitivity in Figure 3.15 tapers off at the largest distances. On the very largest scales the Universe is beginning to look homogeneous, which we shall quantify with the power spectrum.

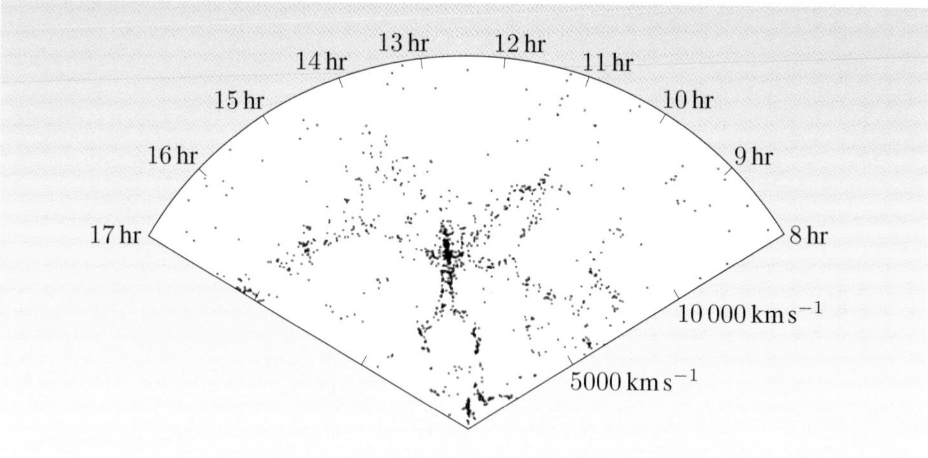

Figure 3.14 The famous 'stick man' in the CfA galaxy redshift surveys. The angular axis is right ascension, and the distance from the apex is galactocentric.

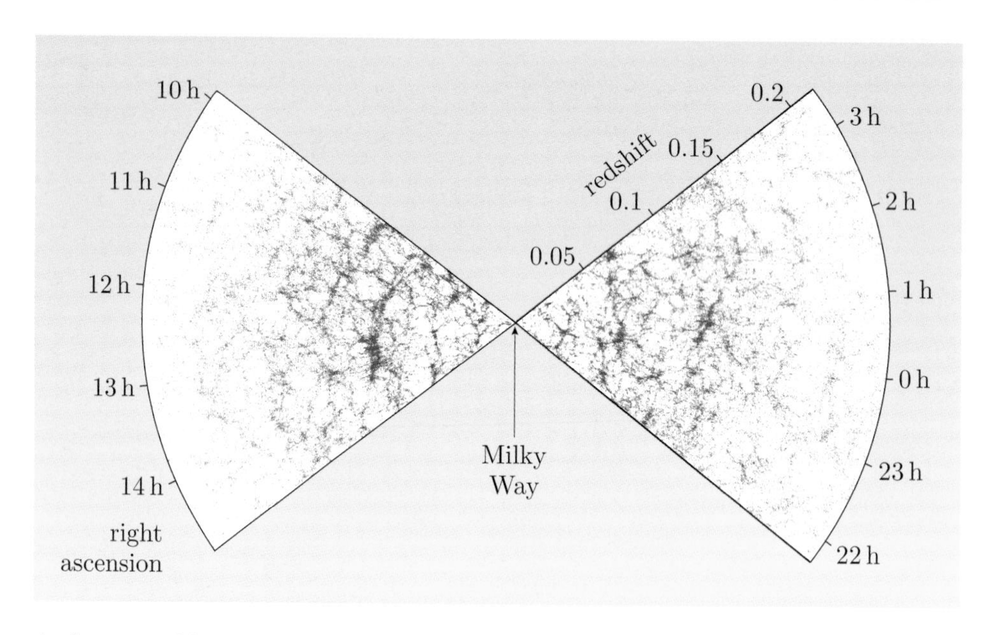

Figure 3.15 The distribution of galaxies seen in the 2dF galaxy redshift survey.

A sharp-eyed look at Figure 3.15 reveals several structures apparently pointed directly at the Earth! These are known as **fingers of God**. To understand how these come about, imagine how a galaxy cluster affects the redshifts of its galaxies. These galaxies are decoupled from the Hubble flow and acquire the typical redshift of the cluster, plus or minus its velocity dispersion. The length of the 'finger' simply reflects the velocity dispersion in the cluster. On bigger scales there is another effect, known as the **Kaiser effect**, caused by galaxies beginning to fall into the cluster. Galaxies between us and the cluster (but near the cluster in space) will tend to be falling in. This peculiar velocity (Section 1.4) adds a Doppler component to the redshift and makes the galaxies appear further away. Similarly, galaxies on the other side of the cluster appear closer. Together, the Kaiser effect and fingers of God are known as **redshift space distortions**, and the clustering analysis done with redshifts (rather than independently-derived distances) is known as redshift space clustering. We'll return to this in Chapter 4 and see at which point the Kaiser effect turns into a finger of God.

This is illustrated in Figure 3.16. This shows the amplitude of the redshift space correlation function from 2dFGRS as a function of transverse separation on the sky (symbolized σ) and redshift-axis separation (symbolized π). Note the finger of God and the Kaiser effect flattening. This flattening is determined by the value of $\Omega_{\rm m}^{0.6}/b$, where b is known as the bias parameter and expresses how much stronger the clustering of galaxies is compared to dark matter (see also Chapter 4).

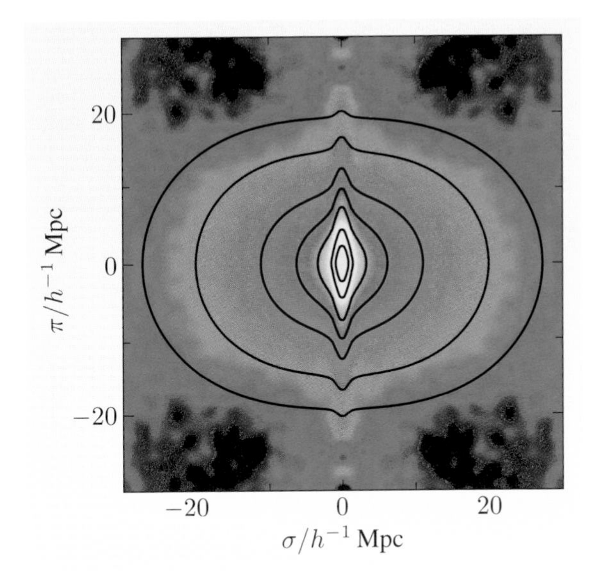

Figure 3.16 The amplitude of the redshift space correlation function of galaxies, $\xi(\sigma,\pi)$, as a function of the transverse (σ) and radial (π) separations. The data from the first quadrant have been repeated in the other three quadrants, to highlight the deviation from circular symmetry. The 'fingers of God' effect is seen at small scales, and the flattening on larger scales is due to the Kaiser effect. Contours are plotted at $\xi = 10, 5, 2, 1, 0.5, 0.2, 0.1$ with the contour of highest ξ in the centre.

The dimensionless power spectrum of galaxies is shown in Figure 3.17. This diagram can be used to make a rough estimate of how big the fluctuations from large-scale structure would be in any galaxy survey. We'll show examples of this in Chapter 4.

Another way of measuring galaxy clustering is through the **correlation function** $\xi(r)$, which is defined as follows. What is the probability that a galaxy will have a neighbour at a distance r from it, in a small volume δV? If galaxies are unclustered, the answer will be $n\,\delta V$, where n is the number of galaxies per unit volume. However, if galaxies are clustered, we could write this as

$$\Pr(r) = \{1 + \xi(r)\}\, n\,\delta V, \tag{3.8}$$

where $\xi(r)$ just expresses the excess over what's expected in a non-clustered case. Slightly more generally, the probability of finding one galaxy in the volume $\mathrm{d}V_1$ and another in $\mathrm{d}V_2$ separated by a vector $\boldsymbol{r}$ is

$$\mathrm{d}^2 \Pr(\boldsymbol{r}) = \{1 + \xi(\boldsymbol{r})\}\, n^2\, \mathrm{d}V_1\, \mathrm{d}V_2. \tag{3.9}$$

The correlation function can be estimated by counting the numbers of galaxy pairs as a function of their separation, and comparing the results for random unclustered populations. This measure of galaxy clustering is sometimes seen as more intuitive since it avoids Fourier series.

It turns out to be closely related to the galaxy power spectrum. We'll spare you the details, but if we make a Fourier expansion of $\xi(\boldsymbol{r})$, it turns out that it can be expressed as

$$\xi(\boldsymbol{r}) = \frac{V}{(2\pi)^3} \int |\delta_{\boldsymbol{k}}|^2 \mathrm{e}^{-\mathrm{i}\boldsymbol{k}\cdot\boldsymbol{r}}\, \mathrm{d}^3 k. \tag{3.10}$$

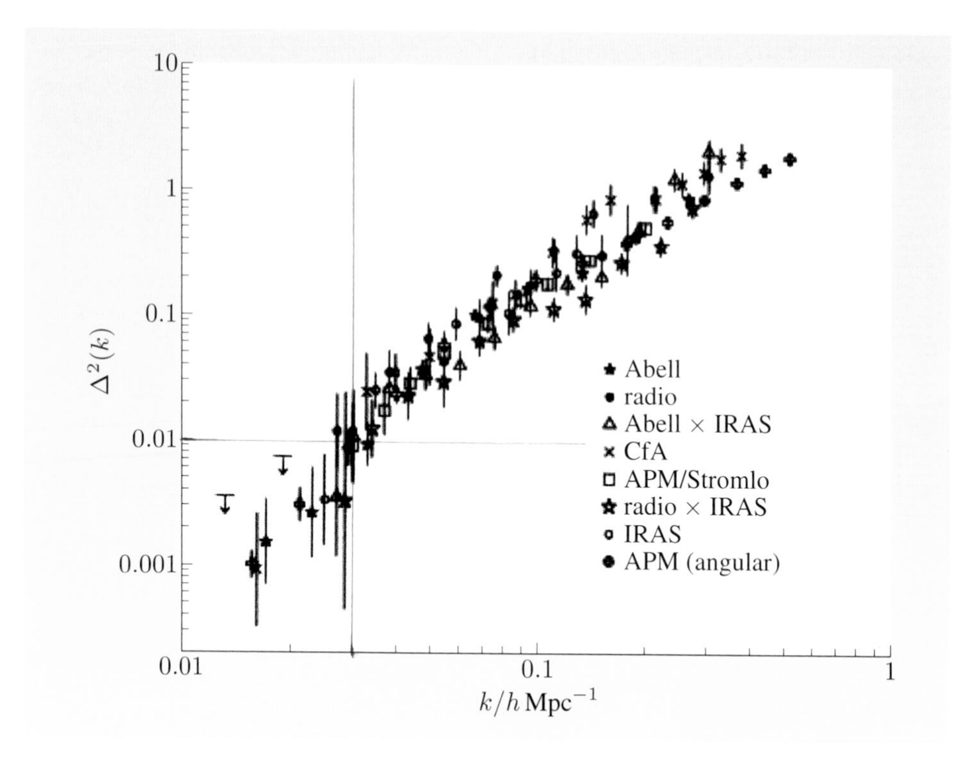

Figure 3.17 The dimensionless power spectrum of galaxies, selected through a variety of means. The bias parameter depends on the nature of the selection, so the power spectra have been normalized to $b = 1$.

In other words, the correlation function is the Fourier transform (see the box in Section 2.9) of the power spectrum. If the density field is isotropic, then the power spectrum will depend only on the magnitude of the $\boldsymbol{k}$ vector, not its direction: $\langle|\delta_{\boldsymbol{k}}|^2\rangle = |\delta_k|^2(k)$. Since $\xi(r)$ must be real, we can replace $e^{i\boldsymbol{k}\cdot\boldsymbol{r}}$ with $\cos(kr\cos\theta)$, and integrating over all angles in three dimensions can be shown to give

$$\xi(\boldsymbol{r}) = \frac{V}{(2\pi)^3}\int P(k)\,\frac{\sin kr}{kr}\,4\pi k^2\,\mathrm{d}k. \tag{3.11}$$

Similarly, we can write the power spectrum as a transform of the correlation function:

$$\Delta^2(k) = \frac{V}{(2\pi)^3}\,4\pi k^3\,P(k) = \frac{2}{\pi}k^3\int_0^\infty \xi(r)\,\frac{\sin kr}{kr}\,r^2\,\mathrm{d}r. \tag{3.12}$$

In summary, the power spectrum $P(k)$ (or $\Delta^2(k)$) is very closely related to the correlation function $\xi(r)$.

So far we've talked about three-dimensional power spectra and the correlation function of galaxies in three dimensions. To measure this we need the positions of galaxies on the sky, and the distance (e.g. redshift) to the galaxies. But what if we don't have redshifts? Can we determine anything about the clustering? It turns out that we can, though we need to make some assumptions about the number of galaxies that we see per unit redshift. This might evolve for example because of galaxy merging, and needs to take account of the fact that less-luminous galaxies can be seen only nearby while the rarer bright galaxies can be seen further away.

We can write down an **angular correlation function** $w(\theta)$ on the sky in a similar way to the spatial correlation function, as the excess neighbours that you'd see over an unclustered population:

$$\mathrm{d}^2\,\Pr(\theta) = n^2(1 + w(\theta))\,\mathrm{d}\Omega_1\,\mathrm{d}\Omega_2, \tag{3.13}$$

where $d\Omega_1$ and $d\Omega_2$ are the solid angles of nearby patches of sky, and n is the number of galaxies per unit area on the sky. We won't prove the relationship here between $\xi(r)$ and $w(\theta)$ (it is known as *Limber's equation*), but if $w(\theta)$ is a power law with $w(\theta) = A\theta^{1-\gamma}$ (where A is some constant), then it turns out that $\xi(r)$ is also a power law with $\xi(r) = (r/r_0)^{-\gamma}$ (where r_0 is a scale length known as the clustering length scale). The constants A and r_0 can be related with a knowledge of the number of galaxies per unit redshift.

See, for example, Gonzalez-Solares, E.A. et al., 2004, *Monthly Notices of the Royal Astronomical Society*, **352**, 44.

The clustering of galaxies was different at high redshifts. We can add an evolution term into the correlation function:

$$\xi(r) = \left(\frac{r}{r_0}\right)^{-\gamma} (1+z)^{-(3+\varepsilon)}. \qquad (3.14)$$

If $\varepsilon = 0$, then the galaxy clustering is constant in *physical* coordinates, i.e. it's unaffected by the expansion of the Universe. If $\varepsilon = 3 - \gamma$, then the clustering is constant in *comoving* coordinates. Equation 3.14 still yields a power law angular correlation function $w(\theta) = A\theta^{1-\gamma}$. Just so you have it, in this case **Limber's equation** (which we won't prove) is

See, for example, Phillips, S. et al., 1978, *Monthly Notices of the Royal Astronomical Society*, **182**, 673.

$$A = C r_0^{\gamma} \frac{\int d_A^{1-\gamma}\, g^{-1}(z)\, (1+z)^{-(3+\varepsilon)} (dN/dz)^2\, dz}{\left(\int (dN/dz)\, dz\right)^2}, \qquad (3.15)$$

where d_A is the angular diameter distance, $g(z)$ is the derivative of proper distance with redshift, and dN/dz is the number of galaxies per unit redshift. The normalization C is a function of γ and is given by $C = \sqrt{\pi}\,\Gamma((\gamma-1)/2)/\Gamma(\gamma/2)$, where Γ is a standard function (usually calculated numerically) known as the gamma function.

We'll end our tour of the large-scale structure of the Universe with the clustering of quasars, shown in Figure 3.18. Quasars, as we'll see in Section 4.6, are supermassive black holes in the centres of galaxies. The accretion discs around these black holes can outshine the rest of the galaxy, as we'll discuss in Chapter 4. This extraordinary luminosity makes them visible in large numbers throughout most of the Hubble volume (Section 1.9). The Universe appears far more homogeneous on these scales, but some clustering signal is still present. The strength of quasar clustering today is comparable to the predicted dark matter clustering, but in the early Universe the amplitude of the quasar correlation function was higher (unlike the dark matter). In general, virialized dark matter haloes (when considered as objects in their own right) cluster more strongly than the dark matter distribution as a whole, so the enhanced clustering strength of quasars could reflect the haloes that they inhabit. The observed quasar clustering could be accounted for by assuming that quasars inhabit dark matter haloes with a typical mass of around $10^{13}\,M_\odot$. The phenomenon can be characterized by the **bias parameter** b, mentioned briefly above:

$$\left(\frac{\delta\rho}{\rho}\right)_{\text{QSOs}} = b \left(\frac{\delta\rho}{\rho}\right)_{\text{dark matter}}, \qquad (3.16)$$

where the bias parameter can be a function of redshift. There isn't much theoretical basis for assuming this relationship; it's more of an empirical rule of thumb.

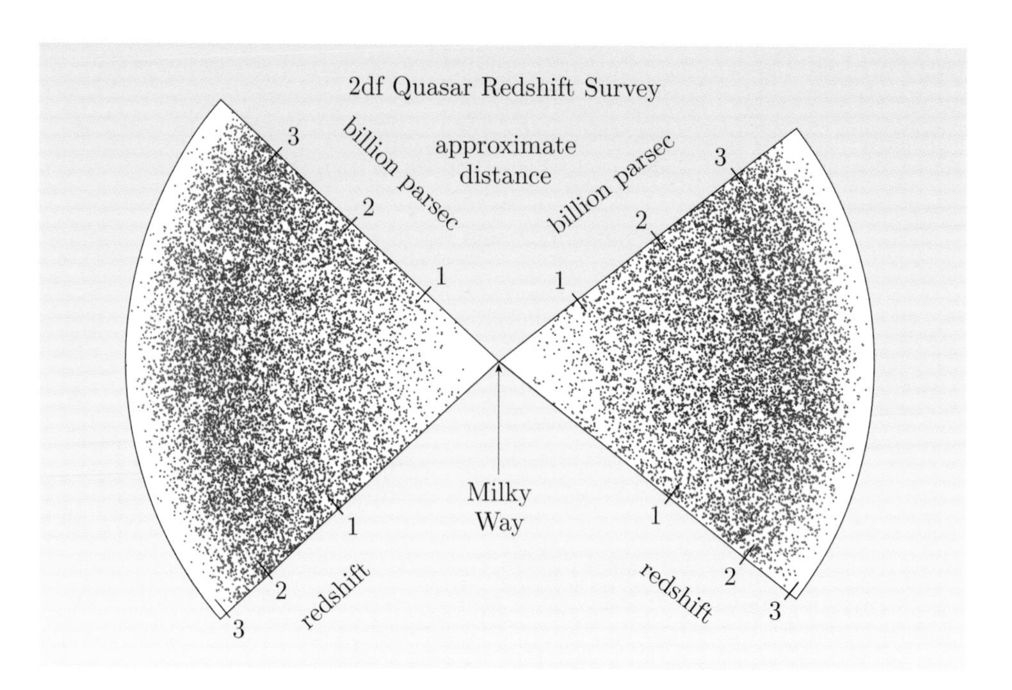

Figure 3.18 The spatial distribution of quasars in the 2dF quasar survey, which covered two $75^\circ \times 5^\circ$ strips on the sky, one in each Galactic hemisphere. The approximate distances marked are comoving, assuming $\Omega_\mathrm{m} = 1$ and $\Lambda = 0$.

3.12 Baryon wiggles

The acoustic peaks in the matter power spectrum did not disappear after recombination — they are still there in the power spectrum of galaxies! These appear as wiggles in the galaxy power spectrum and are sometimes referred to as **baryon wiggles** or more fully as **baryonic acoustic oscillations** (BAOs). The detection of these features is difficult but there have been hints of detections from SDSS and the 2dF galaxy redshift survey, shown in Figure 3.19.

At the time of writing this work is at its early stages, but the crucial advantage of detecting BAOs is that the positions of the acoustic peaks are a standard rod that can be used right back to the time of the CMB. (There is a phase difference between the photon and matter oscillations, but this is calculable.) This is ideal for measuring changes in the expansion rate of the Universe, i.e. constraining the cosmological parameters and the evolving dark energy equation of state. This goal is so attractive that many international groups are planning ambitious sky surveys with enough photometric and spectroscopic data for baryon wiggles and other dark energy constraints such as cosmic shear (Chapter 7) and high-redshift supernovae. In the further reading section, we've included a report on the future of dark energy surveys in the coming decade; we'll meet some of these ambitious future surveys briefly in Chapter 7.

BAOs are not just detectable in the angular correlation function of galaxies at a particular redshift. They could also be detected along the redshift axis, which as the following exercises show can give us a measure of the evolving Hubble parameter $H(z)$. This in turn can be a measure of the dark energy equation of state, because the generalization of Equation 1.33 for a flat Universe with dark energy is

$$H(z) = H_0\sqrt{\Omega_{\mathrm{m},0}(1+z)^3 + (1-\Omega_{\mathrm{m},0})(1+z)^{3(1+w)}}. \tag{3.17}$$

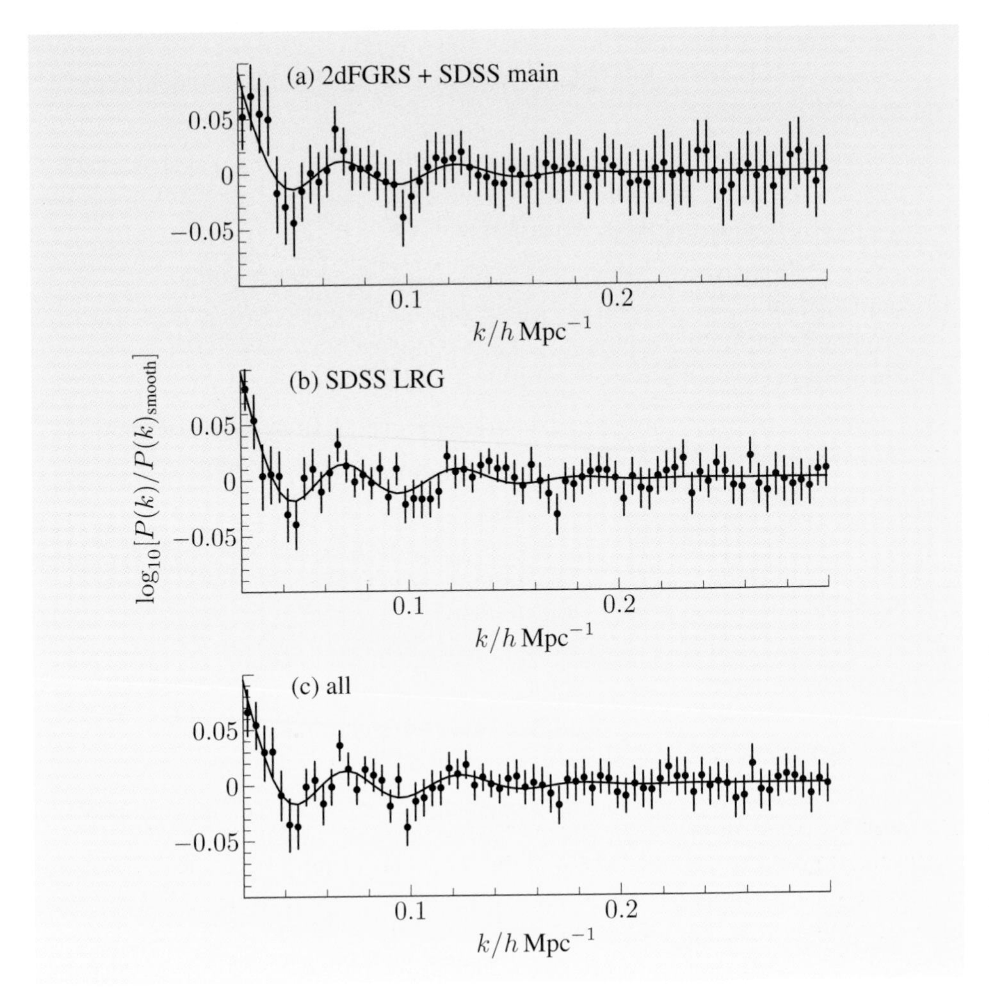

Figure 3.19 The deviations from a smooth power spectrum seen in the galaxy power spectrum, compared to the predictions for baryon wiggles. The galaxy surveys are the 2dFGRS, the whole SDSS galaxies, and the subset of luminous red galaxies from SDSS.

Exercise 3.4 If $L_{\rm BAO}$ is the *comoving* size of the BAOs at some redshift z, and $\theta_{\rm BAO}$ is the observed angular size in a *flat* universe, show that the comoving distance to redshift z is $d_{\rm comoving} = L_{\rm BAO}/\theta_{\rm BAO}$.

Exercise 3.5 If δz is the size along the redshift axis of the BAOs, show that $H(z) = c\,\delta z/L_{\rm BAO}$, where $L_{\rm BAO}$ is the comoving size of the BAOs.

Exercise 3.6 Is the scale length of baryon wiggles dependent on galaxy bias? ■

Summary of Chapter 3

1. One piece of evidence in favour of dark matter is galaxy rotation curves.
2. Local galaxies can be classified by eye (or by software) along the Hubble tuning fork. Irregular galaxies lie outside this framework, and low surface brightness galaxies were unknown in Hubble's time. Local elliptical galaxies tend to occur in galaxy clusters, while local spiral galaxies tend to occur in the field (the morphology–density relation).
3. There is experimental support for the Tolman test (that surface brightness decreases as $(1+z)^4$) once stellar evolution is taken into account.

4. Spiral galaxies obey the Tully–Fisher relation, that luminosity L scales with velocity width Δv as $L \propto (\Delta v)^\alpha$, with $\alpha \simeq 3$–4. This can also be used as a distance indicator.
5. Elliptical galaxies also have a similar relation, known as the Faber–Jackson relation: $L \propto \sigma_\text{v}^\alpha$, where σ_v is the velocity dispersion and $\alpha \simeq 3$–4, though the physical underpinnings are different.
6. There is a more general relationship for ellipticals, known as the fundamental plane: $L \propto I_0^x \sigma_\text{v}^y$, where L is the luminosity, I_0 is the surface brightness within a given radius, and σ_v is the velocity dispersion. This can also be used as the basis of distance indicators.
7. Clusters of galaxies show evidence for dark matter, from the virial theorem (kinetic energy equals minus one half times the potential energy for a virialized system).
8. If we assume that the baryonic mass fraction of galaxy clusters doesn't depend on redshift, then galaxy cluster sizes and X-ray luminosities can be used to constrain cosmological parameters.
9. The Sunyaev–Zel'dovich effect is the change of temperature of CMB photons passing through a galaxy cluster, due to Compton scattering. It can be used to find the line-of-sight size of a cluster, and assuming spherical symmetry (true on average) gives an angular diameter distance estimate.
10. Galaxy surveys to determine the large scale structure of the Universe reveal redshift space distortions known as the fingers of God and the Kaiser effect. These are due to the velocity dispersion of galaxies within a cluster and the peculiar velocities of galaxies falling into clusters respectively.
11. The clustering of galaxies can be measured with the correlation function $\xi(r)$, which is a measure of the number of neighbours that a galaxy has in excess of what's expected in an unclustered population. This is proportional to the Fourier transform of the galaxy power spectrum. It's also related to the angular correlation function, i.e. the clustering of galaxies as they appear on the sky.
12. The acoustic peaks imprinted into the cosmological density field in the early Universe are still present in the present-day clustering of galaxies. These features, known as baryonic acoustic oscillations or simply baryon wiggles, are a standard rod that can be used to derive angular diameter distances and also the variation in the Hubble parameter $H(z)$. This is a route to constraining the dark energy equation of state parameter w.

Further reading

- The Galaxy Zoo website is currently www.galaxyzoo.org.
- Gaitskell, R.J., 2004, 'Direct detection of dark matter', *Annual Reviews of Nuclear and Particle Science*, **54**, 315.
- For more on the virial theorem, see, for example, Chapter 2 of Ryan, S.G. and Norton, A.J., 2010, *Stellar Evolution and Nucleosynthesis*, Cambridge University Press.

- Young, J.S. and Scoville, N.Z., 1991, 'Molecular gas in galaxies', *Annual Review of Astronomy and Astrophysics*, **29**, 581.
- Reese, E.D., 2004, 'Measuring the Hubble constant with the Sunyaev–Zel'dovich effect', in Freedman, W.L. (ed.) *Measuring and Modeling the Universe*, Carnegie Observatories Centennial Symposia, Cambridge University Press (also available at level 5 of http://nedwww.ipac.caltech.edu).
- Freeman, W.L., 2001, 'Final results from the Hubble Space Telescope key project to measure the Hubble constant', *Astrophysical Journal*, **553**, 47.
- For an accessible account of recent developments in the cosmological distance ladder, see, for example, Rowan-Robinson, M., 2008, 'Climbing the cosmological distance ladder', *Astronomy and Geophysics*, **49**, 3.30–3.33.
- An accessible review of the WiggleZ baryon oscillation survey is in Blake, C., 2008, *Astronomy and Geophysics*, **49**, 5.19–5.24.
- The classic graduate-level text for galaxy clustering (e.g. galaxy correlation functions and power spectra) is Peebles, P.J.E., 1980, *The Large-scale Structure of the Universe*, Princeton University Press.
- Albrecht, A. et al., 2006, *Report of the Dark Energy Task Force*, available at astro-ph/0609591.

Chapter 4 The distant optical Universe

It's one of nature's ways that we often feel closer to distant generations than to the generation immediately preceding us.

Igor Stravinsky

Introduction

Perhaps one of the most astonishing surprises about the Universe is that it's possible to build telescopes that can observe galaxies throughout almost all the Hubble volume, most of the way back to the Big Bang. In this chapter we'll discuss some of the profound insights that optical telescopes have brought us on the evolution of galaxies in our Universe.

4.1 Source counts

We started Chapter 1 with a derivation of the Euclidean source count slope $\mathrm{d}N/\mathrm{d}S \propto S^{-5/2}$, where $\mathrm{d}N/\mathrm{d}S$ is the number of objects per unit flux S on the sky. You may be surprised to hear that something as simple as counting galaxies in this way has led to profound insights and questions.

First, however, let's derive the Euclidean slope in a simpler and quicker way by considering instead the **integral source counts** (or **integral number counts**), defined as the number of objects *brighter* than a flux S. This is often given the absurdly non-mathematical symbol $N(> S)$, and it can be related to the number per unit flux as $N(> S) = \int_S^\infty (\mathrm{d}N/\mathrm{d}S)\,\mathrm{d}S$. For this reason $\mathrm{d}N/\mathrm{d}S$ is sometimes called the **differential source counts** (or **differential number counts**).

Suppose that all galaxies have the same luminosity L, and have a constant number per unit volume in a static Euclidean space. The number of objects brighter than a flux S will be given by all the objects within a radius r, where $S = L/(4\pi r^2)$. But we've assumed that the space density (number per unit volume) of galaxies is a constant, say ρ, so the number of galaxies brighter than S must be $N(> S) = \rho \times \frac{4}{3}\pi r^3 \propto r^3$. But $S \propto r^{-2}$, so $N(> S) \propto S^{-3/2}$. Differentiating this gives $\mathrm{d}N/\mathrm{d}S \propto S^{-5/2}$. As in Chapter 1, objects with different luminosities will still give power-law number counts with the same slope, so a mix of galaxies with different luminosities will still give the same slope. This argument is very old and dates back at least as far as John Michell in 1767!

Michell, J., 1767, *Philosophical Transactions of the Royal Society of London*, **57**, 234–64; a reference url is given in the *Further reading* section at the end of this chapter.

As early as the 1970s it became clear that the number counts of galaxies in the B-band optical (blue) filter were strongly inconsistent with an unevolving population of galaxies almost regardless of Ω_m ($\Lambda = 0$ was implicitly being assumed). There seemed to be too many faint blue galaxies, which became known as the **faint blue galaxies problem**. We'll see below that galaxies should evolve in luminosity, and adding some luminosity evolution to the models helped but did not resolve the discrepancy on its own. The only solution appeared to be a new population of recently-formed dwarf galaxies, and theorists struggled to explain this population in an unforced way.

This changed with the advent of precision cosmology and the concordance model (Chapters 1 and 2) with $\Omega_{m,0}\, h^2 = 0.1326 \pm 0.0063$, $\Omega_{\Lambda,0} = 0.742 \pm 0.030$ and $h = 0.72 \pm 0.03$. A flat universe with a cosmological constant has both more time and more volume at high redshifts than a flat $\Lambda = 0$ universe. Both these effects increased the expected numbers of galaxies at the faint end of the B-band number counts. This resolved most of the faint blue galaxies problem, but perhaps it's a pity that optical galaxy number counts are no longer demanding such strong changes to our picture of galaxy formation and evolution. (The number counts of galaxies at submm wavelengths nevertheless turned out to be an unexpectedly strong constraint, as we'll see.)

Although this chapter is mainly about optical astronomy, it's worth just adding that radio source counts were found as early as 1959 to be steeper than an $S^{-5/2}$ power law. This was one of the first pieces of evidence that we don't live in a flat, steady-state universe, and it predated the discovery of the CMB.

4.2 Cold dark matter and structure formation

Why is dark matter sometimes described as 'cold'?

The key to this is the time when the dark matter particles decouple, i.e. the time when interactions cease with other matter. If the dark matter was relativistic at decoupling, the mean energy of a particle would be $\langle E \rangle = 2.7kT$ for bosons, or $3.1kT$ for fermions. Also, the average momentum $\langle |p| \rangle$ will satisfy $\langle E \rangle = \langle |p| \rangle\, c$. At later times, the particle's momentum will be reduced by redshifting (the de Broglie wavelengths will redshift like that of a photon) until the motion is non-relativistic and $\langle |p| \rangle = m \, \langle |v| \rangle$. However, the shape of the momentum distribution will still be the same, so we still apply $\langle E \rangle = \langle |p| \rangle\, c = m\, |v|\, c$. A neutrino background would have a present-day temperature of $1.95\,\mathrm{K}$ (this is cooler than the CMB temperature because of the additional later contribution from e^+e^- annihilation to the photons), which leads to $|v| = 159\,(mc^2/\mathrm{eV})^{-1}\,\mathrm{km\,s^{-1}}$ for the present-day velocities of a hypothetical neutrino of mass m. However, if the dark matter particles decoupled when non-relativistic, they can be arbitrarily slow. These relative speeds lead to the terms 'hot' and 'cold' dark matter (**HDM** and **CDM**, respectively), or compromises such as 'warm' or 'mixed' dark matter.

Several effects bend the matter power spectrum over (see Figure 4.1) from the nearly scale-invariant fluctuations imprinted by inflation. For example, if a matter perturbation Fourier mode enters the horizon (i.e. when the horizon becomes bigger than the scale of the Fourier mode) before the epoch of matter–radiation equality, then this Fourier mode can't grow through gravitational collapse, because the dominant energy of radiation drives the Universe to expand so fast that matter has no time to self-gravitate, i.e. the Fourier mode is 'frozen' at a constant value. This is known as the *Mészáros effect*. Also, if the dark matter particles are fast-moving as in HDM, their free-streaming movement can erase small-scale structure in the early Universe, until the particles become non-relativistic. However, HDM models do not reproduce the observed large-scale galaxy power spectrum. In fact, the upper limit on the cosmological neutrino density (Equation 3.1) comes not from particle physics constraints but from the evolution of large-scale structure in the Universe.

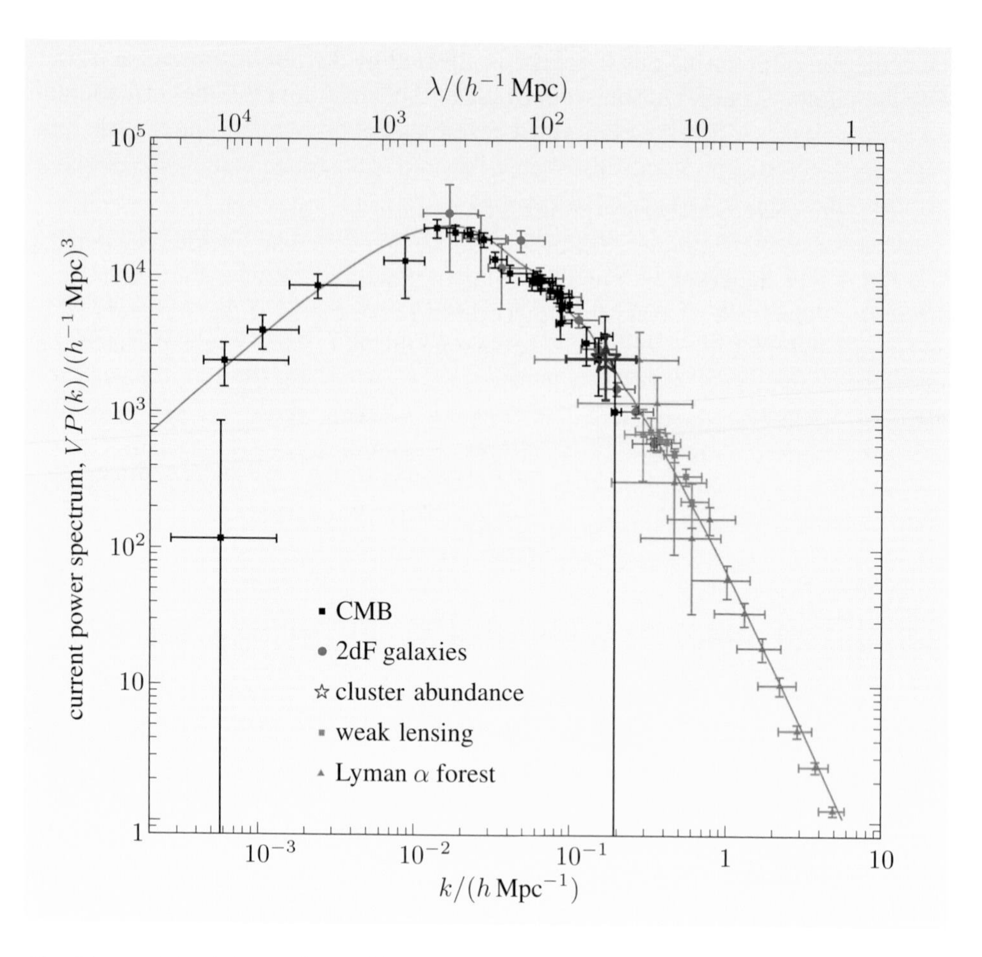

Figure 4.1 Matter power spectrum from the largest to the smallest scales, derived from multiple methods. This figure pre-dates WMAP. The solid red curve is a model based on CDM in the concordance cosmology.

The **hierarchical formation** of large-scale structure in the Universe is shown schematically in Figure 4.2. Small virialized dark matter haloes merge to form larger dark matter haloes. This is known as **bottom-up** structure formation (as opposed to the top-down formation that results in HDM from the erasing of small-scale structures). Note, however, that what happens to the baryons (e.g. galaxies, stars, gas) is another matter altogether. In the early Universe,

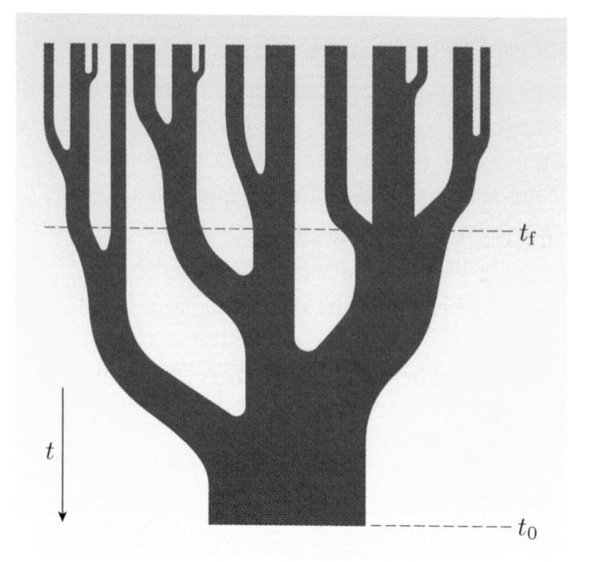

Figure 4.2 Schematic representation of the hierarchical growth of dark matter haloes. Time increases downwards in this figure. Haloes at a nominal formation time t_{f} in the figure merge and form a single halo in the present Universe at time t_0.

the evolution of perturbations in matter is affected by the interaction of baryons with the photons (via Thomson scattering). In the later Universe, the formation of stars and galaxies is a very complex process, as we explore later in this book.

The evolution of density perturbations can be very complicated mathematically, if you take into account the spacetime curvature perturbing the Robertson–Walker metric. It's relatively easy to derive the collapse of a spherical overdensity in a flat $\Lambda = 0$ universe, however, assuming that the motion is non-relativistic. First, by Birkhoff's theorem (that you can neglect the external matter if the system is spherically-symmetric — we shall demonstrate this in Chapter 6) the perturbation must behave in exactly the same way as an identical region in a homogeneous closed universe, which we'll calculate in the next exercise.

Exercise 4.1 Verify that in a $\Lambda = 0$ universe with $k = +1$, the Friedmann equation with $\mathrm{d}R/\mathrm{d}t$ (Equation 1.7) can be written as

$$\left(\frac{\mathrm{d}R}{\mathrm{d}\theta}\right)^2 = R_{\max} R - R^2, \tag{4.1}$$

where $R_{\max}$ is the radius where $\mathrm{d}R/\mathrm{d}t = 0$, and we've introduced a new parameter θ defined with $\mathrm{d}t/\mathrm{d}\theta = R/c$.

Also verify that the solution satisfies the cycloid equations

$$R(\theta) = \frac{R_{\max}}{2}(1 - \cos\theta), \tag{4.2}$$

$$t(\theta) = \frac{R_{\max}}{2c}(\theta - \sin\theta). \tag{4.3}$$

■

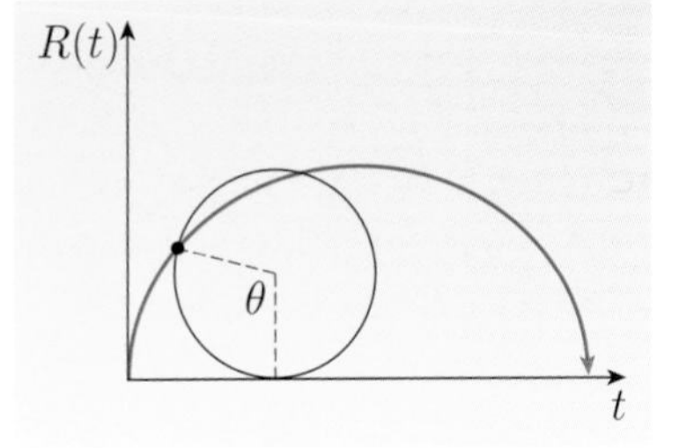

Figure 4.3 Schematic representation of the cycloid solution for a $k = +1$, $\Lambda = 0$ universe.

Incidentally, you may be curious why this is called the 'cycloid' solution. It's because Equations 4.2 and 4.3 describe the path swept out by an object on a wheel, known as a cycloid, shown in Figure 4.3.

In the cosmological case, $R_{\max} = 8\pi G\rho_0 R_0^3/(3c^2)$. In the case of a spherical overdensity in a flat $\Lambda = 0$ universe, the same equations apply but R this time stands for the radius of the overdensity. The mass of the overdensity is

$$M = \tfrac{4}{3}\pi R^3 \rho, \tag{4.4}$$

and since mass is conserved we can use the value at time t_0, which is $M = \frac{4}{3}\pi R_0^3 \rho_0$, so we can write

$$R_{\max} = \frac{2GM}{c^2}. \tag{4.5}$$

(This is curiously similar to the Schwarzschild radius, but remember that we're working in the Robertson–Walker metric, not the Schwarzschild metric.) In the cosmological case, the universe begins at $\theta = 0$, reaches a maximum size at $\theta = \pi$, and ends at $\theta = 2\pi$, while in the overdensity the moment $\theta = \pi$ is called the **turn-around time** and $\theta = 2\pi$ is the time when it would collapse to a point. In practice, we suppose that the clump would become virialized before then.

Rearranging Equations 4.2 and 4.3 to find $R(t)$ is not possible analytically, but we can find the early behaviour of the fractional overdensity δ using Taylor series. Combining Equations 4.4 and 4.5 gives us the density of the spherical clump:

$$\rho = \frac{3M}{4\pi R^3} = \frac{3R_{\max}\, c^2}{8\pi G\, R^3(\theta)}. \tag{4.6}$$

The density of the surrounding universe can be found by using the results of Exercises 1.4 and 1.5 in Chapter 1, and applying $\Omega_m = 1$ (Equation 1.15). The density of this Einstein–de Sitter universe comes out as

$$\rho_{\rm EdS} = \frac{1}{6\pi G t^2}. \qquad (4.7)$$

Taking the ratio of these two densities gives

$$\begin{aligned}\frac{\rho}{\rho_{\rm EdS}} &= \frac{3R_{\rm max}\,c^2}{8\pi G\,R^3(\theta)} 6\pi G\,t^2(\theta) = \frac{9}{4}R_{\rm max}\,c^2\frac{t^2(\theta)}{R^3(\theta)} \\ &= \frac{9}{4}R_{\rm max}\,c^2\frac{R_{\rm max}^2}{4c^2}\frac{8}{R_{\rm max}^3}\frac{(\theta-\sin\theta)^2}{(1-\cos\theta)^3} \\ &= \frac{9}{2}\frac{(\theta-\sin\theta)^2}{(1-\cos\theta)^3}.\end{aligned}$$

The function of θ is singular at $\theta = 0$, but by writing a Taylor expansion around a point arbitrarily close to $\theta = 0$, it can be shown that

$$\frac{(\theta-\sin\theta)^2}{(1-\cos\theta)^3} \simeq \frac{2}{9} + \frac{1}{30}\theta^2 + \cdots$$

and therefore

$$\frac{\rho}{\rho_{\rm EdS}} \simeq \frac{9}{2}\left(\frac{2}{9} + \frac{1}{30}\theta^2 + \cdots\right) = 1 + \frac{3}{20}\theta^2 + \cdots.$$

This is still in terms of θ, but if we now Taylor-expand Equation 4.3, we find that

$$t(\theta) \simeq \frac{R_{\rm max}}{2c}\frac{\theta^3}{6} + \cdots$$

and so

$$\frac{\rho}{\rho_{\rm EdS}} \simeq 1 + \frac{3}{20}\left(\frac{12ct}{R_{\rm max}}\right)^{2/3}.$$

We can therefore identify the fractional overdensity as

$$\delta \simeq \frac{3}{20}\left(\frac{12ct}{R_{\rm max}}\right)^{2/3}. \qquad (4.8)$$

Despite the high orders of θ that we've used, this is known as the **linear theory** approximation (we'll see why in Exercise 4.2). In this approximation, density perturbations grow as $t^{2/3}$, i.e. proportional to the dimensionless scale factor a. However, by the time the overdensity reaches the turn-around radius $R_{\rm max}$, linear theory has broken down. This is at $\theta = \pi$, i.e. $t = \pi R_{\rm max}/(2c)$ (Equation 4.3). The overdensity will then have a turn-around density of exactly $\rho = 3M/(4\pi R_{\rm max}^3)$. Using Equations 4.5 and 4.7 we find that the density ratio is

$$\begin{aligned}\frac{\rho}{\rho_{\rm EdS}} &= \frac{3M}{4\pi R_{\rm max}^3} 6\pi G t^2 = \frac{9}{2}\frac{MG}{R_{\rm max}^3}\left(\frac{\pi R_{\rm max}}{2c}\right)^2 = \frac{9\pi^2}{8c^2}\frac{MG}{R_{\rm max}} \\ &= \frac{9}{16}\pi^2 \simeq 5.552,\end{aligned}$$

so the overdensity is $\delta = 5.552 - 1 = 4.552$. Linear theory would predict $\delta = (3/20)(6\pi)^{2/3} \simeq 1.062$ at this point (Equation 4.8).

If the cycloid solution applied at $\theta = 2\pi$, then the overdensity would collapse to a point of infinite density. The linear theory prediction (now infinitely wrong) is $\delta = \frac{3}{20}(12\pi)^{2/3} \simeq 1.686$ at this point. However, it's more likely that the overdensity will stabilize at the virial equilibrium, where the potential energy equals -2 times the kinetic energy. It's usually assumed that virialization happens around the time $\theta = 2\pi$. It's not too hard to calculate the final virialized density. Total energy is conserved, and since there's no kinetic energy at turn-around, the total energy is just the binding energy at turn-around ($E \propto 1/R_{\text{max}}$, Equation 3.6). But since the virial binding energy E_{v} must be related to the virial kinetic energy $E_{\text{K,v}}$ by $E_{\text{v}} + 2E_{\text{K,v}} = 0$, and the total energy is $E = E_{\text{v}} + E_{\text{K,v}}$, we must have $E = E_{\text{v}}/2$. This means that the virialized radius must be half the turn-around radius (Equation 3.6), i.e. the virialized density is eight times the turn-around density. The collapse time is at $\theta = 2\pi$, which is twice the turn-around time (proved either by symmetry in Figure 4.3 or using Equation 4.3), so the density of the surrounding Universe has gone down by a factor of 4 (Equation 4.7). The overdensity is therefore $8 \times 4 = 32$ times bigger than the turn-around overdensity, i.e. about 146. This collapse also sheds light on the fingers of God. Figure 4.4 shows how the collapse of a spherical overdensity would appear in redshift space.

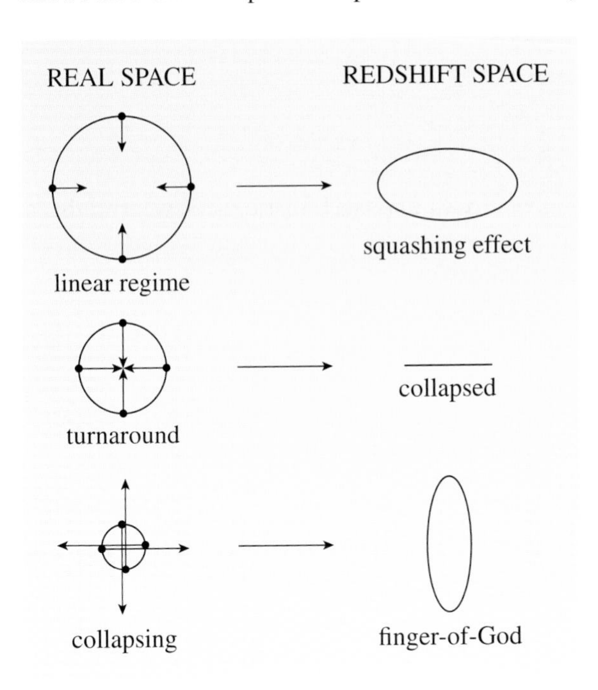

Figure 4.4 A schematic view of how a collapsing overdensity appears in real space and in redshift space.

Of course, you might quite reasonably object to this analysis, because real overdensities are much more likely to be highly aspherical. The growth of these density perturbations can be modelled with numerical ***N*-body simulations**. However, the spherical collapse model contains more elements of the underlying physics than you might expect, as we'll see. We've also assumed an $\Omega = 1$ universe — but what happens more generally?

In general, the growth of dark matter perturbations in the Universe is governed by fluid dynamics equations. Taking the linear approximation of these equations gives the relation

$$\ddot{\delta} + 2\frac{\dot{a}}{a}\dot{\delta} = 4\pi G \rho_{\text{m}} \delta, \tag{4.9}$$

Deriving Equation 4.9 would take us too far off-topic, and introduce mathematical machinery that we won't use elsewhere, but if you're not satisfied with having this used without proof, you can find proofs in Chapter 15 of Peacock, 1999, or Chapter 11 of Coles and Lucchin, 1995, in the *Further reading* section below.

where δ is the shorthand for the matter overdensity $(\delta\rho)/\rho_0$, and $a = R/R_0$ is the dimensionless scale factor (Chapter 1).

Exercise 4.2 Show using Equation 4.9 that in a flat matter-dominated universe with no cosmological constant, matter perturbations grow following the power law $\delta(t) \propto t^{2/3}$, i.e. $\delta(t)$ is just proportional to the scale factor a. (You may assume that $H(t) = 2/(3t)$ in this universe.) ■

See Peacock, 1999, in the *Further reading* section.

The solution for general cosmologies can be expressed as

$$\frac{\delta(z=0,\Omega_{\mathrm{m},0},\Omega_{\Lambda,0})}{\delta(z=0,\Omega_{\mathrm{m},0}=0,\Omega_{\Lambda,0}=0)} \simeq \tfrac{5}{2}\Omega_{\mathrm{m},0}\left[\Omega_{\mathrm{m},0}^{4/7} - \Omega_{\Lambda,0} + \left(1+\tfrac{1}{2}\Omega_{\mathrm{m},0}\right)\left(1+\tfrac{1}{70}\Omega_{\Lambda,0}\right)\right]^{-1}. \tag{4.10}$$

A reasonable approximation to this equation for a flat universe is $\Omega_{\mathrm{m},0}^{0.65}$, while for a $\Lambda = 0$ universe a good approximation is $\Omega_{\mathrm{m},0}^{0.23}$. So, for fixed initial conditions, open universes or $\Lambda > 0$ universes have weaker large-scale structure δ than the flat $\Lambda = 0$ case. This is because the increased expansion rate at late epochs suppresses the growth of large-scale structure. The suppression is weaker in the flat $\Lambda > 0$ case because the Λ term becomes dynamically relevant only at later epochs. A related analysis to the proof of Equation 4.9 shows that density perturbations gravitationally induce peculiar velocities given approximately by $\delta v \simeq H_0 r\, \Omega_{\mathrm{m},0}^{0.6}\,(\delta\rho)/\rho$ (Section 3.11).

In the linear theory, we have that $\delta \propto t^{2/3}$ in a flat matter-dominated $\Lambda = 0$ universe, so that $P(k) \propto t^{4/3}$ in this case. In the linear regime, each Fourier mode evolves independently. In general this is expressed as the **transfer function** $T(k)$, which depends on the Fourier mode k: $\Delta^2(k, z = 0) = T^2(k)\, f^2(a)\, \Delta^2(k, z)$, where $f(a)$ is the linear growth factor which we found is $t^{2/3}$ in a matter-dominated $\Lambda = 0$ universe. The calculation of the transfer function is in general quite complicated and includes the Mészáros effect, Silk damping and other related effects in the early Universe. Finally, though, the (linear) matter power spectrum can be expressed as $\Delta^2(k) \propto k^{3+n_s}\, T^2(k)$. How do we specify the constant of proportionality? The galaxy spatial correlation function is about $\xi(r) \simeq 1$ on a comoving scale of $r = 8h^{-1}$ Mpc. However, galaxy clustering may be biased, so it's conventional to **convolve** (see the box below) the matter density distribution with a spherical kernel with $8h^{-1}$ Mpc, and use the standard deviation of the resulting distribution known as σ_8 to define the constant of proportionality in the galaxy power spectrum. (A further complication is that σ_8 is usually measured from the filtered *linear-theory* density field.) We could then use the observed value of σ_8 as an equivalent definition of the galaxy bias ($b = \sigma_8(\mathrm{galaxies})/\sigma_8(\mathrm{mass})$), provided that the bias isn't scale-dependent.

Convolution

If you've ever experimented with image software like Photoshop, you'll know that it's possible to smooth or blur an image. How does this work? Well, at any particular point in the image, the software will consider a little box around that point and take the average value in that little box. Perhaps all pixels in that little box will be given the same weight, or perhaps it'll be a weighted average with more weight given to pixels in the middle of the little box than the edge. The software repeats this little-box-averaging around all the positions in the image, and the result is a blurred or smoothed image.

It turns out that Fourier analysis has a very simple and beautiful way of expressing this. Let's suppose that the pixels are small enough that we can consider an integral instead of a sum. In one dimension, the blurring could be expressed as $B(x) = \int_{-\infty}^{\infty} I(x)\, P(x+\alpha)\, \mathrm{d}\alpha$, where $I(x)$ is the un-blurred data, P is the weighting function (e.g. 1 in a range $-L$ to $+L$, representing the little box size, and 0 otherwise), and $B(x)$ is the blurred data. The weighting function P is sometimes called the **kernel**. In two or more dimensions, we'd change x into a vector $\boldsymbol{x}$ and α into a vector $\boldsymbol{\alpha}$, but otherwise it's the same. We call this the **convolution** of I with P, and it's sometimes written as $B = I \otimes P$. Now, it turns out that the Fourier transforms of B, I and P, which we can write as $\widetilde{B}$, $\widetilde{I}$ and $\widetilde{P}$, are related by $\widetilde{B} = \widetilde{I} \times \widetilde{P}$. In other words, a convolution in real space can be expressed as just a multiplication in Fourier space. The reverse is true too: a convolution in Fourier space is the same as a multiplication in real space.

An example is the treble control on a stereo. High musical notes have high frequencies and short wavelengths, and since the Fourier wave number k is proportional to the reciprocal of wavelength, the high musical notes are in the high Fourier k modes. Now suppose that we don't want the treble (high) notes to be so loud, so we rig up some electronics that suppresses these high k modes, with (say) an $\exp(-\frac{1}{2}(k/k_0)^2)$ factor, where k_0 is some constant. We picked this factor because it's a Gaussian, and it turns out that the Fourier transform of a Gaussian is another Gaussian. Multiplying the Fourier transform of our sound with a Gaussian to suppress the high treble notes is therefore the same as convolving our sound with another Gaussian. So, when you're turning down the treble control on a stereo, you're smoothing out the sound waves coming out of the speaker, just like blurring an image in Photoshop by convolving it with a Gaussian.

The number density of dark matter haloes per unit mass, known as the **mass function**, can be predicted by assuming that the density fluctuations are Gaussian. First we convolve the density field with a spherical kernel with radius R, which corresponds to a mass scale

$$M = \tfrac{4}{3}\pi R^3 \rho_0. \tag{4.11}$$

The probability that a random point has an overdensity $\delta_{\text{convolved}}$ is just

$$\Pr(\delta_{\text{convolved}})\, \mathrm{d}\delta_{\text{convolved}} = \frac{1}{\sqrt{2\pi\,\sigma^2_{\text{convolved}}}} \exp\left(\frac{-\delta^2_{\text{convolved}}}{2\sigma^2_{\text{convolved}}}\right) \mathrm{d}\delta_{\text{convolved}}, \tag{4.12}$$

i.e. just a Gaussian distribution. It turns out that the variance $\sigma^2_{\text{convolved}}$ can be calculated from the power spectrum $P(k) \propto k^{n_s}$ (Equation 2.44): $\sigma_{\text{convolved}} \propto M^{-(3+n_s)/6}$. In particular, the probability that a region is above the

See, for example, Coles and Lucchin, 1995, in the *Further reading* section.

overdensity for virialized collapse $\delta_c = 1.686$ is just

$$\Pr(>\delta_c, M) = \int_{\delta_c}^{\infty} \Pr(\delta_{\text{convolved}})\,\mathrm{d}\delta_{\text{convolved}}. \tag{4.13}$$

We've written this as a function of M because the smoothing scale R is equivalent to a mass scale M (Equation 4.11). This isn't quite enough to calculate the mass function, because we need the clump to be isolated, i.e. surrounded by a less dense region, so we need to subtract the probability $\Pr(>\delta_c, M + \mathrm{d}M)$. Also, there is the **cloud-in-cloud problem**: could an object that's a virialized clump on one mass scale M also be later contained in another larger clump on a bigger mass scale? Moreover, doesn't this blob counting miss all the mass in the underdense regions? Perhaps we should multiply this probability by a factor of two, since we're missing the half of the matter that's in underdense regions? A careful but lengthy analysis of the cloud-in-cloud problem has shown that such a factor of two correctly solves both problems, so the mass function is

$$\begin{aligned} n(M)\,M\,\mathrm{d}M &= 2\rho_0 \left\{\Pr(>\delta_c, M) - \Pr(>\delta_c, M + \mathrm{d}M)\right\} \\ &= 2\rho_0 \left|\frac{\mathrm{d}\Pr(\delta_c)}{\mathrm{d}\sigma_{\text{convolved}}}\right| \left|\frac{\mathrm{d}\sigma_{\text{convolved}}}{\mathrm{d}M}\right| \mathrm{d}M, \end{aligned}$$

thus

$$n(M) \propto \left(\frac{M}{M_*}\right)^{-\alpha-2} \exp\left(-\left(\frac{M}{M_*}\right)^{2\alpha}\right), \tag{4.14}$$

where we've defined $\sigma_{\text{convolved}} \propto M^{-\alpha}$ and M_* is a constant that depends on δ_c, α and the normalization of the $\sigma_{\text{convolved}}$–$M$ relationship. This is known as the **Press–Schechter model**, and is a surprisingly good fit to the results of N-body simulations (an agreement that is not entirely understood).

The evolution of dark matter can be approximated by the Press–Schechter approach and tracked more accurately with N-body simulations, but the evolution of baryons is much more complicated because the physics is not purely gravitational, but includes gas cooling and heating from photons and shocks, energy and momentum input from winds, and multiple phases in the interstellar medium. Understanding the formation of galaxies and the stars within them is one of the key goals in observational cosmology. Much recent work has been done on combining N-body simulations with simplified numerical models of the chemical/dust/stellar content evolution of particular galaxies or classes of galaxy. These so-called **semi-analytic models** have had many successes in reproducing observed galaxy properties but currently still contain many adjustable parameters that need to be tuned to match observations. There is, as yet, no consensus between the semi-analytic interpretations. Different semi-analytic models make different assumptions about, for example, the number of stars forming per unit stellar mass (the **initial mass function**) and the feedback effects of energy and momentum input from black hole accretion and star formation.

We also shouldn't assume that there's a one-to-one correspondence between dark matter haloes and galaxies, and the **halo occupation distribution** describes the probability distribution of the number of galaxies that a halo contains. This distribution is one of the key predictions of models of galaxy evolution. It can be constrained by the shape and amplitude of the spatial or angular correlation function on *small* scales (e.g. less than one comoving Mpc), where this function reflects the way that galaxies populate the haloes.

Figure 4.5 shows why baryons must behave differently to dark matter: if they didn't, galaxies would have far more satellites, just as galaxy clusters have many satellite clumps containing galaxies. Recent discoveries of Milky Way satellites have alleviated this problem (Section 3.11), but there is still a factor of $\sim\times 4$ deficit at some mass ranges. It's possible to explain this deficit if the star formation in the lowest-mass haloes is suppressed, which numerical simulations of feedback (Chapters 5 and 6) suggest may be the case. If so, our Milky Way galaxy is surrounded by puddles of 'failed dwarf galaxies' that have yet to undergo their first star formation. Another dark matter puzzle that baryonic physics might solve is the lack of sharp density cusps at the centres of dark matter haloes: N-body simulations predict that haloes should have a Navarro–Frenk–White profile — which we shall meet in detail in Chapter 7, and which predicts a density profile varying as $\rho(r) \propto 1/r$ at the centre. Local galaxy observations suggest otherwise, perhaps because winds from supernovae or black hole accretion drive out matter and smooth out the density profile.

Figure 4.5 The dark matter density within a galaxy cluster halo with a mass $5 \times 10^{14}\,M_\odot$ (top), compared to a galaxy halo with a mass $2 \times 10^{12}\,M_\odot$ (bottom). Note the similar number of neighbouring objects. So why do galaxies not look like mini galaxy clusters?

4.3 Population synthesis

We saw in Section 4.1 that having evolution in galaxy luminosities helped to resolve the faint blue galaxies problem. Can we predict how galaxies evolve from first principles? Most of the optical light from galaxies comes from stars. There's a very well-developed theory for the evolution of stars along the main sequence, along the giant branch and (for more massive stars) to supergiants and ultimately to white dwarfs, neutron stars or black holes. If we make some assumption about how the galaxies started, we could derive how the colours and luminosities of galaxies evolve.

This technique is known as **population synthesis** (or sometimes 'pop synth', colloquially). Figure 4.6 shows synthetic spectra of a galaxy evolving from a burst of star formation that lasted 1 Gyr. There are several assumptions that one needs to make, such as: whether star formation is ongoing or whether it occurred in a single burst; the initial distribution of stellar masses, known as the **initial mass function** (IMF); and what effect dust has on the spectra. This astronomical dust is quite different from household detritus: it's created in the winds of red giants and in supernova explosions, and is composed of graphites and/or silicates. The grain sizes of astronomical dust vary from sub-micron sizes to clusters of just a few atoms. Large interstellar grains can penetrate the Solar System despite the solar wind, and interplanetary/interstellar dust grains have been collected in the upper atmosphere (see Figure 4.7). These are also the materials that comprise protoplanetary discs and ultimately form asteroids and planets.

The amount of star formation per unit time is an important input to population synthesis models. Depending on the context, an initial starburst might be assumed, or the star formation history of a galaxy might be taken from predicted merger rates. If there is no star formation, the change in colours purely from stellar evolution is known as **passive stellar evolution**. In general this involves a reddening and dimming of the galaxy.

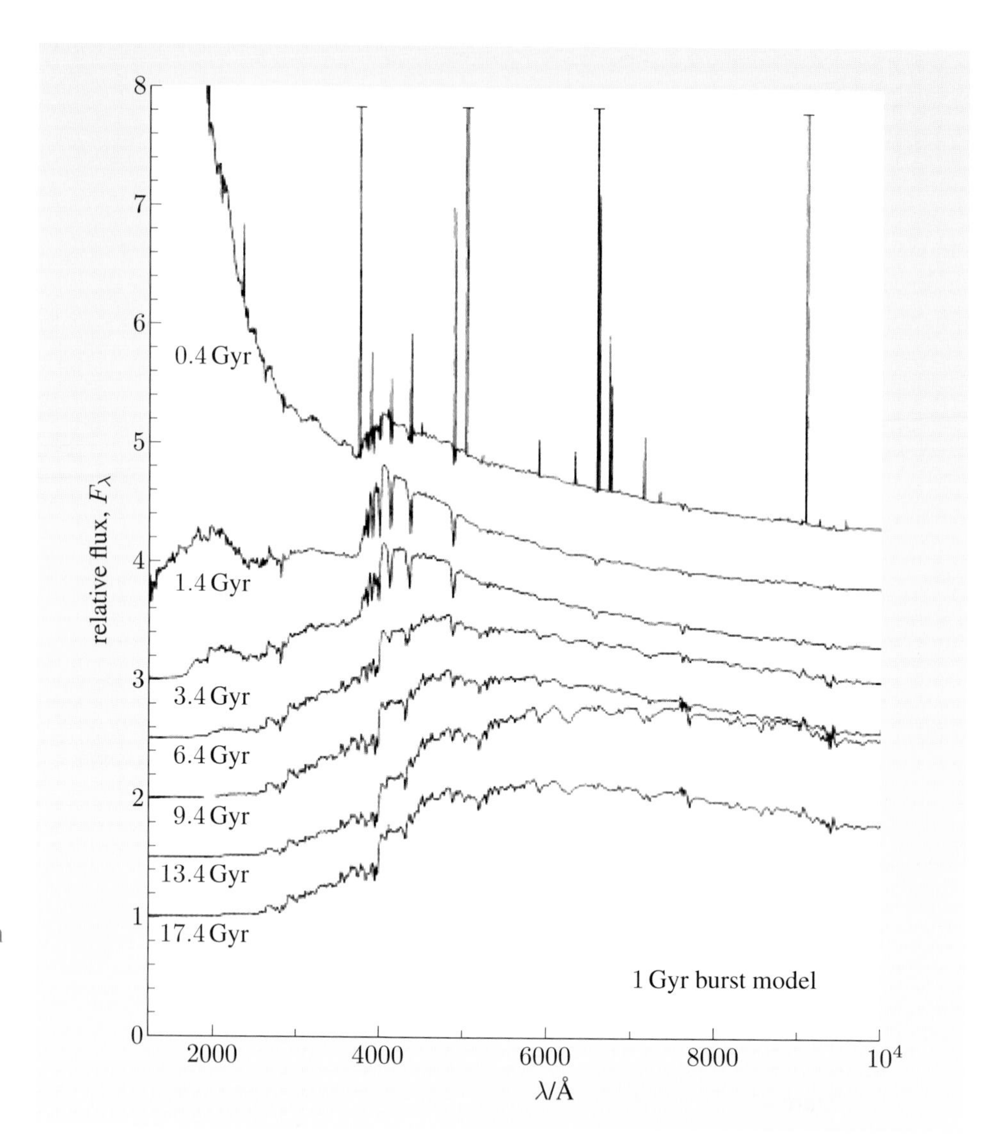

Figure 4.6 Simulated galaxy spectra following a 1 Gyr-long burst of star formation. Ages are marked in Gyr on each spectrum. The spectra have been normalized to 1 and offset vertically for clarity. Some spectra have lines or continua off the top of the figure, and have been truncated for clarity.

The initial mass function is one of the key unknowns in observational astronomy. Often the initial number of stars per unit mass, $\mathrm{d}N/\mathrm{d}m$, is assumed to have a simple form such as $\mathrm{d}N/\mathrm{d}m \propto m^{-2.35}$ over a range 0.1–100 $\mathrm{M}_\odot$ or so, which matches inferences from present-day mass distributions of Galactic stars. However, it's not at all clear whether the IMF varies from place to place within a galaxy, or between galaxies. We'll see in Chapter 5 that when star formation rates of galaxies are estimated, the estimates all depend on light generated ultimately by the most massive stars. In order to find the total number of stars being formed, we need to extrapolate from these largest stars to the smallest, which is very sensitive to the shape of the IMF.

Dust reddening is another unknown. Dust preferentially absorbs blue optical light compared to red. This changes the (B–V) colour of a background star by an amount symbolized as E(B–V). We can write the observed value of (B–V) as $(\mathrm{B–V})_{\mathrm{obs}}$, and it's related to the true value $(\mathrm{B–V})_{\mathrm{true}}$ by

$$(\mathrm{B–V})_{\mathrm{obs}} = (\mathrm{B–V})_{\mathrm{true}} + E(\mathrm{B–V}). \tag{4.15}$$

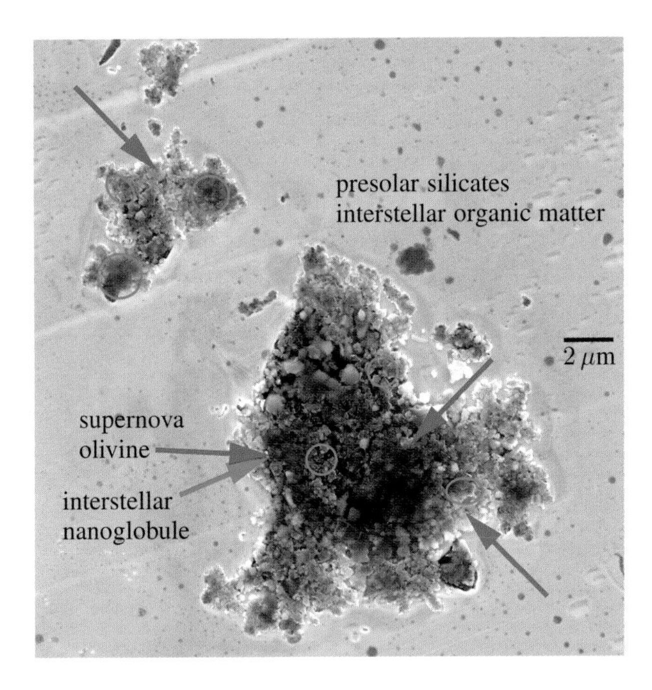

Figure 4.7 A large interplanetary dust grain collected from the upper atmosphere of the Earth by a NASA aircraft in 2003, during the Earth's passage through the dust stream from comet 26P/Grigg-Skjellerup. Parts of the dust grain appear to be pre-solar, and other parts originated in the interstellar medium.

Exercise 4.3 Which is redder: (B–V) = 0 or (B–V) = 1? Remember that (B–V) is the B-magnitude minus the V-magnitude, and magnitudes m are related to fluxes S by $m = -2.5 \log_{10} S + \text{constant}$. ■

If we imagine a dust screen in front of a star, the amount of V-band attenuation in magnitudes A_V is related to the $E(\mathrm{B–V})$ reddening by

$$A_\mathrm{V} = R_\mathrm{V}\, E(\mathrm{B–V}), \tag{4.16}$$

where R_V is a constant that depends on the composition of the dust. For typical Galactic dust, $R_\mathrm{V} = 3.1$, though it can range from 2.75 to 5.3. Different dust compositions also change how absorbent the dust is at other wavelengths, as shown in Figure 4.8. The LMC and SMC appear to be less enriched with heavy elements than our Galaxy, and the dust in the LMC and SMC appears to have different extinction properties. The differences in absorption at 217.5 nm are also about dust composition; this feature is probably caused by small graphite grains. We'll revisit dust composition in Chapter 5.

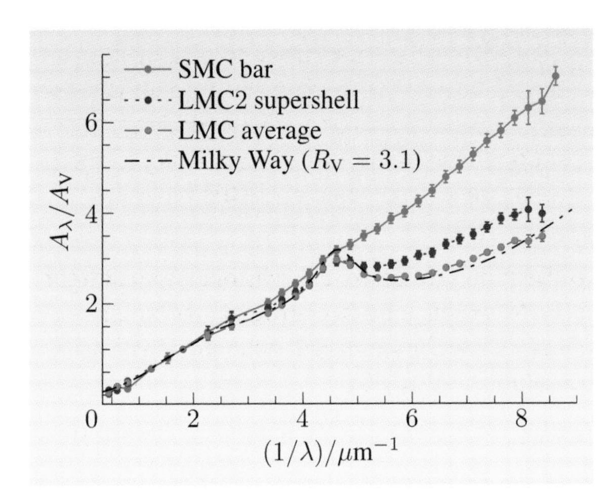

Figure 4.8 The relative extinction as a function of wavelength, for dust found in the LMC in two locations, in the SMC and in our Galaxy.

One way of measuring the amount of dust attenuation is the **Balmer decrement**. The hydrogen Balmer lines Hα (656.3 nm) and Hβ (486.1 nm) are emitted by gas that's ionized by hot stars, with a characteristic ratio that's calculable from atomic physics. We'll meet these lines again in Chapters 5 and 8. If we write the optical depth to Hα photons as $\tau_{\mathrm{H}\alpha}$, the Hα fluxes are by definition attenuated by $\mathrm{e}^{-\tau_{\mathrm{H}\alpha}}$. The optical depth turns out to be related to A_{V} by $\tau_{\mathrm{H}\alpha} \simeq 0.7A_{\mathrm{V}}$. (It shouldn't be a surprise that it's a linear relation, since magnitudes are a logarithmic system and the optical depth appears in an exponential, $\mathrm{e}^{-\tau_{\mathrm{H}\alpha}}$.) However, as we see in Figure 4.8, the extinction and optical depth to Hβ will be greater. For Galactic dust, $\tau_{\mathrm{H}\beta} \simeq 1.45\,\tau_{\mathrm{H}\alpha}$. If we compare the observed flux ratio of Hα and Hβ, $S_{\mathrm{H}\alpha}/S_{\mathrm{H}\beta}$, and compare the predicted ratio of 2.8 from atomic physics, we can infer $\tau_{\mathrm{H}\alpha}$ and hence A_{V}. The extinction is also empirically related to the gas column density (which we shall meet in subsequent chapters) through $N_{\mathrm{H}}/E(\mathrm{B{-}V}) = 4.93 \times 10^{21}$ atoms cm^{-2} mag^{-1}.

This assumes that the gas is optically thick to the Lyman continuum — see, for example, Osterbrock, D.E. and Ferland, G.J., 2005, *Astrophysics of Gaseous Nebulae and Active Galactic Nuclei*, University Science Books.

See, for example, Diplas, A. and Savage, B.D., 1994, *Astrophysical Journal*, **427**, 274.

It's not just the type of dust that affects the colours, it's the location of the dust, as the following (optional) exercise shows.

Exercise 4.4 The previous discussion of Balmer decrements assumed that the dust is placed in a screen in front of the emission-line-emitting gas. Suppose instead that the dust is evenly mixed in with this gas. Calculate the Balmer decrement $S_{\mathrm{H}\alpha}/S_{\mathrm{H}\beta}$ that you'd measure in this situation. Now suppose that you observed this but *wrongly* assumed that the dust was in a screen in front of the stars. What A_{V} would you wrongly infer? (This is a more difficult and slightly more open-ended exercise than most in this book, so for quantities not supplied in the question, improvise!) ■

Another way of looking at this exercise is that the optical depth of extinction to a given position depends on the wavelength, so Hα photons would have a lower optical depth than Hβ photons or ultraviolet photons. However, the light that we receive will always be dominated by the regions with optical depths of $\tau < 1$, so the observed light at shorter and shorter wavelengths will be dominated by regions with lower and lower extinctions. The modern approach in population synthesis is to include assumptions or predictions for the dust location, density variation and composition.

4.4 Photometric and spectroscopic redshifts

All these effects complicate the spectra of galaxies, but this is also an advantage because there is a lot of useful information in optical spectra. For example, emission lines imply the presence of ionizing radiation from young, hot O stars and B stars, or from an active nucleus, which we shall meet in Section 4.6. Blue spectra also suggest the presence of young hot stars (Figure 4.6), while red spectra can be the signature of old stellar populations (since the blue O and B stars are short-lived). Elliptical galaxies and spiral bulges tend to have spectra consistent with old stellar populations, while the bluer spectra of spiral discs are attributable to more recent star formation.

The emission and absorption lines in Figure 4.6 are used to measure the redshift z of galaxies, since each line will be shifted in wavelength by a factor $(1 + z)$. We need to know which emission line we're observing in order to know the emitted wavelength $\lambda_{\rm em}$, and hence infer the redshift from the observed wavelength $\lambda_{\rm obs}$ and $1 + z = \lambda_{\rm obs}/\lambda_{\rm em}$ (see Equation 1.10). But how can we find which emission line is which? Often in practice certain features are characteristic enough to be instantly recognizable, but this is not always the case.

One way is to use the wavelength ratios. Suppose that we have two emission lines observed at wavelengths $\lambda_{\rm obs,1}$ and $\lambda_{\rm obs,2}$. If they're at the same redshift, then

$$\frac{\lambda_{\rm obs,1}}{\lambda_{\rm obs,2}} = \frac{\lambda_{\rm em,1}(1+z)}{\lambda_{\rm em,2}(1+z)} = \frac{\lambda_{\rm em,1}}{\lambda_{\rm em,2}}. \qquad (4.17)$$

There are a limited number of astrophysically-plausible emission lines in galaxies, so the observed wavelength ratios can quickly identify the emitted wavelengths. Even when a galaxy doesn't have emission lines, the absorption lines can be used to find redshifts. Another absorption feature is the 'break' in the spectra in Figure 4.6 at about 400 nm, known as the **4000 Å break**, caused by Balmer continuum absorption in the atmospheres of stars.

Spectroscopy is not always available, because it's harder to get good signal-to-noise ratios in spectra (where you need enough photons in each of hundreds of pixels along the wavelength axis) than in broad-band imaging (where all the photons passing through the filter are directed onto a few pixels in the image). No matter how big your telescope, it's always possible to make images of galaxies too faint to find redshifts from spectra with that telescope. The next best approach is to use the broad-band colours of galaxies to estimate the redshift. If a galaxy has photometric measurements in many filters (e.g. UBVRIJHK — see Section 3.3), this is effectively a spectrum with a very coarse wavelength resolution. Instead of using emission lines, this approach uses the broad shape of the spectrum to estimate the redshift. With good enough photometric measurements in enough filters, it's possible to distinguish redshifting from the reddening from dust or the presence of old stars. The most likely redshift is usually found from the minimum-χ^2 fit to the photometric data using template spectra. Figure 4.9 shows that this works on the whole, but it's hard to avoid a few strong outliers known sometimes as 'catastrophic failures' of the fitting. Reducing these outliers is one of the main challenges in calculating photometric redshifts. The Sloan Digital Sky Survey (SDSS) used a custom-made set of broad-band filters named u, g, r, i, z. The sharper boundaries of the filter curves help with photometric redshifts.

See, for example, Bolzonella, M., Miralles, J.-M. and Pelló, R., 2000, *Astronomy and Astrophysics*, **363**, 476.

As we'll see in Chapter 8, high redshift galaxies can be identified from the fact that intervening neutral clouds cause absorption of Lyman series lines. This means that at observed wavelengths shorter than the redshifted Lyman α line (the $n = 2 \rightarrow 1$ hydrogen transition at an observed wavelength of $121.6(1 + z)$ nm), and particularly those below the redshifted Lyman limit ($91.2(1 + z)$ nm, $n = \infty \rightarrow 1$), the continuum from the high redshift galaxy is strongly suppressed.

This means that $z > 4$ galaxies can be found by searching for galaxies with detections in the B or g filters and longer wavelengths, yet non-detections in the U or u filters. This photometric redshift technique is known as **U-band dropouts** or **Lyman break galaxies**. Searches for galaxies at higher redshifts tend to focus on dropouts at longer wavelengths, though it becomes progressively more difficult because we start running out of cosmological volume, since $\mathrm{d}V/\mathrm{d}z$ tends to zero as z increases (the Hubble volume has a finite comoving size). Follow-up

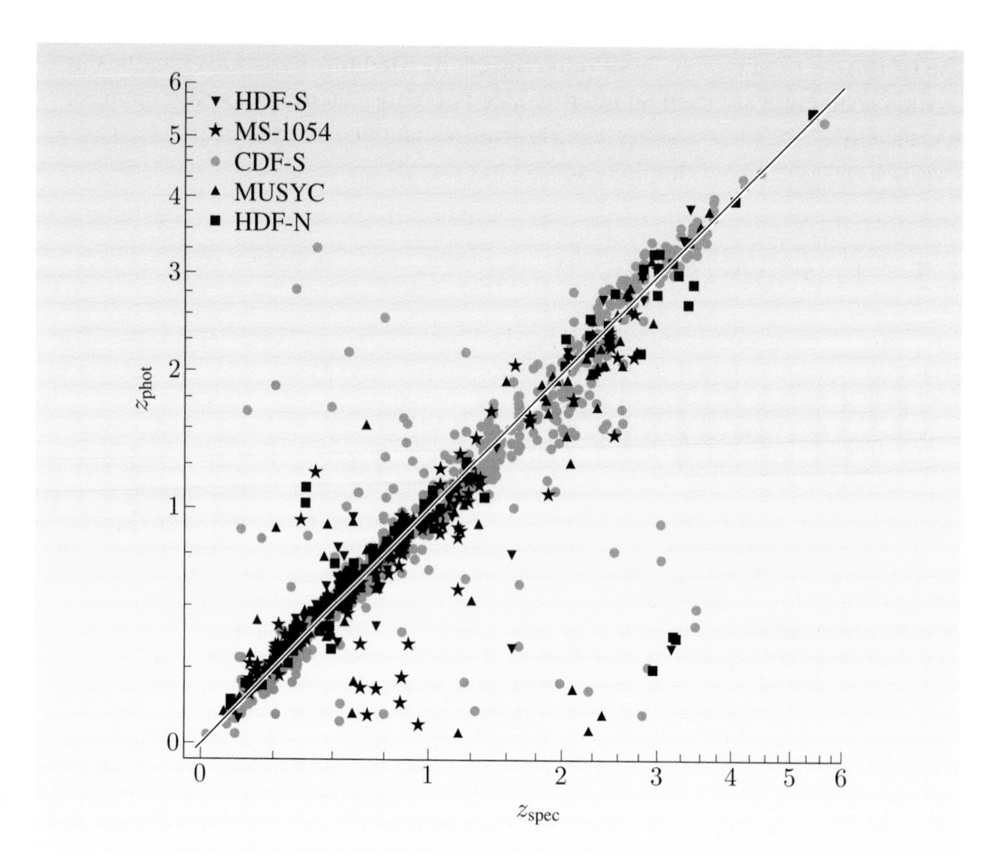

Figure 4.9 A comparison of photometric redshift estimates with spectroscopic redshifts (which have much higher precision). Note that while the technique works for most objects, there are some strongly outlying points known as 'catastrophic' failures. Five survey data sets are used and are marked in different symbols.

spectroscopy often finds a Lyman α emission line (rest wavelength 121.6 nm), as shown in Figure 4.10. Note the weak continuum at λ above Lyman α, and the lack of continuum at λ below. However, the presence of the rest-frame 4000 Å break in lower-redshift galaxies and the nearby [O II] 372.7 nm emission line has resulted in some high-z galaxy claims that failed under closer examination.

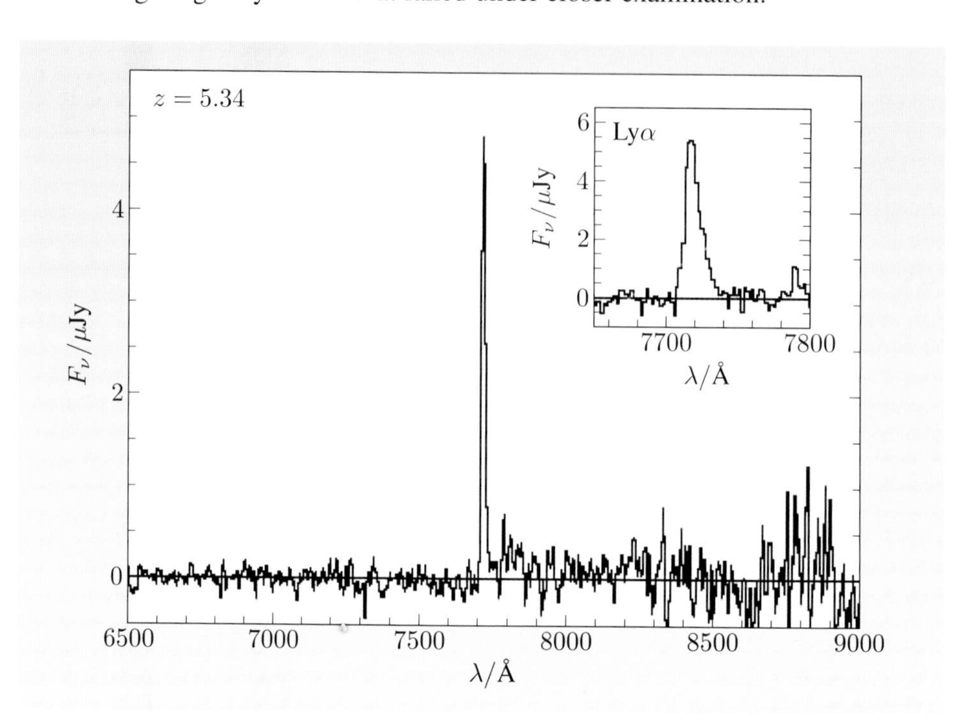

Figure 4.10 The spectrum of a $z = 5.34$ galaxy.

The COMBO-17 survey (Classifying Objects by Medium-Band Observations in 17 filters) ingeniously used an intermediate approach, taking images in many intermediate-width filters (Figure 3.3), from which photometric redshifts accurate to $\delta z/(1+z) \simeq 0.02$ could be found.

4.5 Luminosity functions

What are the most common galaxies in the present-day Universe? If we picked a galaxy at random, what sort of galaxy would it be? The Milky Way is sometimes described as a 'typical' spiral galaxy, so you might think that the answer is Milky-Way-sized galaxies, but they turn out to be enormously outnumbered by dwarf galaxies.

The number of objects per unit volume, per unit luminosity is known as the **luminosity function**, sometimes given the symbol $\phi(L)$. Figure 4.11 shows the SDSS g-band luminosity function of all galaxies, and for early-type and late-type galaxies separately. For higher redshift galaxies it becomes important to account for the expansion of the Universe, and the usual convention is to use comoving volume (Chapter 1) when calculating luminosity functions. The shape is generally quite simple: a shallow slope, which changes to a steeper slope or exponential decline around some characteristic luminosity L_* known as the **break luminosity**, or equivalently at some characteristic magnitude M_*. This is very often parameterized as a **Schechter function**, which fits many galaxy luminosity functions:

$$\phi(L) = \phi_* \left(\frac{L}{L_*}\right)^{-\alpha} \exp\left(\frac{-L}{L_*}\right), \tag{4.18}$$

where the normalization ϕ_*, the faint-end slope α and the break luminosity L_* are free parameters in fitting the luminosity function. This is suggestively similar to the Press–Schechter form above (identical if $\alpha = 1/2$, i.e. a white-noise spectrum), giving some theoretical motivation for this form. However, the underlying assumptions of Press–Schechter are overly simplistic, as we've seen. Also, the observed galaxy luminosity functions have different faint-end slopes to the predicted dark matter halo distributions, implying that there is no simple linear relationship between galaxy luminosity and the mass of its halo. The physics of dark matter halo formation is (relatively) simple, but the physics of the baryons (star formation, winds, shocks, radiation) is much more complicated. Understanding the formation and evolution of galaxies is one of the major challenges of modern cosmology.

Exercise 4.5 Figure 4.11 shows the number density of galaxies per *absolute magnitude*. How is this related to the number density of galaxies per unit luminosity? ■

Calculating the number density of galaxies in principle is very simple: count the number of galaxies N within a volume V (we'll assume that the galaxies aren't evolving), to find the number density $\rho = N/V$. This can be done if we know *all* the galaxies within a particular volume V_0, known as a **volume-limited sample**.

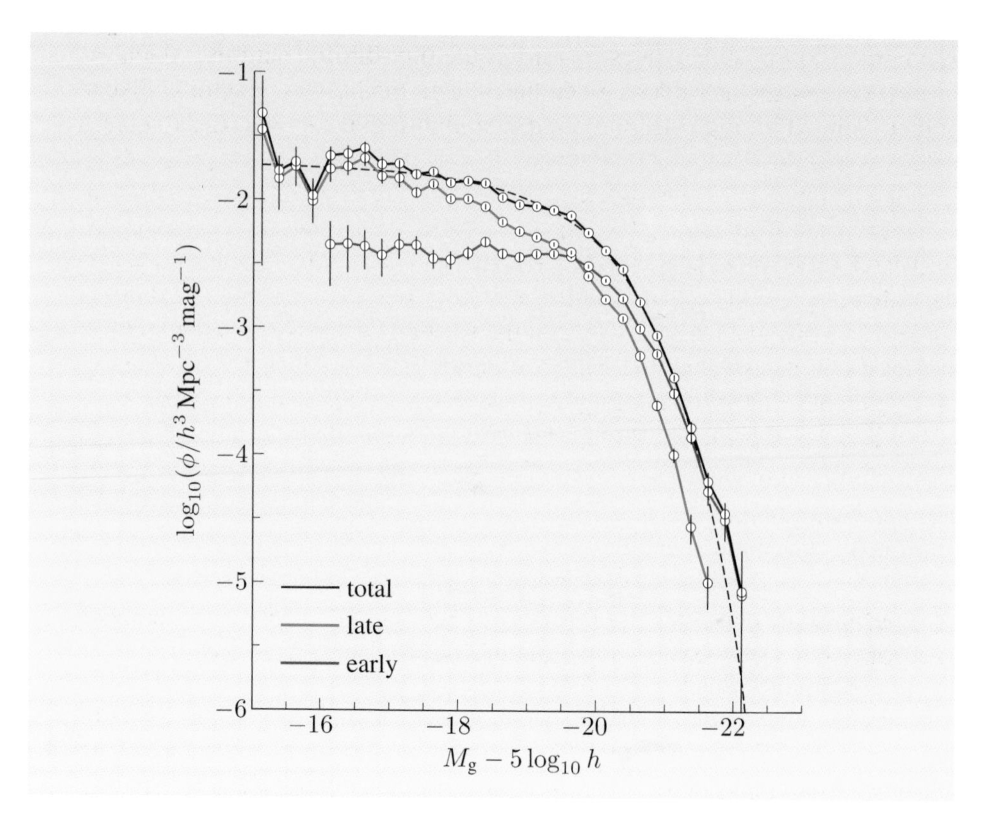

Figure 4.11 The SDSS g-band luminosity functions for galaxies. (The notation mag^{-1} means 'per magnitude'.) Late-type and early-type (i.e. spiral and elliptical) galaxies are also shown separately.

We could write this as

$$\rho = \frac{N}{V_0} = \frac{1}{V_0}\sum_{i=1}^{N} 1. \tag{4.19}$$

In practice, though, it's slightly trickier, because high-luminosity galaxies can be seen to greater distances than low-luminosity galaxies. Suppose that galaxy i has probability p_i of being detected in the volume. (Again, we're assuming that the galaxies don't evolve.) The number density would then be

$$\rho = \frac{1}{V_0}\sum_{i=1}^{N} \frac{1}{p_i}. \tag{4.20}$$

For example, if there's only a probability of $1/10$ that a particular sort of galaxy is in the volume, then each time we see that type of galaxy in the volume, there must be nine more we've missed, so each one we see is 'worth' ten. If it's a volume-limited sample, then $p_i = 1$ for every galaxy, and it reduces to Equation 4.19. The root-mean-square (RMS) estimate of the uncertainty in ρ from Equation 4.20 is just

$$\sigma_\rho = \frac{1}{V_0}\sqrt{\sum_{i=1}^{N} \frac{1}{p_i^2}}. \tag{4.21}$$

The probabilities or their equivalents are in general known as the **selection function** ('function' because in general it depends on properties of the galaxy such as luminosity and redshift). As long as the selection function is known, the number density can be calculated, as can the number density per unit luminosity (the luminosity function).

A common sort of survey in astronomy is the **flux-limited sample**, which lists all objects brighter than a given flux at a particular wavelength. Unlike a volume-limited sample, this one can detect high-luminosity galaxies to greater distances than low-luminosity galaxies. To calculate the selection probabilities p_i, imagine that we could push galaxy i to greater distances. When this galaxy is sufficiently far away that its flux equals the flux limit of the survey, this is the furthest distance at which it can be seen. We can write this distance as d_{max}, and the volume enclosed by that distance as $V(d_{\text{max}})$ or V_{max}. In static Euclidean space, V_{max} will just be $\frac{4}{3}\pi\, d_{\text{max}}^3$ times the fraction of the sky covered by the survey. Let's choose a volume V_0 that's bigger than the biggest V_{max} of any of the galaxies. The probability that galaxy i is above the flux limit will be $p_i = V_{\text{max},i}/V_0$ (again we're assuming no evolution), where $V_{\text{max},i}$ is the V_{max} for galaxy i. The number density is therefore

$$\rho = \frac{1}{V_0}\sum_{i=1}^{N}\frac{V_0}{V_{\text{max},i}} = \sum_{i=1}^{N}\frac{1}{V_{\text{max},i}}. \tag{4.22}$$

We could also calculate the number density per unit luminosity (the luminosity function) $\phi(L)$ by counting the number of galaxies in intervals ΔL:

$$\phi(L) = \frac{1}{\Delta L}\sum_{i \text{ in } L\to L+\Delta L}\frac{1}{V_{\text{max},i}}, \tag{4.23}$$

where this time the sum is over only those galaxies with luminosities in the interval L to $L + \Delta L$. This is known as the $1/V_{\text{max}}$ method for calculating luminosity functions. Luminosity functions from more complicated selection functions can be found by writing down the detection probabilities p_i. For example, a wide shallow flux-limited survey with a central deep part would have the p_i dropping discontinuously as an increasingly-distant galaxy crosses the shallow flux limit, and then dropping to zero as the galaxy passes below the deeper flux limit. In these cases we use the p_i to calculate the accessible volumes for each galaxy, sometimes written as V_{a} rather than the V_{max} used for a single flux-limited sample.

There are a number of assumptions built into these estimates. First, we're assuming that there are no types of galaxies that could not be detected anywhere in the survey, i.e. p_i is never 0. Second, we're assuming that we have a good estimate of the detection probabilities p_i. Third, we're assuming no evolution. It's worth checking to see if these conditions hold. The first is tricky, unless you are measuring quantities that deliberately exclude objects outside your selection function. For example, the g-band luminosity function of g-band-selected galaxies is not prone to this problem, but (say) the g-band luminosity function of radio-flux-selected galaxies might be. Spectroscopic redshifts can sometimes be difficult to obtain, and the second assumption can fail in practice if only galaxies in a biased subset of the sample have redshifts (e.g. galaxies that happen to already have redshifts from other research programmes). Surveys without such biases are known as **complete samples**. Sometimes only a random subset of galaxies is targeted for redshifts, a technique known as **sparse sampling**. This reduces the effective sky coverage of the survey, which can be folded into the p_i estimates. Note that a survey can be sparse sampled yet still be considered 'complete'.

If we have a wrong estimate of the selection function, it would make a non-evolving population look like it's evolving. In local galaxy populations, the key test is therefore whether they seem to evolve. There are many ways of testing this, but one method is particularly free of other assumptions: the $\langle V/V_{\text{max}}\rangle$ test.

The trick here is to use the cumulative probabilities. For any probability distribution $p(x)$, a sample x will be in the 10th percentile exactly 10% of the time, in the 50th percentile 50% of the time, and so on. These percentiles are essentially the cumulative probabilities. If we plotted a histogram of the values of x, the shape would be $p(x)$ (within the uncertainties). However, if we plotted a histogram of the *percentiles* associated with each x, the histogram would be uniform (again within the uncertainties). In other words, the cumulative probability distribution $c(x) = \int_0^x p(x')\,\mathrm{d}x'$ is uniform from 0 to 1.

In the case of our flux-limited sample, the cumulative probability that a galaxy at distance d_i is seen at any distance from 0 to d_i is $V_i/V_{\text{max},i}$, where V_i is the volume enclosed by the distance d_i, and $V_{\text{max},i}$ is, as before, the volume enclosed by $d_{\text{max},i}$. The test that there's no evolution is therefore that $V_i/V_{\text{max},i}$ should be uniformly distributed from 0 to 1. In particular, this means that the average value should be $\langle V/V_{\text{max}}\rangle = 1/2$. To test the null hypothesis of no evolution, we need a measured value of $\langle V/V_{\text{max}}\rangle$ and its uncertainty. You might think that the best approach is to propagate the uncertainties on the redshifts, but in practice these are effectively negligible. However, even if the null hypothesis holds and there is no evolution, there would still be some variation in $\langle V/V_{\text{max}}\rangle$, as Exercise 4.6 shows. The usual procedure is to use this expected variation (in the case of no evolution) as the uncertainty $\langle V/V_{\text{max}}\rangle$, which is then used to test whether the sample data are consistent with no evolution.

Exercise 4.6 Show that the variance of a uniform distribution from 0 to 1 is $1/12$, then use the central limit theorem to show that the standard deviation of $\langle V/V_{\text{max}}\rangle$ from a sample of N galaxies is $1/\sqrt{12N}$, assuming that the null hypothesis of no evolution holds. (Recall that the standard deviation is the square root of the variance, and the variance is the mean of the squares minus the square of the mean.)

Exercise 4.7 Does no evolution imply $\langle V/V_{\text{max}}\rangle = 1/2$?

Exercise 4.8 Does $\langle V/V_{\text{max}}\rangle = 1/2$ imply no evolution? ■

For more complicated selection functions, the idea is to use the p_i to calculate the enclosed volumes V_{e} and compare them to the accessible volumes V_{a}, so the $\langle V/V_{\text{max}}\rangle$ test is generalized to $\langle V_{\text{e}}/V_{\text{a}}\rangle$. In surveys of high-redshift galaxies and quasars (Section 4.6) the populations evolve strongly, so $\langle V/V_{\text{max}}\rangle$ is no longer useful for testing the knowledge of the selection function. Instead, it's sometimes used to demonstrate the presence of evolution.

A related and possibly equivalent concept to the selection function is **selection effects**. This can occasionally have a pejorative overtone if there is some debate over whether an effect is really present in a population or whether it is just due to the selection of the sample of objects. For example, suppose that we had a galaxy survey with both a radio flux limit and an optical flux limit, so a galaxy has to have enough radio flux *and* enough optical flux to be present in the sample. If we plotted the optical *luminosity* against the radio *luminosity* in this sample of objects, there could well be a broad correlation, but this would be because both luminosities would correlate with distance (because the numerous faint objects can only be seen nearby). This correlation would therefore be attributable to selection effects.

Having said that, there is no such thing as an 'unbiased' survey without selection effects. *All* astronomical surveys are defined by what they include and what they exclude. Even if you set out to measure every luminous object in the observable Universe, from galaxies to the smallest star, you would still have chosen to exclude many classes of object (planets, dark matter haloes without galaxies, and so on). Ideally you would choose a survey selection function that is tailored to answer a particular scientific question; these inevitable selection effects can work in one's favour.

Finally, there *is* a sense in which our Galaxy can be considered typical, as the following exercise shows.

Exercise 4.9 Assume that the galaxy luminosity function has a Schechter function form, with a faint-end slope $\alpha < 1$. Show that most of the light emitted by galaxies per unit volume is dominated by galaxies near L_* (also near the Milky Way luminosity). ■

4.6 Active galaxies

We've already seen that the discovery of non-Euclidean radio source counts was very strong evidence that the early Universe was very different to the present, even before the discovery of the CMB. The radio source population surprised in all sorts of other ways. In 1954, Baade and Minkowski discovered that the bright radio source Cygnus A (Figure 4.12) is identified with a faint optical galaxy. Suddenly it became apparent that radio sources sample a much more cosmologically distant population than anyone had considered before. But what generated these giant radio-emitting lobes? Baade wrote of the optical imaging:

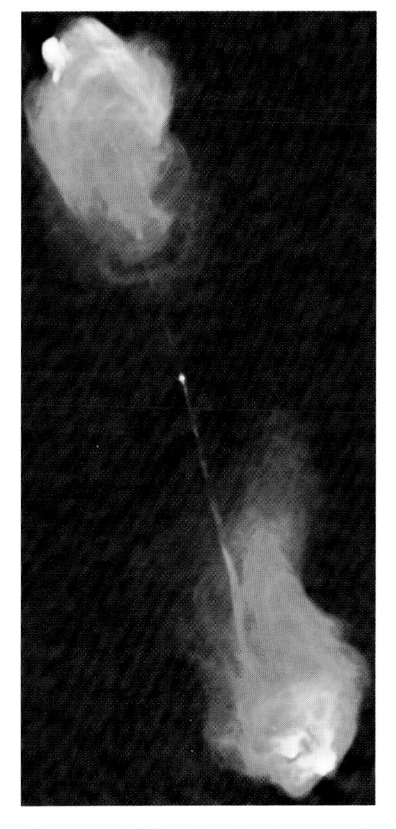

Figure 4.12 Radio image of the radiogalaxy Cygnus A, taken with the Very Large Array at a frequency of 5 GHz at 0.5″ resolution. This image is about 4.3′ from top to bottom.

> I knew something was unusual the moment I examined the negatives. There were galaxies all over the plate, more than two hundred of them, and the brightest was at the center. It showed signs of tidal distortion, gravitational pull between the two nuclei. I had never seen anything like it before. It was so much on my mind that while I was driving home for supper, I had to stop the car and think.
>
> Pfeiffer, J., 1956, *The Changing Universe*, Victor Gollancz.

Then in 1963, Schmidt and Greenstein discovered that the radio source 3C273 was identified with an object at a (then gigantic) redshift of $z = 0.158$. The optical luminosity implied by this was astounding: the V-band absolute magnitude of -26.3 is about 120 times larger than the absolute magnitude of -21.1 of the entire Andromeda galaxy.

- ● Why is a magnitude of -26.3 some 120 times larger than -21.1?
- ❍ Magnitudes m are related to fluxes S by $m = -2.5\log_{10} S + k$ (where k is some constant), so the difference between two magnitudes m_1 and m_2 is

related to their respective fluxes S_1 and S_2 by

$$\begin{aligned} m_1 - m_2 &= -2.5 \log_{10} S_1 + k + 2.5 \log_{10} S_2 - k \\ &= -2.5(\log_{10} S_1 - \log_{10} S_2) \\ &= -2.5 \log_{10}(S_1/S_2). \end{aligned}$$

In other words, $S_1/S_2 = 10.0^{-0.4(m_1-m_2)}$. We're given a magnitude difference of $26.3 - 21.1 = 5.2$ magnitudes, which corresponds to a flux ratio of just over 120.

Figure 4.13 SDSS image of the quasar 3C273.

These prodigious luminosities meant that quasars could be found at far greater distances than galaxies; the astronomer J.H. Oort wrote that 'quasars gave great expectations for cosmology'. In optical photographic plates the object looked like a star, so 3C273 was referred to as a **quasi-stellar object** (abbreviated **QSO**) or a **quasar** for short. Figure 4.13 shows an optical image from SDSS, and Figure 4.14 shows an optical spectrum of 3C273.

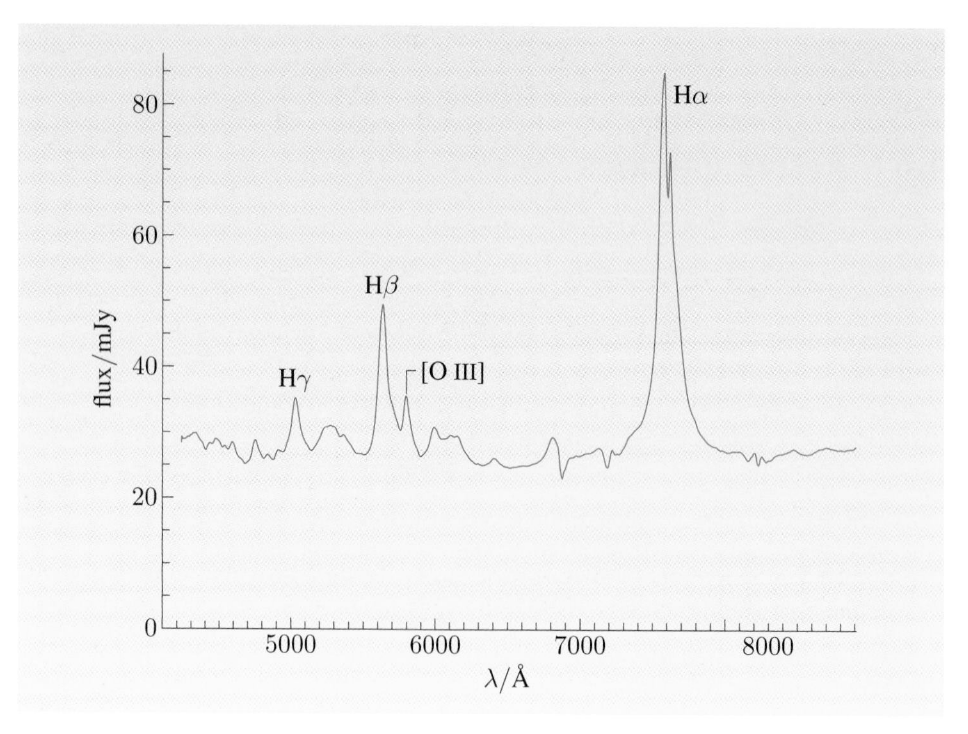

Figure 4.14 Optical spectrum of the quasar 3C273. Note the broad optical emission lines from the quasar's broad-line region (e.g. Hα).

Quasars typically (though not always) have very blue optical–ultraviolet spectra and broad emission lines, sometimes with widths suggesting Doppler motions of thousands of $\mathrm{km\,s^{-1}}$. The optical–ultraviolet continuum varies on timescales of weeks or less, suggesting physical sizes of light-weeks. These observations suggest accretion onto a central massive object, and the size and velocity constraints make it likely that this object is a **supermassive black hole**, i.e. one with a mass $\gtrsim 10^6\,\mathrm{M_\odot}$, if only because any astrophysical alternatives would very rapidly evolve into a single black hole.

We'll cover some of the detailed physics of quasars in Chapter 6 (see also the further reading section, especially Kolb's book *Extreme Environment Astrophysics*). For now, we'll state a few general results to define the terminology. The modern picture of quasars and radiogalaxies is of a dusty torus that makes the optical appearance depend strongly on orientation. Figure 4.15 shows a schematic diagram of these models of **active galactic nuclei** (AGN).

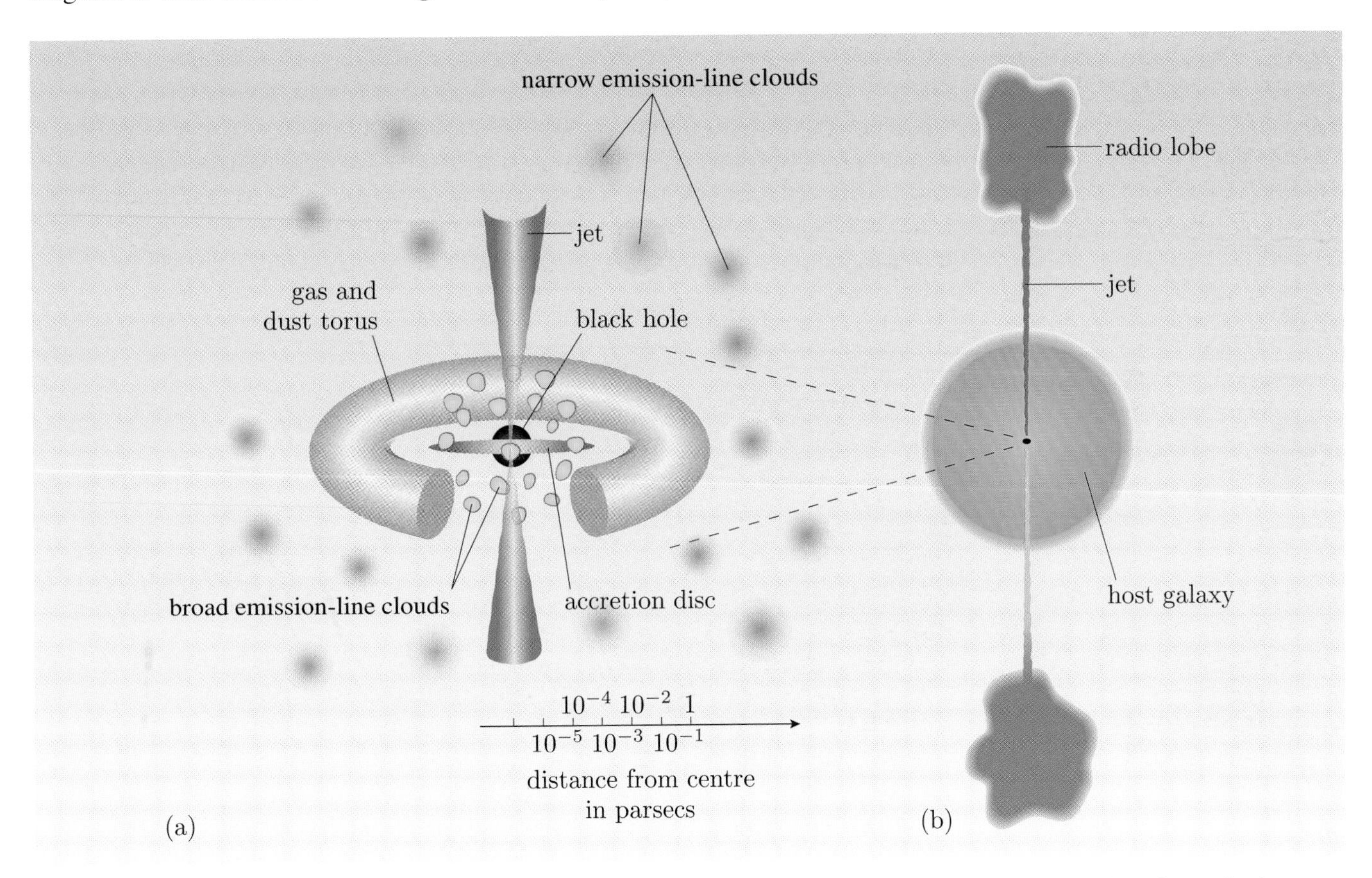

Figure 4.15 Schematic view (not to scale) of the dust-torus-based unified model of radio-loud active galaxies. Where the broad-line region is visible, the active galaxy is seen as a type 1 object (i.e. with broad and narrow lines), such as a Seyfert 1 or quasar. When the torus obscures the line of sight to the broad-line region, the active galaxy is seen as a type 2 object (i.e. with narrow lines only), such as a Seyfert 2 or radiogalaxy. When the jet is pointed directly along the line of sight, the jet luminosity can sometimes swamp the rest of the active nucleus; these examples are referred to as blazars.

When the central black hole accretion (sometimes called the **central engine**) is visible along the line of sight, the broad ($> 1000\,\mathrm{km\,s^{-1}}$) emission lines and blue optical–ultraviolet continuum are visible; when the host galaxy is visible, these are known as Seyfert 1 galaxies, and in general these are type 1 AGN. When the dusty torus obscures the line of sight to the central engine, only narrow emission lines are visible and the continuum is dominated by starlight from the host galaxy; these are Seyfert 2 or type 2 AGN. The broad line region and narrow line region are shown in Figure 4.15. These AGN also show evidence for gas with higher ionization than found in starbursts, such as having high ratios of [O III] $495.9 + 500.7$ nm to [O II] 372.7 nm or Hβ 486.1 nm, or a high ratio of [N II] $654.8 + 658.4$ nm flux to Hα 656.3 nm flux.

Figure 4.16 illustrates how star-forming galaxies and AGN emission lines differ. About 10% of active galaxies are **radio-loud**, i.e. they have luminous radio-emitting lobes. It's not clear why some active galaxies have radio lobes while others (the **radio-quiet** AGN) do not, but the lobes appear to be caused by particle jets emanating from the central engine. These impact on the interstellar/intergalactic medium and create a cocoon of plasma, in which electrons spiral along magnetic fields lines and emit synchrotron radiation at radio wavelengths. Even the radio-loud objects come in at least two distinct types: Fanaroff–Riley (FR) types I and II (not to be confused with Seyfert type). FR-I radiogalaxies have less luminous radio lobes that taper off in brightness towards the edges ('edge-darkened'), while FR-II radiogalaxies have more luminous, edge-brightened lobes. About half the energy output of an FR-II radiogalaxy comes out as jet kinetic energy — an astonishing output given that a quasar's luminous energy can exceed that of the rest of the galaxy combined.

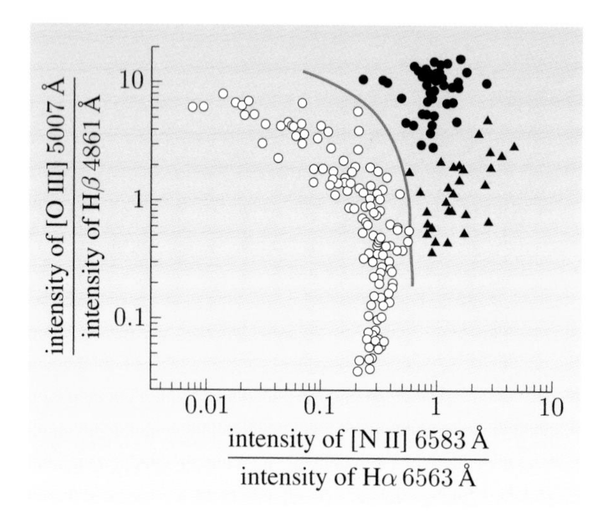

Figure 4.16 The relative strengths of narrow emission lines can be used to diagnose whether black hole accretion or star formation is present in a galaxy. This diagram is known as the 'Baldwin–Phillips–Terlevich diagram' or sometimes 'BPT diagram'. Both wavelength pairs are close in wavelength, so are insensitive to dust reddening. Open circles represent galaxies with emission lines from H II regions (i.e. star formation), while the closed symbols are active galaxies. (Filled circles are Seyfert 2 galaxies, and triangles are weaker AGN known as 'low-ionization nuclear emission regions' or LINERs.) AGN can be separated from star-forming galaxies using the curved line.

Baade's original question on the triggers of active galaxies is still with us. As we'll see in Chapter 6, the formation of supermassive black holes and their growth through accretion are closely related to the formation of stars in their host galaxies. The luminosity function of quasars is not in general well fit by the Schechter function; instead, researchers have opted for an arbitrary double-power-law parameterization:

$$\phi(L) = \frac{\phi_*}{(L/L_*)^\alpha + (L/L_*)^\beta}, \tag{4.24}$$

where ϕ_*, L_*, α and β are free parameters to be determined from the data. This model doesn't have any underlying physical motivation, but if $\alpha > \beta$ then it has the property that $\phi \propto L^{-\beta}$ at luminosities far below the break luminosity (i.e. $L \ll L_*$), and $\phi \propto L^{-\alpha}$ at $L \gg L_*$. The overall shape is therefore of a shallow power law at faint luminosities, steepening at luminosities around L_* to a steeper power law at high luminosities.

Quasars held another surprise: the luminosity function of quasars evolves very strongly. Figure 4.17 shows the evolution of bright quasars from SDSS.

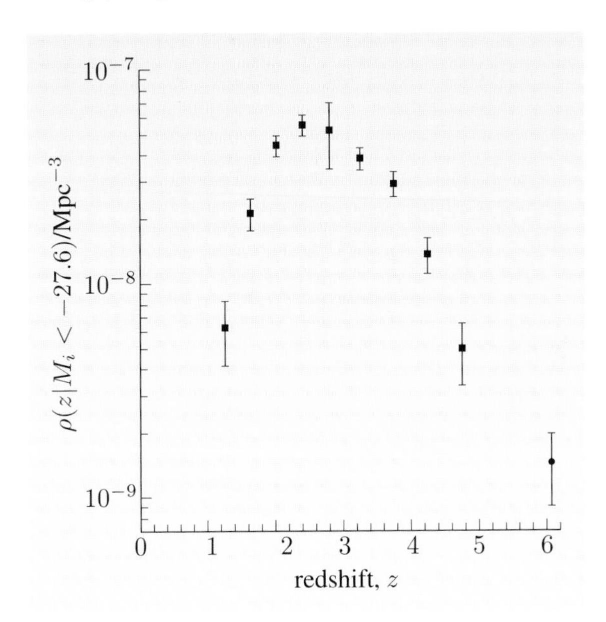

Figure 4.17 The total comoving number density of SDSS quasars with i-band absolute magnitudes brighter than -27.6.

It appears that quasars were far more common at a redshift of $z = 2$ than at the present. The first measurements of this evolution parameterized these changes as evolution in ϕ_* only (**pure density evolution** or PDE) or in L_* only (**pure luminosity evolution** or PLE). Initial indications were that PLE fitted better, suggesting a long-lived population of quasars that dimmed over cosmological times; however, the radio-loud subset also showed PLE, while the electron energy loss from synchrotron radiation in their radio lobes implied ages of only tens of millions of years. The PLE decline in quasar number density from $z = 2$ to $z = 0$ varies approximately as $(1 + z)^3$. We'll return to the underlying causes of this sudden increase and decline in quasar activity in the cosmos in Chapters 5 and 6, but an interesting clue comes from comparing the evolution in galaxy–galaxy major merger rates (i.e. mergers of similarly-sized galaxies) inferred from optical galaxy surveys: the merger rate varies as $(1 + z)^{2.7\pm0.6}$ at least up to $z = 1$. Numerical simulations have shown that galaxy–galaxy mergers could drive gas to the final common centre, so a link is plausible. At the highest redshifts, the quasar number density appears to drop quickly (Figure 4.17), known as the **redshift cut-off** of quasars. Again the underlying physical causes of this change are still debated.

Exercise 4.10 How would PDE and PLE translate a luminosity function in the $(\log \phi, \log L)$ plane? ■

Exercise 4.11 (This is a more open-ended and difficult exercise than most in this book.) The luminosity L of radio lobes depends on the kinetic energy output per unit time of the radio jet, Q, on the density of the surrounding medium ρ (which can be inferred independently from other observations), and on time. As the radio jet burrows deeper into the surrounding medium, the cocoon of ionized plasma increases in size with time, so the linear size r of the radio lobes increases. It can be shown that the jet power output Q is related to these observables roughly as $Q \propto L^{6/7} r^{-4/7} \rho^{-1/2}$. Comment on whether radio lobe surface brightness is suitable for the Tolman test. ■

See, for example, Miller, P., Rawlings, S. and Saunders, R., 1993, *Monthly Notices of the Royal Astronomical Society*, **263**, 425.

One advantage that radiogalaxies have over radio-loud quasars is that the observed-frame optical luminosities are not dominated by the central active nucleus, so the host galaxy is visible. The K-band light is much less sensitive to young stars than the ultraviolet (Figure 4.6), so should be a measure of the assembled stellar mass in the host galaxy. How does the K-band luminosity of radiogalaxies evolve? It turns out that the K-band Hubble diagram (apparent magnitude versus redshift) has a tight scatter, consistent with just passive stellar evolution in the host galaxies that locally are giant ellipticals. However, this scatter increased considerably above a redshift of around 2. Is this the formation epoch of giant elliptical galaxies? Another effect conspired to increase the observed scatter: at $z > 2$ the [O III] 500.7 nm emission line redshifts into the K-band window at 2–2.5 μm, so the added dispersion in the K–z relation might just reflect variations in the emission line contribution.

4.7 Deep-field surveys and wide-field surveys

In 1996 the Hubble Space Telescope (HST) made an ultra-deep survey that was to revolutionize the study of the evolution of galaxies. The idea was to invest 150 orbits (about 5% of a year's total) using the HST's Wide Field Planetary Camera 2 instrument (WFPC2), mainly making a deep image of a single camera field of view in several broad-band filters, in a region in the constellation of Ursa Major. Many of the galaxies detected would be too faint for optical spectroscopy, which in part drove the development of better photometric redshift estimates. This project was known as the **Hubble Deep Field** (HDF). A later HST survey performed a similar WFPC2 map in the southern hemisphere with a $z = 2.24$ quasar centred in the HST's ultraviolet imaging spectrograph STIS, and the two are now known as the HDF North and HDF South (HDF-N and HDF-S). The images of HDF-N are shown in Figure 4.18. Unusually, the data were made public immediately and the astronomical community pounced on them, metaphorically. The HDF-N and it successors contained many surprises. One is that the sky is mostly black — it's not a hopeless jumble of overlapping objects (a further answer to Olbers' paradox). Most distant galaxies appear quite small, despite the fact that angular diameter distance has a maximum around $z \simeq 2$ (Figure 1.18). Another surprise was the evolution of the galaxy merger rate (see above) and galaxy morphologies. While there are plenty of examples of spirals and ellipticals as in the local Universe, the familiar Hubble sequence disintegrates at high enough redshift. There are more irregular morphologies at faint apparent magnitudes. The U-band dropouts in particular tended to have the very disturbed morphologies. Figure 4.19 shows an example of a type of U-band dropouts that became known as **chain galaxies**.

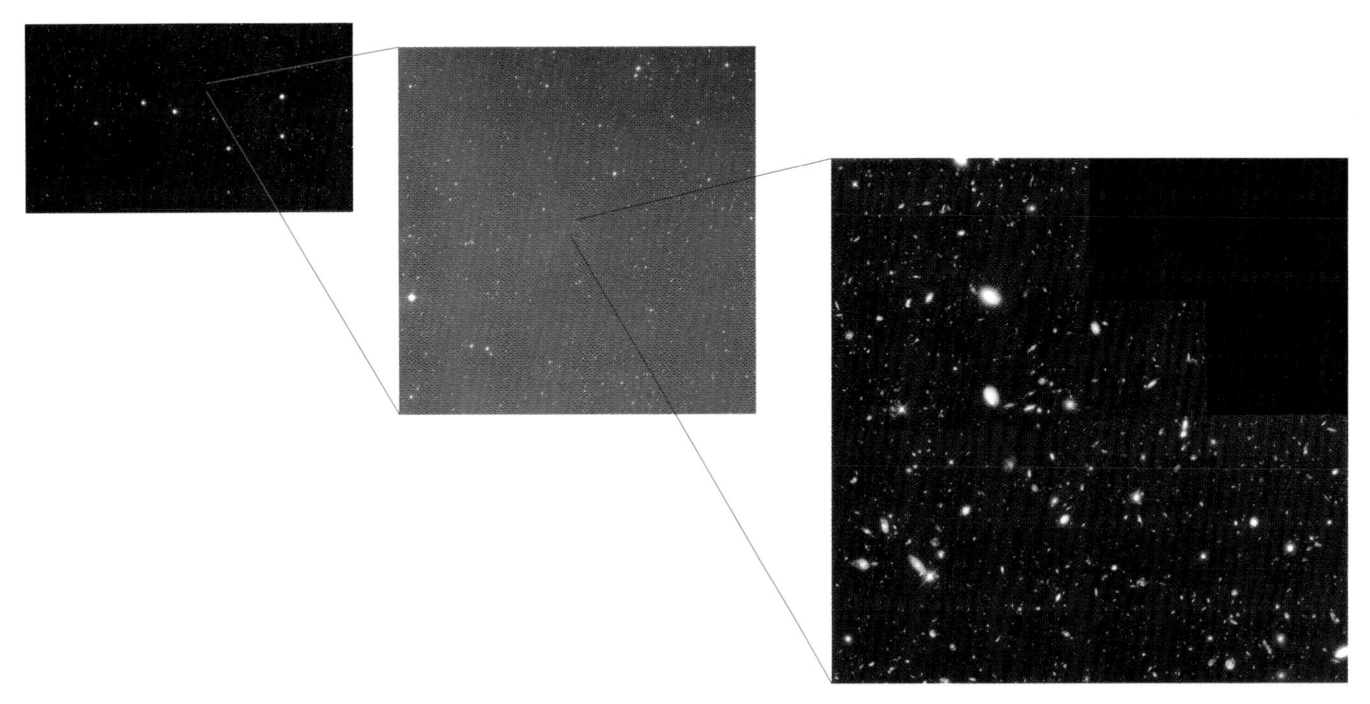

Figure 4.18 The Ursa Major constellation (top left), with a square degree region marked in red. A zoom into this region shows the location of the Hubble Deep Field North (HDF-N) in red. Note that the region of sky is unremarkable. The location was chosen to avoid bright objects at all wavelengths (e.g. bright stars in the optical, bright galaxies in the far-infrared or radio); not every patch of sky is suitable for a blank-field survey. The panel to the right shows the final zoom into the HDF-N itself.

Figure 4.19 Details from the Hubble Deep Field North. Alongside elliptical and spiral morphologies familiar from the Hubble sequence, there are many disturbed and/or irregular galaxies, including blue elongated 'chain galaxies'.

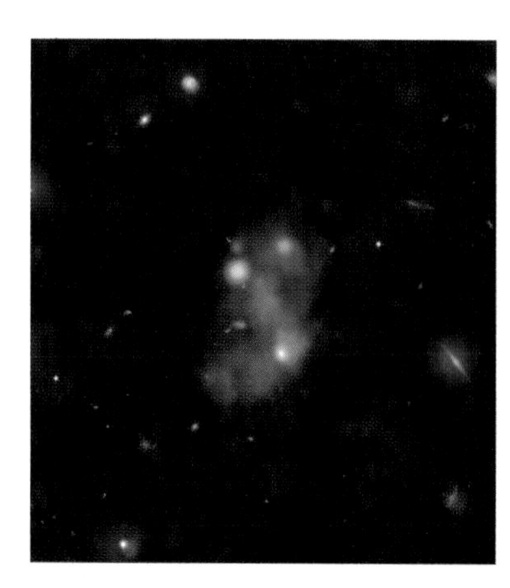

Figure 4.20 A composite image of a Lyman α blob, with Lyman α coloured yellow, Spitzer infrared data marking the sites of dust-shrouded star formation coloured red, and Chandra X-ray data marking the site of supermassive black hole accretion marked in blue. This may be an example of the feedback processes in action (Chapters 5 and 6).

See Daddi, E. et al., 2004, *Astrophysical Journal*, **617**, 746.

These are part of what was originally called the 'faint blue galaxy' population. These Lyman break galaxies cluster strongly. Combined with their number density, it became clear that the bias parameter for these galaxies was high, effectively cancelling the weaker clustering of dark matter at earlier epochs, so the overall clustering strength resembled present-day galaxies.

Searches for U-band dropouts yielded another surprise: a population of galaxies with a strong Lyman α emission line but weak continuum. (See, for example, Steidel, C.C. et al., 2000, *Astrophysical Journal*, **532**, 170.) These have become known as **Lyman α blobs**. The widths of the lines suggest that the ionization causing the Lyman α line is from star formation, though at least one candidate has been found with an obscured X-ray core (Figure 4.20), suggesting hidden AGN. We'll return to the coupled formation of black holes and galaxies in later chapters. It's not clear what fraction of the ionizing radiation escaped from galaxies at high redshifts; we'll return to this in Chapter 8.

Other new galaxy populations have been discovered in these deep field surveys. The **Extremely Red Objects** (ERO) were found to be very faint in optical wavelengths, but very bright in the near-infrared, with colours (R–K) > 5. It was not initially clear whether these were very dusty star-forming galaxies or redshifted old stellar populations. In fact, both appear to be present in the ERO population. EROs cluster strongly, suggesting that they are associated with more massive dark matter haloes than Lyman break galaxies. We'll return to this in the next chapter. A related population is the **BzK galaxies**, selected in the (B–z) versus (z–K) colour–colour plane to satisfy (z–K)–(B–z) > -0.2 (when the magnitude zero points are given in the AB system). This appears to select $z > 1$ star-forming galaxies independently of reddening because reddening moves galaxies *parallel* to the (z–K)–(B–z) $= -0.2$ threshold.

Like quasars, the galaxy luminosity function evolves strongly at all wavelengths investigated so far. Figure 4.21 shows the evolution in M_* and ϕ_* in the best-fit Schechter functions at several wavelengths. The deepest survey so far is the **Hubble Ultra Deep Field** (UDF). This is a further ultra-deep survey with the more sensitive Advanced Camera for Surveys on the upgraded HST, some details of which are shown in Figure 4.22. The most recent (and possibly final) refurbishment of the HST introduced the WFC3 instrument, which has yielded $z \sim 10$ galaxy candidates in the UDF using the Lyman dropout technique, shown in Figure 4.23. A surprise from these observations has been that the luminous output from these galaxies is declining so quickly that these galaxies could not have reionized the Universe (see Chapter 8).

We can use the evolution in the luminosity function to estimate the cosmic star formation history, i.e. the amount of mass being turned into stars per unit comoving volume per year, as a function of redshift. Because heavy elements are synthesized in stars, this also traces the metal enrichment of the Universe throughout its history. Recall that astronomers use 'metal' to refer to any elements heavier than helium.

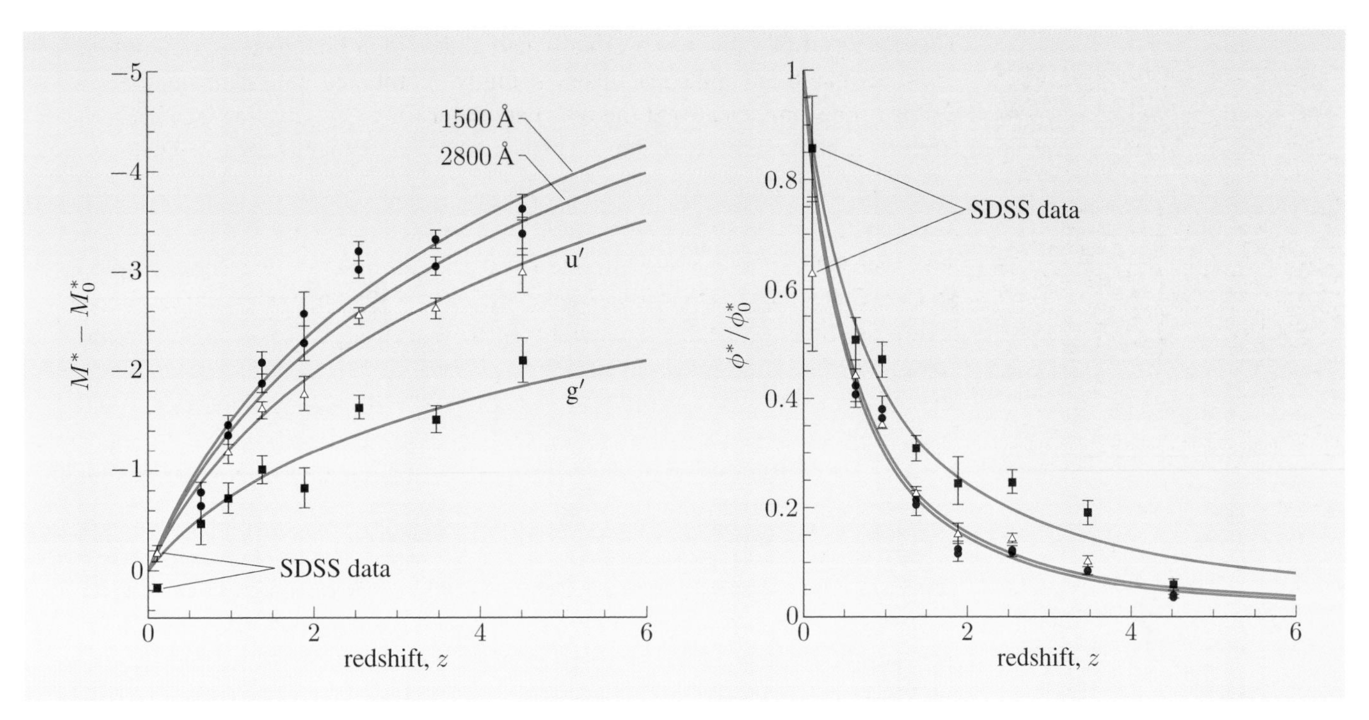

Figure 4.21 The evolution of ϕ_* and M_* of the best-fit Schechter function for galaxies, at various wavelengths, determined from the FORS Deep Field survey. Also shown are the local SDSS galaxy constraints.

Figure 4.22 Details from the Hubble Ultra Deep Field, taken with the HST's Advanced Camera for Surveys.

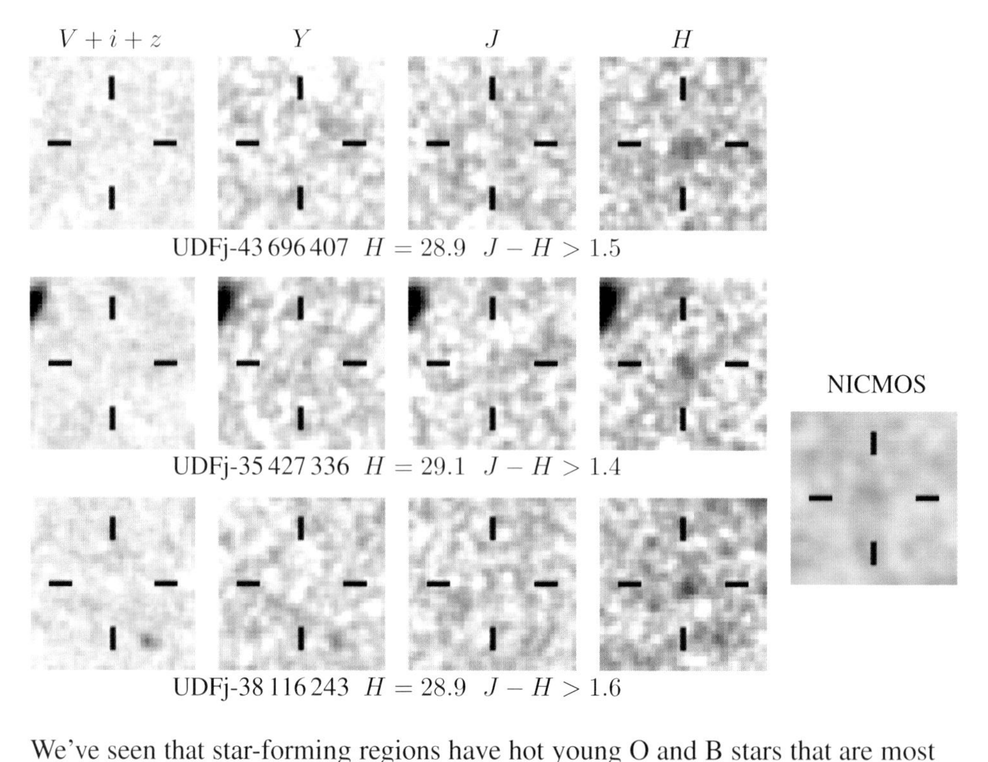

Figure 4.23 Candidate $z \simeq 10$ galaxies discovered using the Lyman break technique as J-band (1.2 μm) dropouts, using the new WFC3 on the HST. Confirmation of the H-band (1.6 μm) detections comes from an average of the H-band images taken with the HST's earlier observations using the less-sensitive NICMOS instrument. These galaxies are seen just 500–600 Myr after recombination. Each image is $2.4'' \times 2.4''$.

We've seen that star-forming regions have hot young O and B stars that are most luminous at ultraviolet wavelengths. The amount of ultraviolet light being emitted per unit comoving volume could therefore be used as an estimator of the volume-averaged star formation rate. If we integrate $L\,\phi(L)$ (where L is the ultraviolet luminosity and $\phi(L)$ is the luminosity function at this ultraviolet wavelength), we can calculate the ultraviolet luminosity density. The next step is to extrapolate from the O and B star formation to calculate the formation of stars of all types. For this, one needs to assume an initial mass function for the stars. The next step is to account for dust obscuration, to which ultraviolet luminosity is particularly prone. This is also difficult. As we've seen, the extinction depends on both the geometry of the dust and the dust composition. One approach (known as the Calzetti extinction law) is to use empirical correction derived from a comparison of models with optical–ultraviolet rest-frame spectra of galaxies. Figure 4.24 shows the cosmic star formation history derived from ultraviolet observations with this extinction law. This was first known as the 'Madau plot' after the 1996 Madau et al. paper that first appeared to detect the high-redshift decline. Later usage changed to refer to the **Madau diagram** or **Madau–Lilly diagram** (the latter acknowledging earlier work that detected the initial increase from $z = 0$ to $z = 1$). Equivalently, the term **cosmic star formation history** is used. This diagram has been enormously influential, but it's wise to keep in mind the uncertainties in the underlying assumptions. We'll see in Chapter 5 other approaches to constraining this diagram. It is superficially similar to the evolution of quasars, which immediately suggested a physical connection between the formation of stars and the growth of black holes.

See Calzetti, D., Kinney, A.L. and Storchi-Bergmann, T., 1994, *Astrophysical Journal*, **429**, 582.

Madau, P. et al., 1996, *Monthly Notices of the Royal Astronomical Society*, **283**, 1388.

Another way of constraining the cosmic star formation history is with the present-day optical spectra of nearby galaxies. Population synthesis models can be used to determine (or at least constrain) when the bulk of the stars formed, i.e. the location of the peak in the Madau diagram. The result of this analysis applied to the SDSS spectra of about 3×10^5 galaxies shows a decline from $z = 1$ to

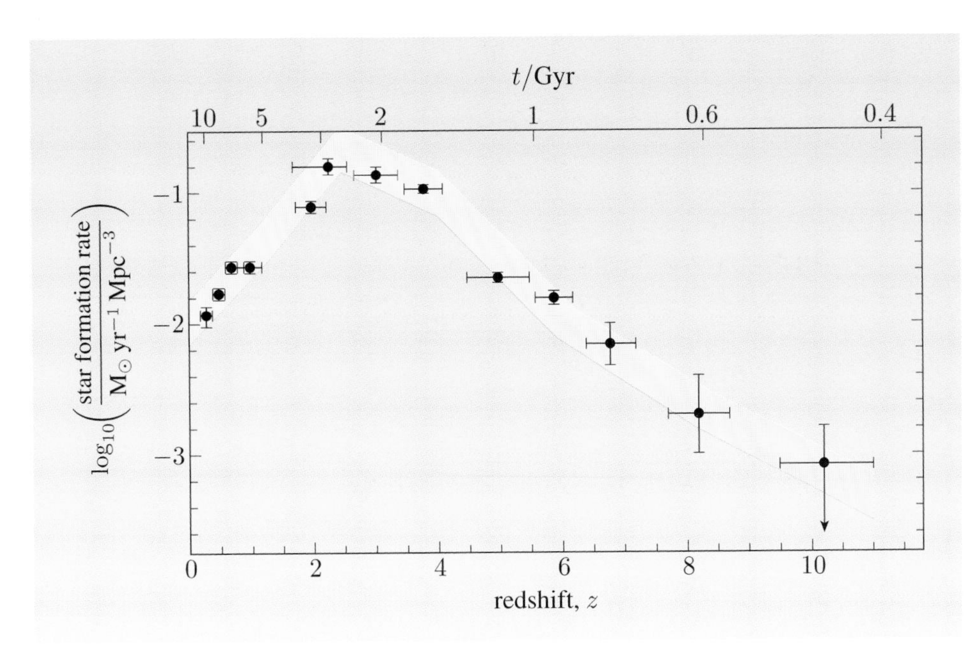

Figure 4.24 The cosmic star formation history (also known as the Madau diagram) derived from rest-frame ultraviolet luminosities of galaxies. The data have had a correction for dust extinction applied.

$z = 0$, but the analysis doesn't have the redshift resolution to find the location of the peak.

In the previous chapter we spent some time discussing local galaxy scaling relationships, so it's worth saying briefly how these relationships change at higher redshift. At redshifts $z < 1$, ellipticals show some evolution in the fundamental plane, at least some of which may be due to passive stellar evolution, though there are discussions of how the selection effects of magnitude-limited samples affect the result.[3] Meanwhile, spiral galaxies show evolution in the Tully–Fisher relation, suggesting 'differential' evolution,[4] meaning that different types of galaxies evolve differently: low-mass galaxies appear to have undergone more star formation more recently than higher-mass galaxies. The $z \simeq 1$ edge-on spiral discs appear to have thicker widths than their $z = 0$ counterparts, and disturbances such as warps were more common[5] at $z = 1$.

The HST seems to be a long way from being limited by galaxy–galaxy overlaps, but other telescopes have not been so fortunate. There is a threshold known as the **confusion limit** beyond which one can't rely on being able to separate individual objects. Even the HST can reach this limit when observing a dense star cluster. Intuitively, there has to come a point where the RMS fluctuations of the image are no longer dependent on the noise in the image (which would reduce as $1/\sqrt{\text{time}}$) but instead are dominated by the fluxes of the faint sources lying below the detection threshold.

The confusion limit is usually defined as three or five times the fluctuations from background objects. The location of the confusion limit varies depending on the slope of the source counts, and is surprisingly high, as the following worked example shows. We shall use the term **beam** to mean the area on the sky occupied

[3] See, for example, Treu, T., 2003, astro-ph/0307281, and Almedia, C., Baugh, C.M. and Lacy, C.G., 2007, *Monthly Notices of the Royal Astronomical Society*, **376**, 1711.

[4] See, for example, Böhm, A. and Ziegler, B.M., 2007, *Astrophysical Journal*, **668**, 846.

[5] See, for example, Reshetnikov, V.P., Dettmar, R.-J. and Combes, F., 2003, *Astronomy and Astrophysics*, **399**, 879.

by a point source, i.e. an object that's spatially unresolved by that telescope. For example, if a point source has a Gaussian shape with a (standard deviation) width of r arcseconds, then the beam area Ω could be regarded as πr^2 square arcseconds (though conventions differ slightly on this choice). We'll use a detection limit of five times the noise, because then less than one beam in 3.5 million will have a random detection. This may seem excessively cautious, but remember that megapixel cameras are now very common and it's quite possible to have many millions of beams in an image, particularly in wide-field astronomical mosaics.

Worked Example 4.1

Find the noise caused by point sources below the detection limit $S_{\rm lim}$, assuming a source count slope of $N(>S) = kS^{-\alpha}$ and a beam area of Ω. At what surface density of sources are these fluctuations one-fifth of $S_{\rm lim}$? (You might predict a bit less than one source per beam, but this is very wrong.) You can make use of the fact that the variance in the number of galaxies equals the number of galaxies (Poisson statistics).

Solution

In an interval $S \rightarrow S + \mathrm{d}S$, the total number of sources per unit area will be $\mathrm{d}N = \alpha k S^{-\alpha-1}\,\mathrm{d}S$. We can write this as $\mathrm{d}N = \alpha(N(>S)/S)\,\mathrm{d}S$. Therefore the number of sources in one beam will be $\Omega\,\mathrm{d}N = \alpha\,\Omega(N(>S)/S)\,\mathrm{d}S$. This will be subject to Poisson statistics, so the variance will equal the mean, i.e. $\alpha\,\Omega(N(>S)/S)\,\mathrm{d}S$. The variance of the *flux* (as opposed to the number) will be S^2 times the variance in the number, i.e. $\alpha S\Omega\, N(>S)\,\mathrm{d}S$. Integrating this from $S = 0$ to $S = S_{\rm lim}$ gives

$$\begin{aligned}
\mathrm{Var}(S) &= \int_0^{S_{\rm lim}} \alpha S \Omega k S^{-\alpha}\,\mathrm{d}S \\
&= \frac{\alpha}{2-\alpha} k\, S_{\rm lim}^{2-\alpha}\,\Omega \\
&= \frac{\alpha}{2-\alpha} \Omega\, N(>S_{\rm lim})\, S_{\rm lim}^2.
\end{aligned}$$

Therefore the noise $\sigma = \sqrt{\mathrm{Var}(S)}$ will be

$$\sigma = \sqrt{\frac{\alpha}{2-\alpha}\Omega\, N(>S_{\rm lim})}\; S_{\rm lim}.$$

Now, the quantity $\Omega\, N(>S_{\rm lim})$ is the number of sources per beam. We'll find it more useful to work in the number of beams per source, $(\Omega\, N(>S_{\rm lim}))^{-1}$. If we want the detection limit $S_{\rm lim}$ to be five times the noise level, i.e. $S_{\rm lim} = 5\sigma$, then

$$\left(\frac{\alpha}{2-\alpha}\Omega\, N(>S_{\rm lim})\right)^{-1/2} = 5,$$

which rearranges to give

$$(\Omega\, N(>S_{\rm lim}))^{-1} = \frac{25\alpha}{2-\alpha}.$$

If the source count slope is Euclidean, then $\alpha = 1.5$, so we'd need one source per 75 beams! In practice the source counts flatten at the very faintest fluxes, and one source per 20–40 beams is usually considered as the confusion limit.

The confusion limit severely constrains what can be said about individual objects, but it may still be possible to constrain the shape of the source count slope from the histogram of pixel values in a confusion-limited map. This is known as a ***P*(*D*) analysis**, where D is the deflection from the mean in the map, and P is the probability of that deflection. The observed histogram is compared to the pixel value distribution predicted by a given source count model. We've also assumed that the underlying point sources are not clustered, but clustering will increase the confusion limit. It may also be possible to constrain the clustering of the sources below the confusion limit by measuring the angular power spectrum of the distribution of pixel values, known as a **fluctuation analysis**.

Exercise 4.12 You might think that a limit of five times the RMS fluctuations (a one in 3.5 million chance of a random noise spike for Gaussian noise) is extraordinarily conservative. For a beam of one square arcsecond, which is typical in some ground-based optical imaging, calculate how big an image has to be in square degrees to have one random 5σ noise spike on average, assuming Gaussian noise. (For comparison, there are several cameras on world-class optical telescopes with fields of view of at least a half a degree along each side.) ■

The HDF-N and HDF-S are examples of **pencil-beam surveys**: very deep but narrow surveys, covering a long and thin volume of the Universe. The opposite strategy is a **wide-field survey**: shallower, but wider area. On the widest scales, we've already met the SDSS survey, which covers a quarter of the sky. Digitized versions of photographic plates taken by Schmidt survey telescopes are available online for the whole sky, known as the **Digitized Sky Survey** (DSS).

The DSS is available online from several sites, such as http://archive.eso.org/dss.

Exercise 4.13 For many observations, including HST imaging, the depth of an image is proportional to the square root of the time spent integrating, i.e. the faintest flux S is proportional to $1/\sqrt{t}$. Suppose that the galaxy source counts are Euclidean. Would you detect more galaxies in a pencil-beam survey or a wide-field survey, for a fixed amount of observing time? What source count slope would tip the balance in the opposite direction? ■

The HDF has perhaps had such an enormous impact on cosmology that time allocation committees on the HST and other telescopes have been emboldened. In any case, the HDF-N and HDF-S have been supplemented by various wider-area HST surveys, most notably COSMOS (Cosmological Evolution Survey), which covered $2\,\text{deg}^2$. We shall meet some of its key results in Chapter 7.

Figure 4.25 shows the redshift histogram in the HDF-N. The non-uniformity is quite striking. This illustrates one disadvantage of pencil-beam surveys: large-scale structure fluctuations can have a big effect on some measurements such as the redshift distribution. This was part of the motivation for making a second Hubble Deep Field, HDF-S. The similarity of the galaxy populations in HDF-N and HDF-S is sometimes cited as confirmation of the cosmological principle, though the redshift histograms differ. We saw in Chapters 2 and 3 how the variance of the galaxy distribution depends on scale, and we can use this to give an order-of-magnitude estimate for the fluctuations in our survey, e.g. by defining $k = 1/V^{1/3}$, where V is the comoving volume of the survey. (This is, in fact, an underestimate of the fluctuations — see the further reading section.) The clustering of dark matter is expected to be weaker at high redshift, but to first order this is cancelled by the evolving bias parameter of galaxies, as we've seen.

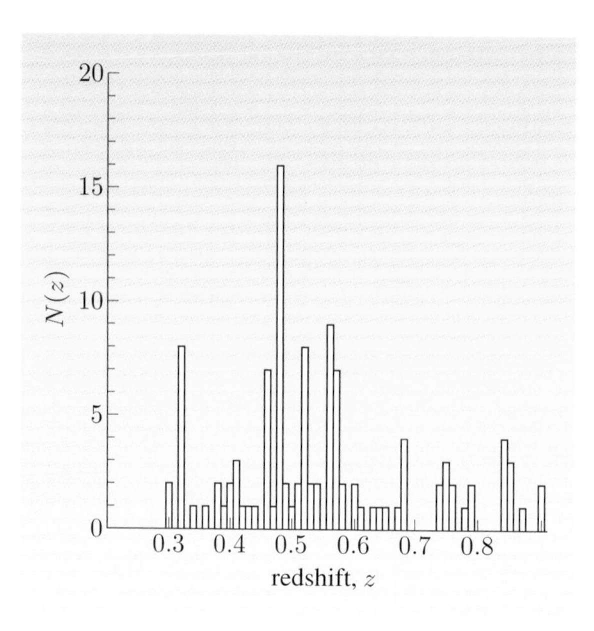

Figure 4.25 Redshift histogram in the HDF-N estimated from a sparse sample of 140 objects. Note the peaks in the redshift distribution, caused by the pencil-beam survey passing through large-scale structures.

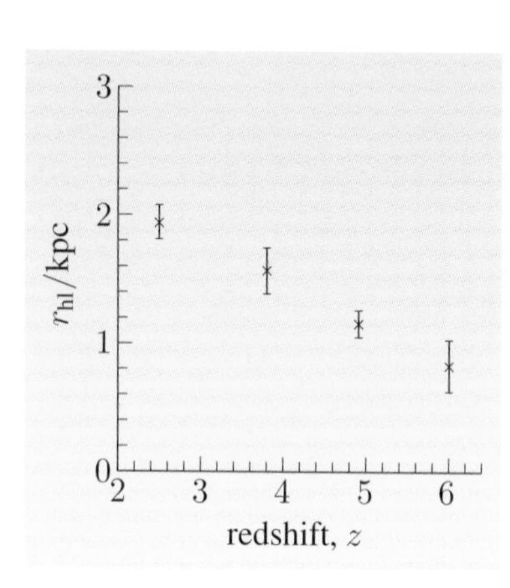

Figure 4.26 The half-light radii (the radii containing 50% of the light) of galaxies in the Hubble UDF.

Figure 4.26 shows how the sizes of galaxies evolve in the Hubble UDF. Galaxies of a given luminosity are consistently smaller at higher redshifts, with a size-dependence scaling approximately as $(1+z)^{-1.1\pm0.2}$. What could cause this size evolution? One clue comes from the **phase space density** of a galaxy. This is the volume that a galaxy (or a portion of a galaxy) occupies in an imagined six-dimensional space of three space dimensions (x, y, z) and three velocity axes (v_x, v_y, v_z). One can estimate it by dividing the mass density by the volume of an ellipsoid with the axes equal to the velocity dispersions in each of the three velocity axes. This is a useful quantity because numerical simulations of galaxy–galaxy mergers have shown that phase space density decreases by a factor of a few during a merger, and it can be shown by Liouville's theorem (Chapter 7) that phase space density cannot be increased, unless the stars' kinetic energy is dissipated into, for example, gas motions or radiation. Some elliptical galaxies have phase space densities consistent with being the merger product of spiral galaxy collisions, and numerical simulations predict that the final product would have an elliptical morphology. On the other hand, the cores of giant elliptical galaxies have phase space densities much higher than those of spiral galaxies, so they cannot be formed by (dissipationless) mergers. However, Lyman break galaxies at $z > 5$ have phase space densities similar to the cores of present-day massive ellipticals. Could these be the progenitors of today's giant ellipticals? Perhaps. We'll meet another population of galaxies making a similar claim in Chapter 5: the submm galaxies, which have giant starbursts as expected in the original monolithic collapse model (Chapter 3). It seems that we're still some way from resolving the monolithic collapse versus disc–disc merger debate on the origin of elliptical galaxies.

The sizes of high-redshift elliptical galaxies have recently thrown up another fundamental puzzle, which at the time of writing is unresolved. There is a very numerous population of small, passively-evolving (i.e. not star-forming) elliptical galaxies at $z > 1$ whose luminosities imply large stellar masses of $> 10^{11}\ M_\odot$. These are sometimes called **red nuggets**, and they have no local counterparts. Could their luminosities be a misleading measure of the underlying stellar mass, for some reason? Extremely deep spectroscopy of one example suggests not:

See van Dokkum, P.G., Kriek, M. and Franx, M., 2009, *Nature*, **460**, 717, and references therein.

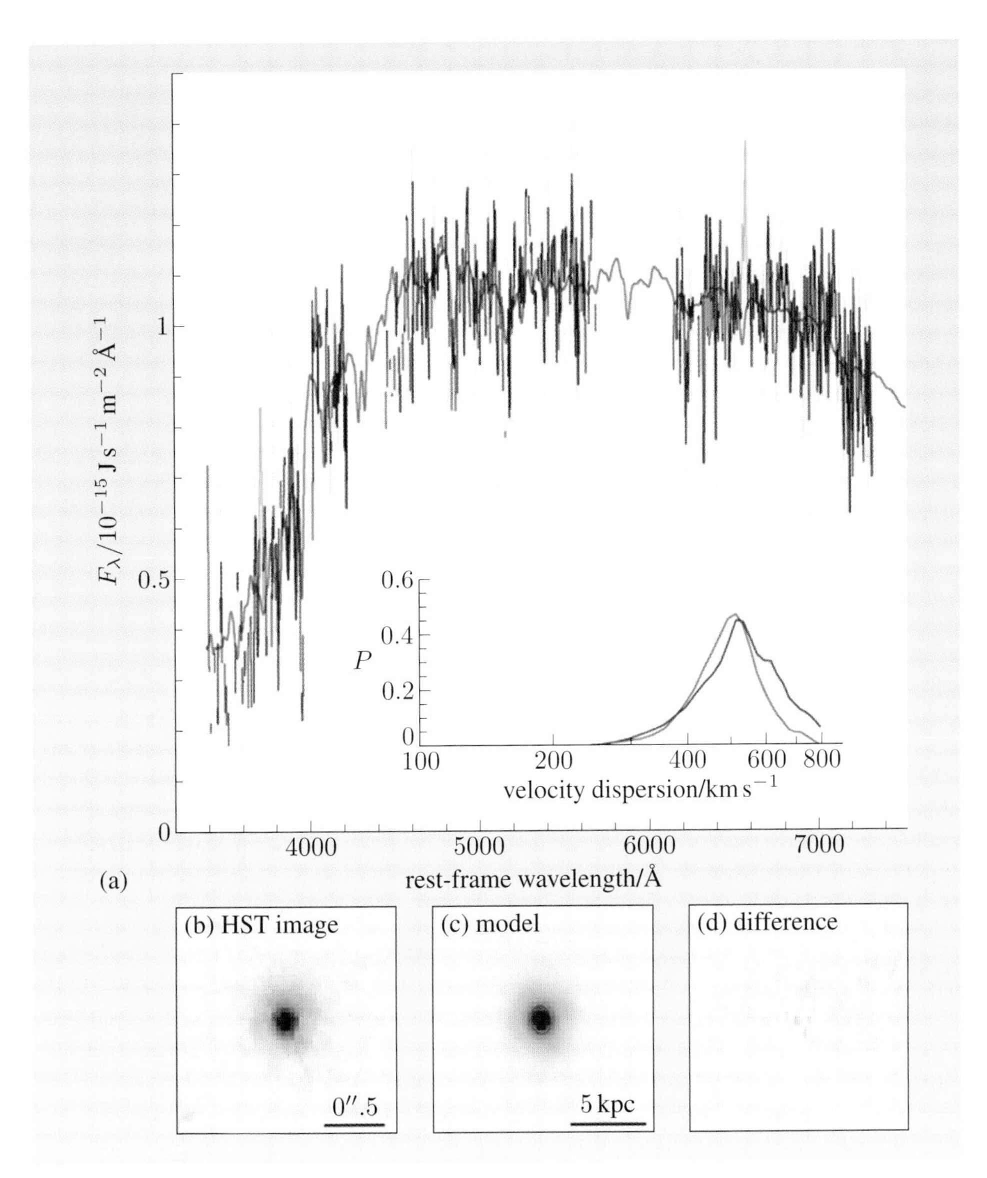

Figure 4.27 The spectrum of galaxy 1255-0 (grey), with a smoothed version shown in black and the best-fit population synthesis model shown in red. The wavelengths of some absorption lines are marked in yellow. The insert shows the likelihood distribution of the galaxy's velocity dispersion, using two different methods of estimating the noise. The panels to the right show the HST 1.6 μm image, a model, and the difference between the two.

despite a small effective radius of just 0.78 ± 0.17 kpc, this galaxy has an enormous velocity dispersion of 510^{+165}_{-95} km s^{-1} (Figure 4.27), implying a stellar mass of around 2×10^{11} M$_\odot$. How do these galaxies transform into present-day massive ellipticals? Major galaxy–galaxy mergers don't appear to puff up the sizes by enough, though a large number of accretion events of small galaxies might work, as might a significant energy input from supermassive black hole accretion, which we'll discuss in Section 5.8. The implication of having many minor mergers has been seen by some as significant evidence against monolithic collapse models for massive ellipticals; whatever mechanisms are involved, a single monolithic collapse cannot be the whole story. But with so many minor mergers, how can the tight dispersion in the fundamental plane be maintained?

What about elliptical galaxies that form later — why do they conform to the same fundamental plane? One possibility is that red nuggets are the *cores* of giant elliptical galaxies at $z > 1$, and we simply haven't imaged deeply enough to see the diffuse faint outer regions. Clearly, there are still many unanswered questions about the evolution of massive galaxies.

See Mancini, C. et al., 2009, arXiv:0909.3088.

4.8 Morphological K-corrections

Figure 4.28 The Andromeda galaxy, M31, as seen in the ultraviolet.

At high redshifts, optical imaging samples the rest-frame ultraviolet light in galaxies. Might this affect the observed morphologies of high-redshift galaxies?

Examples in the local Universe strongly suggest that it might. Figure 3.10 showed the familiar optical image of the Andromeda galaxy, M31. Compare this to Figure 4.28, which is an ultraviolet (135–275 nm) image taken with NASA's Galaxy Evolution Explorer (GALEX) space telescope. As we saw in Figure 4.6, young hot O stars and B stars emit most of their light in the ultraviolet, so we shouldn't be too surprised to see the ultraviolet images dominated by star formation in H II regions. The rest-frame ultraviolet morphology of M31 is quite different to the rest-frame optical morphology. Could the unusual high-redshift morphologies in the Hubble Deep Fields, such as chain galaxies, similarly be due to the different rest-frame wavelengths at high redshifts?

This effect is sometimes known as **morphological K-corrections** by analogy to the K-correction effect on fluxes discussed in Chapter 1. The key test is whether the rest-frame optical morphologies (observed-frame near-infrared) in the Hubble Deep Fields are the same as those in the rest-frame ultraviolet (observed-frame optical). Some of the highest angular resolution near-infrared images have been made with the HST's NICMOS camera, and the NICMOS imaging of the Hubble Deep Field North found very similar morphologies, as shown in Figure 4.29. At the moment it appears that morphological K-corrections are not a strong effect for most galaxies, though counter-examples in a minority have been found.

4.9 The blue cloud and red sequence

The evolution of stars is completely characterized by their tracks on the Hertzsprung–Russell diagram, also known as the colour–magnitude diagram. A star's position on the colour–magnitude diagram can reveal a great deal about its composition and internal structure, past and future. Galaxies are made (partly) of stars, so can we characterize their evolution on the galaxy colour–magnitude diagram? This approach was first made in 2004 by Eric Bell and generated some useful insights. However, keep in mind that much less of the information about galaxies is encoded in the galaxy colour–magnitude diagram than is the case with stars, because galaxy properties are not uniquely determined by the average luminosity and colour. Figure 4.30 shows the colour–magnitude diagram of galaxies from the COMBO-17 survey.

Bell, E.F. et al., 2004, *Astrophysical Journal*, **608**, 752.

The distribution of galaxies in the colour–magnitude plane is typically segregated into the **red sequence** of red, passively-evolving galaxies and the **blue cloud** of more actively star-forming galaxies. Broadly speaking, the red sequence is occupied by early-type galaxies and the blue cloud by late-type galaxies.

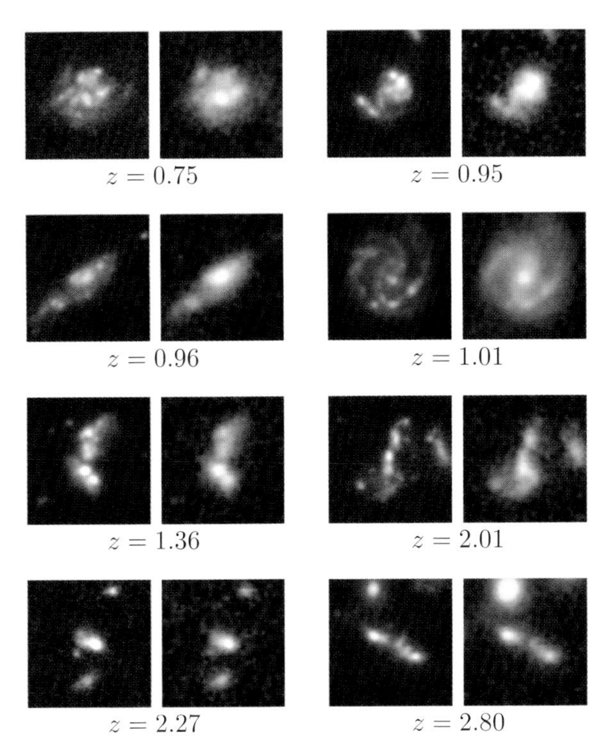

Figure 4.29 Optical and near-infrared (0.8, 1.1, 1.6 μm) morphologies of galaxies in the Hubble Deep Field North, at a variety of redshifts. Most optical and near-infrared morphologies are similar, so morphological K-corrections are not responsible for observed morphological evolution.

The underpopulated narrow region between these two has come to be known as the **green valley**. Some authors have used the red sequence as a *de facto* morphology-independent definition of early-type galaxies at $z < 1$; in fact, one result of the Galaxy Zoo project is that the **colour–density relation** is stronger than the morphology–density relation (Chapter 3) in the local Universe. The red sequence and blue cloud both evolve with redshift (Figure 4.30). The blue cloud becomes slightly redder with time, perhaps because of stellar ageing or increasing dust content. There are also far more luminous blue cloud galaxies at $z > 0.5$ than in the local Universe. Similarly, the red sequence becomes redder on average with time, consistent with passive stellar evolution. Taking into account the effects of passive stellar evolution, the numbers and magnitudes of galaxies in the red sequence imply a build-up of stellar mass in early-type galaxies by a factor of 2 since $z = 1$.

Bamford, S. P. et al., 2009, *Monthly Notices of the Royal Astronomical Society*, **393**, 1324.

These observations could be explained if some star-forming galaxies in the blue cloud stop forming stars, then move to the red sequence and evolve passively. Galaxy evolution in general would then be a story of formation in the blue cloud (or merger-induced starbursts moving a system into the blue cloud), then migration across the green valley to join the red sequence. This, however, begs the question of what mechanism stopped the star formation. There have been suggestions that active galaxies are more common in the green valley, suggesting that AGN activity is somehow responsible for truncating star formation (because the green valley objects might be expected to be transition objects). Furthermore, ultraviolet estimates of the star formation rate in (morphologically) early-type galaxies find that the amount of star formation anti-correlates with the velocity dispersion of the galaxy, but not with the overall galaxy luminosity. We'll see in Chapter 6 that this velocity dispersion is closely linked to the mass of the central supermassive black hole. The truncation of star formation does appear to have something to do with the central black hole.

Schawinski, K. et al., 2006, *Nature*, **442**, 888.

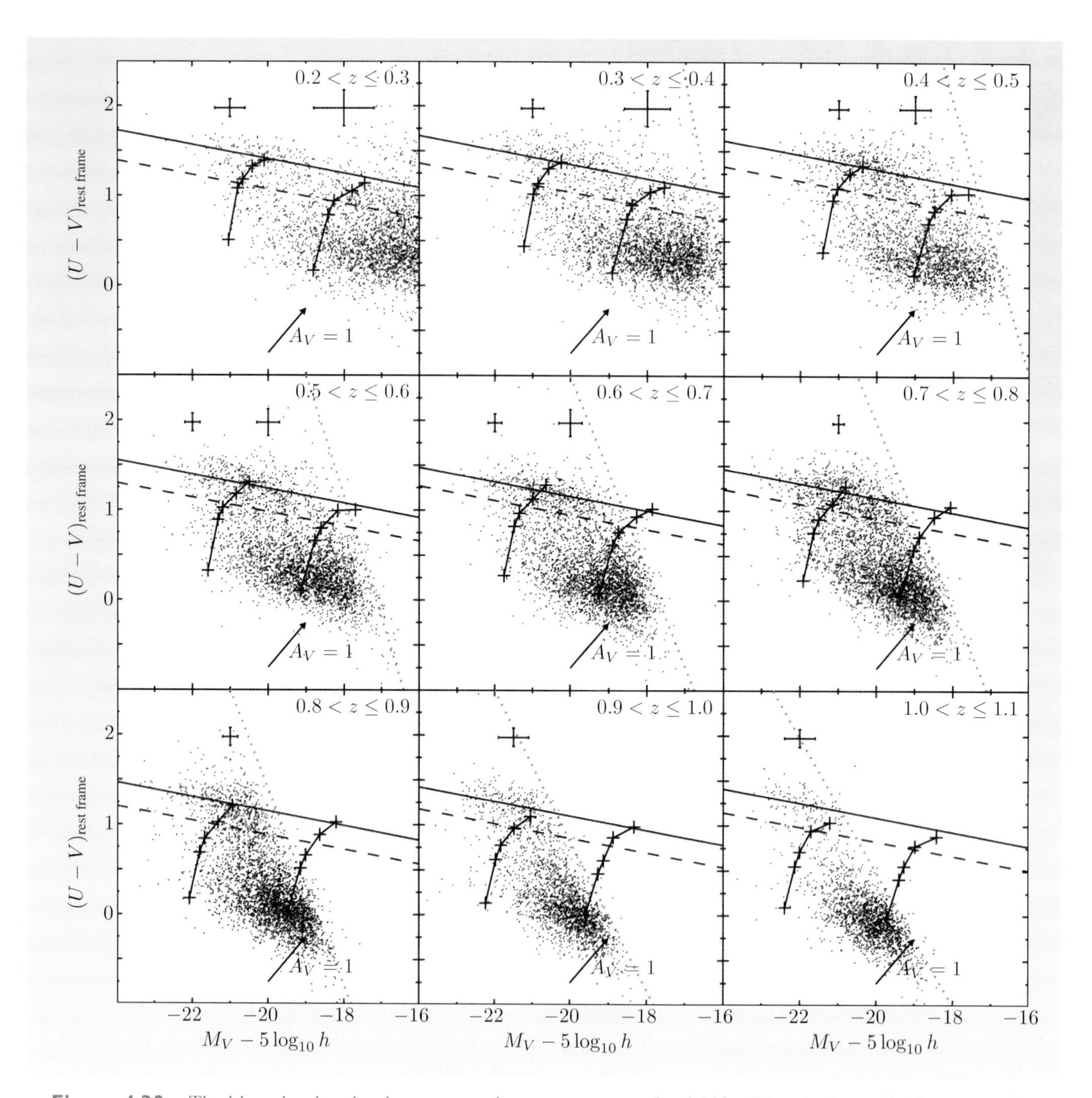

Figure 4.30 The blue cloud and red sequence, shown at a range of redshifts. The sloping solid lines are a fit to the red sequence location, and the dashed line marks the distinction between clouds regarded as 'blue' or 'red'. The sloping dotted line is the approximate apparent magnitude limit of the survey. Simulated evolutionary tracks of some galaxies are shown with lines and crosses. The predicted movement of a galaxy undergoing a reddening of $A_V = 1$ is shown as a vector. Only representative error bars are shown.

A detailed comparison of galaxy evolution models with the evolving blue cloud and red sequence, together with the requirements of needing to reproduce the Madau diagram and the evolution of the total stellar mass density Ω_*, revealed that the number densities of the most massive early-type galaxies could not be reproduced. One possibility is that these most massive early-type galaxies are the result of **dry mergers**, i.e. mergers of galaxies with very little gas, so no star formation results. However, there is currently debate in the community as to whether dry mergers could account for the **mass–metallicity** relation in early-type galaxies, and the structure and sizes of early-types. The mass–metallicity relation is the correlation between stellar mass and the metal enrichment as measured by (for example) emission line ratios in star-forming galaxies. There is much that we have yet to understand about the formation of the most massive elliptical galaxies.

Summary of Chapter 4

1. Dark matter is described as 'cold', 'hot' or 'warm' according to the relative speeds of the dark matter particles.
2. The hierarchical formation model of large-scale structure describes the merger of dark matter haloes to make progressively larger dark matter haloes.
3. Population synthesis models describe the evolution of galaxy spectra by modelling the evolution of stars (and dust) within them.
4. Dust extinction in the V-band is expressed as A_V, measured in magnitudes. This can be measured with the Balmer decrement, among other ways, though the estimates are sensitive to assumptions about the dust distribution.
5. Redshifts can be determined from galaxy emission lines. Redshifts can also be estimated by modelling the changes of observed colours with redshift, a technique known as photometric redshifts. An important example of a photometrically-selected high-redshift galaxy population is the Lyman break galaxy population.
6. The luminosity function of galaxies is the number per unit luminosity (or per decade luminosity), per unit comoving volume. It can be measured using the $1/V_{max}$ statistic.
7. The V/V_{max} values can be used to test for evolution, or in local galaxy samples to test for incompleteness.
8. Type 1 active galaxies have a direct view of the quasar's broad-line region, whereas type 2 active galaxies do not. The type 2 systems can be distinguished from star-forming galaxies using emission line ratios.
9. Quasars evolve strongly, with a peak in number density around $z \simeq 2.5$ and a decline at higher redshifts.
10. The rest-frame ultraviolet luminosity density can be used to measure the comoving star formation density of the Universe, known as the Madau–Lilly diagram or the Madau diagram. This is similar, but it seems not identical, to quasar evolution.
11. The Madau diagram can also be inferred from the ages of stars in local galaxies.
12. The confusion limit restricts how deep a given telescope can image. This limit is conventionally set at 3 or 5 times the noise per beam from background objects.

13. Pencil-beam surveys such as the Hubble Deep Fields have redshift histograms that show peaks due to the large-scale structures through which the surveys pass.
14. Although the rest-frame ultraviolet morphologies of local galaxies can be quite different to the rest-frame optical morphologies, it appears that these differences are not responsible for most of the changing appearance of galaxies in deep Hubble Space Telescope (observed-frame) optical surveys.
15. High-redshift galaxy surveys have used galaxy colour–magnitude diagrams to generalize the local division of early-type and late-type galaxies into the red sequence, the blue cloud and the green valley.

Further reading

- For more on the evolution of large-scale structure, see the graduate-level text Peacock, J.A., 1999, *Cosmological Physics*, Cambridge University Press.
- Alternatively, try Coles, P. and Lucchin, F., 1995, *Cosmology*, Wiley.
- For more on gamma-ray bursts and active galaxies, see Kolb, U., 2010, *Extreme Environment Astrophysics*, Cambridge University Press.
- Antonucci, S., 1993, 'Unified models for active galactic nuclei and quasars', *Annual Review of Astronomy and Astrophysics*, **31**, 473.
- Kennicutt, R.C., 1998, 'Star formation in galaxies along the Hubble sequence', *Annual Review of Astronomy and Astrophysics*, **36**, 189.
- There is a curious analogy between the clustering of galaxies or CMB fluctuations, and the uncertainty principle in quantum mechanics: see Tegmark, M., 1995, *Astrophysical Journal*, **455**, 429, and Tegmark, M., 1996, *Monthly Notices of the Royal Astronomical Society*, **280**, 299.
- For an accessible review on the conundrums posed by red nugget galaxies, see Glazebrook, K., 2009, *Nature*, **460**, 694.
- Binney, J. and Tremaine, S., 2008, *Galactic Dynamics*, Princeton University Press.
- To view John Michell's famous paper of 1767 go to http://adsabs.harvard.edu/abs/1767RSPT...57..234M.

Chapter 5 The distant multi-wavelength Universe

The distant view is not always the truest view.

Nathaniel Hawthorne

Introduction

We've seen how strongly dust can affect optical observations — but what happens hidden behind the dust? One of the great surprises in cosmology in the past decade has been the tremendous amount of star formation and black hole accretion hidden behind heavy dust extinction. This is invisible to optical telescopes but not to other telescopes, as we'll see in this chapter.

5.1 The extragalactic optical and infrared background light

What colour is the sky in space? It would appear black to us, but with a sensitive enough camera we'd be able to detect the background light from all the stars and galaxies that have ever existed. This would appear, if we had the eyes to see it, as a dull reddish colour. Of course, this background is not restricted to optical light: Figure 5.1 shows the extragalactic background at X-ray to radio wavelengths.

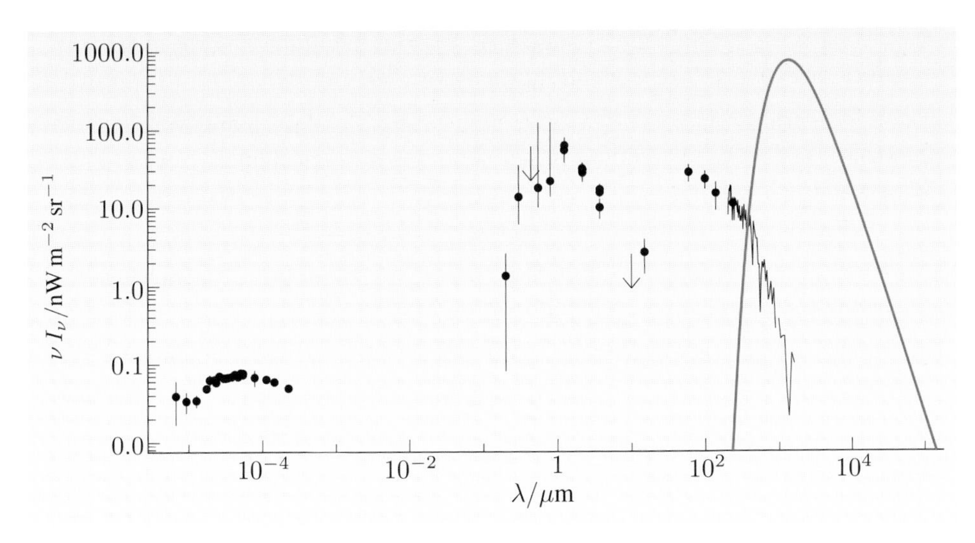

Figure 5.1 The spectrum of the extragalactic sky. To the right is the CMB, peaking around $10^3\,\mu$m, which dominates the energy density of photons in our Universe. Moving to the left, there is the contribution from the far-infrared extragalactic background light peaking at around $10^{1.5}\,\mu$m. Around $10^1\,\mu$m is a minimum. (There are stringent upper and lower limits at $15\,\mu$m, and for clarity we've just drawn this as a single data point, though it's not a direct measurement.) At shorter wavelengths still there is the optical and near-infrared extragalactic background light, peaking around $10^0\,\mu$m. At the shortest wavelengths we have the cosmic X-ray background.

The most striking features of Figure 5.1 are that there's a bump in the optical and near-infrared, a valley at around $15\,\mu$m, and another bump in the far-infrared. The far-infrared bump is believed to be thermal radiation from dust in galaxies. As Exercise 5.1 shows, the energy output in the far-infrared bump is roughly the same as the output in the optical/near-infrared bump. The consequences are profound: roughly speaking, for every two photons created by stars or by black hole accretion, one photon has been absorbed by dust and its energy has been re-radiated as thermal emission by the dust.

Exercise 5.1 The background intensity I_ν per unit frequency is sometimes measured in $\mathrm{W\,m^{-2}\,Hz^{-1}\,sr^{-1}}$. Show that the quantity νI_ν is proportional to the background intensity per decade in frequency (i.e. the intensity per factor of 10 interval in frequency), and hence that there's about as much energy output in the far-infrared bump as in the optical/near-infrared bump. ■

The light from all the stars, galaxies and dust that have ever existed is only a small part of the cosmic photon energy budget. Figure 5.1 compares the cosmic microwave background to the extragalactic far-infrared and optical backgrounds. The energy density from the luminous output of every object in the history of the Universe is only about 5% of the energy density of the CMB.

Another way to think of the extragalactic background from objects is to integrate the contributions as a function of flux. We'll write the flux as S_ν, to refer to the flux density (in, for example, $\mathrm{W\,m^{-2}\,Hz^{-1}}$) for the energy received from a galaxy, per unit time and frequency, measured at a frequency ν. The number of objects in a flux interval $S_\nu \rightarrow S_\nu + \mathrm{d}S_\nu$ is $(\mathrm{d}N/\mathrm{d}S_\nu)\,\mathrm{d}S_\nu$, so the intensity from them must be $S_\nu(\mathrm{d}N/\mathrm{d}S_\nu)\,\mathrm{d}S_\nu$. The extragalactic background must therefore be

$$I_\nu = \int_0^\infty S_\nu \frac{\mathrm{d}N}{\mathrm{d}S_\nu}\,\mathrm{d}S_\nu. \tag{5.1}$$

As we've already seen, a Euclidean source count slope has an integrated background light that diverges at small fluxes (Chapter 1). Therefore the source counts must at some faint flux eventually be less steep than the Euclidean slope. The galaxies that dominate the extragalactic background light per unit interval in flux will be the ones that have the largest values of $S_\nu\,\mathrm{d}N/\mathrm{d}S_\nu$. Nevertheless, it's common in extragalactic astronomy to plot the **Euclidean-normalized differential source counts**, which means $\mathrm{d}N/\mathrm{d}S_\nu$ multiplied by $S_\nu^{2.5}$. If the source counts are Euclidean, then a plot of $S_\nu^{2.5}\,\mathrm{d}N/\mathrm{d}S_\nu$ is a horizontal line. These diagrams are used to illustrate the deviations from the Euclidean slope. Figure 5.2 shows some example Euclidean-normalized counts in the mid-infrared. This figure uses a non-SI unit that is very common in astronomy, namely the **jansky** (symbol Jy), after the radio astronomer Karl Jansky:

$$1\,\mathrm{Jy} = 10^{-26}\,\mathrm{W\,m^{-2}\,Hz^{-1}}. \tag{5.2}$$

- In this plot, a no-evolution curve is plotted. Why is it not horizontal?
- The Euclidean (horizontal) slope assumes a flat *unexpanding* space with no redshift, while the plotted no-evolution model includes these effects.

The source count diagram in Figure 5.2 shows a Euclidean slope, a pronounced bump, and a shallower-than-Euclidean slope at the faintest fluxes. The bumps are broad, so the galaxies in these bumps tend also to be the ones that dominate the extragalactic backgrounds per unit interval in flux at those wavelengths (i.e. $S_\nu\,\mathrm{d}N/\mathrm{d}S_\nu$ is maximum).

Exercise 5.2 We've just seen that the flux interval $\mathrm{d}S_\nu$ that contributes the most background will be the one in which $S_\nu\,\mathrm{d}N/\mathrm{d}S_\nu$ is a maximum. Which logarithmic flux interval $\mathrm{d}\ln S_\nu$ contributes the most background? ■

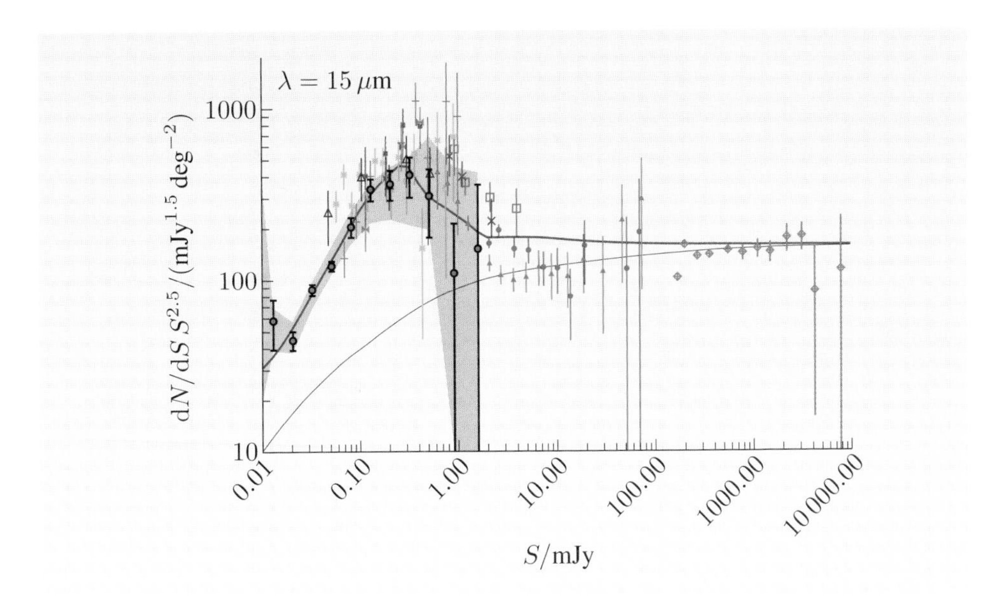

Figure 5.2 Galaxy differential source counts from a variety of sources, normalized to the Euclidean prediction, at an observed wavelength of 15 μm. Also shown are a no-evolution model (black) and a galaxy evolution model that better fits the data (red).

Another way of thinking about these backgrounds is as an integral over redshift of the galaxy populations. Suppose that the comoving luminosity density at some redshift z and some frequency ν is $E_\nu = \int L_\nu\,\phi(L_\nu)\,\mathrm{d}L_\nu$ (so E_ν could be measured in, for example, W per Hz per cubic comoving Mpc). In general, this luminosity density E_ν will depend on redshift (because galaxies evolve) and on frequency (because galaxies don't have flat featureless spectra). We'll write this as $E_\nu = E_\nu(\nu, z)$. The subscript ν is there as a reminder that it's per unit frequency. To find the background energy density $B_\nu(\nu_0)$ at some observed wavelength ν_0, we can just add up the contributions to the comoving intensity throughout time, taking account of the fact that the rest-frame wavelength is different to the observed wavelength by a factor of $(1+z)$:

$$B_\nu(\nu_0) = \int_{t=0}^{t_0} E_\nu((1+z)\nu_0, z)\,\mathrm{d}t, \tag{5.3}$$

where t_0 is the present-day age of the Universe, as in Chapter 1. To convert this to an energy flux density, measured in (for example) $\mathrm{J\,s^{-1}\,Hz^{-1}\,sr^{-1}}$, we multiply this by $c/(4\pi)$:

$$I_\nu(\nu_0) = \frac{c}{4\pi}\int_{t=0}^{t_0} E_\nu((1+z)\nu_0, z)\,\mathrm{d}t. \tag{5.4}$$

It turns out that this can also be expressed quite simply as an integral over comoving distance. From Equation 1.42 we have that

$$\mathrm{d}d_{\mathrm{comoving}} = \frac{-c\,\mathrm{d}z}{H},$$

and using Equation 1.28 we can write this as

$$\mathrm{d}d_{\mathrm{comoving}} = \frac{c\,\mathrm{d}z}{(1+z)^{-1}\,\mathrm{d}z/\mathrm{d}t} = c(1+z)\,\mathrm{d}t.$$

So Equation 5.4 becomes an integral of the comoving luminosity density over comoving distance:

$$I_\nu(\nu_0) = \frac{1}{4\pi}\int E_\nu\,\frac{\mathrm{d}d_{\mathrm{comoving}}}{(1+z)}. \tag{5.5}$$

Now, we've already seen how the galaxies that dominate the ultraviolet luminosity density can dominate the (optically-derived) cosmic star formation history. What this means is that the galaxies that dominate the cosmic star formation history at some redshift are necessarily the same ones that contribute the most to the extragalactic background, at that frequency and redshift. Finding out which galaxies dominate the extragalactic background light is a very similar research problem to measuring the cosmic star formation history.

Reproducing the extragalactic background light is therefore a key objective of **source count models**, one of which is plotted in Figure 5.2. These models aim to account for the observed number counts and (where available) redshift distributions, at all wavelengths, and reproduce the present-day stellar mass density Ω_*. There are many approaches. One approach is to vary the numbers of galaxies of different types with redshift, and find the best-fit evolution for this assumed population mix. Another approach is semi-analytic modelling, in which the locations of dark matter haloes are given by a numerical model, and the haloes are populated by galaxies based on some physical assumptions with adjustable free parameters. Ideally, one would like to simulate a cosmological volume right down to the scales of the formation of stars in molecular clouds, but this is a very long way from being computationally possible, so many source count models rely on template galaxy **spectral energy distributions** (SEDs), i.e. template spectra from the ultraviolet to the far-infrared and beyond, to extend the galaxy number count predictions to different wavelengths. These templates are sometimes predictions from numerical radiative transfer models of dust and stars in particular galaxies, or sometimes taken directly from observations.

5.2 Submm galaxies and K-corrections

Imagine that you had some spectacular power to push a galaxy to higher redshift. Ordinarily, you'd expect the galaxy to appear fainter. We saw in Chapter 1 that the difference between observed and rest-frame wavelengths changes the observed flux of a high-redshift galaxy — an effect known as the K-correction. At submm wavelengths and mm wavelengths, galaxy spectra are dominated by the steep Rayleigh–Jeans slope of the thermal dust emission. The K-correction is therefore extremely strong and can make a more distant galaxy *brighter* than a closer identical galaxy! Figure 5.3 shows a redshifted galaxy template spectrum, keeping its luminosity constant. At most observed wavelengths the galaxy is fainter at high redshifts, except at wavelengths around 1 mm. Figure 5.4 shows how the observed flux of a dusty galaxy is predicted to vary with redshift.

But in order to make use of this K-correction, we need cameras that operate at submm or mm wavelengths. The first was the Submillimetre Common User Bolometer Array (SCUBA) on the Anglo–Dutch–Canadian James Clerk Maxwell Telescope (JCMT) in Hawaii. Prior to SCUBA, the best available models of galaxy evolution predicted zero or at most one galaxy detectable in even the deepest single SCUBA exposures. However, it's often the way that opening up a new wavelength regime in astronomy brings unexpected discoveries, and SCUBA's image of the HDF-N field and of galaxy cluster lenses (Figure 5.5; see also Chapter 8) revealed a completely unanticipated population of far-infrared-luminous high-redshift galaxies!

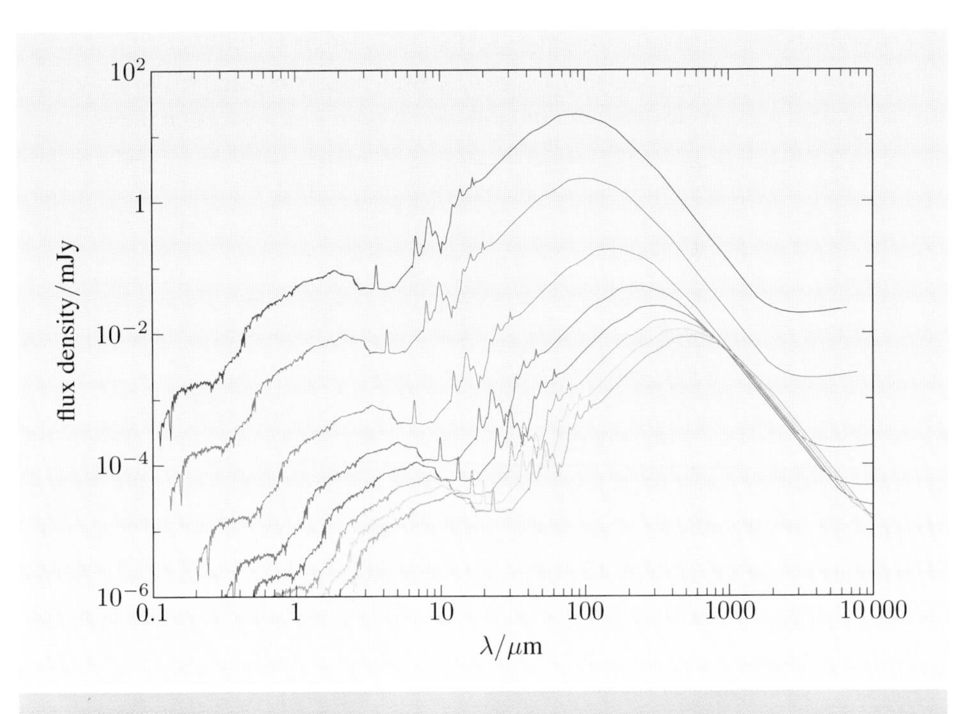

Figure 5.3 The spectral energy distribution of the star-forming galaxy M82, as it would appear at different redshifts. The colours from blue to red refer to redshifts of 0.1, 0.3, 1, 2, 3, 4, 5 and 6. Note that at submm-wave and millimetre wavelengths, the K-corrections are strong enough to counter the cosmological dimming. The same galaxy is shown on the cover of this book. M82 is often used as a template SED for a starburst galaxy.

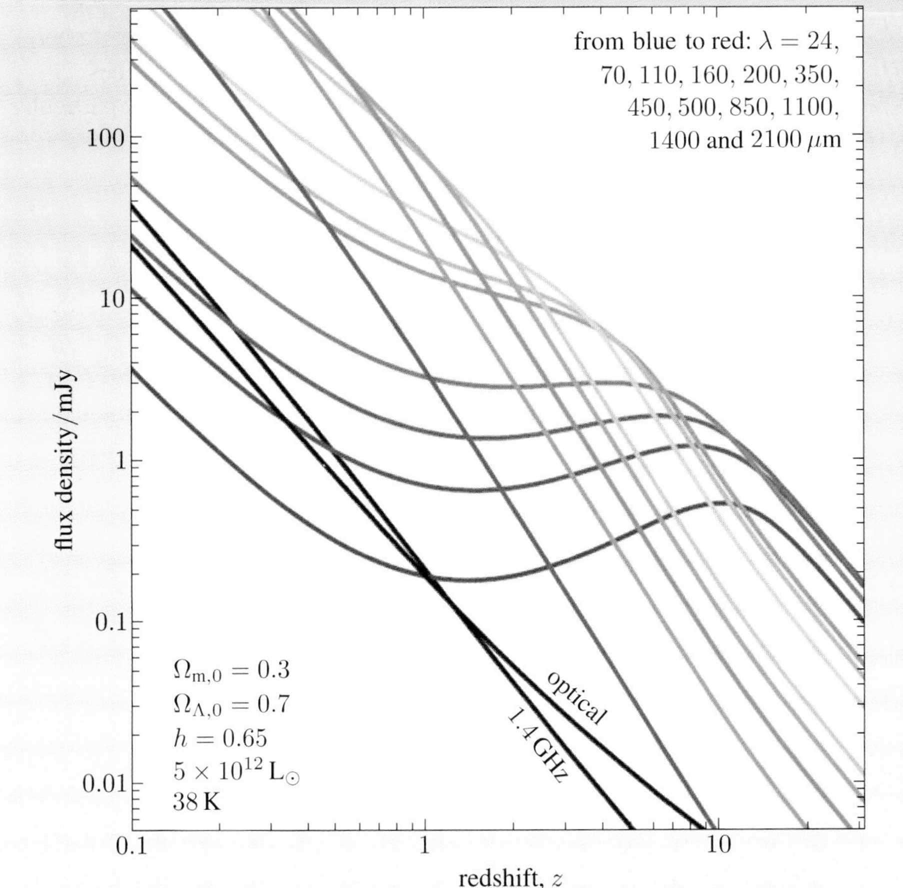

Figure 5.4 The flux variation with redshift of a typical star-forming galaxy with a fixed luminosity, at a variety of observed-frame wavelengths.

These galaxies were soon known as 'SCUBA galaxies', but as other cameras became available (with names such as MAMBO, BOLOCAM, AzTEC, SHARC-II, BLAST, LABOCA) the more generic term of **submillimetre galaxies** (or **SMGs**) became current. The selection function of SMGs (or their mm-wave counterparts MMGs) is strikingly uniform, as you can see in Figure 5.4: an SMG at $z = 10$ would have almost the same brightness as an identical galaxy at $z = 1$.

- Would the histogram of redshifts of SMGs also be uniform?
- No. Even if the number density of SMGs didn't evolve, we'd still be sampling different amounts of comoving volume at different redshifts, i.e. $\mathrm{d}V/\mathrm{d}z$ is not constant.

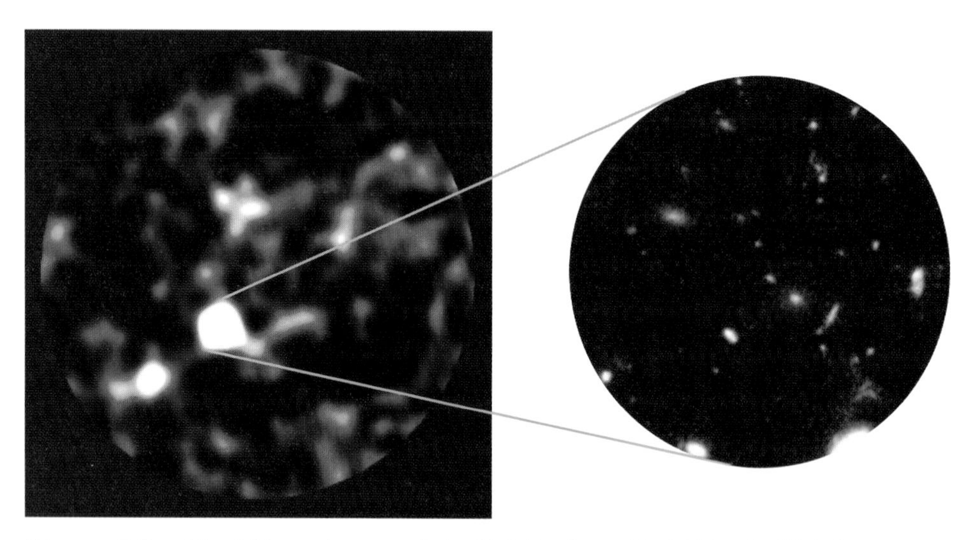

Figure 5.5 The 850 μm image of the Hubble Deep Field North. This image has a radius of 100 arcseconds. The zoom shows the corresponding Hubble Space Telescope data in the region of the brightest SMG.

An intense campaign of optical spectroscopy with the Keck telescopes found a median redshift of around $z = 2.2$ for SMGs.[6] Even without redshifts, the far-infrared luminosities suggested star formation rates of around $1000\,\mathrm{M}_\odot$ per year. (Because submm flux is more or less independent of redshift at $1 < z < 10$, the luminosities can be estimated even without redshifts.) We'll discuss in Section 5.7 how SMGs changed our physical picture of galaxy formation and evolution.

Bigger submm- and mm-wave cameras have led to larger surveys of submm- and mm-wave galaxies. One daring experiment to survey parts of the sky at submm wavelengths involved dangling a telescope with a 2 m primary mirror from a weather balloon! In 2006 the Balloon-borne Large Aperture Submm Telescope (BLAST) flew for 11 days around the South pole and mapped the Chandra Deep Field South (CDF-S) field (among other fields), shown in Figure 5.6. Highly redshifted galaxies should be visible at the longest wavelengths, but less visible at the shorter wavelengths because the peak of the emission has redshifted past (Figure 5.3). BLAST used the same detector technology as the SPIRE instrument on the ESA Herschel Space Observatory, which launched in 2009. The first images from Herschel have been spectacular. Figure 5.7 shows the local spiral galaxy M74. Even in this short exposure, there are many background galaxies, which appear to be clustered.

[6]Chapman, S.C. et al., 2005, *Astrophysical Journal*, **622**, 772.

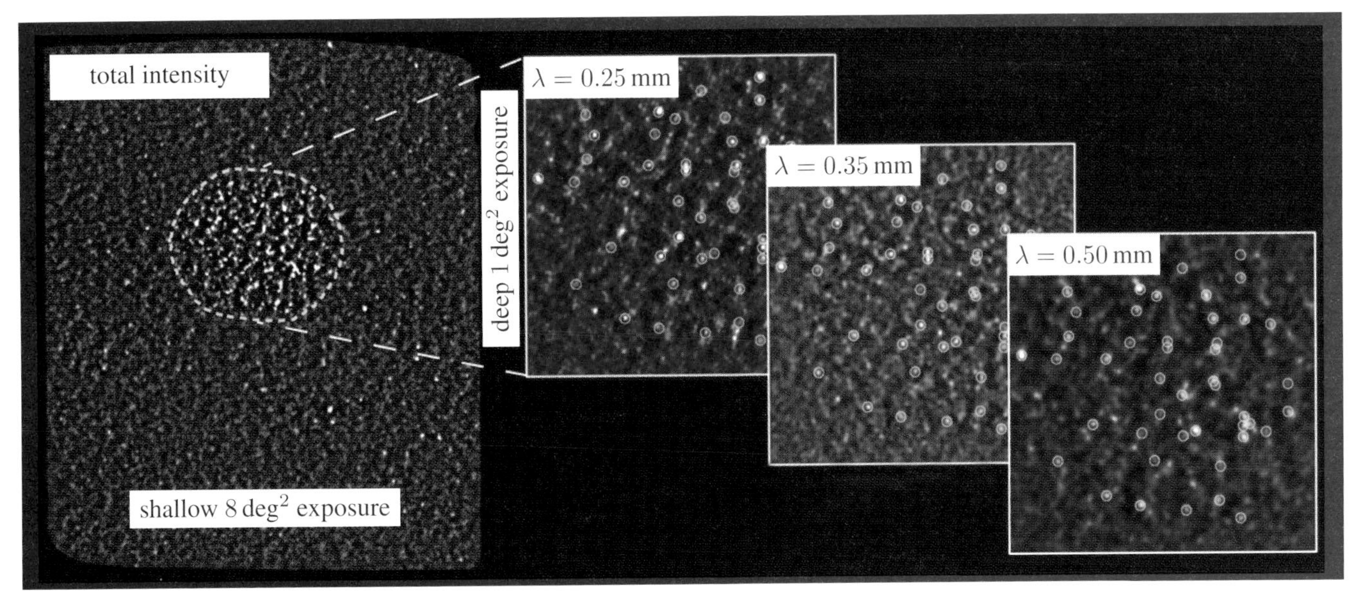

Figure 5.6 Submm-wave maps of the Chandra Deep Field South region from the BLAST mission. For reference, the same features have been circled in each of the images. Total intensity = sum of the three wavelengths.

But what are these SMGs, apart from being galaxies detected at submm wavelengths? To find out, we need to cross-match the submm-wave objects with images or catalogues at other wavelengths, including the optical if we want to take optical spectroscopy of the optical counterpart. However, the **diffraction limit** of telescopes makes this difficult: the angular resolution of a circular aperture is $1.22\lambda/D$ radians, where λ is the wavelength of the light and D is the diameter of the aperture.

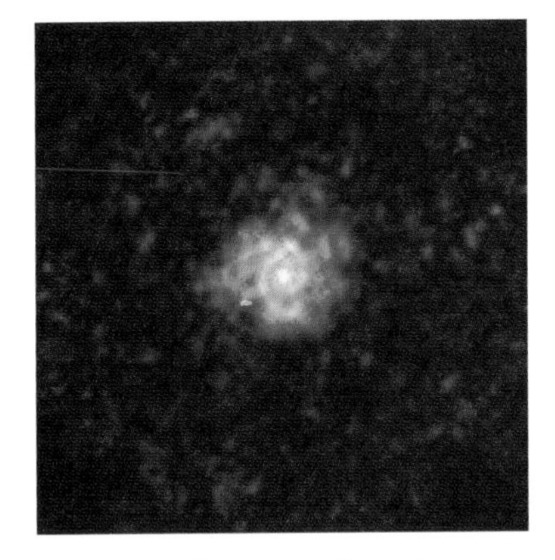

Figure 5.7 First-light image from the Herschel Space Observatory at $250\,\mu$m, of the nearby galaxy M74. There are also many background SMGs visible.

Exercise 5.3 Calculate the angular resolution in arcseconds of the Herschel Space Observatory (diameter 3.5 m) at its longest imaging wavelength of $500\,\mu$m. ■

Figure 5.5 illustrates the problem. Submm-wave imaging has the benefit of negative K-corrections, but imaging at other wavelengths does not. The positions of SMGs are typically only accurate to a few arcseconds or tens of arcseconds, but within this area there can be many optical galaxies. Ultra-deep radio mapping has managed to detect around half of the SMGs but the rest were unidentified in even the deepest 1.4 GHz images of the sky. It's possible to improve the positions using submm- or mm-wave interferometry, but this takes very long exposures and is currently feasible only for the brightest SMGs.

This will change dramatically with the Atacama Large Millimeter/Submillimeter Array (ALMA), a huge interferometer in Chile that is scheduled to start full operations no earlier than 2012. In the meantime, the Spitzer Space Telescope achieved a breakthrough in identifying SMGs: soon after its launch in 2003, it found that it could identify SMGs in only ten-minute snapshots. There does appear to be some considerable overlap between the populations of galaxies seen by Spitzer, and (except for a minority at very high redshifts) the population of SMGs. There are also good indications from stacking analyses that the galaxies that dominate the extragalactic background light at Spitzer's mid-infrared wavelengths are largely the same as the populations that dominate the submm-wave extragalactic background at $< 500\,\mu$m. However, much of the background at 850–1100 μm is still unaccounted for.

We've already met the population of Extremely Red Objects (EROs), some of which appear to be dusty starbursts, particularly at fainter K-band apparent magnitudes. It turned out that approximately 10–20% of SMGs are also ERO galaxies, and early indications are that a similar fraction appear to be BzK starbursts. There is now a considerable variety of definitions of various types of red galaxies, with often partly overlapping memberships, such as **Distant Red Galaxies** (DRGs) with (J–K) > 2.3, or **Dust Obscured Galaxies** (DOGs) with $S_{24}/S_{\rm R} > 1000$ (where $S_{\rm R}$ is the R-band flux, and S_{24} is the 24 μm flux). DRGs and BzK galaxies contribute tens of percent to the cosmic submm background light.

See, for example: Pope, A. et al., 2008, *Astrophysical Journal*, **689**, 127; Knudsen, K.K. et al., 2005, *Astrophysical Journal Letters*, **632**, 9; Takagi, T. et al., 2007, *Monthly Notices of the Royal Astronomical Society*, **381**, 1154.

One final subtlety is that selection effects may have an insidious effect on the types of galaxies seen in the far-infrared and submm. If there are populations of galaxies with lots of cool dust radiating predominantly at longer wavelengths, we might expect these galaxies to be over-represented in SMG samples. Similarly, galaxies selected in the far-infrared at, say, 70 μm, may tend to have warmer colour temperatures. It is too early to say definitively if this is the case, and to what extent these biases operate, but several observations have been found consistent with the presence of these subtle biases.

5.3 Ultraluminous and hyperluminous infrared galaxies

SMGs were not the first tremendous surprise in infrared extragalactic astronomy. The US/UK/Dutch Infrared Astronomy Satellite (IRAS) mapped 98% of the sky at 12, 25, 60 and 100 μm, and as well as detecting many tens of thousands of star-forming galaxies, it soon became clear after its launch in 1983 that there were many galaxies with luminosities of $\simeq 10^{12}\,{\rm L}_\odot$ or more, about 1–2 orders of magnitude more luminous than the Milky Way, yet with most of the energy output in the infrared. It was not immediately clear what caused these high luminosities, but violent starbursts or dust-shrouded active nuclei were suspected. All-sky surveys are also very useful for finding very rare populations, and 1991 saw the spectacular discovery by Rowan-Robinson et al. of the galaxy IRAS FSC 10214+4724. (Here FSC stands for 'faint source catalogue' and the numbers refer to approximate right ascension and declination coordinates.) Most IRAS galaxies were found at redshifts of < 0.1, and even the $10^{12}\,{\rm L}_\odot$ objects rarely exceeded $z = 0.2$, but this galaxy was found at the (then) tremendously high redshift of $z = 2.286$ (Figure 5.8), with a derived far-infrared luminosity of $3 \times 10^{14}\,{\rm L}_\odot$. This was the most luminous object known so far in the Universe, and given the predictions of monolithic collapse galaxy formation models, this galaxy excited a great deal of interest as a candidate 'protogalaxy'.

Houck, J.R. et al., 1985, *Astrophysical Journal Letters*, **290**, 5.

Rowan-Robinson, M. et al., 1991, *Nature*, **351**, 719.

Astronomy has a tendency as a field to be fond (perhaps overly fond) of classifying. These discoveries of infrared-luminous galaxies led to the classes of **luminous infrared galaxies** (or **LIRGs**) with 10^{11}–10^{12} solar luminosities, **ultraluminous infrared galaxies** (or **ULIRGs**) with 10^{12}–$10^{13}\,{\rm L}_\odot$, and

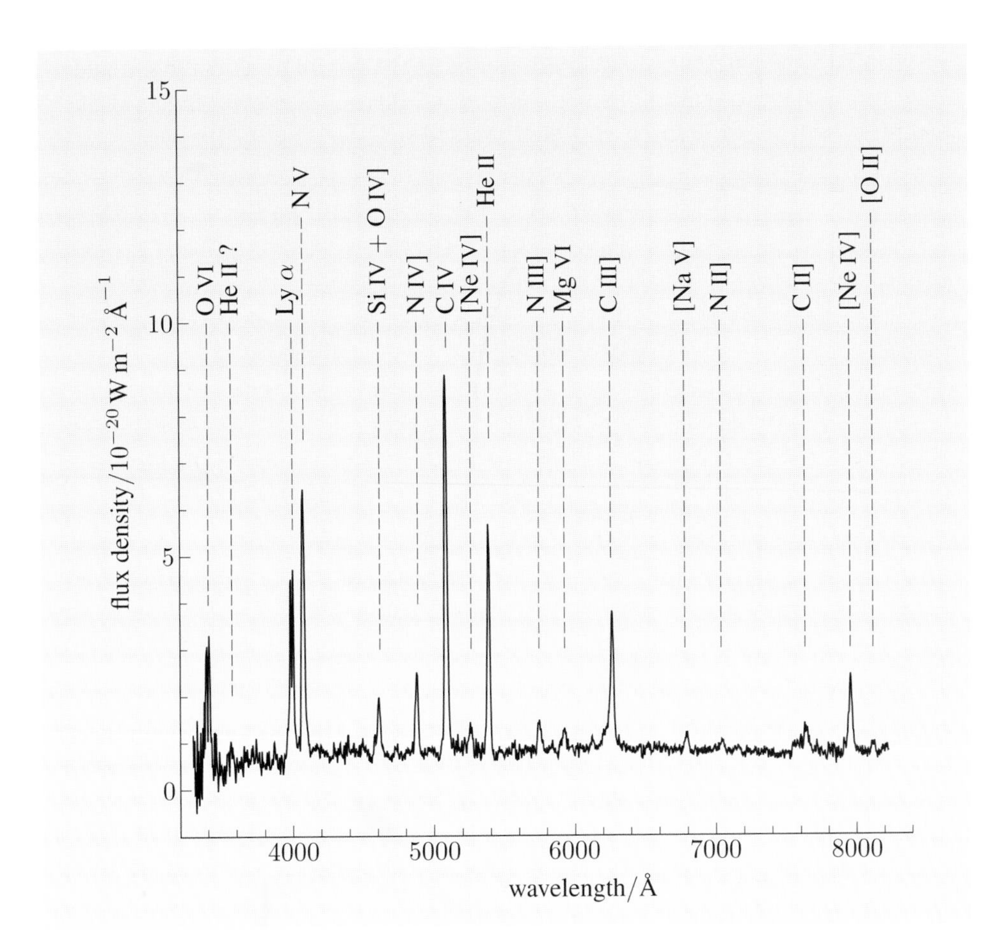

Figure 5.8 Spectrum of the redshift $z = 2.286$ hyperluminous galaxy IRAS FSC 10214+4724. Note the numerous emission lines from ionized gas, some of which are characteristic of starburst galaxies, and others characteristic of active galaxy narrow line regions.

hyper-luminous infrared galaxies (or **HLIRGs**) with $> 10^{13}\,L_\odot$. Some models predicted different physical mechanisms driving the evolution of these classes, so these divisions were not without physical motivation, though these interpretations were by no means unique. Figure 5.9 shows HST morphologies of ULIRGs; it appears that major galaxy–galaxy mergers are important in the local Universe in triggering ultraluminous starbursts. There was a twist to the story of IRAS FSC 10214+4724: the enormous apparent luminosity turned out to be in part due to gravitational lensing, as we'll see in Chapter 7, though it remains a prototypical HLIRG, albeit a lensed one.

As a joke I once tried to coin the term 'überluminous' in a paper describing hypothetical $\sim 10^{14}$–$10^{15}\,L_\odot$ galaxies, but the referee (perhaps quite rightly) would have none of it!

IRAS was followed by the ESA Infrared Space Observatory (ISO) in 1995. As an observatory rather than a sky survey, it specialized in a few deeper surveys and follow-ups of individual objects. NASA's Spitzer Space Telescope, launched in 2003, had a primary mirror with a similar diameter to the ISO, but enormously more sensitive detectors. Both the ISO and Spitzer resulted in the discovery of many new ULIRGs and HLIRGs, as well as shedding (metaphorical) light on many star-forming galaxies. Figure 5.10 shows the Antennae galaxies, a pair of colliding galaxies observed with the HST and with Spitzer. Some of the heavily-extincted regions in the HST image are strongly luminous at mid-infrared wavelengths in the Spitzer image. In the local Universe, ULIRGs contribute a negligible amount to the cosmic star formation density, but the discovery of SMGs and the surveys by Spitzer have demonstrated that the fractional contribution from ULIRGs to the cosmic star formation history increases strongly with redshift, as we'll see.

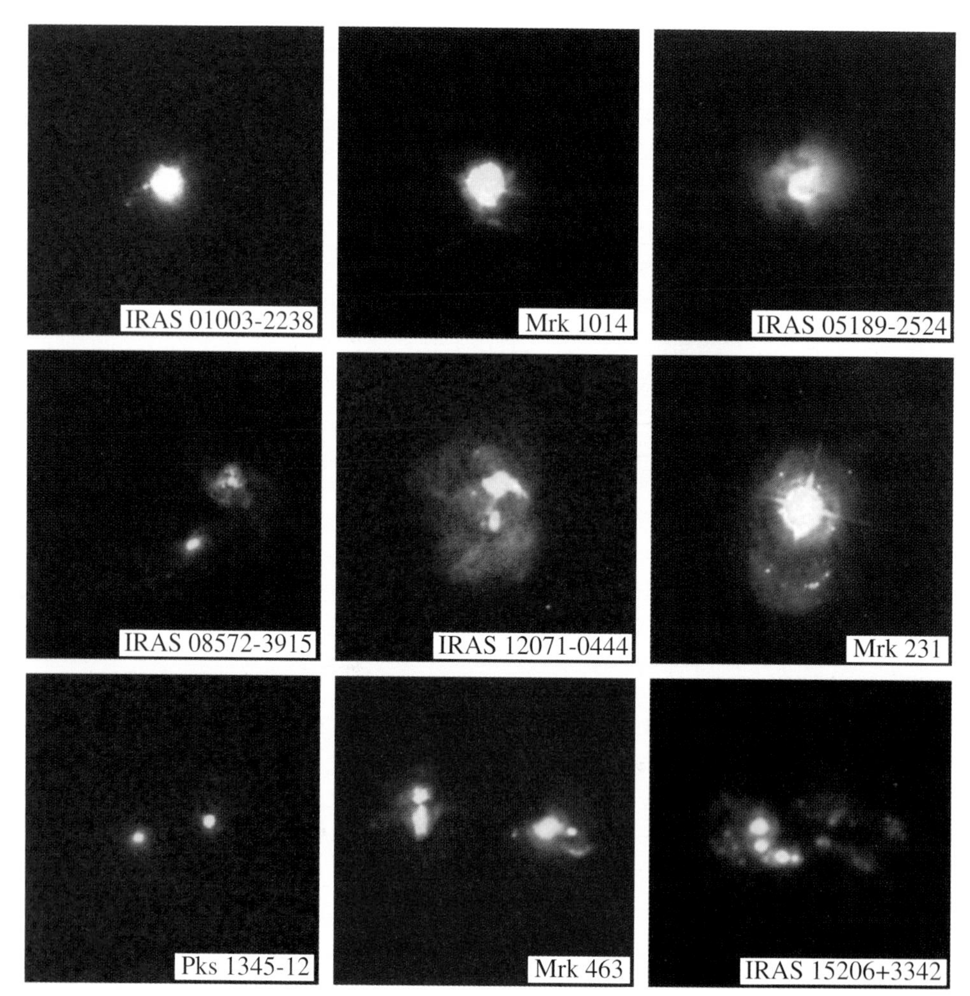

Figure 5.9 HST imaging of ultraluminous infrared galaxies in the local Universe. Many contain quasars, which appear as bright point sources, but all have at least some evidence of a violent galaxy–galaxy merger.

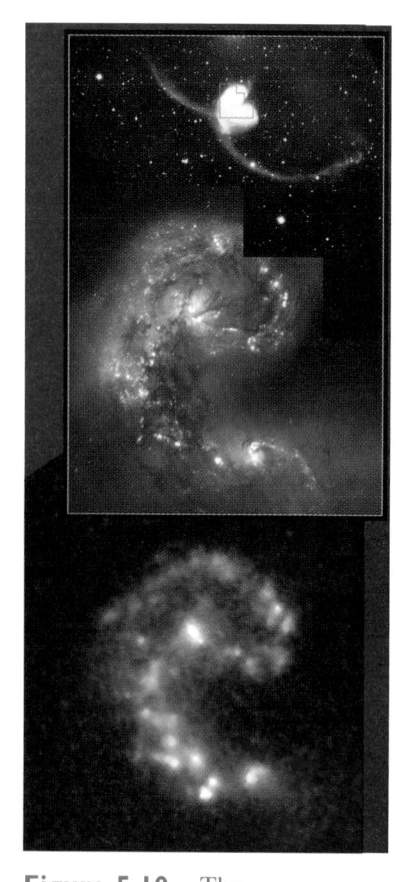

Figure 5.10 The Antennae galaxies, a pair of colliding galaxies. The top black-and-white image shows a ground-based optical image, with a green overlay that shows the area of the HST optical imaging (centre). The lower image is the same field seen by the Spitzer Space Telescope at wavelengths of 3.6–8 μm. There is a large dusty region that is dark to the HST, but luminous at longer wavelengths. The widths of the two lower images are about 2.3 arcminutes, or about 13 kpc at the distance of these galaxies.

The thermal emission from dust is not entirely black body, because the wavelength of the light emitted can sometimes be of the same order as the dust grain size. As a result, the black body spectrum is modified by a wavelength-dependent factor k_d, typically in the range $k_\mathrm{d} \propto \lambda^{-1}$ to λ^{-2}. This index (1 to 2 in this case) is often given the symbol β and is known as the **grey body emissivity index**. It is usually found empirically by observing the Rayleigh–Jeans tail. Knowledge of this index is important in calculating dust masses from the thermal spectra of galaxies. If we are observing dust with a *single* temperature, the dust mass M for a given flux S_ν at an observed frequency ν would be

$$M = \frac{1}{1+z}\,\frac{S_\nu d_\mathrm{L}^2}{k_\mathrm{d}(\nu_\mathrm{rest})\,B(\nu_\mathrm{rest},T)}, \tag{5.6}$$

where d_L is the familiar luminosity distance (Chapter 1), $\nu_\mathrm{rest} = \nu/(1+z)$, and $B(\nu,T)$ is the Planck black body function.[7] However, it's physically implausible that the single-temperature approximation holds in practice, so some research

[7]See, for example, Hughes, D.H., Dunlop, J.S. and Rawlings, S., 1997, *Monthly Notices of the Royal Astronomical Society*, **289**, 766.

groups opt instead to use sophisticated numerical radiative transfer models of the three-dimensional heating of dust. (Sometimes these models impose a simplifying symmetry, such as cylindrical or spherical symmetry, for more rapid computations.) The single-temperature approximation may sometimes nevertheless still be useful for order-of-magnitude estimates of dust masses.

Exercise 5.4 The factor k_d has a large uncertainty. Estimates range from $0.04\,m^2\,kg^{-1}$ to $0.3\,m^2\,kg^{-1}$ at a rest-frame wavelength of $800\,\mu m$, i.e. a range of about a factor of 7, though more typical values are $0.15 \pm 0.09\,m^2\,kg^{-1}$. For an $850\,\mu m$ observation of an SMG at a redshift of $z = 3$, what would the fractional range be in the possible dust mass assuming the following?

(a) $k_d = 0.15 \pm 0.09\,m^2\,kg^{-1}$, a grey body emissivity index of $\beta = 1.5$ and a fixed temperature.

(b) The same as (a), except also allowing a grey body emissivity index of $\beta = 1$–2.

(c) The same as (b), but also allowing a range of assumed temperature of $T = 20$–$40\,K$ (a wide but not unreasonable range for galaxies).

What advantages are there to measuring fluxes at more wavelengths than just $850\,\mu m$? ■

Although undoubtedly difficult to measure, the dust masses are often key predictions for galaxy evolution models, especially those in which giant elliptical galaxies form at high redshifts, converting gas to stars in high star formation rates and generating large dust masses.

5.4 Measuring star formation rates

We've already met how to use the ultraviolet comoving luminosity density to estimate the star formation history of the Universe, and discussed the systematic uncertainties from the IMF and from dust obscuration. There are other ways of making these estimates that are less prone to dust obscuration. We'll see in Section 5.6 that this leads to an astonishing conclusion that is similar to the result from the extragalactic background light.

The short-lived but luminous O stars and B stars ionize their environments and cause strong Balmer emission lines, as we found in Chapter 4, as well as [O II] 372.7 nm emission and Lyman α 121.6 nm emission. The Hα 656.3 nm Balmer line, [O II] line and Lyman α line, once corrected for extinction, could therefore be used to estimate star formation rates. Longer wavelengths are less prone to dust extinction, as we saw in Chapter 4, giving Hα an advantage, but it redshifts out of the observed-frame optical (i.e. $> 1\,\mu m$) above a redshift of a half or so. Infrared spectra can be measured but so far not in such large numbers as optical spectra, because of technological limitations. As we saw in Chapter 4, shorter wavelength observations are dominated by lower extinction regions, so longer rest-frame wavelengths directly sample more of the star formation, but Hα is still a long way from being an extinction-independent measure of star formation.

However, the energy absorbed by the dust in star-forming regions does not go away; it is re-radiated as thermal radiation by the dust. The peak emission from dust in galaxies is roughly around 70–130 μm, as shown in Figures 5.1 and 5.3. Could we use this as our star formation rate indicator? How do we know that the dust is heated by star formation, and not (for example) just by the ambient interstellar radiation or from an active nucleus (Chapter 6)? In our Galaxy, the ambient interstellar light heats the ambient dust, and the thermal radiation from that dust has been detected in the all-sky far-infrared surveys from IRAS and the Japanese AKARI space telescope. This dust has come to be known as **cirrus** owing to its wispy appearance. (This foreground cirrus structure also places a similar limit to some deep-field observations to point source confusion, known as **cirrus confusion noise**. The power spectrum of cirrus is approximately $P(k) \propto k^{-3}$, where k is inverse angle, so cirrus is smoother on smaller scales, thus observations with larger beams are more susceptible to cirrus confusion.)

See, for example, Gautier, T.N. III et al., 1992, *Astronomical Journal*, **103**, 1313.

Ultimately, the case for using the far-infrared luminosity to measure star formation (like all other estimators) rests on astrophysical plausibility. Radiative transfer models predict that the cirrus contribution is cooler than star-forming giant molecular clouds (Orion, for example, is warmer than Galactic cirrus, shown in Figure 5.11), and in star-forming galaxies the cirrus component is predicted to be lower luminosity than the far-infrared emission from star formation. Supermassive black hole accretion (Chapter 6) in a galaxy's active nucleus also heats dust in the circumnuclear torus, but most models predict this contribution to dominate in the mid-infrared rather than at longer wavelengths. Far-infrared measurements capture the obscured star formation but still aren't free of IMF assumptions, because the massive stars (above $5\,M_\odot$ or so) dominate the dust heating.

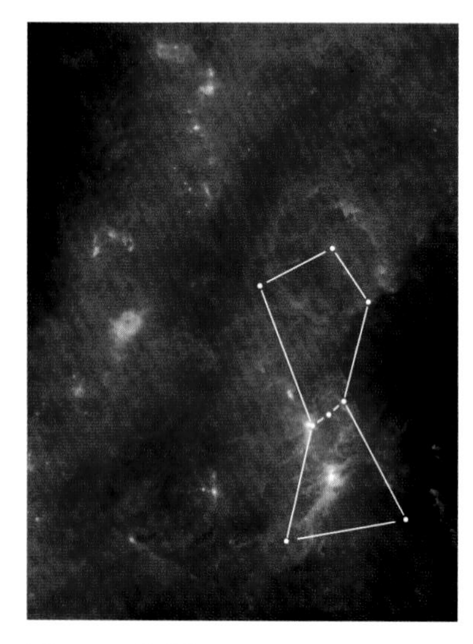

Figure 5.11 The Orion nebula, as seen by the AKARI space telescope at 140 μm. The constellation itself is marked in white. The bright far-infrared knot at the location of the nebula (inside the bottom of the constellation) is caused by dust-shrouded star formation.

Another star formation rate indicator relies on an entirely different physical process, and is completely independent of dust obscuration: the radio luminosity. Supernovae from massive stars accelerate charged particles, which spiral along the field lines of the galaxy's magnetic field, emitting synchrotron radiation. The synchrotron luminosity should therefore be proportional to the recent supernova rate in the galaxy. This synchrotron radiation dominates at radio wavelengths (e.g. metre-scale) with a power-law spectrum $S_\nu \propto \nu^{-\alpha}$ with $\alpha \simeq 0.7$ to 1.0 depending on the energy distribution of the charged particles (which itself depends on time). Dust clouds are transparent to this radiation, and synchrotron-emitting regions are themselves optically thin to synchrotron radiation, so the radio luminosity is ostensibly obscuration-independent. However, it's still subject to the IMF. Figure 5.12 shows schematically how stars of different masses contribute to the radio, far-infrared and ultraviolet luminosities. All three are sensitive to massive stars, to varying degrees, but none covers stars less massive than $5\,M_\odot$.

These three star formation rate indicators are sensitive to different parts of the galaxy too: the ultraviolet traces the unobscured regions, the far-infrared traces the dust-shrouded regions, and the radio traces the total. Perhaps it's not too surprising, then, that the radio and far-infrared luminosities of galaxies correlate, as in Figure 5.13. This correlation was discovered by George Helou, Tom Soifer and Michael Rowan-Robinson in 1985. However, the tightness of the **radio–far-infrared correlation** over nearly four orders of magnitude in luminosity is an unsolved puzzle, as is the physical origin of the normalization. The radio–ultraviolet and far-infrared–ultraviolet correlations, meanwhile, are less tight.

Helou, G., Soifer, B.T. and Rowan-Robinson, M., 1985, *Astrophysical Journal Letters*, **298**, 7.

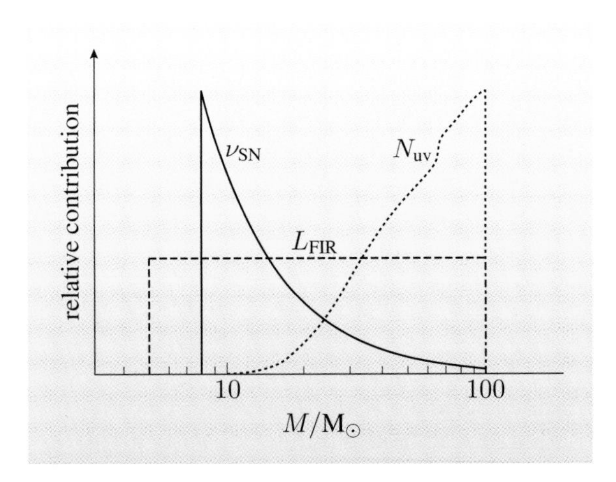

Figure 5.12 The relative contributions to the far-infrared luminosity, the supernova rate and the ultraviolet ionizing radiation, per logarithmic interval of stellar mass, assuming an IMF varying as $M^{-5/2}$ up to $100\,M_\odot$.

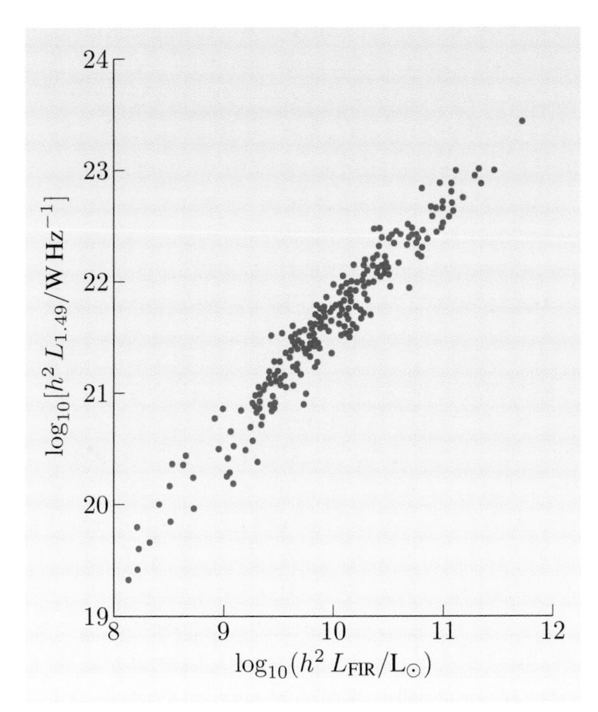

Figure 5.13 The correlation between far-infrared and radio (1.49 GHz) luminosities of star-forming galaxies. Galaxies with active nuclei have been excluded from this plot.

Ideally, we'd like to track the cosmic star formation history with far-infrared, radio and ultraviolet tracers at the same time. However, we've seen that the **diffraction limit** of telescopes limits far-infrared observations. Moreover, the atmosphere is opaque in most far-infrared wavelengths, and the Earth's atmosphere is strongly luminous in the few transparent windows.

Exercise 5.5 Why would a high background flux from (for example) the Earth's atmosphere limit astronomical observations? Demonstrate your answer using Poisson statistics. ■

Radio observations get around the diffraction limit with interferometry, but this is technically very challenging in the far-infrared partly because of the more stringent timing requirements at higher frequencies. A solution to the terrestrial sky background is space telescopes, but it's difficult to launch large primary mirrors into space. The largest so far is Herschel (Exercise 5.3). The proposed Japanese SPICA space telescope will have a similar aperture, but will cool the

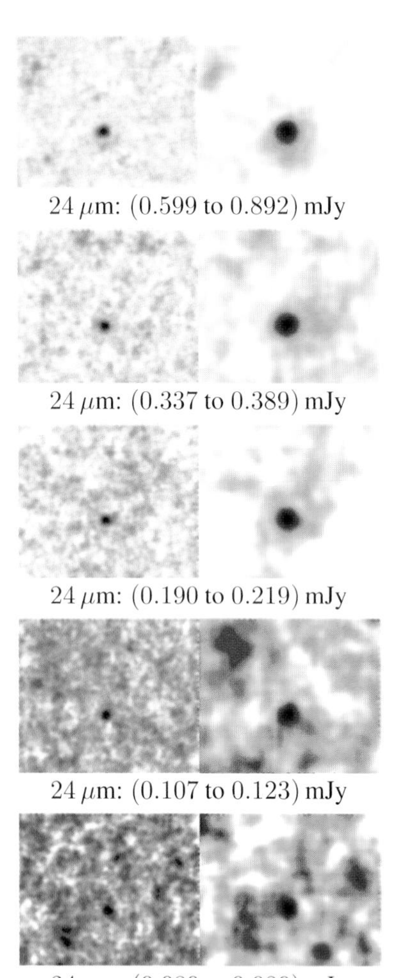

Figure 5.14 The *average* 70 μm (left) and 160 μm (right) images of 24 μm-selected galaxies, for a variety of ranges of 24 μm flux. Most of these 24 μm galaxies are *individually* undetected at longer wavelengths. For fainter 24 μm fluxes, the detections in the average images are noisier, but nonetheless significant.

optics to reduce the background from the telescope and increase the sensitivity. At the time of writing, there are also ambitious proposals with both NASA and ESA for future far-infrared space-based interferometers.

The coarse angular resolution of far-infrared images makes it difficult to identify which optical galaxy is the far-infrared emitter: the angular size of the submm image in Figure 5.5 is approximately the same as the whole Hubble Deep Field North. This also increases the confusion noise. However, there are some possible shortcuts. One method is to use stacking analyses (Chapter 2), such as averaging together the far-infrared images of galaxies detected at other wavelengths, in order to measure the average far-infrared flux for those galaxies. Another method is to see what else the far-infrared luminosities correlate with. In the local Universe, the mid-infrared luminosities correlate with bolometric luminosities (i.e. total luminosities). The shorter wavelengths in the mid-infrared lead to higher angular resolutions. Figure 5.14 shows stacked far-infrared images of 24 μm-selected galaxies, implying that this correlation exists at higher redshifts too.

But why should mid-infrared luminosities correlate with far-infrared? The dust emitting the mid-infrared light is physically distinct from the far-infrared emission. In order to be radiating at these shorter wavelengths, the dust must be hotter. It turns out that the dust grains responsible are small (less than 0.1 μm, or even as small as tens of atoms), and are transiently heated sometimes by a single photon. The mid-infrared spectra of star-forming galaxies have strong spectral signatures (see, for example, Figure 5.3), often (but not always) attributed to **polycyclic aromatic hydrocarbons** (or **PAHs**). This small-grained and PAH dust is heated by very short-wavelength light, because wavelengths much longer than the grain size are largely unaffected by these small dust particles, so the mid-infrared spectra measure the dust-shrouded ultraviolet light from O and B stars. The PAH emission line ratios can also be used to investigate the dust composition, and at around 10 μm there is an additional absorption feature from silicate dust grains. These mid-infrared spectral features do, however, make the K-corrections quite complicated. This can make it difficult to estimate the mid-infrared luminosities, but the Japanese AKARI space telescope turned this to its advantage and made deep surveys in many mid-infrared filters, in order to make mid-infrared photometric redshifts possible.

One disadvantage of using the mid-infrared for estimating star formation rates is that black hole accretion can also contribute. The dust tori in active galactic nuclei (Chapter 4) are predicted to emit the peak of their radiation at exactly these wavelengths. However, the thermal spectra from dust tori are largely featureless (notwithstanding a 10 μm absorption feature) so mid-infrared spectra can in principle distinguish star formation from black hole accretion.

The proposed successor to the HST, the **James Webb Space Telescope** (JWST), will have an expected collecting area of 25 m^2, and will operate from 0.6 μm to 28 μm. SPICA should still out-perform the JWST in the mid-infrared, despite SPICA's smaller mirror, because of its cooled optics. The proposed ESA planet-finding mission Darwin may also be able to take high-resolution mid-infrared images, using mid-infrared space-based interferometry.

One more star formation rate measure deserves a brief mention. **High-mass X-ray binary** stars (HMXBs) have one massive star, emitting a wind that is accreted by a companion neutron star, and the accretion is responsible for the

X-ray radiation. Since massive stars are shorter-lived, the numbers of HMXBs must be a measure of the recent star formation rate. In practice the high-redshift X-ray luminosities from star formation are often overwhelmed by that from supermassive black hole accretion, though this star formation rate estimator can be useful in some systems. For more details, see the further reading section.

5.5 Multi-wavelength surveys

With so many approaches to measuring star formation rates in galaxies, each with different advantages and disadvantages, what's the best way to study galaxy evolution? In practice, astronomy at different wavelengths uses different observing techniques and different detector technologies, and so has developed specialist research communities; each community would at one time undoubtedly have promoted their own as 'best'. However, the more modern approach is to take a coordinated multi-wavelength view, acknowledging the complementary insights at different wavelengths.

A classic example of this is the Great Observatories Origins Deep Survey (GOODS), which used three of NASA's Great Observatories: the HST, the Spitzer Space Telescope and the Chandra X-ray space telescope. GOODS made or uses the deepest pencil-beam surveys with all three facilities: the Chandra Deep Field North (CDF-N) in the HDF-N field and the Chandra Deep Field South (CDF-S); the HST HDF-N, HDF-S and the Ultra-Deep Field in CDF-S; and deep Spitzer imaging in HDF-N and CDF-S, now also known as GOODS-N and GOODS-S. Note the deliberate choices to map the same sky areas (though HDF-S and CDF-S are in different locations).

It isn't just pencil-beam surveys that have taken this approach; many other surveys of various depths and areas have taken a multi-wavelength approach. Nevertheless, there are only around a dozen or so well-studied fields in extragalactic astronomy. This is partly because of the competing requirements of multi-wavelength astronomy, such as:

- avoiding bright foreground objects at optical, mid/far-infrared and X-ray wavelengths;
- avoiding bright radio sources in the field or nearby for radio interferometers such as the Very Large Array (VLA) or the UK's MERLIN array (or the forthcoming upgraded versions e-VLA and eMERLIN);
- having low Galactic neutral hydrogen column density for X-ray observations;
- having low cirrus intensity and hence low cirrus confusion noise for far-infrared observations;
- having high visibility for the space-based observatories such as Chandra or ESA's XMM-Newton X-ray space telescope.

We've met the play-offs between depth and area in pencil-beam surveys and wide-field surveys. Some facilities have taken a deliberate approach of combining a variety of deep and wide surveys, sometimes called a 'tiered' approach or a 'wedding cake'. Figure 5.15 shows a simulation of tiered surveys with the Herschel Space Observatory. Each tier is a flux-limited survey, which means that the survey includes all the objects brighter than a given flux, within the survey area. Note that in any single tier, there is a strong correlation of luminosity with redshift (Section 4.5), which as a selection effect is also known as **Malmquist bias**. However, by combining surveys of different depths, there is better coverage of the luminosity–redshift plane.

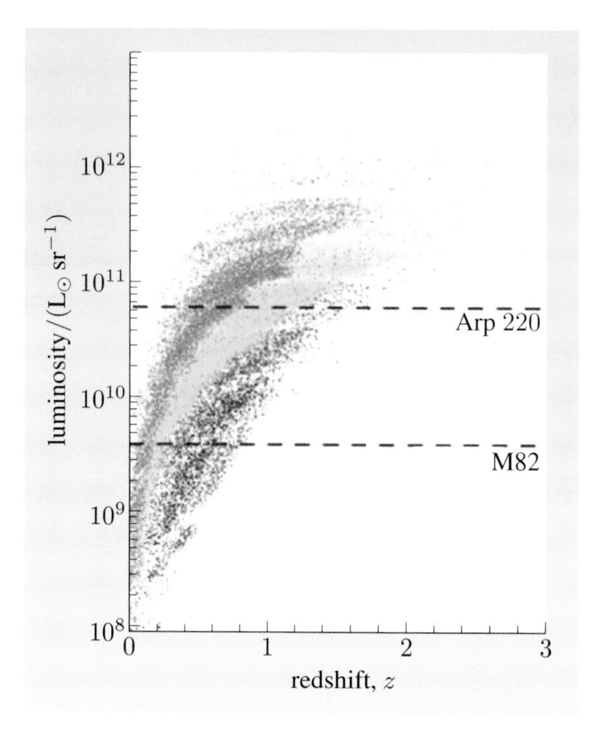

Figure 5.15 Simulated catalogues for planned surveys with the Herschel Space Observatory. Luminosity per unit solid angle is plotted against redshift, with objects in each survey colour-coded. The luminosities of the local starbursts Arp 220 and M82 are marked with dashed lines. The wide and shallow surveys cover the upper parts of this figure, while the deep and narrow pencil-beam surveys cover the lower parts of the figure. In each survey in isolation, redshift is correlated with luminosity, due to the Malmquist bias selection effect. Taking the surveys as an ensemble improves the coverage of the luminosity–redshift plane. The slight horizontal banding is an artefact of the simulation, but the paucity of high-luminosity objects at low redshifts is real and due to Malmquist bias.

- Why are there few high-luminosity objects at low redshift in any of the surveys?

- This is partly because the luminosity function evolves, so high-luminosity objects are intrinsically more common at higher redshifts. However, it's also because the amount of comoving volume sampled per unit redshift is smaller at low redshift than at high redshift (Figure 1.19).

Exercise 5.6 Draw (or otherwise indicate) the approximate line of any flux limit in Figure 5.15. ■

5.6 Cosmic star formation history and stellar mass assembly

The cosmic star formation histories derived from obscured star formation indicators and obscuration-independent indicators show just as much evolution as the rest-frame ultraviolet luminosity density. There's a strong increase from the local Universe to redshift $z = 1$ seen in radio surveys, but it's difficult to extend this to higher redshifts because of the difficulty in obtaining reliable redshift estimates for the radio-selected starbursts.

However, this has been achieved for some submm-wave surveys. In these, the negative K-corrections give a very uniform selection function over most of the Hubble volume, as we've seen. Figure 5.17 shows the cosmic star formation history of SMGs. While this seems similar to the ultraviolet star formation history, remember that the observed optical identifications of SMGs are often faint or unprepossessing optical galaxies, and not strongly luminous in the rest-frame ultraviolet. Therefore this star formation is *in addition* to the ultraviolet-derived star formation rate!

Meanwhile, the Spitzer Space Telescope has made it possible to conduct wide-field and deep extragalactic surveys at 3–8 μm. In the near-infrared, the luminosities of galaxies are *in*sensitive to star formation (see Chapter 4) and relatively insensitive to dust obscuration, notwithstanding the possibility of a large proportion of old stars in dust clouds. We could therefore use the near-infrared luminosity density to measure the total density in stars, Ω_*. By definition, this must be an integral of the cosmic star formation history. These constraints are shown in Figure 5.16. Differentiating this with respect to time gives the cosmic star formation history in Figure 5.17. Surprisingly, these constraints turn out not to be consistent with the observed cosmic star formation history!

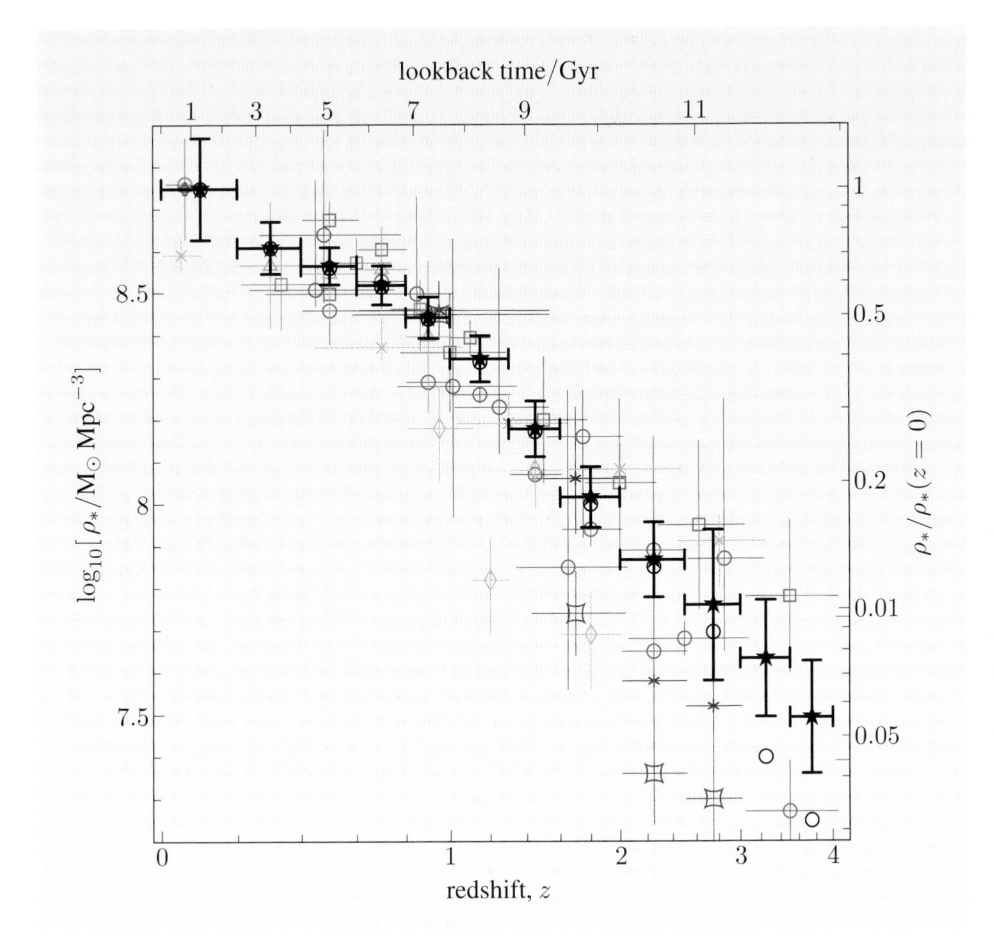

Figure 5.16 The stellar mass density of the Universe as a function of time and of redshift. A compilation of earlier estimates is shown in coloured symbols, while the most precise determination so far is shown in black symbols. Open circles show the direct measurements, while filled stars incorporate an additional correction for galaxies below the flux limits of the data.

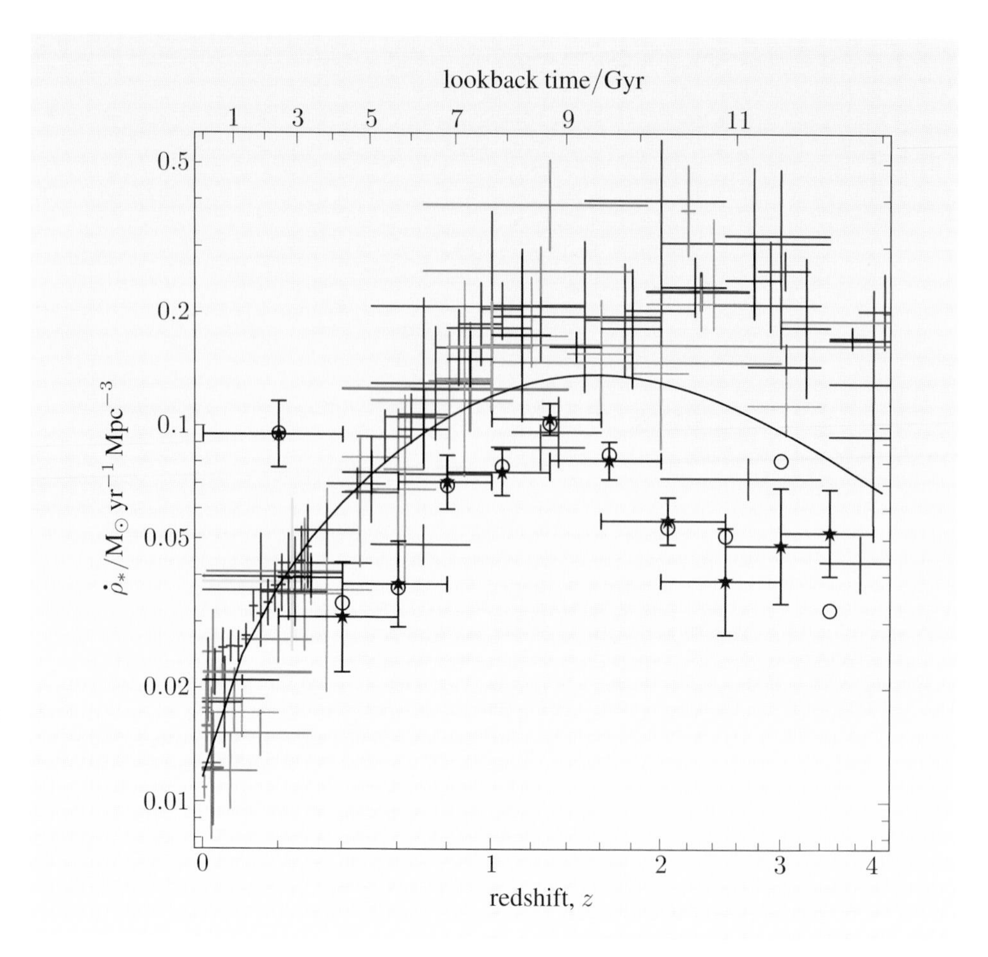

Figure 5.17 The comoving star formation density history of the Universe, derived from mid-infrared luminosities (blue error bars) and submm luminosities (magenta error bars). These estimates have been averaged together to make the black error bars. The error bars with filled stars and the open circles show the result of differentiating Figure 5.16, and are substantially lower.

What could cause this discrepancy? There are many potential sources of systematic uncertainties that we've discussed in this chapter and the previous one. Perhaps, for example, the dust extinction corrections need revision, or perhaps some of the star formation rate indicators have additional unrecognized contributions from AGN. A more fundamental underlying cause could be changes in the IMF. Indeed, some semi-analytic models assume a strongly top-heavy IMF in order to be able to reproduce the SMG population without overproducing Ω_*. The revisions to the IMF are radical, though. Many observations point to the same or very similar IMFs, but there is evidence that the IMF is strongly top-heavy in the central parsec of our Galaxy.

See, for example, Larson, 2005, in the further reading section, and Bartko, H. et al., 2009, arXiv:0908.2177.

Another surprise has been the high-redshift reversal of the relationship between star formation and environment. In the local Universe, star-forming galaxies avoid rich environments such as the cores of galaxy clusters (see Section 3.7). However, there is increasing evidence that at $z \simeq 1$, star-forming galaxies detected at 24 μm are more common in richer environments. This is already a challenge to semi-analytic models, and might suggest that mergers were more important in triggering star formation at higher redshifts. There are also hints that SMGs are found in richer environments than non-starbursting galaxies. The natural generalization of the Butcher–Oemler effect (Chapter 3) is to measure the cosmic star formation history in different environments, and this is the subject of much active research.

5.7 Downsizing

When did galaxies form? Up to the 1980s, the tendency was to speak of an epoch of galaxy formation at some (as yet unfathomed) redshift, perhaps involving monolithic collapse (Chapters 3 and 4). The advent of deep HST imaging changed the terminology and the thinking, with galaxy formation then being seen as an ongoing process. The growing sophistication of N-body and semi-analytic simulations led to a growing acceptance of hierarchical galaxy formation, following the schematic picture in Figure 4.2.

The discovery of SMGs at redshifts $z > 2$ with enormous star formation rates of thousands of solar masses per year therefore came as a tremendous surprise. Evidence mounted that present-day massive galaxies formed *earlier* than present-day less-massive galaxies. This is exactly the opposite to what you might expect from hierarchical structure formation (Figure 4.2). Figure 5.18 shows the fraction of stellar mass assembled in present-day galaxies as a function of redshift. A similar trend is seen in direct measurements of the star formation history at an observed wavelength of 24 μm (Figure 5.19).

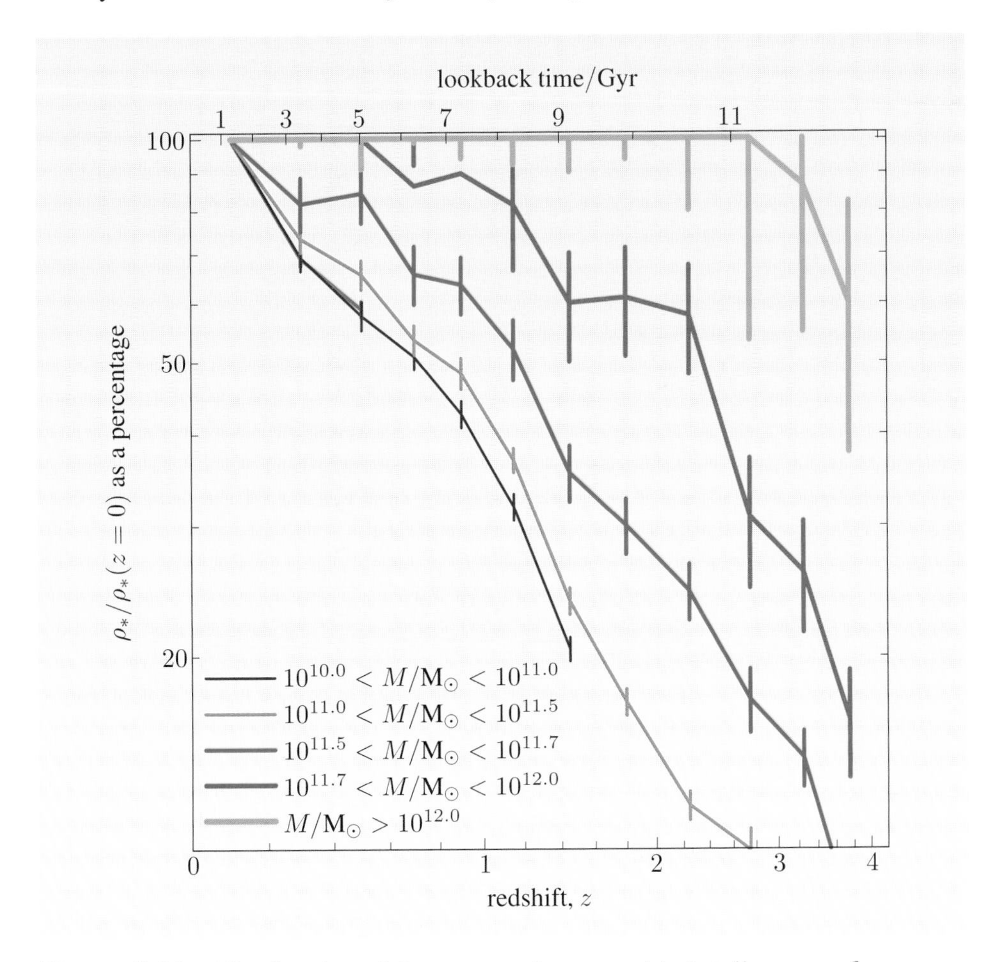

Figure 5.18 The fraction of the present-day assembled stellar mass, for a variety of galaxy masses. Note that the most massive galaxies formed their stars earlier in the history of the Universe.

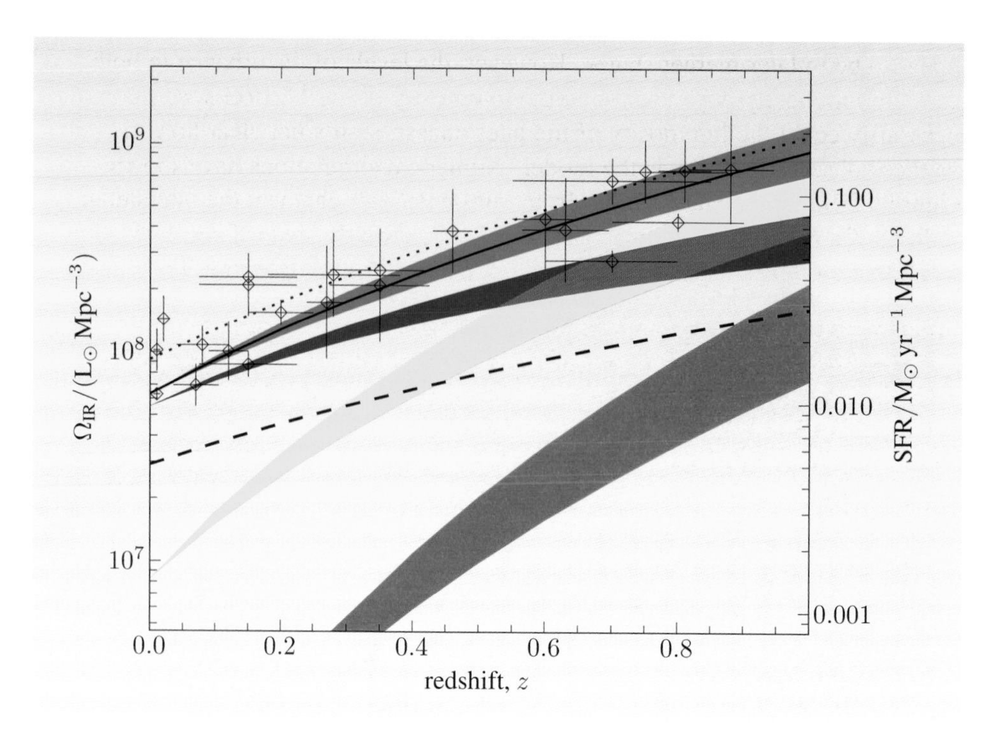

Figure 5.19 The cosmic star formation history for all galaxies (green), for $< 10^{11}\,L_\odot$ galaxies (blue), and for galaxies with $> 10^{11}\,L_\odot$ (yellow) and $> 10^{12}\,L_\odot$ (red). The contribution from higher-luminosity galaxies evolves more quickly, making it negligible in the present-day Universe but important by $z = 1$.

Note that massive starbursts comprised a greater proportion of the cosmic star formation rate at higher redshifts. This is so striking that it was sometimes called 'anti-hierarchical'. Nevertheless, hierarchical semi-analytic models were ultimately able to account for these populations, though needing to invoke non-standard IMFs (Section 5.6) or particular feedback mechanisms, which we shall meet in Section 5.8 and in Chapter 6. A more apt and widely-used term to describe this top-down galaxy formation is **downsizing**.

If SMGs are the progenitors of giant elliptical galaxies, then we'd expect them to be in massive dark matter haloes, and (according to semi-analytic models) strongly biased tracers of the underlying dark matter distribution. They should therefore cluster strongly. There have been hints of strong clustering in the SCUBA surveys, but one of the key aims of the new SCUBA-2 camera on the JCMT aims to measure the clustering of SMGs. The BLAST balloon-borne telescope has already inferred a clustering signal in SMGs from fluctuations in its background measurements, consistent with $z \sim 1$ SMGs having typical halo masses of $\sim 10^{13}\,M_\odot$. Also, it appears that the central brightest cluster galaxies (BCGs) in $z > 1$ galaxy clusters have already formed $> 90\%$ of their stellar masses by $z = 1.5$, compared to their $z = 0$ counterparts. This suggests that the formation of BCGs is more akin to monolithic collapse than resembling the result of repeated mergers. The forthcoming wide-field surveys by SCUBA-2 and Herschel may find the rare violent starbursts that accompanied this collapse.

See Collins, C.A., 2009, *Nature*, **458**, 603.

5.8 Feedback in galaxy formation

What do ULIRGs evolve into? It's very hard to know what any galaxy evolves into without an impossibly long wait, but it's been proposed that ULIRGs evolve into quasars. Many local ULIRGs appear in optical imaging to be in the late stages of a galaxy–galaxy merger, while there are at least some quasars that appear to be in later merger stages. However, the levels of disturbance in both cases are luminosity-dependent, and the luminosity of the ULIRG need not necessarily equal the luminosity of the later quasar, so it's not clear how to use these observations to test the model. Numerical simulations have shown (Figure 5.20) that the energy input from quasar activity can heat the interstellar medium of a galaxy and/or expel gas, which can abruptly shut off star formation (recall that the Jeans mass is temperature-dependent — see Exercise 3.3).

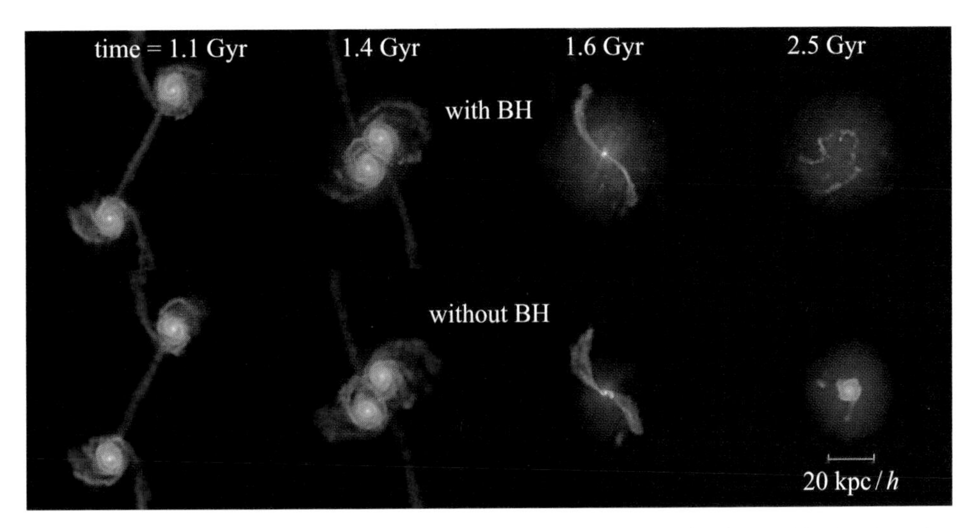

Figure 5.20 A simulation of a collision between two galaxies, with (top) and without (bottom) energy input from accretion round a supermassive black hole. The gas temperature distribution is colour-coded, blue to red. The maximum in the star formation and black hole accretion is at 1.6 Gyr, when the galaxies merge. The energy output from black hole accretion expels the gas from the inner regions of the merged galaxy after this stage. However, without this energy input, the result would be very different.

Another source of energy input and a cause of gas expulsion is supernovae. Local starburst galaxies often have supernova-driven 'superwinds' of gas being expelled from the galaxy. The amount of star formation (or black hole accretion) can therefore affect the future star formation; in general, this is known as **feedback**. Both star formation and black hole accretion are very complex phenomena and are so far best addressed with large numerical simulations, the results of which are used in generic ways in semi-analytic models. The role of feedback is perhaps the single greatest unknown in our understanding of galaxy evolution.

The intra-cluster medium in galaxy clusters is enriched with heavy elements, which is good evidence that galactic winds have played an important role in galaxy evolution. The figure on the cover of this book shows an example galactic wind observed with the Chandra X-ray telescope. Numerical simulations show that supernova-driven winds succeed in driving out most of the gas only in dwarf galaxies ($< 10^8\,M_\odot$). However, the lack of resolution in these simulations means that they don't account for the Rayleigh–Taylor instability (which could help the hot wind escape) or the Kelvin–Helmholtz instability. Alternatively, an outflow driven by the energy from black hole accretion could also lead to a bigger wind from the galaxy than supernovae can generate on their own.

AGN feedback could also solve the cooling flow problem in galaxy clusters (Section 3.9). Figure 5.21 shows a smoothed X-ray image of the Hydra A galaxy cluster (greyscale) tracing the hot gas of the intra-cluster medium, superimposed on contours of radio flux density from the radio lobes. The X-ray gas temperature and profile is consistent with a cooling flow, but the AGN radio lobes have cleared out cavities. Perhaps AGN radio lobes are the mechanism by which cooling flows are stopped or regulated? A further source of mechanical energy input from the AGN was found in Chandra X-ray images of the Perseus galaxy cluster, shown in Figure 5.22 (the cavities are similarly due to radio lobes). After an image processing technique called 'unsharp masking' (which amplifies high-frequency Fourier components in the image, while suppressing low-frequency components), large-scale ripples are seen (Figure 5.23). The lack of temperature changes in these oscillations led to them being interpreted as acoustic waves with a wavelength of about 11 kpc and a period of 9.6×10^6 years, or a frequency about 57 octaves below middle C.[8] AGN can therefore inject energy into the surrounding medium in two ways: radiation energy and mechanical energy. These are sometimes called **radiative mode** and **kinetic mode**.

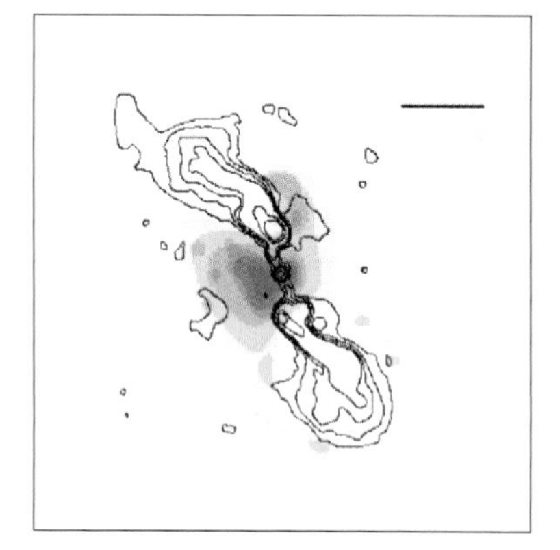

Figure 5.21 Smoothed X-ray image of the galaxy cluster Hydra A (grey shading), compared to the radio lobes from the active nucleus (contours). The bar marks a distance of 20 arcseconds.

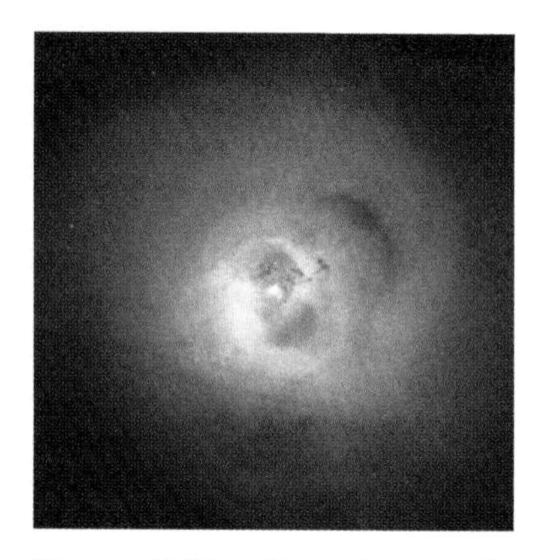

Figure 5.22 X-ray image of the Perseus galaxy cluster, colour-coded by temperature. The image is 350 arcseconds across, or 131 kpc.

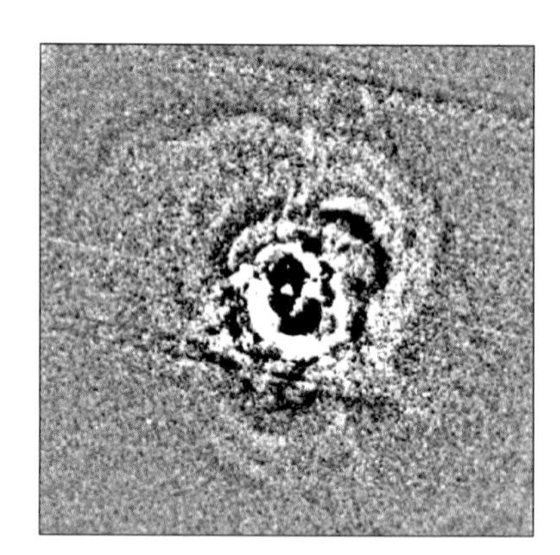

Figure 5.23 The result of applying the *unsharp masking* technique to Figure 5.22, which removes large-scale features and accentuates small-scale features, revealing subtle ripples.

AGN may also play a more complex role than simply shutting down star formation and expelling gas. Radio jets from radio-loud AGN have been observed to trigger star formation in some systems, and there is at least one quasar in which this **jet-induced star formation** has been argued[9] to pre-date the formation of the quasar host galaxy.

[8]The press referred to this as the 'deepest bass' note in the Universe.
[9]See Elbaz, D. et al., 2009, arXiv:0907.2923, and references therein.

Summary of Chapter 5

1. The energy density in the cosmic optical/near-infrared background is about the same as the energy density in the cosmic far-infrared background.
2. The observed cosmic backgrounds from optical to far-infrared wavelengths are closely linked to the evolving comoving luminosity densities:
$$I_\nu(\nu_0) = \frac{1}{4\pi} \int E_\nu \, \frac{\mathrm{d}d_{\text{comoving}}}{(1+z)}. \quad \text{(Eqn 5.5)}$$
At any redshift, the populations that contribute most to the background will be the same as those that contribute most to the corresponding luminosity density. This links the extragalactic background light to the cosmic star formation history.
3. A common unit used in astronomy is the jansky (symbol Jy), where $1\,\text{Jy} = 10^{-26}\,\text{W}\,\text{m}^{-2}\,\text{Hz}^{-1}$.
4. No-evolution source count models don't follow the Euclidean slope at the faint end, partly because of K-corrections and partly because of the redshift-dependence of luminosity distance. In other words, the source counts of a non-evolving population in a flat *non-expanding* universe are very different to no-evolution counts in a flat *expanding* universe.
5. K-corrections often strongly affect the redshift distributions of surveys.
6. The angular resolution of a telescope in radians is given by $\theta = 1.22\lambda/D$, where λ is the observed wavelength and D is the diameter of the telescope's primary mirror. The large diameters of submm-wave telescopes are not enough to compensate for the large wavelengths, so optical telescopes have sharper images. Many optical galaxies can sometimes be found within the observed position of an SMG, and it's not always obvious which optical galaxy is responsible for the submm flux. Other information is needed in these cases.
7. Infrared-luminous galaxies are sometimes described as LIRGs (10^{11}–$10^{12}\,\text{L}_\odot$), ULIRGs (10^{12}–$10^{13}\,\text{L}_\odot$) and HLIRGs ($> 10^{13}\,\text{L}_\odot$). Locally, ULIRGs and HLIRGs appear to be galaxy–galaxy mergers.
8. Far-infrared luminosity in galaxies is not always cospatial with optical-ultraviolet light.
9. There are many assumptions in calculating dust masses: the values of k_d, the temperature and the grey body index are needed, even if assuming a single dust temperature.
10. There are many methods of determining star formation rates, such as ultraviolet luminosities from young stars, far-infrared luminosities from star formation in giant molecular clouds, or radio luminosities deriving ultimately from supernovae. However, all measure the numbers of massive stars, so one needs to assume an initial mass function to extrapolate to all stellar masses.
11. The far-infrared and radio luminosities of star-forming galaxies correlate strongly.

12. In a flux-limited survey, the luminosity correlates with redshift. This is a selection effect called Malmquist bias, caused by the effect of the flux limit and the evolution of the objects being studied.
13. Downsizing is the apparently anti-hierarchical behaviour of massive present-day galaxies having formed the bulk of their stars earlier in the history of the Universe than present-day less-massive galaxies.
14. The energy and momentum input to the interstellar medium from an active nucleus, or from supernovae, can have a large effect on the evolution of a galaxy. In general, this is known as feedback, and is a major area of uncertainty in models of galaxy formation and evolution.

Further reading

- Kolb. U.C., 2009, *Extreme Environment Astrophysics*, Cambridge University Press.
- The Space Telescope Science Institute has an online tool for converting between janskys, magnitudes and other systems, currently at http://www.stsci.edu/hst/nicmos/tools/conversion_form.html.
- Condon, J.J., 1992, 'Radio emission from normal galaxies', *Annual Review of Astronomy and Astrophysics*, **30**, 575.
- De Zotti, G. et al., 2009, 'Radio and millimeter continuum surveys and their astrophysical implications', arXiv:0908.1896.
- Hauser, M.G. and Dwek, E., 2001, 'The cosmic infrared background: measurements and implications', *Annual Review of Astronomy and Astrophysics*, **39**, 249.
- Kennicutt, R.C., 1998, 'Star formation in galaxies along the Hubble sequence', *Annual Review of Astronomy and Astrophysics*, **36**, 189.
- Larson, R.B., 2005, 'Thermal physics, cloud geometry and the stellar initial mass function', *Monthly Notices of the Royal Astronomical Society*, **359**, 211.
- For more about the BLAST mission, including its crash and the heroic recovery of its data, see the BLAST web page currently at blastexperiment.info.

Chapter 6 Black holes

Black holes ... are the most perfect macroscopic objects there are in the universe: the only elements in their construction are our concepts of space and time.

S. Chandrasekhar

Introduction

Where are the biggest black holes in the Universe? Why does every galaxy contain a giant black hole at its centre, and how can we tell? And where did they come from?

We'll see in this chapter that black holes have been extremely important in galaxy evolution. Most of the light that's generated in the Universe has, ultimately, two main origins: the release of nuclear binding energy through nuclear reactions in stars, and the release of gravitational binding energy through accretion onto black holes. This accretion luminosity is extraordinarily efficient — black holes turn out to be not so black after all.

6.1 What are black holes?

Dark matter is an older idea than you might suppose. In the late eighteenth century, John Michell speculated that some stars might be so massive that light could not escape their surfaces. It was then very unusual for astronomers to use statistics, but Michell used a statistical argument to show that most close superpositions of stars on the sky (such as the Pleiades) are in fact neighbours, 'to whatever cause this may be owing, whether to their mutual gravitation, or to some other law or appointment of the Creator'.[10] In a brief but fascinating speculation[11] in 1784, he pointed out that if many stars are in binary systems, then there could be stars seen orbiting an invisible massive companion.

The modern counterpart to Michell's ingenious suggestion of 'dark stars' is **black holes**. In general relativity, the exterior of any spherically-symmetric non-rotating (uncharged) mass is described[12] by the Schwarzschild metric

$$\mathrm{d}s^2 = \left(1 - \frac{R_\mathrm{S}}{r}\right) c^2\,\mathrm{d}t^2 - \frac{\mathrm{d}r^2}{1 - R_\mathrm{S}/r} - r^2(\mathrm{d}\theta^2 + \sin^2\theta\,\mathrm{d}\phi^2), \qquad (6.1)$$

where the **Schwarzschild radius** is $R_\mathrm{S} = 2GM_\mathrm{BH}/c^2$, and M_BH is the black hole mass.[13] The derivation assumes only spherical symmetry, a vacuum at the radii of interest, and Einstein's field equations, and so also proves Birkhoff's theorem

[10] Michell, J., 1767, *Philosophical Transactions of the Royal Society of London*, **57**, 234–64; available at http://adsabs.harvard.edu/abs/1767RSPT...57..234M. This paper also proves the $N(>S) \propto S^{-3/2}$ Euclidean integral source counts in Chapter 4.

[11] In Michell, J., 1784, *Philosophical Transactions of the Royal Society of London*, **74**, 35–57; available at http://adsabs.harvard.edu/abs/1784RSPT...74...35M. This paper also makes the first prediction of gravitational redshift, using Newtonian arguments.

[12] This is derived, for example, in *Theoretical Cosmology* by R. Lambourne.

[13] Some texts use $M_\bullet$ to represent the black hole mass.

(mentioned in Chapter 4). General relativity therefore shares two important results with Newtonian gravity: that there is no gravitational field inside a spherical shell, and that the gravitational field outside a spherically-symmetric matter distribution is the same as the field from a central point mass. Another corollary is that spherically-symmetric pulsation cannot produce gravitational waves, of which more later in this chapter.

The surface $r = R_S$ is known as the **event horizon**, and black holes have their matter within this surface. A light ray (i.e. $\mathrm{d}s = 0$) on a radial trajectory in a Schwarzschild metric (i.e. $\mathrm{d}\theta = \mathrm{d}\phi = 0$) has $\mathrm{d}r/\mathrm{d}t = \pm c(1 - R_S/r)$. As $t \to \infty$, $r \to R_S$ and $\mathrm{d}r/\mathrm{d}t \to 0$, so the light ray never crosses $r = R_S$ as seen by a distant observer. Nevertheless, it can be shown that an infalling observer would still cross the event horizon in a finite *proper* time τ. Seen by a distant observer, an infalling watch would tick increasingly slowly as it approached the event horizon, but would not cross it; seen by an observer wearing that watch, time would not dilate and he or she would cross the horizon in a finite time. Once the horizon is crossed, this infalling observer is out of communication with the rest of the Universe.

There had been some speculation that some physical processes could prevent matter inside a black hole collapsing to the singularity at $r = 0$. The Penrose–Hawking singularity theorems put an end to these hopes, by showing that the collapsing matter is trapped inside a volume whose surface shrinks to zero. General relativity makes no predictions for the conditions at the singular point, so it's a theory with the unusual property that it demonstrates its own incompleteness.

Black holes are also surprisingly simple: it turns out that they can be described completely by their mass, spin and electric charge. Once the event horizon is formed and the metric eventually settles down to become time-independent, all other information is lost. (The saying 'black holes have no hair' has led to this being called the 'no-hair theorem'.)

This is described in detail in graduate-level texts such as *Gravitation* by Misner, C. W., Thorne, K. S. and Wheeler, J. A. (published by W. H. Freeman).

The metric of a rotating (uncharged) black hole is the **Kerr metric**, which we shall describe only briefly here. It is completely determined by the mass M_{BH} and angular momentum J, often expressed in terms of angular momentum per unit mass $a = J/(M_{\mathrm{BH}}\, c)$. Provided that the spin is less than a maximal value of $a = GM/c^2$, there are *two* astrophysically-relevant critical surfaces, at

$$r_+ = \frac{GM_{\mathrm{BH}}}{c^2} + \sqrt{\left(\frac{GM_{\mathrm{BH}}}{c^2}\right)^2 - a^2}$$

and

$$s_+ = \frac{GM_{\mathrm{BH}}}{c^2} + \sqrt{\left(\frac{GM_{\mathrm{BH}}}{c^2}\right)^2 - a^2 \cos^2\theta}\,.$$

The radius r_+ is an event horizon, but the outer horizon is the boundary of the 'ergoregion' in which all matter and light rays are forced to co-rotate with the black hole — an extreme case of a relativistic phenomenon known as **frame dragging**. (There is also a third surface — see the further reading section.)

It's not clear if black holes can have spins at or above the maximal value $a \geq GM/c^2$. No mode of matter accretion can take a sub-maximally-spinning black hole and increase the spin above its maximal value — it turns out that any angular momentum gained is more than balanced by an increase in M_{BH}. From the equations above, black holes spinning above the maximal rate have no event horizons. It turns out that it then becomes possible for particle world-lines around the black hole to be closed time-like loops, i.e. the black hole would be a time machine! Roger Penrose has proposed a 'cosmic censorship hypothesis' that there are no naked singularities in nature, i.e. no singularities without event horizons. At the time of writing, this remains unproven in classical gravity (unless naked singularities are taken as pre-existing). The hypothesis is the subject of a bet between Stephen Hawking, who contends that it is correct, and Kip Thorne and John Preskill, who contend that it is not. The stake of the bet (re-formulated in 1997 to eliminate possible loopholes) is: 'The loser will reward the winner with clothing to cover the winner's nakedness. The clothing is to be embroidered with a suitable, truly concessionary message.' A related conjecture by Stephen Hawking is the 'chronology protection conjecture' that fundamental physics forbids closed time-like loops. This conjecture would be addressed by, or could form part of, a future theory of quantum gravity.

This view has recently been challenged by Jacobsen, T. and Sotiriou, T. P., arXiv:0907.4146.

General relativity is time-symmetric, so one could conceive of the time-reverse of black holes, sometimes called 'white holes'. Perhaps entropic reasons forbid them, for a reason similar to why molecules in a lake do not suddenly conspire to throw pebbles out. We shall not dwell on this, except to say that many deep problems in classical and quantum gravity involve entropic aspects of gravity. White holes also occur in discussions of the Kerr metric. The curvature singularity in the Kerr metric is ring-shaped, unlike the point-like Schwarzschild singularity. Infalling matter in a Kerr metric could pass through a further inner horizon and though the ring, and, if the same metric holds past these points, would emerge out of a white hole — where? This has been the subject of much speculation in science fiction. But an infalling traveller attempting this journey would receive an infinite flux of radiation that fell into the hole from the other side, infinitely blueshifted — an infinitely ferocious gauntlet to run. Clearly, there is still much to be understood about these singular regions.

Black holes can form as the end-point of the most massive stars' evolution, when nuclear reactions or degeneracy pressure can no longer support a stellar core against gravity. This conclusion was first reached by Chandrasekhar and infamously opposed by Sir Arthur Eddington, who said: 'I think there should be a law of nature to prevent a star from behaving in this absurd way.' Hindsight has shown Chandrasekhar to be correct, but ironically Eddington's name has become associated with black hole accretion, as we shall see in the next section. One might nevertheless share a similar reaction to the inevitable curvature singularities inside black holes.

6.2 The Eddington limit

How quickly can black holes grow? There must come a point where the outward photon pressure from the accretion luminosity balances the inward gravitational attraction towards the black hole.

Suppose that the luminosity of the central object is L. The rate of energy output from this central object is just L, i.e. $\mathrm{d}E/\mathrm{d}t = L$. For photons, the momentum p and energy E are related by $E = pc$, so the *momentum* flux from the central object is $\mathrm{d}p/\mathrm{d}t = L/c$. The photon pressure P at any distance r from the centre will be the momentum output per unit area, i.e.

$$P = \frac{L}{4\pi c r^2}, \tag{6.2}$$

because the area of the surface of a sphere with radius r is $4\pi r^2$. Suppose that the infalling gas is hot enough to be a plasma (which is true for most astrophysical black holes). The force felt by an electron in this plasma from this photon flux will be this pressure P times the cross section σ_T for Thomson scattering of photons by electrons:

$$F_{\text{photon}} = \sigma_T P = \frac{\sigma_T L}{4\pi c r^2}. \tag{6.3}$$

Atomic nuclei also present a Thomson scattering cross section, but $\sigma_T \propto q^2/m^4$, where q is the particle charge and m is its mass, so the electron Thomson scattering cross section provides the dominant outward force on the infalling gas.

The inward force from gravity on a hydrogen nucleus is just

$$F_{\text{gravity}} = \frac{G\, M_{\text{BH}}\, m_{\text{p}}}{r^2}, \tag{6.4}$$

where m_p is the mass of a hydrogen ion, i.e. a proton, and M_{BH} is the mass of the black hole. Although we've assumed that the infalling matter is a plasma, the electron gas would still be strongly coupled electrostatically to the gas of positive nuclei or ions. The outward force on the infalling plasma is exerted mainly on the electrons, while the inward force is exerted mainly on the protons and neutrons, but the plasma nevertheless responds as a whole rather than separating out by charge.

Exercise 6.1 Demonstrate this, using some quantified argument of your own invention. (Like Exercises 4.4 and 4.11 in Chapter 4, this is a more open-ended exercise.) ■

The inward and outward forces balance when $F_{\text{photon}} = F_{\text{gravity}}$, so

$$\frac{\sigma_T L}{4\pi c r^2} = \frac{G\, M_{\text{BH}}\, m_{\text{p}}}{r^2}. \tag{6.5}$$

Remarkably, the r^2 terms cancel, so the balance is independent of radius. The luminosity at this balance is

$$L_E = \frac{4\pi G c\, M_{\text{BH}}\, m_{\text{p}}}{\sigma_T}, \tag{6.6}$$

which is known as the **Eddington luminosity**, after Sir Arthur Eddington who (among many contributions in astronomy and physics) first made these calculations in the context of stellar opacity.

In black hole accretion, this luminosity is generated by the accretion disc, so $L \propto \mathrm{d}M_{\text{BH}}/\mathrm{d}t$. Ultimately, some of this energy output comes from the release of gravitational binding energy of matter falling towards the black hole. This happens because friction in the accretion disc leads to energy losses through thermal radiation, so the orbital radius of the matter decreases. As the accreting matter approaches the black hole, some of the combined mass-energy is converted to luminosity. The accretion luminosity can be expressed in terms of the mass-energy accretion rate ($E = mc^2$ so $\dot{E} = \dot{m}c^2$):

$$L = \eta \frac{\mathrm{d}M_{\mathrm{BH}}}{\mathrm{d}t} c^2, \tag{6.7}$$

where η (which is ≤ 1) is the conversion efficiency from mass-energy accretion to luminosity. The maximum black hole growth rate will happen when the accreting matter reaches the Eddington luminosity, known as the **Eddington limit**.

Exercise 6.2 The Eddington limit is also the highest luminosity that a gravitationally-bound object can have without photon pressure blowing it apart. How close is the Sun to the Eddington limit? (The value of σ_{T} is $6.65 \times 10^{-29}\,\mathrm{m}^2$. The mass of a proton, and the mass and luminosity of the Sun, are given in Appendix A.)

Exercise 6.3 How close is a light bulb to the Eddington limit? Comment on your answer. ■

We can also define a characteristic timescale t_{E}, sometimes called the **Eddington timescale** or the **Salpeter timescale**:

$$t_{\mathrm{E}} = \frac{M_{\mathrm{BH}}\, c^2}{L_{\mathrm{E}}} = \frac{\sigma_{\mathrm{T}} c}{4\pi G m_{\mathrm{p}}} \simeq 4 \times 10^8\ \mathrm{yr}. \tag{6.8}$$

This is the time that an object would take to radiate away all its rest mass, at the Eddington limit. It's also the e-folding timescale for Eddington-limited black hole growth, i.e. the time to increase by a factor of $\mathrm{e} = 2.718\,28\ldots$ if the efficiency is $\eta = 1$. (This can be proved by combining Equations 6.6, 6.7 and 6.8, then solving the resulting differential equation to find $M \propto \exp(t/t_{\mathrm{E}})$.) If the efficiency isn't 100%, the e-folding growth timescale is just t_{E}/η.

It's worth keeping in mind that we've assumed spherical symmetry in deriving the Eddington luminosity and accretion rate. Non-spherical accretion is complicated to describe analytically, and is often studied in numerical simulations. Accreting matter in astrophysics usually settles into a disc geometry, for which there is a well-developed theory (see the further reading section). Nevertheless, the Eddington limit provides an important order-of-magnitude reference for accretion flows in general. It's also possible for accretion rates above the Eddington rate (called **super-Eddington accretion**) to occur under certain circumstances. If the accreting matter is optically-thick and also has a very high accretion rate, it could trap the radiation and drag it down with it into the black hole — a process known as **advection**. These **advection-dominated accretion flows** (**ADAFs**) have been conjectured in the rare class of narrow line Seyfert 1 AGN, which appear to lack a broad line region yet have a strong non-thermal continuum. Super-Eddington ADAFs are not spherically symmetric, and strong gas flows along the angular momentum axis are driven out by radiation pressure. In the opposite limit, in which the accretion rate is much lower than the Eddington limit, advection can again play a role. If the accreting gas has a very low density, it may be inefficient at radiating and the matter may be advected onto the central black hole. These low accretion rate ADAF models reproduce the spectral energy distribution of Sgr A* at the centre of our Galaxy, and have been conjectured to be present in some low-luminosity AGN.

6.3 Accretion efficiency

As matter orbits a black hole, it gradually loses energy through radiative losses from, for example, friction. One could think of this as a conversion of gravitational binding energy or mass-energy to luminosity. We'll find that accretion onto black holes can be extraordinarily efficient at converting mass-energy into luminosity — much more efficient than, say, a nuclear bomb. Once matter has fallen inside the event horizon, no light signals can reach the outside world, but before that point the accretion process can be prodigiously luminous. Paradoxically, this makes black holes the best candidates for some of the most luminous objects in the Universe.

To calculate the accretion efficiency of black holes, we'll first have to calculate the orbital motion. We calculated this for radial motion of light rays in Section 6.1, but for a massive particle $\mathrm{d}s^2 \neq 0$, so we need more information to find the motion. Freely-falling particles in general relativity follow geodesics, which are extremal paths in the relativistic sense, i.e. $\int \mathrm{d}s$ along the path is a maximum. This means that the integral

$$\ell = \int \sqrt{g_{\mu\nu} \frac{\mathrm{d}x^\mu}{\mathrm{d}\tau} \frac{\mathrm{d}x^\nu}{\mathrm{d}\tau}}\, \mathrm{d}\tau = \int \mathrm{d}s \tag{6.9}$$

In space (as opposed to spacetime) geodesics are paths of shortest *distance* between two points, but in the spacetime of general relativity, geodesics are paths between two events along which the *proper time* is a *maximum*.

is a maximum for the path taken by the particle $\boldsymbol{x}(\tau)$, where τ is the proper time, and x^μ means the μth component of the vector $\boldsymbol{x}$ rather than x-to-the-power-μ (see Appendix B). Calculus conventionally finds the value of a *variable* that maximizes a function, but here we want to find the *function* $\boldsymbol{x}(\tau)$ that maximizes the value of a *variable* ℓ.

This is the subject of a branch of calculus known as the *calculus of variations*. The key result for our purposes is that if $\int \mathcal{L}\, \mathrm{d}\tau$ is to be minimized, and $\mathcal{L}$ does not depend on a parameter y, then

$$\frac{\partial \mathcal{L}}{\partial \dot{y}} = \text{constant} \tag{6.10}$$

is a conserved quantity (note the dot in the denominator). For example, if $\mathcal{L}$ doesn't depend on the μth component of the vector $\boldsymbol{x}$, i.e. x^μ, then $\partial\mathcal{L}/\partial\dot{x}^\mu$ is a conserved quantity. The proof takes us away from the main story of this chapter, but in case you're not satisfied with having this pulled out of a hat, there's a proof in the box below.

The principle of least action

The most useful application of the calculus of variations in physics is the principle of least action. From it we'll find a deep and beautiful connection between angular and linear momentum, and a wonderful way of re-stating the fundamental laws of the Universe. It will also indirectly answer this question: why is it that in a plot of linear momentum mv against velocity v, the area under the curve is the kinetic energy $\frac{1}{2}mv^2$? Where's the connection? There's no obvious answer in conventional Newtonian physics.

We define the **action** A as $A = \int \mathcal{L}\, \mathrm{d}t$, where $\mathcal{L}$ (known as the **Lagrangian**) is the *difference* between the kinetic and potential energies, $\mathcal{L} = E_\mathrm{K} - V$. This is not the total energy, which would be $E_\mathrm{K} + V$. As in Fermat's principle in optics, it seems that particles in classical mechanics follow paths that minimize the action (we'll ask why later). Note that some textbooks use the symbol T for kinetic energy in this context.

Suppose that $x(t)$ is the path taken by a particle between two points $x(t_0)$ and $x(t_1)$ that minimizes the action, and that some other neighbouring path $x(t) + \varepsilon(t)$ doesn't do so (see Figure 6.1).

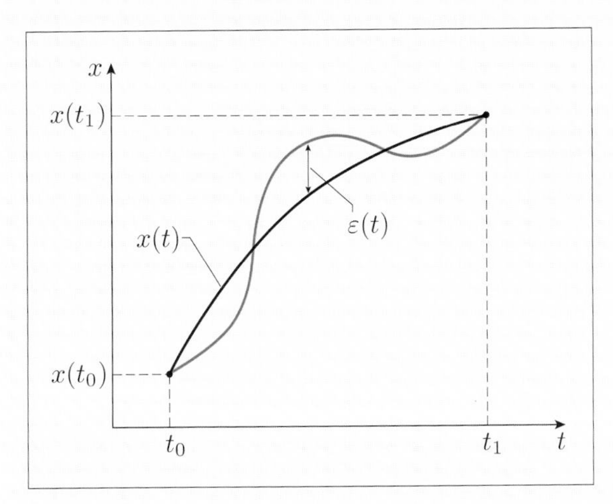

Figure 6.1 The least-action path $x(t)$ and a neighbouring path $x(t) + \varepsilon(t)$.

We set $\varepsilon(t_0) = \varepsilon(t_1) = 0$ so that the paths start and end at the same points. The velocity on the least-action path is $v(t) = \mathrm{d}x(t)/\mathrm{d}t$, while on the neighbouring path it's $v(t) + \dot{\varepsilon}(t)$. To first order, the Lagrangian $\mathcal{L}(x, v)$ on the neighbouring path will be

$$\mathcal{L}(x+\varepsilon, v+\dot{\varepsilon}) = \mathcal{L}(x, v) + \varepsilon(t)\,\frac{\partial \mathcal{L}}{\partial x} + \dot{\varepsilon}(t)\,\frac{\partial \mathcal{L}}{\partial v}, \tag{6.11}$$

so the action on the neighbouring path will be $A + \delta A$, where

$$\delta A = \int_{t_0}^{t_1} \left(\varepsilon(t)\,\frac{\partial \mathcal{L}}{\partial x} + \frac{\mathrm{d}\varepsilon}{\mathrm{d}t}\,\frac{\partial \mathcal{L}}{\partial v} \right) \mathrm{d}t. \tag{6.12}$$

We can integrate the second term in the integral by parts:

$$\int_{t_0}^{t_1} \frac{\mathrm{d}\varepsilon}{\mathrm{d}t}\,\frac{\partial \mathcal{L}}{\partial v}\,\mathrm{d}t = \left[\varepsilon(t)\,\frac{\partial \mathcal{L}}{\partial v} \right]_{t_0}^{t_1} - \int_{t_0}^{t_1} \varepsilon(t)\,\frac{\mathrm{d}}{\mathrm{d}t}\left(\frac{\partial \mathcal{L}}{\partial v} \right) \mathrm{d}t. \tag{6.13}$$

But we set $\varepsilon(t_0) = \varepsilon(t_1) = 0$, so the term in the square brackets is zero, and

$$\delta A = \int_{t_0}^{t_1} \varepsilon(t) \left(\frac{\partial \mathcal{L}}{\partial x} - \frac{\mathrm{d}}{\mathrm{d}t}\,\frac{\partial \mathcal{L}}{\partial v} \right) \mathrm{d}t. \tag{6.14}$$

Now, you already know that if a variable a minimizes a function $y(a)$, then changing a at that minimum doesn't change y to first order (see Figure 6.2).

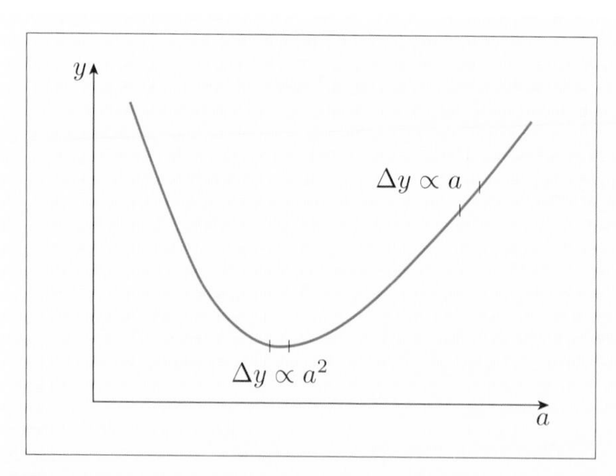

Figure 6.2 Near the minimum of this curve $y(a)$, the displacement Δy is proportional to a^2. This is because $\mathrm{d}y/\mathrm{d}a = 0$ there, which means that the first-order term in a Taylor series expansion around this point is zero, and only the second-order term is non-zero. Elsewhere $\mathrm{d}y/\mathrm{d}a \neq 0$, so $\Delta y \propto a$ to first order.

Similarly, if $x(t)$ is the path that minimizes the action, then putting a small wiggle $\varepsilon(t)$ onto the path won't change the action to first order, i.e. $\delta A = 0$ (sometimes written as $\delta \int \mathcal{L}\,\mathrm{d}t = 0$). But though ε is small, it's arbitrary, so $\delta A = 0$ can happen only if

$$\frac{\partial \mathcal{L}}{\partial x} = \frac{\mathrm{d}}{\mathrm{d}t}\frac{\partial \mathcal{L}}{\partial v}. \tag{6.15}$$

This is known as the **Euler–Lagrange equation**.

We've done this in one dimension for simplicity, but if the Lagrangian depends on many coordinates $q_1, q_2, \ldots, \dot{q}_1, \dot{q}_2, \ldots$ — where, for example, the coordinates could be Cartesians $(q_1, q_2, q_3) = (x, y, z)$ or polars $(q_1, q_2, q_3) = (r, \theta, \phi)$ or indeed any coordinate system — then

$$\frac{\partial \mathcal{L}}{\partial q_i} = \frac{\mathrm{d}}{\mathrm{d}t}\frac{\partial \mathcal{L}}{\partial \dot{q}_i} \tag{6.16}$$

for every q_i.

We can use this to find conservation laws in physics. If the Lagrangian $\mathcal{L}$ is independent of a Cartesian coordinate x, and $v = \mathrm{d}x/\mathrm{d}t$, then

$$\frac{\mathrm{d}}{\mathrm{d}t}\frac{\partial \mathcal{L}}{\partial v} = 0,$$

so $\partial \mathcal{L}/\partial v$ must be a constant. Empty space with no potential ($V = 0$) has this property, and the conserved quantity in Cartesian coordinates turns out to be linear momentum. Similarly, $\mathcal{L}$ in polar coordinates won't depend on θ, and the conserved quantity turns out to be angular momentum. In Newtonian gravitation, $\mathcal{L}$ also doesn't depend on θ, and the conserved quantity gives Kepler's second law. This can also be proved from angular momentum conservation in elliptical orbits, but the Lagrangian trick is much quicker and easier.

These coordinate independencies can also be thought of as symmetries, because if $\mathcal{L}$ is not a function of x (i.e. $\mathcal{L} \neq \mathcal{L}(x)$), then $\mathcal{L}$ is invariant under the transformation $x \to x + \delta x$ for any δx. The fact that symmetries in the Lagrangian imply conservation laws is known as **Noether's theorem**, after the brilliant physicist Amalie Emmy Noether (Figure 6.3), and is very widely used in fundamental physics. A related argument can show that energy conservation in general reflects time-independence. Maxwell's equations and charge conservation are equivalent to Lorentz invariance of the electromagnetic vector potential — or rather, a more general invariance known as *gauge invariance* (see, for example, Ryder, L.H., 1985, *Quantum Field Theory*, Cambridge University Press). Einstein's field equations of general relativity can be found by minimizing an appropriate Lagrangian. All *fundamental* physics can be thought of as simple symmetries and conservation laws. This realization was a crowning achievement of nineteenth and early-twentieth century physics, and is still true today.

What is energy, anyway? What is momentum? Lower-level textbooks tend to describe their effects, but fundamentally these quantities are mostly interesting because they are conserved. If they weren't, we'd find and use other quantities that were. This thinking is useful when particle physics presents you with more abstract quantities such as strangeness or colour charge or lepton number, and you wonder what they are.

You might well ask why Nature obeys the principle of least action — after all, the total energy $E_K + V$ is an obviously physical quantity, but $\mathcal{L} = E_K - V$ isn't. Classically, it's hard to interpret, but the physicist Richard Feynman found that in quantum mechanics the phase of the wave function is $A/\hbar$. He imagined particles in quantum mechanics taking all possible paths simultaneously, but nearby paths would have wave functions that tend to cancel out (because to first order the phases are different), except in the regions where the phases are all the same to first order, i.e. where $\delta A = 0$, which is the path of least action. The smallness of $\hbar$ ensures that this cancellation happens only very close to the least-action path, so macroscopically (i.e. classically) the particle appears to take only the least-action path.

Figure 6.3 Amalie Emmy Noether, 1882–1935.

Geodesics are the paths that maximize the total relativistic spacetime distance $\int \mathrm{d}s$ along the path (i.e. $\delta \int \mathrm{d}s = 0$ in the notation of the box above). In fact, we can pick *any* two points on the path and the geodesic will follow the maximum $\int \mathrm{d}s$ between those points — because if it didn't, we could tweak the path between those points and find a better global $\int \mathrm{d}s$. In general relativity, any free-falling frame is *locally* the metric of special relativity, and the maximum spacetime interval in special relativity between two events is just a straight line in spacetime: $\delta s = \sqrt{(c\,\delta t)^2 - (\delta x)^2 - (\delta y)^2 - (\delta z)^2}$. We can think of a geodesic as a sum of these δs contributions measured in a chain of free-falling reference frames along the path. But if δs is a maximum, then so is $(\delta s)^2$. By instead summing up these $(\delta s)^2$ contributions, we can make a new quantity, say ℓ_2, that's also maximized along geodesics:

$$\ell_2 = \int g_{\mu\nu} \frac{\mathrm{d}x^\mu}{\mathrm{d}\tau} \frac{\mathrm{d}x^\nu}{\mathrm{d}\tau}\, \mathrm{d}\tau. \tag{6.17}$$

The Schwarzschild metric in Equation 6.1 is independent of t and ϕ, so applying the Euler–Lagrange equation (Equation 6.16) gives two conserved quantities, one corresponding to energy and one to angular momentum:

$$\left(1 - \frac{R_S}{r}\right)\frac{dt}{d\tau} = \text{constant} = \frac{E}{mc^2} = \widetilde{E} \tag{6.18}$$

and

$$r^2 \frac{d\phi}{d\tau} = \text{constant} = \frac{J}{mc} = \widetilde{J}, \tag{6.19}$$

where R_S is the Schwarzschild radius, and we *define* the constants of specific angular momentum (i.e. angular momentum per unit rest mass) to be J/mc or $\widetilde{J}$, and the specific energy to be $\widetilde{E} = E/(mc^2)$.

To find the orbital equations in the Schwarzschild metric, we can choose an orbit in the equatorial plane ($\theta = \pi/2$, $d\theta = 0$) without loss of generality. We can then combine the metric (Equation 6.1 with $ds^2 = c^2\,d\tau^2$) with the angular momentum and energy conservation (Equations 6.18 and 6.19) to find that

$$\left(\frac{1}{c}\frac{dr}{d\tau}\right)^2 = \widetilde{E}^2 - \left(1 - \frac{R_S}{r}\right)\left(1 + \frac{\widetilde{J}^2}{r^2}\right). \tag{6.20}$$

We can think of this as $(1/c)^2(dr/d\tau)^2 = \widetilde{E}^2 - \widetilde{V}^2$, where $\widetilde{V}$ is an 'effective potential' per unit mass given by

$$\widetilde{V}(r) = \sqrt{\left(1 - \frac{R_S}{r}\right)\left(1 + \frac{\widetilde{J}^2}{r^2}\right)}. \tag{6.21}$$

Figure 6.4 plots this function for various specific angular momenta $\widetilde{J}$. Most angular momenta have a stable minimum at one radius. This is also the radius of a circular orbit at that $\widetilde{J}$, because for circular orbits, $dr = 0 \Rightarrow dr/d\tau = 0$ $\Rightarrow \widetilde{E} = \widetilde{V}$. But at sufficiently low angular momenta, there is no stable circular orbit. There is therefore a closest possible inner edge of the accretion disc.

We'll use this closest stable circular orbit to calculate the maximum efficiency for converting mass-energy to luminosity in a black hole accretion disc. It's not too hard (but a bit tedious) to show that the smallest stable circular orbit around a Schwarzschild black hole has $\widetilde{J} = \sqrt{3}R_S$ and radius $r = 3R_S$. (Either find where $d^2\widetilde{V}/dr^2 = 0$, or set $d\widetilde{V}/dr = 0$ and require that a finite root exists.) Putting in the numbers, the fractional binding energy of an orbit at this radius is therefore

$$\frac{mc^2 - E}{mc^2} = 1 - \widetilde{E} = 1 - \sqrt{8/9} = 0.0572\ldots \simeq 6\%. \tag{6.22}$$

If a particle spirals in from $r = \infty$, radiating binding energy or mass-energy as luminosity via friction, this is the fraction of energy released. For comparison, $< 0.1\%$ of the uranium rest mass in a fission-based atomic bomb is converted to energy.

The corresponding result for a maximally-spinning Kerr metric is a minimum stable equatorial circular orbit radius of $r = (5 \pm 4)GM_{BH}/c^2$ (use $-$ for prograde orbits, $+$ for retrograde). The fractional binding energy is $1 - \frac{5}{3}\sqrt{\frac{1}{3}} \simeq 4\%$ for retrograde orbits and an astonishing $1 - \sqrt{\frac{1}{3}} \simeq 42\%$ for prograde orbits. (If the black hole has a spin just less than the maximal value, e.g. 0.998 times the maximal spin, then this can drop to 'merely' $\sim$30%, but this is still an astonishingly high efficiency.) The proofs follow similar methods to the Schwarzschild case, but are longer for this more complicated metric (details are in Misner, Thorne and Wheeler). Retrograde accretion orbits seem unlikely, either because of frame dragging or because the black hole and accretion disc both formed from matter with similar angular momentum axes, so spinning black holes are expected to be even more efficient converters of mass–energy to luminosity.

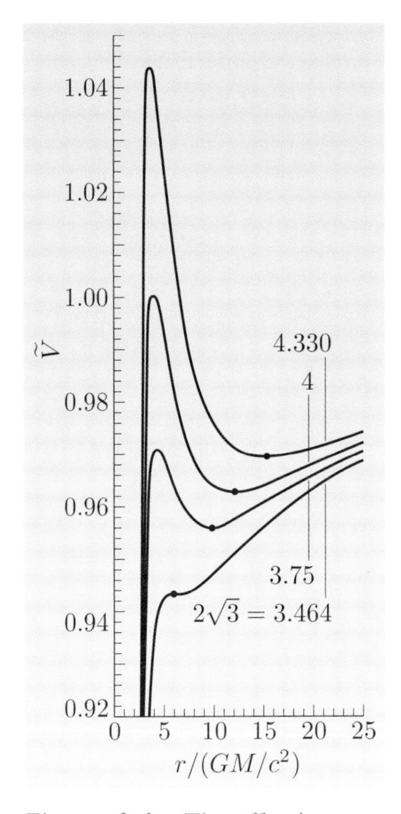

Figure 6.4 The effective potential $\widetilde{V}$ around a Schwarzschild black hole, given by Equation 6.21. The numbers on the curves are the values of the specific angular momentum relative to the black hole mass, $\widetilde{J}c^2/(GM)$, where M is the black hole mass.

6.4 Cosmic mass density of black holes, Ω_{BH}

Almost 200 years after John Michell's suggestion of matter hidden in dark stars, the cosmologist Andrzej Soltan spotted that the number counts of quasars give an ingenious constraint on the present-day total mass of black holes, which is independent of H_0, Ω_m and Ω_Λ. This is done by estimating the total energy output of quasars throughout the history of the Universe, and applying the Eddington limit.

The energy output of quasars per unit comoving volume and per unit time is $E = L\,\Phi(L,z)$, where L is the quasar bolometric luminosity and Φ is the bolometric luminosity function. The bolometric energy output in a time t to $t + \mathrm{d}t$ from quasars with luminosities from L to $L + \mathrm{d}L$ is then

$$E(L,t)\,\mathrm{d}L\,\mathrm{d}t = L\,\Phi(L,z)\,\mathrm{d}L\,\mathrm{d}t,$$

where time t and redshift z are related through Equation 1.34. Now, the number counts are related to the luminosity function by

$$4\pi\, n(S,z)\,\mathrm{d}S\,\mathrm{d}z = \Phi(L,z)\,\mathrm{d}L\,\frac{\mathrm{d}V}{\mathrm{d}z}\,\mathrm{d}z,$$

where $n(S,z)$ is $\mathrm{d}N/\mathrm{d}S$ evaluated at flux S for objects with redshift z, and V is the comoving volume. Luminosity and flux are also related:

$$L = 4\pi d_\mathrm{L}^2 S,$$

so

$$E(L,t)\,\mathrm{d}L\,\mathrm{d}t = (4\pi)^2 S\,n(S,z)\,\mathrm{d}S\,d_\mathrm{L}^2 \left(\frac{\mathrm{d}V}{\mathrm{d}z}\right)^{-1} \mathrm{d}t. \tag{6.23}$$

Now one can show that

$$4\pi d_\mathrm{L}^2 \left(\frac{\mathrm{d}V}{\mathrm{d}z}\right)^{-1} \mathrm{d}t = \frac{1}{c}(1+z)\,\mathrm{d}z. \tag{6.24}$$

Putting this into Equation 6.23 and integrating over L and t, we find that the total energy output of quasars throughout the history of the Universe is

$$\begin{aligned} E_\text{total} &= \int_{L=0}^{\infty} \int_{t=0}^{t_0} E(L,t)\,\mathrm{d}L\,\mathrm{d}t \\ &= \frac{4\pi}{c} \int_{z=0}^{\infty} (1+z)\,\mathrm{d}z \int_{S=0}^{\infty} S\,n(S,z)\,\mathrm{d}S. \end{aligned} \tag{6.25}$$

This does *not* depend on either H_0 or the cosmology!

Exercise 6.4 Soltan's ingenious argument rests in part on Equation 6.24. Prove this relation, by first showing that

$$\frac{\mathrm{d}V}{\mathrm{d}z} = \frac{4\pi c d_\mathrm{L}^2}{(1+z)^2 H(z)} = \frac{c}{H_0} \frac{4\pi d_\mathrm{L}^2}{(1+z)^2 (H(z)/H_0)} \tag{6.26}$$

(in itself a useful relation), and then using other results in Chapter 1. ■

Beware — some textbooks define the comoving volume differential $\mathrm{d}V/\mathrm{d}z$ for a *unit* solid angle, not for the whole sky as we have done, so they miss out the 4π factor. Sometimes this is also referred to as the 'volume element'.

We can use this to estimate the total mass of black holes today. Quasar number counts are rarely measured in bolometric fluxes, but B-band number counts are much more common. We can relate this to the bolometric flux S (where 'bolometric' means the total over all wavelengths):

$$S = k_\mathrm{bol}\, f_\mathrm{B} \nu_\mathrm{B},$$

where k_bol is known as the **bolometric correction**, f_B is the B-band flux, and ν_B is the typical frequency of B-band light. Soltan approximated Equation 6.25 as

$$E_\mathrm{total} \simeq \frac{4\pi}{c}\, k_\mathrm{bol} \int_{f_\mathrm{B}=0}^{\infty} f_\mathrm{B} \nu_\mathrm{B}\, n(f_\mathrm{B})\, \{1 + \langle z|f_\mathrm{B}\rangle\}\, \mathrm{d}f_\mathrm{B}, \tag{6.27}$$

where $\langle z|f_\mathrm{B}\rangle$ is the mean redshift at a given B-band flux f_B. The uncertainty in the B-band number counts $n(f_\mathrm{B})$ is much larger than the uncertainties in $\langle z|f_\mathrm{B}\rangle$.

Once we have the total energy emitted by quasars, we can convert this to a present-day black hole mass density using $\rho_\mathrm{BH}\, c^2 = E_\mathrm{total}\,(1-\eta)/\eta$, where η is the conversion efficiency of black hole mass accretion to luminosity. The factor of $(1-\eta)$, not originally used by Soltan, accounts for the fact that not all the accreted matter falls into the black hole. Putting in the numbers, Soltan found a present-day black hole density of

$$\rho_\mathrm{BH} = (0.1/\eta) \times 8 \times 10^4\, \mathrm{M_\odot\, Mpc^{-3}}$$

for a bolometric correction of $k_\mathrm{bol} = 6.0$ (justified on the basis of quasar spectral energy distributions). This corresponds to a cosmological density of black holes from 'dormant' quasars of

$$\Omega_\mathrm{BH}\, h^2 = 3 \times 10^{-7}\, \frac{1-\eta}{0.9}\, \frac{0.1}{\eta}.$$

For comparison, the present-day mass density in stars has been estimated as $\Omega_* h = (2.9 \pm 0.43) \times 10^{-3}$ for a Salpeter initial mass function, so Ω_BH in dormant quasars is only about $0.01h\%$ of Ω_*. However, we've only counted the type 1 (broad line) active galaxies and not the type 2 (narrow line) ones, so this should be regarded as a lower limit.

Nearby type 1 active galaxies have had black hole masses measured using the widths of the Balmer emission lines. On the assumption that the dynamics of the broad line region is dominated by gravity, $M_\mathrm{BH} \simeq v^2\, R_\mathrm{BLR}/G$, where v is the velocity width and R_BLR is the broad line region radius. This latter parameter can be estimated from models of the ionization within quasars, or from reverberation mapping, which we shall meet in the next section. This yields an integrated mass density of $\simeq 600\, \mathrm{M_\odot\, Mpc^{-3}}$ from local active galaxies. This is two orders of magnitude smaller than Soltan's estimate of Ω_BH. Most of the present-day Ω_BH must therefore be in dormant quasars, which we shall cover in the next section.

6.5 Finding supermassive black holes

6.5.1 Context

There are many lines of evidence that point to *accreting* black holes in other galaxies, such as rapid variability, or non-thermal spectra. We'll meet some of these shortly. We'll also show you some ways to find a black hole that's *not* accreting.

It's not necessary for matter to be compressed to extreme densities to begin the formation of a black hole. Filling the Solar System with liquid water (at standard temperature and pressure, STP, of 0° C and 1 atm) up to the orbit of Jupiter would be more than sufficient. There is not enough water in the Galaxy to do this, but the density of the Sun is only about 1.4 times that of water at STP, and there is no shortage of stars, particularly at the centres of galaxies. In practice the stellar densities are not high enough to generate a supermassive black hole spontaneously; nevertheless, **supermassive black holes** ($M_{\rm BH} > 10^6\,{\rm M}_\odot$) are inferred at the centres of many galaxies.

How can one detect supermassive black holes if they are not accreting matter? Broadly speaking, the trick is to resolve the sphere of influence of the black hole, within which the gravity of the black hole $GM_{\rm BH}/r^2$ dominates over the centripetal acceleration from the galaxy's velocity dispersion σ. These effects balance at a typical radius r_h, where

$$\frac{\sigma^2}{r_h} = \frac{GM_{\rm BH}}{r_h^2} \tag{6.28}$$

so

$$r_h = \frac{GM_{\rm BH}}{\sigma^2} \simeq \left(\frac{M_{\rm BH}}{10^8\,{\rm M}_\odot}\right)\left(\frac{\sigma}{200\,{\rm km\,s^{-1}}}\right)^{-2}\ {\rm pc}. \tag{6.29}$$

For comparison, the Schwarzschild radius can be expressed as

$$R_{\rm S} \simeq 2\left(\frac{M_{\rm BH}}{10^8\,{\rm M}_\odot}\right)\ {\rm AU}, \tag{6.30}$$

where 1 astronomical unit (AU) is the mean distance from the Earth to the Sun, equivalent to about 4.8×10^{-6} pc, so the radius r_h is about 10^5 times bigger than the Schwarzschild radius $R_{\rm S}$ when the galaxy's velocity dispersion is $\sigma \simeq 200\,{\rm km\,s^{-1}}$. So, if we can reach angular resolutions of $\sim 10^5 R_{\rm S}$, we may be able to discern the effects of the black hole.

Proving that the central object is a supermassive black hole, and not some super-dense clump of non-luminous stuff, is another matter. One could imagine, for example, a dense cluster of non-luminous stellar objects such as neutron stars, white dwarfs and stellar-mass black holes. In three galaxies, the limits on the size and density of the central object imply that it would have dispersed within the lifetime of the galaxy, so these could be regarded as providing robust evidence for a black hole. It's still possible to hypothesize exotic alternatives, such as a cluster of very low mass ($\leq 0.04\,{\rm M}_\odot$) black holes, but there is no plausible astrophysical process for making them. It's also very unclear how one would form a central star cluster of that density without creating a supermassive black hole anyway. The evidence for supermassive black holes in galaxies therefore rests on astrophysical plausibility rather than direct detections.

Exercise 6.5 Find the angular size in arcseconds of the sphere of influence of a $10^8\,M_\odot$ black hole in a galaxy with a central velocity dispersion $\sigma = 220\,\mathrm{km\,s^{-1}}$, at a distance of 10 Mpc. Compare this to the best angular resolution typical of optical ground-based telescopes (known as 'seeing') of $\sim 0.5''$ set by turbulence in the Earth's atmosphere. (One arcsecond is $1/3600$th of a degree.) ■

6.5.2 Stellar and gas kinematics

As you've found in the last exercise, you need excellent angular resolution to probe the sphere of influence of a supermassive black hole, and this is far beyond the capabilities of conventional optical ground-based telescopes. One approach is to site your telescope above the Earth's turbulent atmosphere. The Hubble Space Telescope (HST) has been extremely important in making high angular resolution constraints of supermassive black holes. Another approach is to correct for the turbulence in the Earth's atmosphere, by measuring the distortion and correcting telescope optics in real time. This approach, known as **adaptive optics**, uses a nearby bright reference object to monitor the distortions in the wavefront and deforms the telescope's optics in real time to correct for these distortions. The bright reference could be a nearby bright star or a 'laser guide star', i.e. a laser beam sent from the telescope itself.

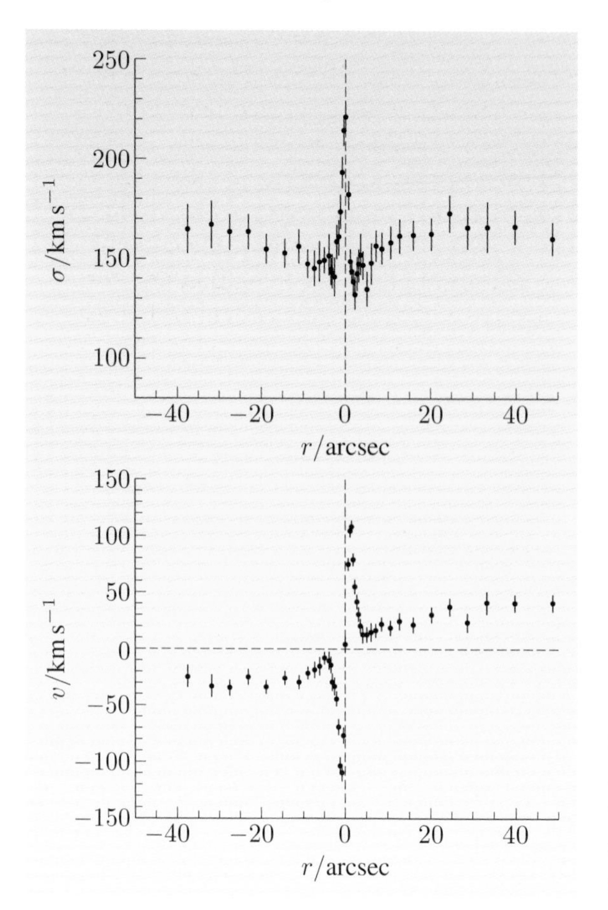

Figure 6.5 The velocity dispersion σ and mean velocity v (both in $\mathrm{km\,s^{-1}}$) along the major axis of the core of M31.

The Andromeda galaxy (M31) is 2.52 million light-years away, or 0.77 Mpc. It is one of our Galaxy's closest neighbours and can even be seen with the naked eye on a dark enough night. At this distance, one arcsecond is about 3.7 pc, so the sphere of influence of the black hole may just be within the capabilities of ground-based telescopes. Figure 6.5 shows the velocities of the stars in the bulge of M31 (measured from the integrated starlight rather than detecting individual stars). The characteristic sharp feature in the centre implies a black hole mass of $\simeq 10^6\,M_\odot$. Subsequent HST spatially-resolved spectroscopy revised this upwards to $(3.0 \pm 1.5) \times 10^7\,M_\odot$.

The HST found evidence for a *much* larger black hole in another nearby galaxy, M87, by measuring the Doppler shifts in ionized gas close to the centre. The spatial sampling of the HST's Faint Object Camera is $0.028''$, corresponding to only about 2 pc in M87. Two sets of observations by the HST yielded $(2.4 \pm 0.7) \times 10^9\,M_\odot$ within 18 pc ($0.25''$), followed by $(3.2 \pm 0.9) \pm 10^9\,M_\odot$ within 3.5 pc ($0.05''$). The HST has since been used for measuring black hole masses in several other galaxies. For comparison, the total mass of the entire Small Magellanic Cloud (Chapter 3), including its dark matter, is about $6.5 \times 10^9\,M_\odot$.

6.5.3 Megamasers

The early HST discoveries received a lot of press coverage, but at about the same time — though finding less press notice — a much stronger constraint on a supermassive black hole came from radio astronomy. The galaxy NGC 4258 (also known as M106) has naturally-occurring masers (microwave lasers) generating coherent radiation through stimulated emission, with H_2O molecules providing the masing medium. These masers are believed to be generated in random directions but with a small random subset lying along the line of sight to the Earth. (This abundant maser activity is sometimes referred to as 'megamasers'.) This masing emission can be detected by radio interferometry, which routinely has milliarcsecond (mas) resolutions or better. Figure 6.6 shows the line-of-sight velocities of masers in NGC 4258 observed with the Very Long Baseline Array (VLBA) of radio telescopes. Currently, observations are consistent with a central mass of $(3.82 \pm 0.01) \times 10^7\,M_\odot$ within a central radius of 0.13 pc (4.1 mas, i.e. $0.0041''$).

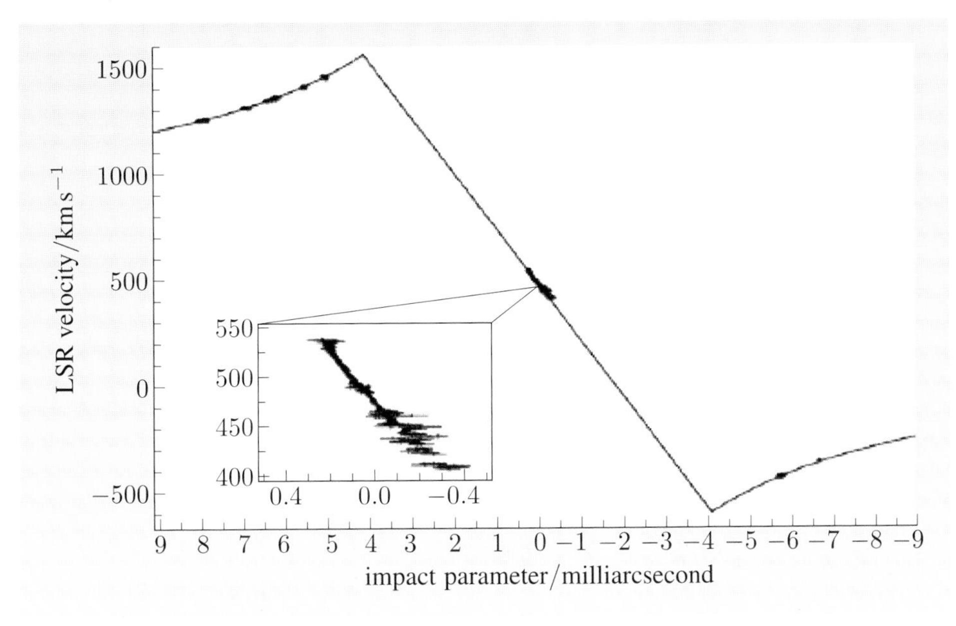

Figure 6.6 The velocities of megamasers around the central supermassive black hole in the galaxy NGC 4258.

6.5.4 Stellar proper motion

You may already have wondered whether the best chance of resolving the sphere of influence of a supermassive black hole could be in our Galaxy. The distance to the Galactic Centre is approximately 8 kpc, so one parsec at this distance corresponds to around $1/8000$th of a radian, or $25''$. This is sufficiently close that we can trace the orbits of individual stars around the central object and make a direct mass estimates. This can only be done in the near-infrared because of the very heavy dust extinction to the Galactic Centre. Ground-based adaptive optics in the 1–$3\,\mu$m wavelength range on telescopes with 8–10 m primary mirrors can be competitive with the HST, and long-term monitoring campaigns of the Galactic Centre have yielded impressive results. For example, Figure 6.7 shows the orbit of a star just over only ten light-hours from the central black hole. The enclosed mass can be derived from the orbits of this star and others, as a function of the distance to the centre. The data are consistent with a central mass of at least $2.6 \times 10^6\,M_\odot$ in our Galaxy. At the time of writing, the current best estimate for the central black hole mass is $(4.5 \pm 0.4) \times 10^6\,M_\odot$.

Ghez, A. M. et al., 2008, *Astrophysical Journal*, **689**, 1044.

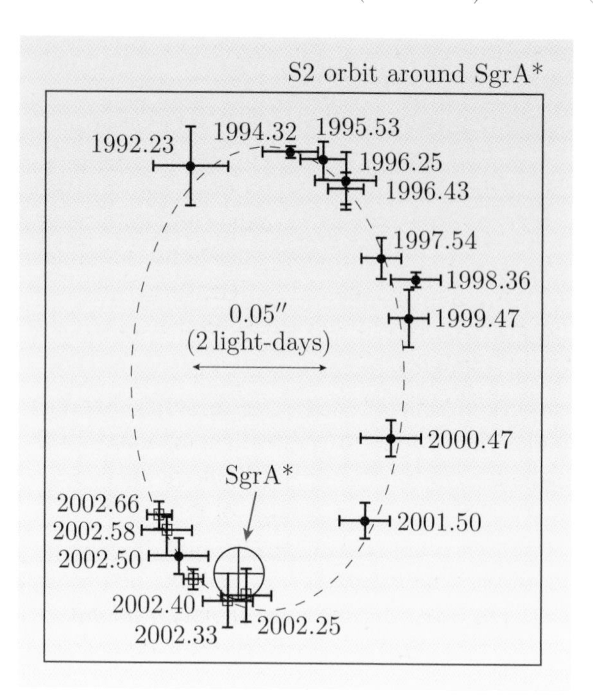

Figure 6.7 The orbit of a star close to the central supermassive black hole in our Galaxy. The numbers indicate the dates of the observations, expressed as decimal years.

6.5.5 Broad iron X-ray emission line

The inner regions of the accretion disc can sometimes give rise to a high-energy emission line in the X-ray from ionized iron, known as the Kα line. A pioneering detection in a nearby active galaxy MGC-6-30-15 has a width of the line implying Doppler motions of $\sim$100 000 km s^{-1}(!), suggesting a location at three to ten Schwarzschild radii. This is a rare chance to sample the relativistic environment very close to a black hole.

6.5.6 Reverberation mapping

Another way to set a limit on the size is the variability of its light output. If an object varies on timescales of weeks, then it's likely to have a physical size of light-weeks or smaller. Quasars vary on timescales from years to (in some cases) days. The extremely high luminosities of quasars and these small inferred physical sizes, together with the very high radiative efficiency of black hole accretion (Section 6.3), led to supermassive black holes being the leading explanation for quasar activity.

However, there are exceptions to these variability arguments: for example, if the light is from a relativistic jet that happens to be pointed at the Earth, the observed variability can appear to be much shorter. This is one of the consequences of combining motion towards the observer with the Lorentz transformation (see Appendix B), known as relativistic beaming, which is covered in more detail elsewhere (see the further reading section).

Does quasar variability reflect light-travel time or is it just a beaming artefact? A variation on this technique, known as **reverberation mapping**, avoids the possibility of relativistic beaming. A flare of emission from the centre will cause a brightening in the broad emission lines, but there will be some time delay because the broad emission line clouds are not as centrally located. Also, the clouds exist in physical conditions that are very different to relativistic jets. By cross-correlating the variation in the continuum with the variations in the emission lines, one can infer how close the emission line clouds are to the central luminous object. Figure 6.8 shows the time delay between continuum variations and broad line luminosity variations in the Seyfert 1 galaxy NGC 5548. The time lag is measured by cross-correlating the line and continuum measurements, i.e. summing $\sum C(t)\, L(t - \Delta t)$ for various supposed lags Δt.

Why are the lag measurements in Figure 6.8 fairly broad? One reason is that if the Δt chosen is close to (but not equal to) the underlying value, then there may still be some positive cross-correlation signal. More importantly, light-travel time effects mean that the broad line clouds are sampled at a range of radii at any fixed time lag, as shown in Figure 6.9. One can use this reverberation mapping in different emission lines to derive the internal structures of the broad line region of active galaxies. High-ionization lines respond most rapidly, implying a stratification of ionization, and they also have the largest widths. This immediately suggests a route to estimating black hole masses: if these widths represent Doppler motion, then material closer to the black hole is moving faster and we can use a formalism analogous to Equation 6.28 to estimate a black hole mass:

$$M_{\mathrm{BH}} \simeq \frac{r(\Delta v)^2}{G}, \tag{6.31}$$

where the radius r is estimated from the reverberation measurements, while Δv is measured from the line widths.

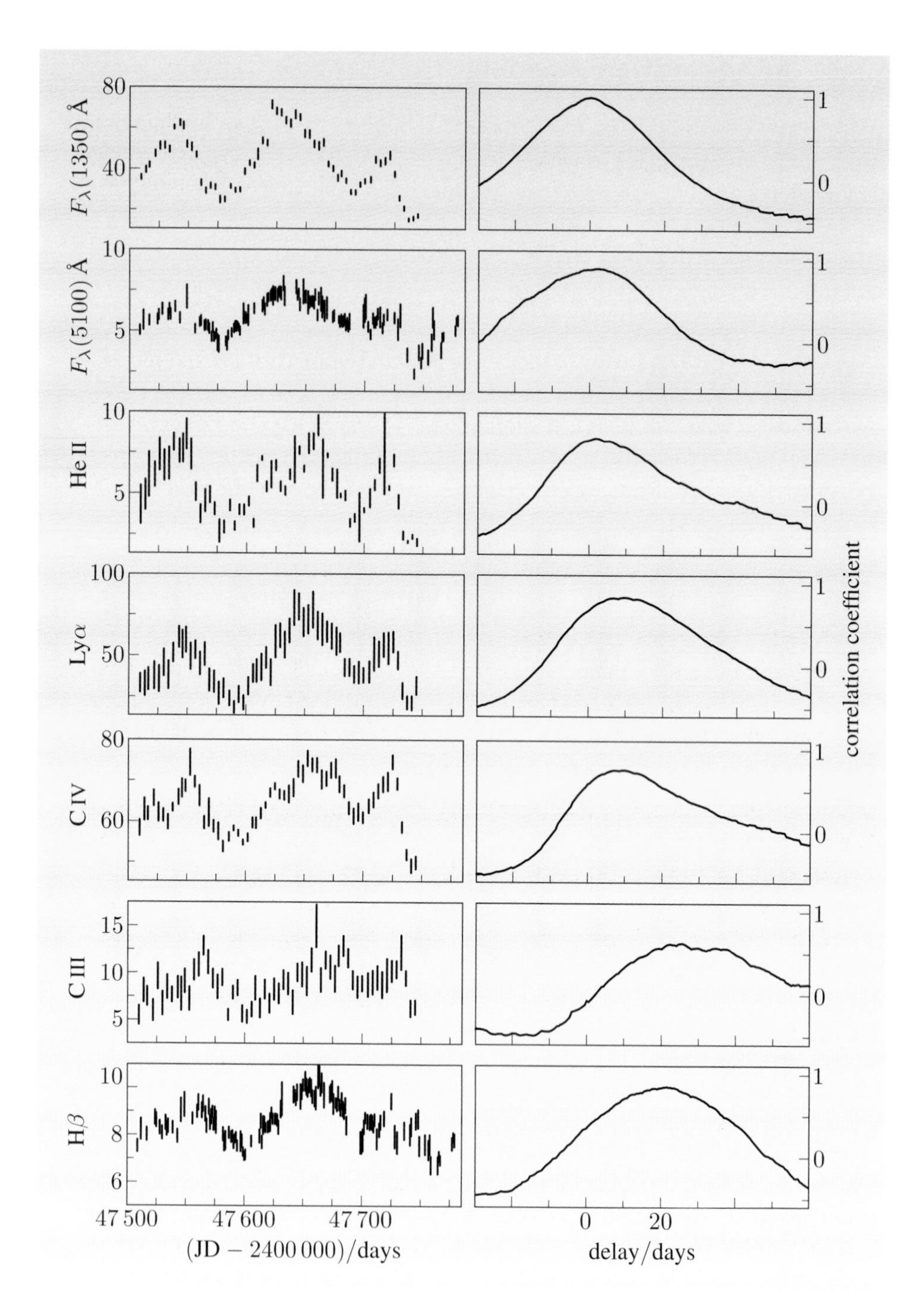

Figure 6.8 The variations in the continuum and in selected emission lines in the Seyfert 1 galaxy NGC 5548. The left-hand panels show the variations themselves, and the right-hand panels show the strength of the lag Δt. This strength is proportional to the cross-correlation $\sum C(t)\, L(t - \Delta t)$, where $C(t)$ are the continuum measurements and $L(t)$ are the line measurements.

Reverberation mapping has also confirmed predictions from photoionization models of the structure of quasar broad line regions. In the simplest such model, the radius r from which an emission line mainly originates scales as $r \propto \sqrt{L}$, where L is the luminosity of the quasar. This follows from the expectation that the number of photons per electron, U, will vary as $U \propto L/(4\pi r^2)$, and any given emission line will be most efficiently produced at a particular U. The parameter U is known as the ionization parameter; we shall meet it again in Chapter 8.

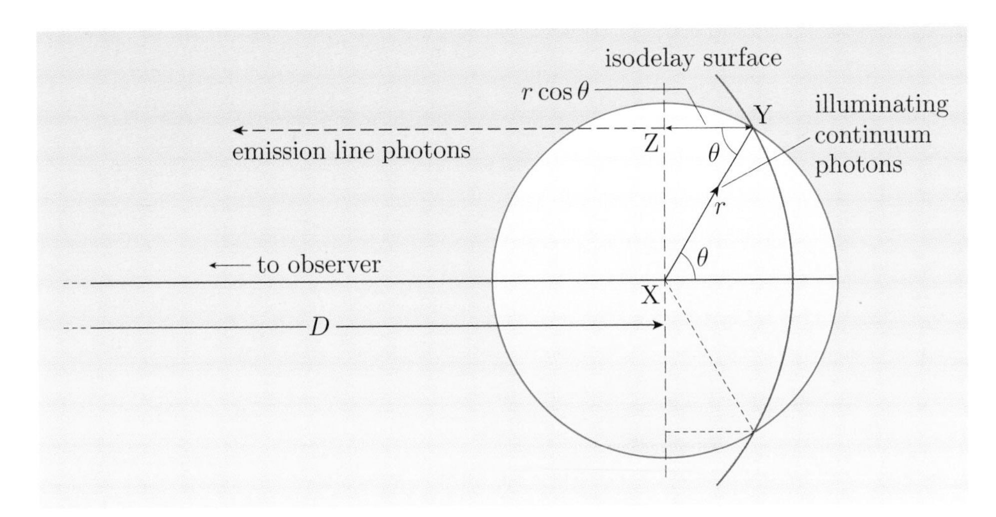

Figure 6.9 Isodelay surface in a quasar broad line region. Broad line clouds lying along the isodelay surface have a time delay of $\Delta t = (r/c)(1 + \cos\theta)$ relative to light from the centre.

In practice the long-term monitoring data are available for only a few quasars, so the reverberation measurements act as calibrators for black hole mass estimators of the form $M_{\rm BH} = kL_\nu^\alpha(\Delta v)^2$, where L_ν is the continuum luminosity measured near a particular emission line, and k and α are constants specific to that emission line. Typically $\alpha \simeq 0.5$, as required if $r \propto \sqrt{L}$.

The quasars in the Sloan Digital Sky Survey (SDSS) have had black hole mass estimates ranging from around $10^7\,{\rm M_\odot}$ to an astonishing $> 10^{10}\,{\rm M_\odot}$ (compare the mass of the entire Small Magellanic Cloud galaxy in Subsection 6.5.2). However, these black hole mass estimates have underlying uncertainties of many tens of per cent at best. It's possible that these largest black holes are in fact lower-mass objects in which the underlying uncertainties happen to have given rise to a higher measurement. Nevertheless, the SDSS quasar data set is consistent with black holes existing up to a maximum mass of $3 \times 10^9\,{\rm M_\odot}$.

6.6 The Magorrian relation

The masses of black holes turn out to have an astonishingly close relationship to certain properties of their host galaxies. This closeness is a very important clue to the formation of supermassive black holes and the galaxies that host them. The next exercise should convince you that it's at least not immediately obvious how supermassive black holes came into existence.

Exercise 6.6 The age of the Universe in the concordance cosmology at redshift $z = 2$ was about 3 Gyr. A $10\,{\rm M_\odot}$ black hole could be made early in the history of the Universe as an end-product of the first stars. Show that even if you start with a $10\,{\rm M_\odot}$ black hole at $z = \infty$, you still cannot create a supermassive black hole by $z = 2$ through Eddington-limited black hole growth. ■

How are black holes related to their host galaxies? One way to test for connections is to correlate the black hole measurements with host galaxy property measurements. A closer physical link should result in a tighter correlation. The important effects are not just the correlation itself, but also the *scatter*.

If we correlate the supermassive black hole masses against the total host galaxy masses, the correlation is somewhat weak. A stronger correlation (shown in Figure 6.10) is between $M_{\rm BH}$ and the *spheroid* luminosity, i.e. the bulges in the case of spiral galaxies, and the entire galaxies in the case of ellipticals.

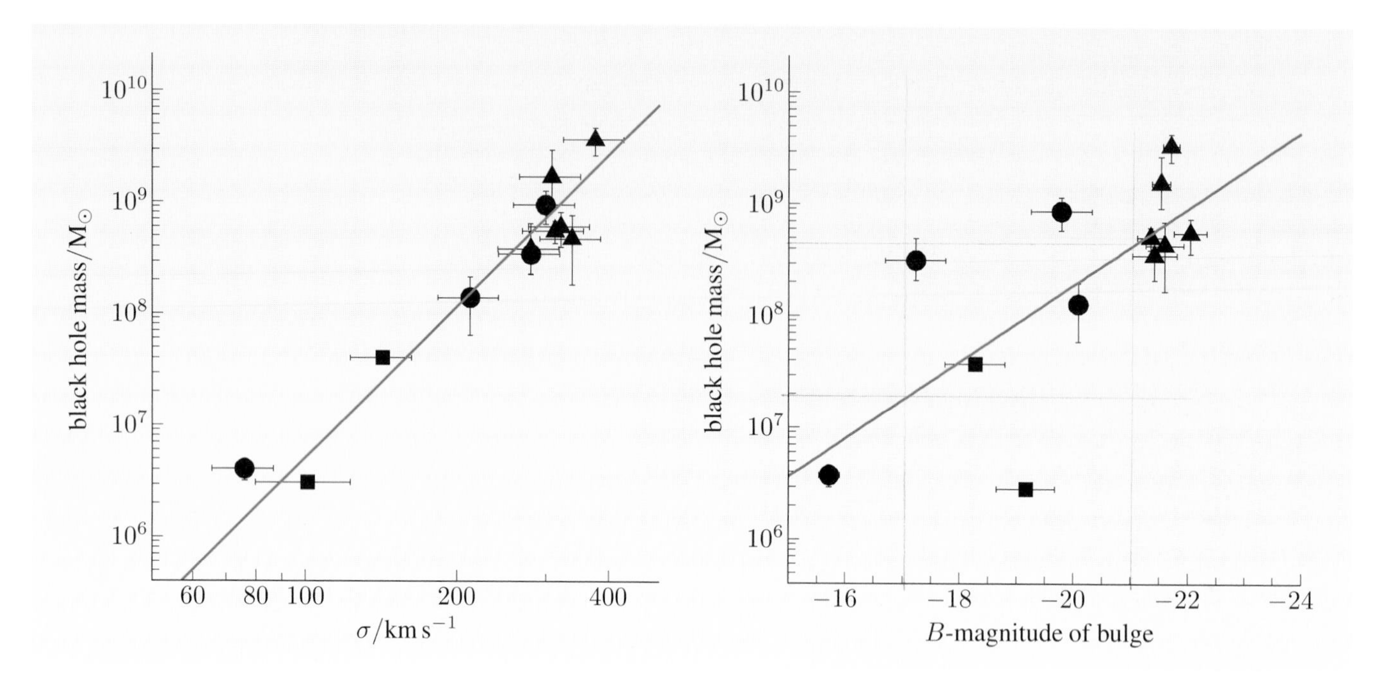

Figure 6.10 Left: correlation between the black hole masses and central velocity dispersions for local galaxies. Right: correlation between black hole masses and bulge B-band luminosities of the same sample of local galaxies. Ellipticals are displayed as circles, spirals as triangles, and the squares represent both lenticulars and compact elliptical galaxies.

This was first discovered in 1998 by a team led by the astronomer John Magorrian, and has become known as the **Magorrian relation**. There is an even stronger correlation between the spheroid velocity dispersions σ and $M_{\rm BH}$. This correlation is also shown in Figure 6.10. The dispersion of the data points is almost entirely attributable to the uncertainties in the measurements — in other words, the measurement of the underlying scatter is almost consistent with zero. The best-fit relationship is

$$M_{\rm BH} = (1.66 \pm 0.32) \times 10^8\,{\rm M}_\odot \left(\frac{\sigma}{200\,{\rm km\,s^{-1}}}\right)^{4.58\pm0.52}. \qquad (6.32)$$

Central supermassive black holes are about 0.6% of the masses of their galactic bulges (or the whole galaxies in the case of ellipticals), with a sphere of influence that is less than a thousand billionth of the volume of the bulge, yet the black hole mass correlates astonishingly strongly with the galaxy velocity dispersion. Clearly, the creation of a central supermassive black hole has somehow been strongly connected to the formation of its galaxy. We shall speculate on how and why later in this chapter.

6.7 The hard X-ray background

In Chapter 5 we turned Olbers' paradox around and made inferences about the evolution of galaxies and star formation from the spectrum of the extragalactic optical and infrared background light. With the cosmic X-ray background, we can make similar inferences about the cosmic history of black hole accretion.

The cosmic X-ray background was discovered in 1962 on a rocket mission designed to study the Moon. The Moon remained undetected in X-rays until 1991, but the discovery of the cosmic X-ray background pre-dates the discovery of the CMB by two years, making it the first known cosmic background. Riccardo Giacconi led this pioneering discovery, and his cumulative contributions to X-ray astronomy were recognized in his Nobel Prize in 2002. Figure 6.11 shows a 1991 image of the Moon by the ROSAT (Röntgensatellit) X-ray satellite.

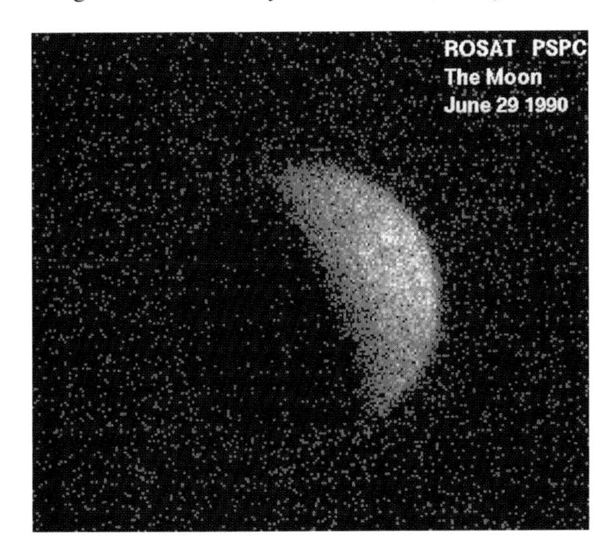

Figure 6.11 Image of the Moon taken with the ROSAT satellite. Note the reflected solar X-rays, the isotropic background that is shadowed by the dark side of the Moon, and the weaker background that appears to emanate from the dark side of the Moon. This latter emission has since been deduced to be local to the Earth and the satellite.

The Moon reflects solar X-rays, but the dark side is clearly obscuring a faint background. The image is grainy because X-ray detectors respond to *individual* photons. The dark side of the Moon isn't quite black because the Earth's geocorona, or extended outer atmosphere, emits X-rays, and the satellite orbits within this geocorona. (The extragalactic X-ray research community sometimes refers to the 'background' as being the proportion that is not yet resolved into point sources, rather than the total flux from the sky. We shall follow the conventions at other wavelengths and take the 'background' to mean the total flux, rather than the unresolved component of it.)

As with the CMB, the X-ray background appears to be fairly isotropic, once the Galaxy has been subtracted. This on its own suggests a cosmological origin. The X-ray background is included in Figure 5.1. One can make inferences about the evolution of black hole accretion from this, but one thing quickly became apparent. The spectrum of the extragalactic X-ray background is very different to the spectrum of a star-forming galaxy or an unobscured quasar, so what could generate this background? This is sometimes known as the **X-ray spectral paradox**.

The shape of the X-ray background therefore requires some objects with **harder** X-ray spectra, i.e. with a greater proportion of higher-energy X-rays. The most likely candidate is type 2 AGN. Figure 6.12 shows the effect of X-ray and optical absorption for various levels of obscuration. As we increase the gas column density (the integral of the gas particle number density along the line of sight), so the optical depth of dust should increase. As the dust content is increased, the optical and near-ultraviolet flux is decreased, with the absorbed energy being re-emitted in the infrared as thermal emission. Dust grains are not effective absorbers of X-rays, but the neutral hydrogen gas associated with the dust does absorb X-rays. The sharp cut-off at low X-ray energies is sometimes known as the photoelectric cut-off. The hardest energy X-rays are the most penetrating, which means that X-ray spectra with absorption have an inverted spectral index. Cosmologically, this has two important observations consequences. First, there is a negative K-correction, as with submm-wave surveys (see Chapter 5). Second, the redshifting causes the *observed* photons to come from *higher*-energy rest-frame X-rays at higher redshifts. Therefore the high-redshift objects will have lower rest-frame obscuration, unlike in optical galaxy surveys.

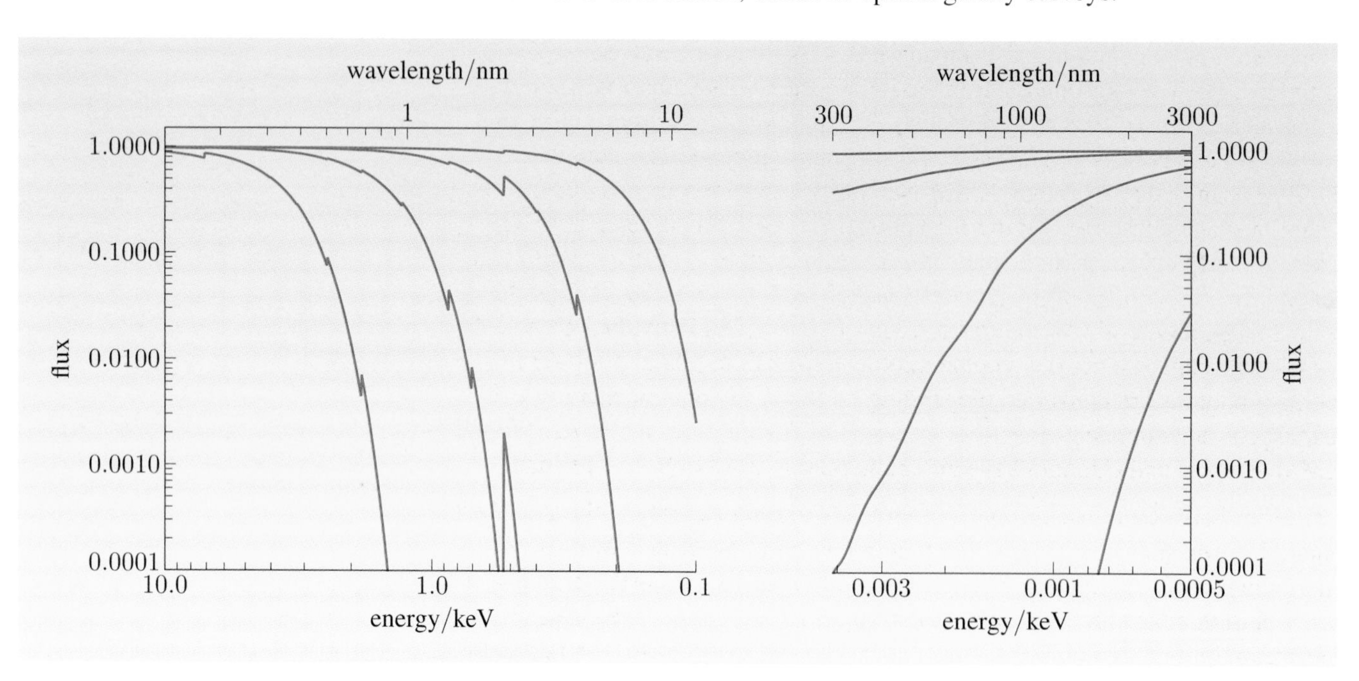

Figure 6.12 Left: the transmitted flux through a neutral hydrogen column density of (from top to bottom) 10^{20}, 10^{21}, 10^{22} and $10^{23}\,\mathrm{cm^{-2}}$. The greater the column density, the less flux is transmitted through. On the right-hand side, the corresponding near-infrared to ultraviolet extinction is shown for $N_{\mathrm{HI}} = A_{\mathrm{V}} \times 1.8 \times 10^{25}\,\mathrm{m^{-2}\,mag^{-1}}$. The right-hand curves show $A_{\mathrm{V}} = 0.055, 0.555, 5.55$ and 55.5 (from top to bottom).

The active galaxies with the highest N_{H} column densities,

$$N_{\mathrm{H}} > \sigma_{\mathrm{T}}^{-1} \simeq 1.5 \times 10^{24}\,\mathrm{cm^{-2}}$$

(where σ_{T} is the Thomson scattering cross section), are optically-thick to Compton scattering of their hard X-ray photons. The scattered X-ray photons have lost some of their energy, which was carried off as kinetic energy by the electron with which they collided. These lower-energy X-rays are much more easily absorbed. These Compton-thick active galaxies are difficult to detect in hard X-ray surveys but their presence can be inferred from interpreting the shape of the hard X-ray background. One could avoid having a Compton-thick population, but (it turns out) only at the cost of failing to reproduce the observed absorption column density distribution. There must therefore have been some significant contribution from Compton-thick active galaxies to the black hole accretion history of the Universe.

If the geometry of the absorption allows it, it may be possible to detect some of the Compton scattered component directly (though faintly). The spectrum of this 'Compton reflected' component is expected to be a broad peak around 20–30 keV, depending on the ionization of the scattering medium. Also, Compton scattering from iron nuclei can excite a strong iron Kα line (Subsection 6.5.5). Both the iron line and the spectral shape can be useful indicators of Compton-thick absorbers, though with current X-ray telescopes this can be done only in luminous and/or local active galaxies.

The softer X-ray background has been mostly resolved into its constituent point sources by ROSAT ($\simeq 75\%$ of the 0.5–2 keV background). It has taken longer to do the same at harder X-ray fluxes where the X-ray spectral paradox suggested new populations. The Japanese ASCA satellite resolved $\simeq 35\%$ of the 2–10 keV background, while the Italian BeppoSAX mission resolved $\simeq 20$–30% of the 5–10 keV background. The breakthroughs in resolving most of the hard X-ray background into its constituent point sources came from the European Space Agency's XMM-Newton space telescope and NASA's Chandra space telescope. Figure 6.13 shows the deep pencil-beam surveys taken by XMM-Newton and Chandra. The point sources in these surveys can account for $\simeq 80$–90% of the 2–6 keV background, but only 50–70% of the 6–10 keV background. Astronomical X-ray CCDs detect individual photons by converting them to electrons via the photoelectric effect (about 10–80% of incident photons are converted, depending on the X-ray photon energies), then reading the accumulated charge in each pixel. The faintest X-ray sources found by XMM-Newton and Chandra have electron count rates of only $\simeq 1$ per day.

The term 'hard X-ray background' often tends to refer roughly to the 2–10 keV range, but one should not forget the higher-energy background. Most of the energy density in the cosmic X-ray background is at 10–100 keV, but only a few per cent of this background has been directly resolved so far. The proposed future European Space Agency X-ray space telescope, currently named the International X-ray Observatory (IXO), will be able to observe 0.1–40 keV and directly probe the Compton-thick populations.

What sort of objects dominate the hard (2–10 keV) X-ray background? Most turn out to be active galaxies, but there is an extraordinary range in the optical properties: the X-ray–optical flux ratios vary by over four orders of magnitude. There are unobscured and obscured AGN, which appear in optical spectra as broad line and narrow line objects, respectively. More surprisingly, there is a minority of objects with obscured X-ray spectra but broad optical emission lines, and others with low X-ray column densities yet narrow line AGN optical spectra (implying that the obscuring gas doesn't follow the same distribution as the obscuring dust in Figure 4.15). Many objects in the XMM-Newton and Chandra pencil-beam surveys are too faint for optical spectroscopy, even with the largest 8–10 m-mirror optical telescopes, though broad-band photometry is consistent with these being mainly distant AGN. Some objects have no optical counterparts in even the deepest ground-based and space-based optical imaging. There are also X-ray bright, optically-normal galaxies (with the delightful acronym XBONGs), which show X-ray evidence for AGN (sometimes obscured, but not always) yet no optical AGN evidence (such as broad or high-ionization emission lines) in the optical spectra. A minority of the hard X-ray background also comes from starburst galaxies (see Chapter 5), galaxy clusters and groups (see Chapters 3 and 7), and Galactic stars.

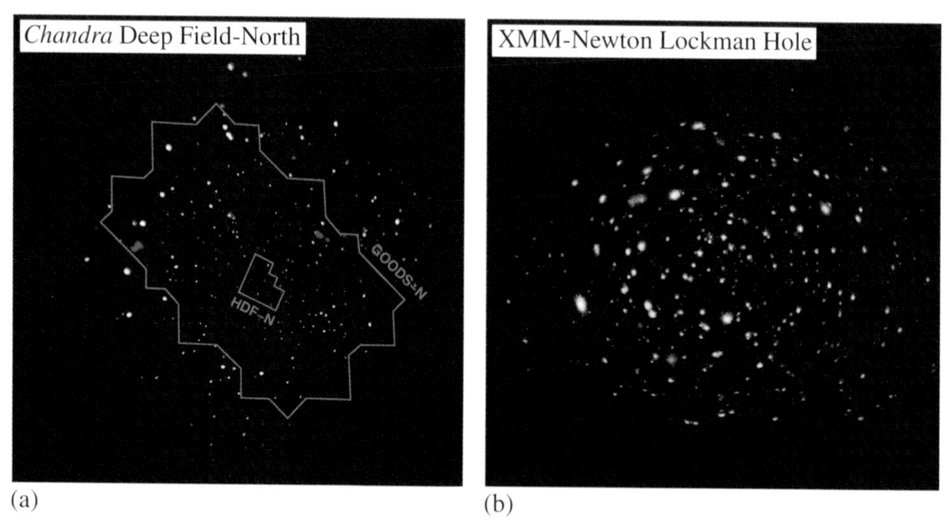

Figure 6.13 Deep fields taken by the XMM-Newton and Chandra space telescopes. (a) The Chandra Deep Field North (CDF-N). This is a 2 Ms image (i.e. an exposure of 2×10^6 seconds) taken by Chandra, in the region of the Hubble Deep Field North (HDF-N, marked in green). The Spitzer GOODS survey field is also shown in green. There are X-ray data over around 448 arcmin^2, i.e. around 60% of the angular size of the Moon. This image represents 0.5–2 keV photons as red, 2–4 keV as green (except for annotations), and 4–8 keV as blue. Confusingly, there is also a Chandra Deep Field South (CDF-S), which does *not* coincide with the Hubble Deep Field South (HDF-S), but the Chandra field is nevertheless the site of the Hubble Ultra Deep Field (UDF). (b) The XMM-Newton deep field in a region of sky known as the Lockman Hole in Ursa Major (named after Felix Lockman who discovered that this region had very low X-ray absorption from Galactic neutral hydrogen). Here, 0.5–2 keV photons are represented as red, 2–4.5 keV as green, and 4.5–10 keV as blue. Objects are broader closer to the edges because the instrumental angular resolution is coarser. This image covers about 1556 arcmin^2.

But where are the Compton-thick objects? A few high-redshift objects are known to be Compton-thick from X-ray observations, such as the hyperluminous galaxy IRAS F10214+4724 (Chapters 5 and 7), but most are very hard to detect in X-rays. An intriguing clue has recently come from infrared surveys. Good candidates for highly dust-shrouded quasars were found through high 24 μm to 3.6 μm flux ratios. (A faint 3.6 μm flux implies that it's not an unobscured quasar, in which case the 3.6 μm flux is dominated by the host galaxy, while the 24 μm excess suggests hot dust as expected for an AGN dust torus.) Furthermore, as we saw in Chapter 5, star-forming galaxies have strong emission and absorption features in the mid-infrared, while the mid-infrared spectra of active galaxies are typically featureless. Mid-infrared spectroscopy of some mid-infrared-bright but optically-faint galaxies in the Chandra Deep Field North (also known as GOODS-North) has found evidence of active nuclei, but the X-ray emission of these galaxies is weak or absent. This suggests that there is a population of 'mid-infrared excess' galaxies having high mid-infrared–optical flux ratios, at least some of which are Compton-thick active galaxies. Could these be the missing galaxies that dominate the highest-energy X-ray backgrounds?

Martínez-Sansigre, A. et al., 2005, *Nature*, **436**, 666.

6.8 Black hole demographics

We can use Equation 6.32 to estimate the present-day cosmological density of black holes, by integrating over the spheroid luminosity function and using the Faber–Jackson relationship between velocity dispersion and luminosity, and its equivalent for spiral bulges (Chapter 3). From this, one finds that

$$\rho_{\rm BH} = 9.4^{+3.9}_{-2.9} h^2 \times 10^5 \,{\rm M_\odot\,Mpc^{-3}}.$$

This is about ten times bigger than Soltan's original estimate of

$$\rho_{\rm BH} = \frac{1-\eta}{0.9}\frac{0.1}{\eta} \times 8 \times 10^4 \,{\rm M_\odot\,Mpc^{-3}}$$

(where η is the accretion efficiency), but Soltan considered only the type 1 (broad line) active galaxies. Locally, there are about four times as many type 2 active galaxies as type 1. In order to bring the Soltan estimate broadly into agreement with this local value, the type 1/type 2 fraction would need to evolve (or the estimated bolometric correction was wrong).

Alternatively, one could make a Soltan-style analysis using the X-ray number counts of active galaxies, since these will include the contributions from type 2 active galaxies, or at least the Compton-thin proportion. This gives

$$\rho_{\rm BH} = (4.7 - 10.6) \times \frac{1-\eta}{9\eta} \times 10^5 \,{\rm M_\odot\,Mpc^{-3}}.$$

Comparing this to the local black hole number density yields a broad constraint on the accretion efficiency of $\eta = 0.04$–1.6. There have also been several attempts to model the fraction of time that quasars spend accreting (known as the **duty cycle**), their accretion efficiencies and their Eddington ratios. The typical inputs to these models are the local black hole mass function (number per unit volume per unit mass), the evolving quasar luminosity function, and the evolution and luminosity dependence of the type 1/type 2 number density ratio. However, as these constraints are model-dependent, we shall not discuss them here.

Figure 6.14 Optical image of the galaxy M74, with the hard X-ray image from the Chandra satellite superimposed in red. The ultraluminous X-ray source is marked as ULX. This is too bright to be a conventional X-ray binary star system, unlike the other X-ray sources in this image.

X-ray observations of local galaxies have revealed a final surprise: a population of 'ultraluminous X-ray sources' with luminosities $> \times 10^{32}$ W (i.e. $> 3 \times 10^5\,{\rm L_\odot}$) that do *not* lie at the centres of the galaxies (see, for example, Figure 6.14). If these are accreting black holes, then their masses are in the range 100–10 000 $\rm M_\odot$, sometimes referred to as 'intermediate mass black holes'. How did they form? Are they formed, for example, at the centres of globular clusters? Some globular clusters do show evidence for central intermediate mass black holes, and they appear to obey the same $M_{\rm BH}$–σ relationship as galaxy spheroids (see, for example, Maccarone, T. J. et al. (2008) *Nature*, **445**, 183). Do intermediate mass black holes eventually sink to the centre of their galaxy to build up the central supermassive black hole? How do intermediate mass black holes participate in the formation of globular clusters, if at all? Are they indeed black holes, or some brief X-ray luminous super-Eddington state of an accreting stellar-mass black hole? At the time of writing, these are still open questions.

6.9 Observations of black hole growth and the effects of feedback

Which came first — the black hole or its galaxy? Can observations constrain which came first? This takes us into what at the moment is uncertain experimental territory. There's a connection to the cooling flow problem in Section 3.9: dense cooling core clusters are a nearby example where we can observe the influence of a black hole suppressing star formation in the central galaxy.

One of the problems is that it's not currently technologically feasible to directly resolve the black hole sphere of influence in any but the most local galaxies. The local M_{BH}–σ relationship is the end result of billions of years of evolution, including multiple galaxy mergers. Intuition suggests that the M_{BH}–σ relationship was somehow imprinted early on, and numerical simulations have confirmed that this relationship, once established, is maintained surprisingly well in galaxy mergers. Unfortunately, the primordial high-redshift links between black holes and their host galaxies are very difficult to observe directly.

However, there are slightly less direct approaches. As we've seen, reverberation mapping can be used to derive the masses of quasar black holes, while the velocity dispersions can be inferred from the widths of absorption lines in the quasars' host galaxies. Provided that these are both reliable measures, we can make some constraint on the evolution of the black hole–host galaxy relationship.

Another approach is to use radio-loud active galaxy unification models. The radio-loud active galaxy population in general has powerful radio jets emitted from the active nucleus, terminating in a bow shock in the intergalactic medium. Figure 4.12 is an example. Only about 10% of active galaxies are radio-loud, but it is not clear why.

In the radio-loud unification model, active galaxies have a dusty torus that obscures the view of the quasar broad lines from some orientations (see Figure 4.15). Quasars and radiogalaxies with the *same* radio lobe luminosities should be members of the same population, though seen with different orientations. Therefore we can measure the host galaxy properties by studying the radiogalaxies, then measure the corresponding quasar properties by comparing the radiogalaxies' counterparts in the quasar population. There is some (admittedly weak but suggestive) evidence for the evolving black hole mass and host galaxy properties inferred from this method.

So which came first, the black hole or its galaxy? The comparison of the Madau diagram (Chapters 4 and 5) to the black hole accretion history suggests that there was plenty of star formation before the quasar epoch. However, the comparison of 3CRR host galaxy masses and black hole masses suggests that the $M_{BH}/M_{spheroid}$ ratio increases with redshift, which in turn suggests that the most massive black holes were pre-existing and spheroids formed around them.

On the other hand, there are hints that the submm-selected galaxy population has a *smaller* M_{BH}/M_{galaxy} mass ratio than these quasars, at least in those submm galaxies with broad optical emission lines (Figure 6.15). What's more, many submm galaxies have been detected in hard X-rays, implying that black hole mass accretion appears to be much more common in those galaxies than in the general galaxy population. Perhaps the most massive starbursts are eventually shut off by the energy input from an exponentially-growing accreting black hole. Quasars at all redshifts are ultraluminous starbursts on average, though this doesn't tell us whether the quasar phase comes at the start of the starburst or at the end. At the time of writing, it seems that the star formation rate in quasars varies roughly as the square root of the quasar luminosity, not linearly:

$dM_*/dt \propto (dM_{BH}/dt)^{0.44\pm0.07}$. This feature has not so far been reproduced by models of quasar feedback.

Serjeant and Hatziminaoglou, 2009, *Monthly Notices of the Royal Astronomical Society*, **397**, 265.

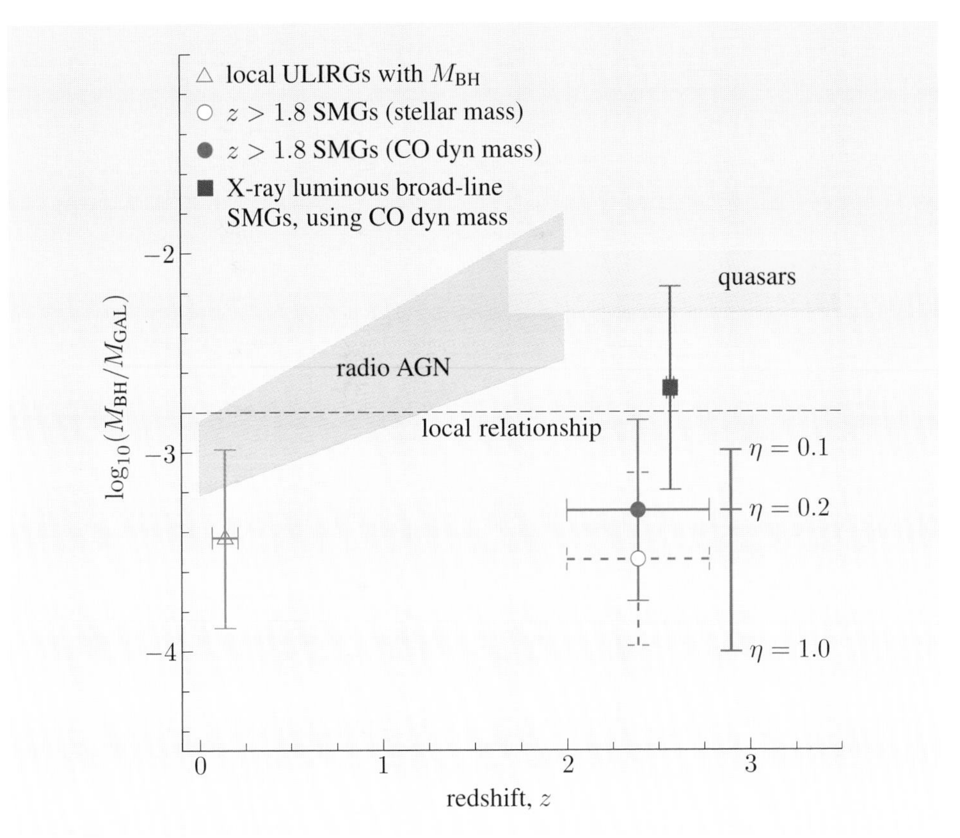

Figure 6.15 Black hole–galaxy mass ratios for the galaxies selected at submm wavelengths (SMGs) that are obscured at X-ray wavelengths with galaxy masses inferred from observed stellar mass and via the width of a carbon monoxide emission line. Also shown is the relationship for local ultraluminous infrared galaxies (ULIRGs), active galaxies, and an indication of the range spanned on average by X-ray luminous QSO SMGs. The effect of changing the assumed value of η (Equation 6.7) for SMGs is also shown. The SMGs have black hole masses that are smaller than those of quasars, for their host galaxy sizes.

The scatter in the M_{BH}–σ relationship appears to be smaller than that of the M_{BH}–M_{halo} correlation, suggesting that the velocity dispersion and not the mass of the dark matter halo is primary. This has been shown (at some length) to be consistent with self-regulated black hole growth, in which the energy output from black hole accretion is enough to unbind the gas, which chokes off the supply of fuel to the black hole. There is currently a great deal of research activity in this area, aiming at inferring the strengths of the physical links from the tightnesses of the correlations. For example, the surprise lack of a black hole in the galaxies M33 and NGC 205, even though their central star clusters obey the same central mass versus spheroid mass relationship, may point at a different fundamental relation. This is still being debated and studied. Also, in nearby active galaxies, there appears to be a different distribution of Eddington ratios for galaxies with recent star formation, compared to those with more quiescent stellar populations.

Wyithe and Loeb, 2005, *Astrophysical Journal*, **634**, 910.

Galaxies with recent star formation also seem to have higher Eddington ratios (Figure 6.16). It's been suggested that this is consistent with self-regulated black hole growth while the gas supply is plentiful (which also fuels the star formation), but when the gas supply runs out, it seems that the only fuel for the black hole comes from mass loss from evolved stars, starving the black hole.

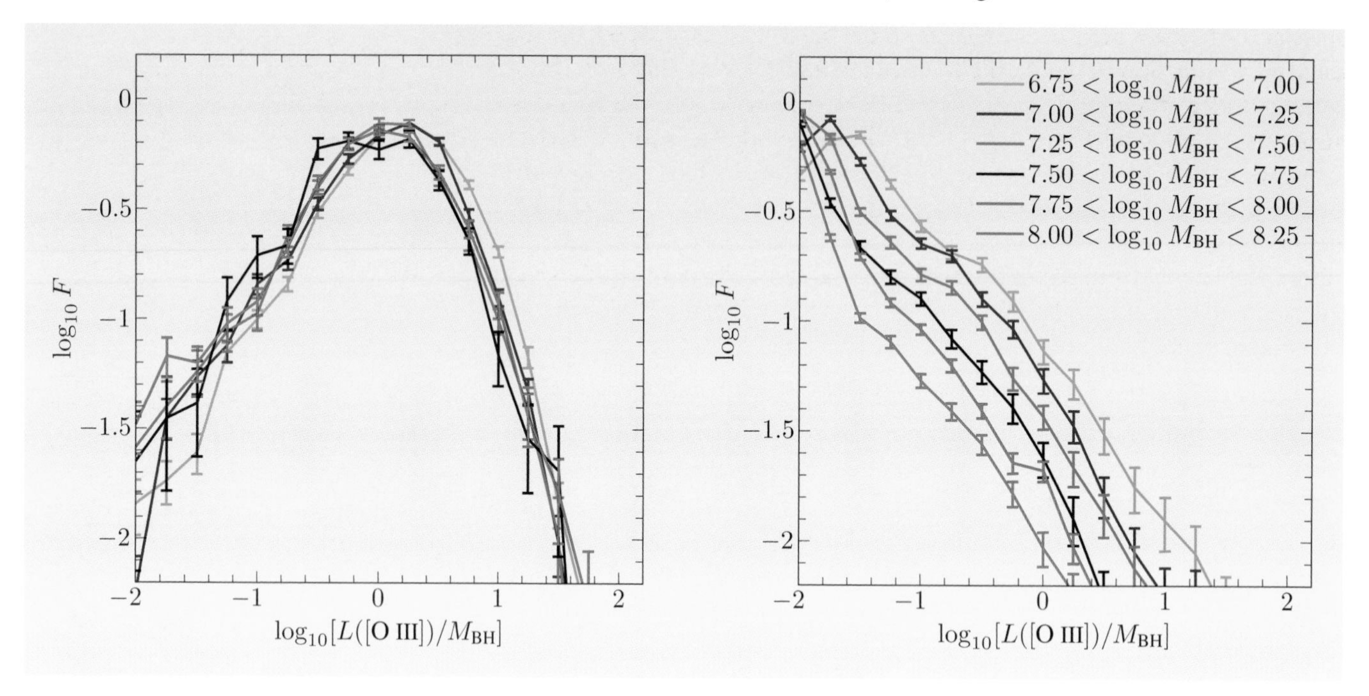

Figure 6.16 Distribution of inferred Eddington ratios L/L_{E} for galaxies with young stellar populations (left), and with old stellar populations (right). The AGN luminosity is taken to be proportional to the luminosity in the [O III] emission line, symbolized as $L[\text{O III}]$, and the y-axis is the logarithm of the fraction of the population. The colours represent black hole mass ranges, as shown in the right-hand panel.

6.10 Merging black holes and gravitational waves

A tremendous new window in observational astronomy may soon open up. We may soon be able to observe the only detectable radiation that comes directly from black holes. (Hawking radiation is too faint ever to be detectable for any known population of black holes.)

In electromagnetism, accelerating charges radiate electromagnetic waves. In general relativity, the analogous process is the radiation of gravitational waves, but in this case the medium of the wave is spacetime itself. We said at the start of this chapter that spherically-symmetric motion does not generate gravitational waves. Dipole gravitational radiation turns out to be impossible because it would violate conservation of momentum: any accelerated mass would be balanced by an equal and opposite change of momentum somewhere else, so any attempt by that mass to radiate dipole gravitational radiation would be cancelled out by the radiation from elsewhere. Only quadrupole-moment motion or above generates gravitational waves.

Gravitational waves took some time to gain wide acceptance, perhaps because they need a very careful treatment of coordinate systems in general relativity.

Einstein himself initially thought that gravitational waves did not exist. Eddington is said to have dismissively quipped that gravitational waves travel 'at the speed of thought', but in truth his remark was directed at a certain spurious subset, and in fact he showed that members of another class of gravitational waves do indeed carry energy.

Eddington, A. S. (1922) *Proceedings of the Royal Society of London A*, **102**, 268–82.

Gravitational waves have been inferred in the binary pulsar PSR B1913+16, in a beautiful verification of the predictions of general relativity that won Russell Hulse and Joseph Taylor the 1993 Nobel Prize (Figure 6.17). The pulsar is in a binary orbit with another star, detectable through subtle variation in the timings of the pulses. (Doppler shifts imply timing variations, in the same way that cosmological redshift implies supernova time dilation — see Chapter 1.) The energy loss from gravitational radiation leads to a gradual spiralling in of the two pulsars, detectable from the timings. Primordial gravitational waves are also expected to contribute to the CMB power spectrum (Chapter 2), though direct detection of primordial gravitational waves will be extremely challenging.

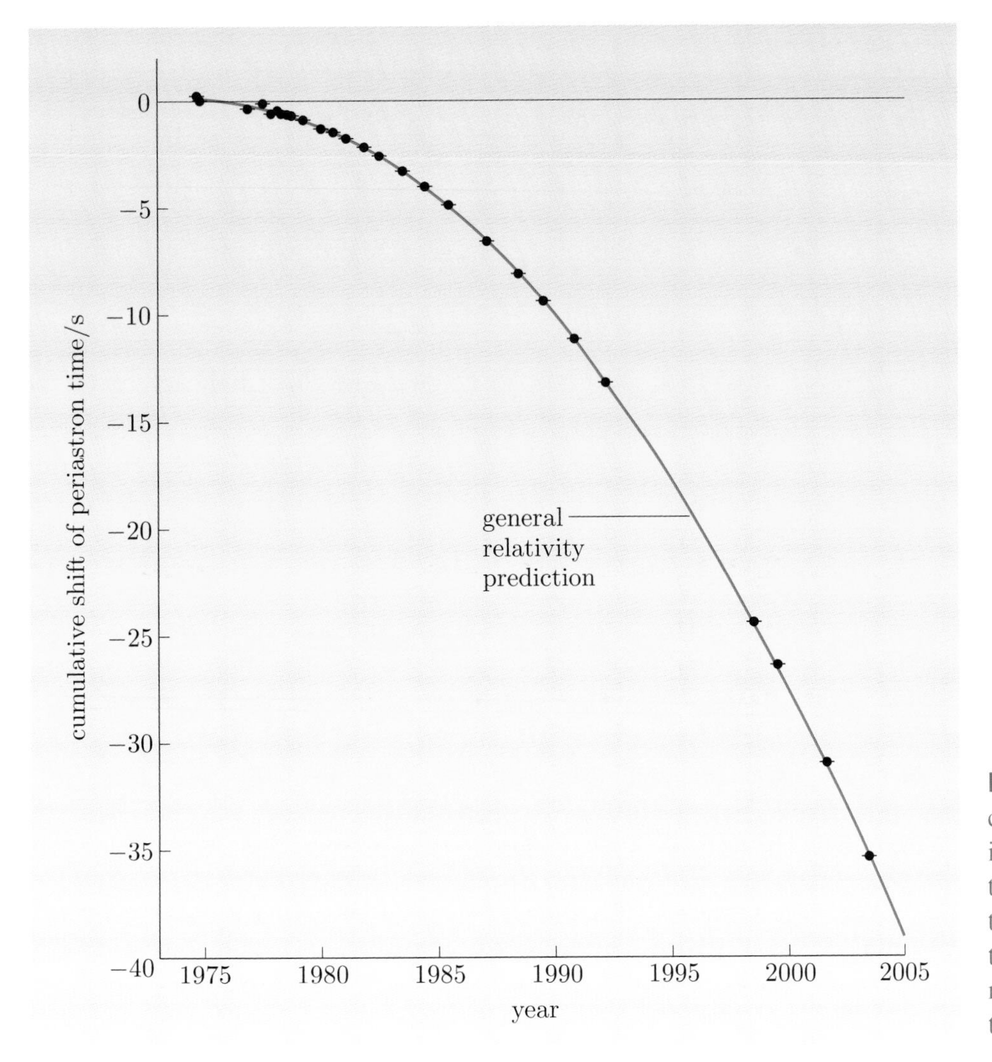

Figure 6.17 The cumulative change of pulsar PSR B1913+16 in the periastron time (the time of closest approach of the two stars), compared to the predictions of general relativity. The data agree with the predictions to 0.2%.

In fact, at the time of writing, no direct detections of gravitational waves have been made, but ambitious experiments may soon succeed and usher in a new era of gravitational wave astronomy. Gravitational waves change the distance between free-falling observers, so gravitational wave observatories seek to monitor distances carefully using laser interferometry. The detectability of gravitational waves is greatly helped by the fact that the *amplitudes* are being measured, rather than the *energies* as with electromagnetic radiation. The energy E varies with amplitude A as $E \propto A^2$. Therefore, while fluxes fall off as $1/r^2$, amplitudes fall off only as $1/r$. Several ground-based gravitational wave detectors are under development at the time of writing, such as LIGO (Laser Interferometer Gravitational-wave Observatory) and GEO, the German–British gravitational wave detector. The European Space Agency also has plans to launch a space-based gravitational wave observatory named LISA (Laser Interferometer Space Antenna), consisting of three free-flying spacecraft linked by laser interferometry. Figure 6.18 shows the expected sensitivity of forthcoming gravitational wave detectors. LIGO has already given useful upper limits to the gravitational waves from a nearby gamma-ray burst.

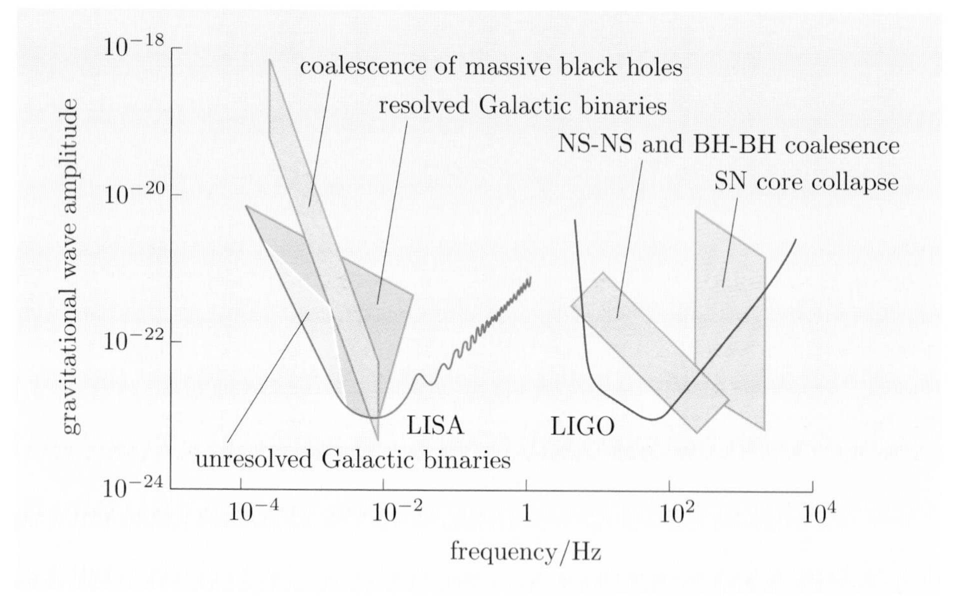

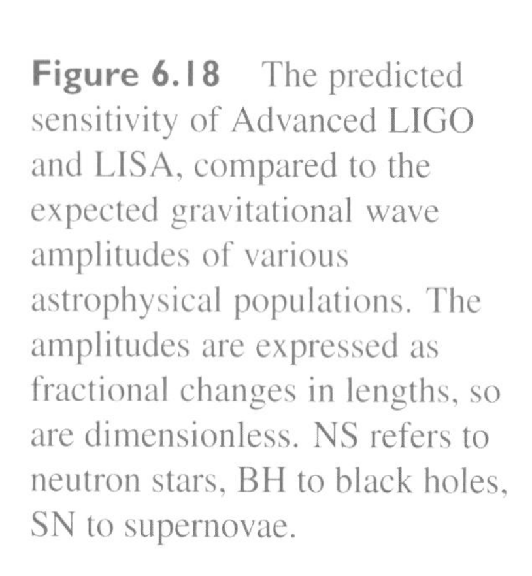
Figure 6.18 The predicted sensitivity of Advanced LIGO and LISA, compared to the expected gravitational wave amplitudes of various astrophysical populations. The amplitudes are expressed as fractional changes in lengths, so are dimensionless. NS refers to neutron stars, BH to black holes, SN to supernovae.

The merger of two black holes would generate copious gravitational waves. Within our Galaxy, many merger events of black holes and neutron stars should be detectable (see Figure 6.18). At cosmological distances, one could detect only the mergers of supermassive black holes. Could this happen? At the time of writing, at least one credible candidate for a binary supermassive black hole has been found in a quasar (Figure 6.19). But supermassive black hole mergers may be much more common than this single example suggests. If quasars and starbursts are triggered by mergers, then galaxy–galaxy merging is common in the history of the Universe. Merging galaxies with pre-existing supermassive black holes will have their supermassive black holes forming a binary system within a million years of the merger, according to numerical simulations. The expectation is of a few tens of supermassive black hole merger events per year detected with LISA.

The gravitational waves from inspiralling black holes are also sufficiently well-understood that they could be treated as standard candles, and they are sometimes referred to as 'standard sirens'. The physical simplicity of such a system, completely determined in practice by two masses and two spins, is very

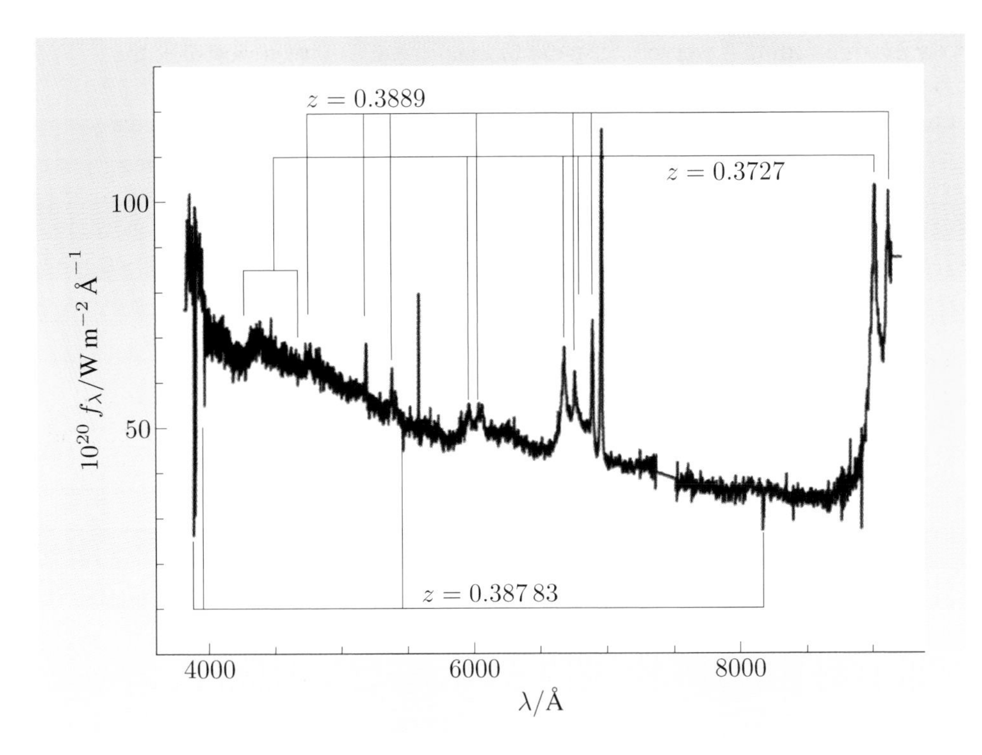

Figure 6.19 The quasar SDSS J1536+0441. The spectrum has evidence of three redshifts. At $z = 0.3889$ are broad hydrogen Balmer emission lines (Hα and Hβ) and narrow emission lines ([O II], [O III], [Ne III], [Ne V]). At $z = 0.3727$ there are further broad Balmer lines and Fe II emission, but no narrow lines. There are also absorption lines at an intermediate redshift of $z = 0.387\,83$ from the host galaxy. This has been interpreted as a binary black hole in the quasar's broad line region.

attractive compared to other standard candles. Luminosity distances can be determined to around 4% accuracy with LISA. If one could independently measure the redshift, one would have an ingenious method to map the geometry of the Universe, determine the evolution of dark energy, and so on. However, the angular resolution of LISA will be only ~ 0.2–$0.5°$, making this a challenging (but not necessarily impossible) experiment. More spectacularly, a merger of black holes at $z = 10$ could be a route to directly measuring the acceleration or deceleration of the expansion of the Universe at $z = 10$.

Summary of Chapter 6

1. Non-rotating uncharged black holes are described by the Schwarzschild metric, and their rotating counterparts by the Kerr metric.
2. The accretion efficiency of a black hole can be calculated by finding the energy released from dropping from infinity to the radius of the smallest stable circular orbit. For a Schwarzschild metric this is 6% of the rest mass at infinity, while for the Kerr metric it is 42%. This is the most efficient conversion process known from mass-energy to luminosity, making black holes prime candidates for powering the central engines of quasars.
3. The present-day contribution to Ω_m from black holes can be estimated from the source counts of quasars, using measurements of the average redshift as a function of apparent quasar magnitude, combined with assumptions of the bolometric correction and the accretion efficiency. This constraint is independent of the Hubble parameter H_0 and the cosmological density parameters.
4. A similar constraint can be made using the hard X-ray background. The background has a harder spectral shape (i.e. more output at higher energies)

than unobscured quasars or starburst galaxies, implying a population of X-ray-obscured quasars. For the most part these can be matched to (optically) type 2 active galaxies but there are exceptions, e.g. optically type 1 active galaxies that are heavily X-ray-obscured, or optically type 2 active galaxies that have low obscuration to hard X-rays.

5. The present-day number density of black holes in active galaxies is at least two orders of magnitude less than the above estimates, implying that most black holes are in dormant quasars in many local galaxies.
6. The centres of all galaxies appear to host supermassive black holes, detected using the kinematics of stars and/or gas or megamasers, by resolving the sphere of influence of the black hole (in which the effects of the black hole's gravity dominate the effects of the galaxy's velocity dispersion). This region is about five orders of magnitude larger than the Schwarzschild radius.
7. The masses of black holes in quasars can also be determined through reverberation mapping.
8. The mass of the central supermassive black hole appears to correlate strongly with the velocity dispersion in the surrounding bulge (for spiral galaxies) or surrounding galaxy (for ellipticals), suggesting a close link between the formation of the black hole and its surrounding galaxy, perhaps through quasar feedback (injection of ionizing flux and/or kinetic energy into the surrounding interstellar medium, affecting both star formation and the amount of infalling gas to the black hole). Various attempts have been made to find evolution in this relationship.
9. There is an additional population of ultraluminous X-ray sources in many local galaxies outside their centres.
10. The inspiralling of black holes and neutron stars releases gravitational waves. These have been inferred in binary pulsars (confirming the predictions of general relativity). Gravitational waves from the merger of black holes may be detectable with the next generation of gravitational wave observatories.

Further reading

- John Michell's 1767 paper 'An inquiry into the probable parallax and magnitude of the fixed stars' is at http://adsabs.harvard.edu/abs/1767RSPT...57..234M. It includes a derivation of integral source counts.
- John Michell's 1784 paper is at http://adsabs.harvard.edu/abs/1784RSPT...74...35M.
- For more on black holes at this level, see Lambourne, R., 2010, *Relativity, Gravitation and Cosmology*, Cambridge University Press.
- At the time of writing, some audio renderings of gravitational waves from inspiralling black holes can be found at http://web.mit.edu/sahughes/www/sounds.html.

- Chongchitnan, S. and Efstathiou, G., 2006, 'Prospects for direct detection of primordial gravitational waves', *Physical Review D*, **73**, 3511; available at http://adsabs.harvard.edu/abs/2006PhRvD..73h3511C.
- One classic graduate-level book on general relativity is Misner, C.W., Thorne, K.S. and Wheeler, J.A., 1970, *Gravitation*, W.H. Freeman.
- For more on the principle of least action (at an accessible level), see Chapter 19 of Feynman, R.P., Leighton, R.B. and Sands, M., 1964, *The Feynman Lectures on Physics*, Vol. II, Addison-Wesley.
- For the history of black hole science, see Ferrarese, L. and Ford, H.C., astro-ph/0411247.
- For more on supermassive black holes (and difficulties if trying to avoid making them), see Begelman, M.C., Blandford, R.D. and Rees, M.J., 1984, 'The theory of extragalactic radio sources', *Reviews of Modern Physics*, **56**, 255–351; for more on alternatives to supermassive black holes, see also Maoz, E., 1998, *Astrophysical Journal Letters*, **494**, 181.
- Brandt, W.N. and Hasinger, G., 2005, 'Deep extragalactic X-ray surveys', *Annual Review of Astronomy and Astrophysics*, **43**, 827.
- Comastri, A., 2004, 'Compton thick AGN: the dark side of the X-ray background', astro-ph/0403693.
- Elvis, M., 2006, 'Quasar structure and cosmological feedback', astro-ph/0606100.
- Colpi, M. and Dotti, M., 2009, 'Massive binary black holes in the cosmic landscape', Invited Review to appear in *Advanced Science Letters*, Special Issue on Computational Astrophysics, edited by Lucio Mayer; available at arXiv:0906.4339.
- For more details on accretion discs and relativistic beaming, see Kolb, U., 2010, *Extreme Environment Astrophysics*, Cambridge University Press.
- For Eddington's 1922 Royal Society article, see http://adsabs.harvard.edu/abs/1922RSPSA.102..268E.

Chapter 7 Gravitational lensing

> Do not Bodies act upon Light at a distance, and by their action bend its Rays; and is not this action (caeteris paribus) strongest at the least distance?
>
> Isaac Newton, *Opticks*

Introduction

Some of the most beautiful images in cosmology are found in gravitational lensing. In these, we see the direct effect that matter has on the curvature of spacetime around it. Most astronomy can investigate only luminous matter, but this is one of the very few opportunities to infer much more. Gravitational lensing effects are created only by the intervening matter distribution, regardless of whether it's luminous or dark, or in equilibrium or not. Lensing can't distinguish between these different sorts of intervening matter, but the positive side of this is that we don't miss anything.

7.1 Gravitational lens deflection

Einstein's theory of general relativity predicts that all mass-energy generates curvature in its surrounding spacetime. The deflection of starlight close to the Sun (e.g. Figure 7.1) was a brilliant confirmation of Einstein's theory (see the *New York Times* report in Figure 7.2).

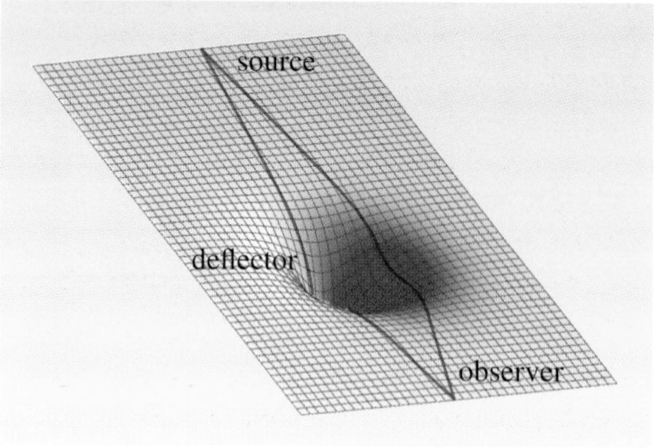

LIGHTS ALL ASKEW IN THE HEAVENS

Men of Science More or Less Agog Over Results of Eclipse Observations.

EINSTEIN THEORY TRIUMPHS

Stars Not Where They Seemed or Were Calculated to be, but Nobody Need Worry.

A BOOK FOR 12 WISE MEN

No More in All the World Could Comprehend It, Said Einstein When His Daring Publishers Accepted It.

Figure 7.2 Headlines on page 17 of the *New York Times*, 10 November 1919.

Figure 7.1 Schematic geometry of gravitational lensing. There can be multiple lines of sight to the background source because of the foreground deflector.

However, gravitational lensing has been found where there are chance alignments of background galaxies with foreground galaxies or clusters of galaxies. Figure 7.3 shows the redshift $z = 0.175$ galaxy cluster Abell 2218, in which higher-redshift galaxies have been distorted into arcs by the curved spacetime around this Abell cluster. We'll show that gravitational lensing conserves surface brightness (i.e. the flux per square degree on the sky), so stretching an image of a background galaxy also magnifies the flux.

Figure 7.3 The galaxy cluster Abell 2218 observed with the Hubble Space Telescope (HST) with the WFPC2 instrument. Note the background galaxies distorted into arcs.

This magnification can also make galaxies appear to have extraordinary luminosities. The Infrared Astronomical Satellite (IRAS) made a surprising detection of a $z = 2.286$ galaxy, IRAS FSC 10214+4724, that appeared to have a tremendous bolometric luminosity of $3 \times 10^{14}\,L_{\odot}$ (see, for example, Figure 5.8 in Chapter 5). This led to a great deal of theoretical speculation about the formation of galaxies, but it later transpired that the galaxy was gravitationally lensed by a $z = 0.9$ interloper (see Figures 7.4 and 7.5). Nevertheless, IRAS FSC 10214+4724 is still one of the most luminous galaxies in the observable Universe, even correcting for lensing. The hunt is on for others like it. Only a handful have been found so far.

How can we tell if an object is a multiply-imaged gravitational lens, rather than some chance pairing of neighbouring objects or an arrangement of galaxies that is curved by chance? In general one looks for a morphology consistent with lensing, a redshift of the candidate background object, a candidate lens and a lens redshift estimate that's a lot lower than the background redshift (e.g. Figure 7.5). The spectra (and variability where one can measure it) should be consistent between candidate multiple images — or at least, any inconsistencies should be small enough to be attributable to differential magnification or microlensing (of which more later). Often, however, only some of these criteria can be met, because of lack of data or difficulty in obtaining data.

Another possible test relies on lensing being purely geometrical. Light travels on geodesics, regardless of the light's wavelength. Therefore lensing must be wavelength-independent or **achromatic**. Different images of the background source must therefore have the same colours, i.e. the same spectra. If the light from the background source is partially obscured by dust in the lensing galaxy, it could give the appearance of different colours, but this achromaticity test can sometimes be done at radio wavelengths where dust extinction has no measurable effect. If the source has some variation in colour and has different magnifications in different parts, then the multiple images could have different colours. One must then carefully model the lens system to see if any observed achromaticity could be due to **differential magnification** (e.g. Figure 7.4).

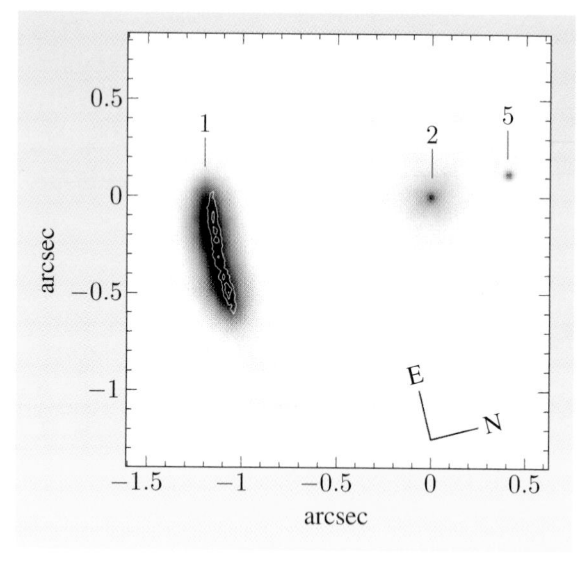

Figure 7.4 HST image of the hyperluminous galaxy IRAS FSC 10214+4724, taken at a wavelength of around 800 nm (just to the red of the visible range). The IRAS galaxy is the arc to the left, gravitationally lensed by the foreground galaxy (marked as 2). There is a second image ('counterimage') of the IRAS galaxy, marked as 5. Objects 2 and 5 have their central pixels boosted artificially in this image for clarity. The contours are HST data at around 400 nm, which surprisingly failed to detect the counterimage; the slight shift in the 400 nm and 800 nm images suggests some colour gradient in the IRAS galaxy and hence differential magnification.

How much does an object deflect light by gravitational lensing? We can make a Newtonian prediction of the gravitational lens deflection angle by a mass M, by treating an incoming photon as being a particle with initial velocity c, as shown in Figure 7.6. In this Newtonian model, the deflection angle ϕ in radians will be $\phi \simeq \tan\phi = v_y/c$, where v_y is the y-axis velocity acquired by the photon as it passes the Sun. We neglect any x-axis change since the imparted velocity will be $\ll c$.

In Newtonian gravity, the photon will move with acceleration $a = GM/r^2$ in the direction towards the Sun. The vertical (y-axis) acceleration in Figure 7.6 will just be $a_y = a\cos\theta = (GM/r^2)\cos\theta$. We can shortcut some tedious algebra by using Kepler's second law (i.e. the conservation of angular momentum): $r^2\,\mathrm{d}\theta/\mathrm{d}t =$ constant. We'll need the value of that constant, and another trick helps: Kepler's laws apply even if the mass M is limitingly small or even zero. Therefore the constant must be bc (where b, known as the **impact parameter**, is shown in Figure 7.6), because that would be the value of $r^2\,\mathrm{d}\theta/\mathrm{d}t$ at the point of closest approach to the mass if the photon were not deflected.

Now imagine a short time interval $\mathrm{d}t$. The change in y-axis velocity in that time will be $\mathrm{d}v_y = a_y(t)\,\mathrm{d}t$, because $a_y = \mathrm{d}v_y/\mathrm{d}t$. But we can rearrange $r^2\,\mathrm{d}\theta/\mathrm{d}t = bc$ to get $\mathrm{d}t = (r^2/bc)\,\mathrm{d}\theta$. Putting this together, we find

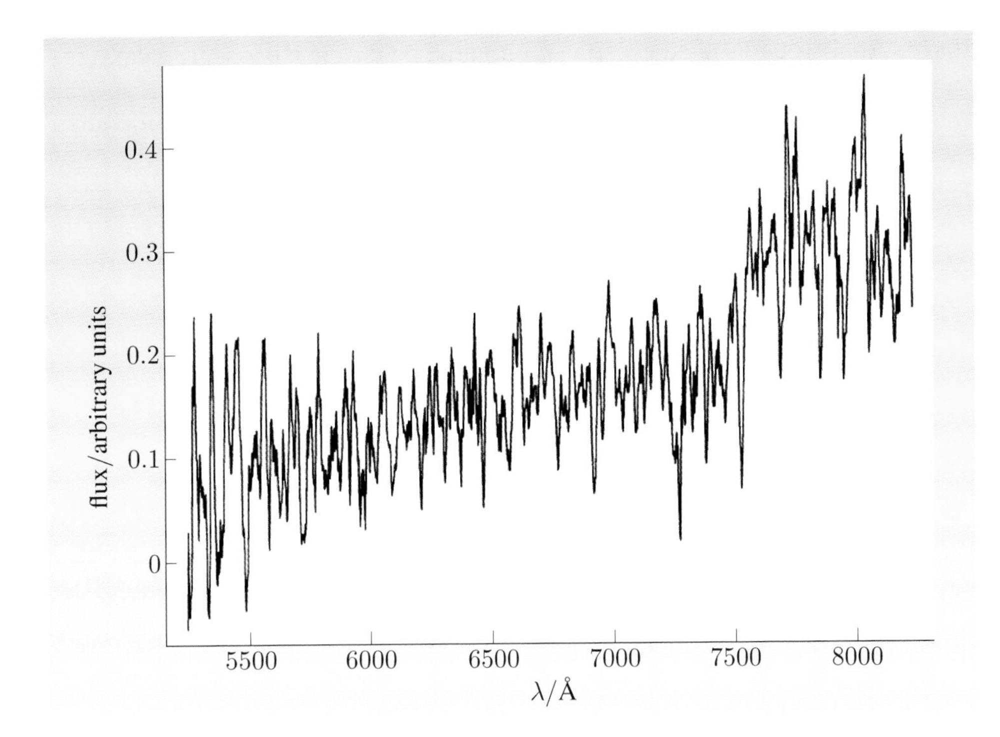

Figure 7.5 Summed spectrum of the two nearest foreground objects that dominate the gravitational lensing of the redshift $z = 2.286$ galaxy IRAS FSC 10214+4724. The discontinuity is the 4000 Å break (Section 4.4), redshifted to about $z = 0.9$. This spectrum is very different to that of the IRAS galaxy (Figure 5.8).

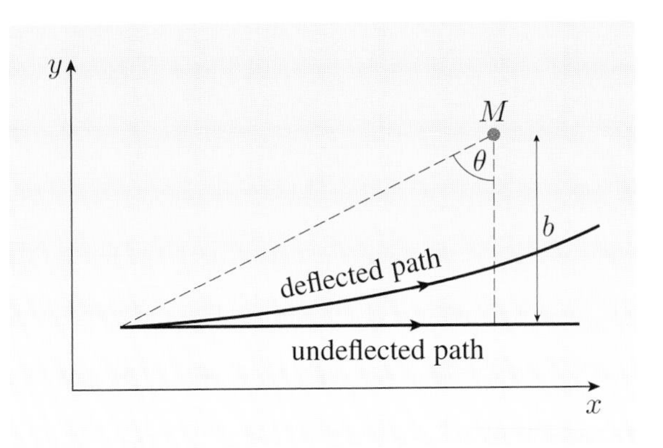

Figure 7.6 The gravitational lensing deflection of a photon by a mass M. Radial distances r are measured outwards from the mass M. The distance b is sometimes known as the impact parameter.

$$\begin{aligned} \mathrm{d}v_y = a_y(t)\,\mathrm{d}t &= \frac{GM}{r^2}\cos\theta\,\mathrm{d}t \\ &= \frac{GM}{r^2}\cos\theta\,\frac{r^2}{bc}\,\mathrm{d}\theta = \frac{GM}{bc}\cos\theta\,\mathrm{d}\theta. \end{aligned}$$

Integrating this from $\theta = -\pi/2$ to $+\pi/2$, we find that

$$v_y = \frac{2GM}{bc},$$

so

$$\phi_{\text{Newtonian}} \simeq \frac{v_y}{c} = \frac{2GM}{bc^2}. \qquad (7.1)$$

In the weak-field limit, the full general relativistic treatment turns out to be exactly a factor of two greater:

$$\phi = \frac{4GM}{bc^2}. \qquad (7.2)$$

Why exactly a factor of two? This is difficult to answer. As you saw in Chapter 6, there is a similar conservation of angular momentum $r^2\,\mathrm{d}\theta/\mathrm{d}\lambda = \text{constant}$, where

λ is a parameter measured along the path of the photon. (For a massive particle we could use $d\lambda = d\tau$, where τ is the proper time, but photons have $ds = 0$ so $d\tau = 0$.) Converting λ to coordinate time t involves a factor also involving GM/c^2 (due to the spacetime curvature), which leads ultimately to the larger deflection angle. We'll return to this in Section 7.5.

7.2 The lens equation

Gravitational lensing has a beguilingly simple geometry, shown in Figure 7.7. The photons spend most of their time travelling between the background source and the lens, or between lens and observer, but they spend very little of their time being deflected. We can therefore use the 'thin lens approximation' and treat the change of direction as instantaneous. The angles are all assumed to be small, so we can write, for example, $\theta \simeq \tan\theta = \xi/D_L$.

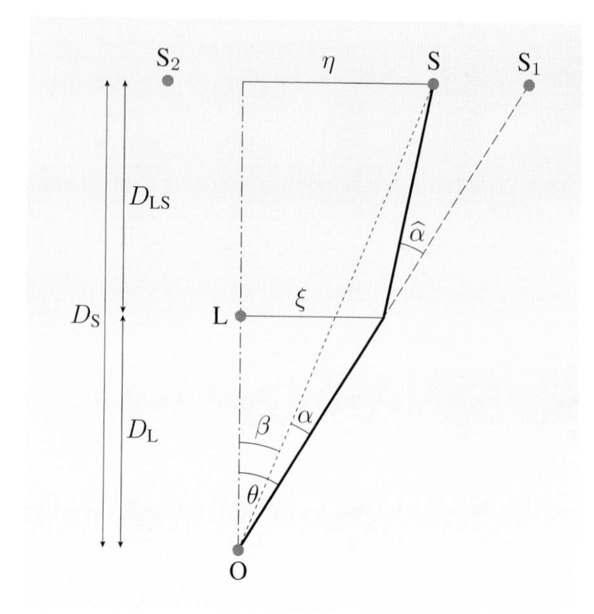

Figure 7.7 The geometry of gravitational lensing of a source S by a lens L, with the angles that are discussed in the text labelled. The apparent positions of the source seen from the Earth are S_1 and S_2. (Light rays for S_2 are not shown, for clarity.) The impact parameter is given the symbol ξ. D_{LS} is the distance to the source as seen from the lens. Note that the distances D_L, D_S and D_{LS} are all *angular diameter distances*, so $D_S \neq D_{LS} + D_L$.

But beware of a subtlety that traps the unwary. The vertical distances in Figure 7.7 are *angular diameter distances* (Chapter 1). For example, D_S is the angular diameter distance from the source to the observer, while D_L is the angular diameter distance from the lens to the observer. But D_{LS} is the angular diameter distance to the source *as seen from the lens*, so D_S does not necessarily equal $D_{LS} + D_L$!

- Do *any* cosmological distances add up, so that Earth-to-source equals Earth-to-lens plus lens-to-source?
- Comoving distances add up in exactly this way.

If the lens, image, background source and the Earth are all in the same plane, as in Figure 7.7, then

$$\beta = \theta - \alpha. \tag{7.3}$$

(To show that this is true while avoiding $D_S \neq D_{LS} + D_L$, compare distances along the top of Figure 7.7.) But what if they aren't in the same plane? This could happen if the lens is not symmetrical, for example. In this case we can treat the angles as *vectors* on the sky, so

$$\boldsymbol{\beta} = \boldsymbol{\theta} - \boldsymbol{\alpha}(\boldsymbol{\theta}). \tag{7.4}$$

This is known as the **lens equation** and is the fundamental equation of cosmological gravitational lensing. Note that we've written $\boldsymbol{\alpha}$ as a function of $\boldsymbol{\theta}$, which is also true in the scalar case. (A subtlety in Equation 7.3 is that α can be negative, i.e. it's not the modulus of $\boldsymbol{\alpha}$.)

Exercise 7.1 Derive a *flat* space expression for D_{LS} involving the comoving distances r_{L} and r_{S} (the comoving distances to the lens and source, respectively), and the lens and source redshifts z_{L} and z_{S}.

Exercise 7.2 Write down a proof of Equation 7.4 by working with vectors on the source plane, keeping in mind that $D_{\text{S}} \neq D_{\text{LS}} + D_{\text{L}}$. ■

So far we've not used any information on the lens mass distribution, or on how much deflection that mass causes. Let's see what happens for a point mass M. Adapting Equation 7.2, a point mass M will cause a deflection of

$$\widehat{\alpha} = \frac{4GM}{c^2 \xi}. \tag{7.5}$$

(By symmetry, the light rays are all confined to a plane in this case, so we don't need to use vectors.) This deflection is related to the observed shift α by

$$\alpha = \frac{D_{\text{LS}}}{D_{\text{S}}} \widehat{\alpha} \tag{7.6}$$

using Figure 7.7, so the lens will cause a visible deflection of

$$\alpha = \frac{D_{\text{LS}}}{D_{\text{S}}} \frac{4GM}{c^2 \xi}. \tag{7.7}$$

We can rewrite the (scalar) lens equation as

$$\beta = \theta - \alpha = \theta - \frac{D_{\text{LS}}}{D_{\text{S}}} \frac{4GM}{c^2 \xi},$$

and using $\theta = \xi / D_{\text{L}}$ (Figure 7.7) we reach

$$\beta = \theta - \frac{D_{\text{LS}}}{D_{\text{L}} D_{\text{S}}} \frac{4GM}{c^2 \theta}. \tag{7.8}$$

Exercise 7.3 What if the background object is exactly behind the lens, so $\beta = 0$? What will this look like? (Give this some thought before looking up the answer!) ■

This angular size is often known as the **Einstein radius** and given the symbol θ_{E}:

$$\theta_{\text{E}} = \sqrt{\frac{4GM}{c^2} \frac{D_{\text{LS}}}{D_{\text{L}} D_{\text{S}}}}. \tag{7.9}$$

It depends only on the source redshift z_{S}, the lens redshift z_{L} and the lens mass M. (Note that θ_{E} isn't just a property of the lens, because it also depends on the distance to the background source.) It's an important quantity in gravitational lensing in general.

When the source position β is around $\theta_{\rm E}$ or less, the magnifications are typically strong. Conversely, if $\beta \gg \theta_{\rm E}$, then there is typically very little magnification. We'll show in Section 7.6 how $\theta_{\rm E}$ can also be a boundary between having multiple images and having only one image. Also, multiple images tend to have separations of roughly $2\theta_{\rm E}$, as we'll show. Figure S7.1 from Exercise 7.3 is an example of an **Einstein ring**.

Substituting in the numerical values and assuming a point mass, we obtain an equation that's useful for cosmological lensing:

$$\frac{\theta_{\rm E}}{\text{arcseconds}} = \left(\frac{M}{10^{11.09}\,{\rm M}_\odot}\right)^{1/2} \left(\frac{D_{\rm L}D_{\rm S}/D_{\rm LS}}{\text{Gpc}}\right)^{-1/2}. \tag{7.10}$$

Typically, galaxy–galaxy lensing gives Einstein radii of the order of an arcsecond, while lensing by a galaxy cluster typically has $\theta_{\rm E}$ about ten times bigger. At the opposite size scale, gravitational microlensing (which we shall meet later in this chapter) can be characterized with

$$\frac{\theta_{\rm E}}{\text{milliarcseconds}} = \left(\frac{M}{1.23\,{\rm M}_\odot}\right)^{1/2} \left(\frac{D_{\rm L}D_{\rm S}/D_{\rm LS}}{10\,\text{kpc}}\right)^{-1/2}. \tag{7.11}$$

For our point mass lens, we can write Equation 7.8 as

$$\beta = \theta - \frac{\theta_{\rm E}^2}{\theta}. \tag{7.12}$$

This quadratic equation has the solution

$$\theta = \frac{1}{2}\left(\beta \pm \sqrt{\beta^2 + 4\theta_{\rm E}^2}\right), \tag{7.13}$$

giving two possible values for θ.

Exercise 7.4 Show that one value of θ in Equation 7.13 is always negative. Is this a physical solution? If it is, then what does it correspond to? If it isn't, then why does it occur in this equation?

Exercise 7.5 In general, is there a unique image position $\boldsymbol{\theta}$ for any given source position $\boldsymbol{\beta}$? Also, is the reverse true — is there a unique source position $\boldsymbol{\beta}$ for every image position $\boldsymbol{\theta}$? ■

7.3 Magnification

Gravitational lensing magnifies not only the *sizes* of distant galaxies, but also their *fluxes*. This is no coincidence: it turns out that surface brightness (flux per unit area on the sky) is conserved in gravitational lensing. We've outlined a proof briefly in the box below, but this is only in case you're unsatisfied with having surface brightness conservation unproven. We won't use the proof later in the book.

Why does lensing conserve surface brightness?

The key idea is the **phase space** density of photons. Phase space is an imagined six-dimensional space that describes both spatial position and momentum. Each photon has a position (x, y, z) and a momentum (p_x, p_y, p_z), and we lump these together and treat any photon's state as being a point (x, y, z, p_x, p_y, p_z) in a six-dimensional space. Now, if we apply a Lorentz transformation along the x-axis, the x-position gets Lorentz contracted by a factor of γ, while the p_x-momentum is increased by the same factor. Therefore the **phase space density** of photons is constant, i.e. the number of photons per unit *phase space volume* is constant. This is sometimes called *Liouville's theorem*.

Next, we imagine that we have a telescope pointed along the z-axis. We put a filter in the optics so that it receives photons only within an energy range $E \to E + \delta E$. Our telescope detector receives N photons from a patch of sky with solid angle area $\Delta\Omega$ (in, say, steradians or square degrees), and our detector itself has an area A (in, say, square cm or square metres). Figure 7.8 shows this schematically.

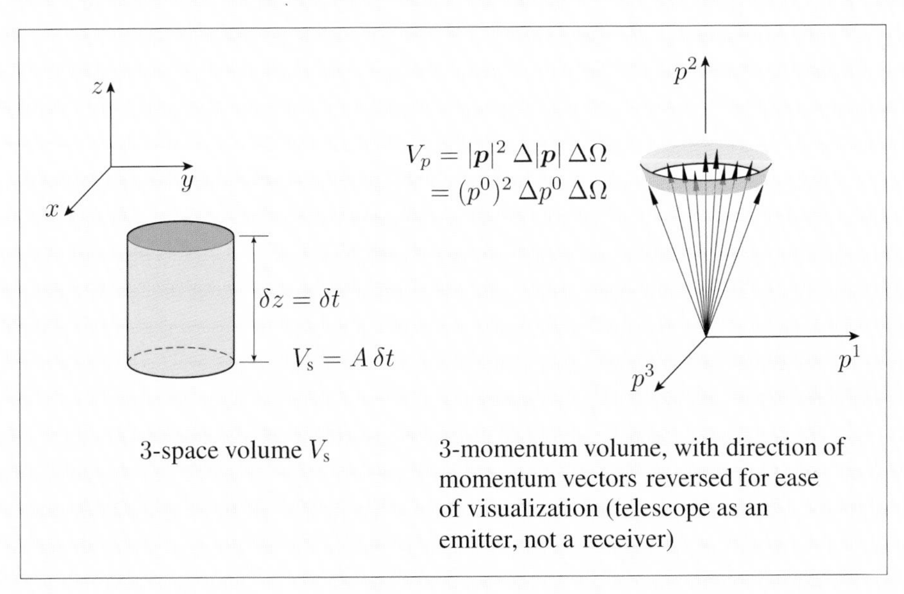

Figure 7.8 The space volume and momentum space volume of photons hitting the detector in a time δt. To make the figure clearer, we've flipped the momentum diagram and shown the detector as an emitter instead of a receiver.

In a short time δt the detector receives the photons from a volume $V_s = A\,\delta t$. Those photons have energy $E \to E + \delta E$, and since $E = pc$ for photons, their z-axis momenta are $E/c \to E/c + \delta E/c$. Their *momentum* space volume is $V_p = (1/c^3)E^2\,\delta E\,\Delta\Omega$ (see Figure 7.8). Putting this together with the spatial volume, we find that our N photons have phase space density

$$\rho_{\text{phase}} = \frac{N}{V_s V_p} = \frac{N c^3}{A\,\delta t\,E^2\,\delta E\,\Delta\Omega} = \frac{N c^3}{h^3 A\,\delta t\,\nu^2\,\delta\nu\,\Delta\Omega} = \text{constant}, \quad (7.14)$$

where for the second step we used $E = h\nu$, with h being Planck's constant and ν the frequency, and the last step is just stating Liouville's theorem.

The surface brightness is the amount of energy per unit area, per unit solid angle, per unit frequency, per unit time:

$$I_\nu = \frac{N h\nu}{A\,\delta t\,\delta\nu\,\Delta\Omega}. \tag{7.15}$$

Combining Equations 7.14 and 7.15, and using the photon phase space density conservation, shows that I_ν/ν^3 has to be constant. But the photons have not gained or lost any energy by moving past the gravitational lens (notwithstanding any Sunyaev–Zel'dovich effect), so the frequencies ν of the photons are the same. Therefore the surface brightness I_ν is the same.

So the flux of a object with a uniform surface brightness I_ν and an area Ω on the sky is $S_\nu = I_\nu \times \Omega$. Lensing increases the area to $\Omega_{\text{lensed}} = \mu\Omega$ (where μ is the magnification factor), and I_ν is the same, so the lensed flux is

$$S_{\nu,\text{lensed}} = I_\nu\,\Omega_{\text{lensed}} = I_\nu\mu\Omega = \mu S_\nu.$$

But hang on — doesn't surface brightness conservation violate energy conservation? We've conserved surface brightness and made the image bigger, so where have the extra photons come from? In fact, it's still consistent with energy conservation. Part of the answer is that photons are being redirected, so in some directions the background source could be demagnified. Another part of the answer is that you must take account of the spatial curvature around the lens: the photons from the background source are now being spread over slightly less than 4π steradians. There is still the same number of photons, but they're being distributed over slightly less space.

The magnification factor of an image is therefore equal to the factor increase of the image's area on the sky. If the lens is circularly symmetric, then the magnification is given by

$$\mu = \frac{\theta}{\beta}\frac{\mathrm{d}\theta}{\mathrm{d}\beta}, \tag{7.16}$$

where θ and β are as given in Figure 7.7.

Exercise 7.6 Show by differentiating Equation 7.12 (or otherwise) that lensing by a point mass (a special case of circular symmetry) gives rise to a magnification

$$\mu = \left[1 - \left(\frac{\theta_{\mathrm{E}}}{\theta}\right)^4\right]^{-1}, \tag{7.17}$$

where, as we've seen, θ has two possible values for any source position β.

Exercise 7.7 If an image is within the Einstein radius, i.e. $\theta < \theta_{\mathrm{E}}$, then the magnification in Equation 7.17 is negative. Is this a physical solution? If it is, what does this correspond to? If it isn't, why does this occur in this equation? (*Hint*: Why would μ in Equation 7.16 be negative?) ■

We can write the total magnification caused by a point mass as $\mu = |\mu_1| + |\mu_2|$, where μ_1 and μ_2 are the magnifications of each of the two images. After a little algebra, it turns out that the total magnification of a point mass is

$$\mu = |\mu_1| + |\mu_2| = \frac{2 + (\beta/\theta_{\mathrm{E}})^2}{(\beta/\theta_{\mathrm{E}})\sqrt{(\beta/\theta_{\mathrm{E}})^2 + 4}}.$$

This has the remarkable property that it is always larger than 1, for any β or θ_E! Again, doesn't this violate energy conservation? Again, it doesn't. Putting a point mass lens into the Universe couldn't change the number of photons that the background source put out, but it would change the *volume* over which they are distributed, because the point mass has a spatial curvature around it. Just like with our discussion of the surface brightness conservation above, the same number of photons is being distributed over slightly less than 4π steradians because of this curvature, so if we compare a universe without the lens (more volume) to one with the lens (less volume), it's possible for the magnification always to be > 1.

Gravitational lens magnification has a curious effect on the source counts of extragalactic objects. We can imagine putting a population of lenses between ourselves and some extragalactic background objects. These lenses will give each extragalactic background object a random magnification $|\mu|$, which has a probability distribution $\Pr(|\mu|)$. Lenses are generally quite sparse on the sky, so $\Pr(|\mu|)$ will have a sharp peak close to $|\mu| = 1$. (We'll ignore any redshift-dependence of this probability for the purposes of demonstration.) The magnification $|\mu|$ can be less than 1, in general, so some objects could be demagnified, while some are boosted in flux.

The *underlying* magnification probability is $\Pr(|\mu|)$, but the *observed* magnification histogram could look very different. Imagine surveying the sky for background objects with an *observed* flux of S_0, and suppose that these background objects have power-law source counts around S_0, i.e. $\mathrm{d}N/\mathrm{d}S \propto S^{-\alpha}$, with α being some constant. There will be a few objects brighter than S_0 that are demagnified, so appear to have flux S_0. However, there will be many *more* objects fainter than S_0, as shown in Figure 7.9, some of which have a high μ so appear to have flux S_0. The net effect is that high magnifications will be over-represented, compared to what you'd expect from the shape of $\Pr(|\mu|)$. The steeper the source counts, i.e. the higher the value of α, the more high-magnification objects you'd find. This is known as **magnification bias** and may be an important new way of finding gravitational lenses, as we'll see in Section 7.11.

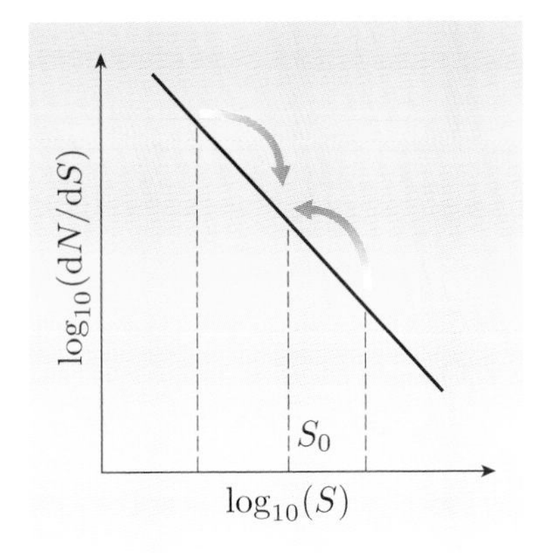

Figure 7.9 The magnification bias effect. At any flux S_0, there are many objects fainter than S_0, some of which will be magnified to the flux S_0. There are far fewer objects brighter than S_0, of which again some will be demagnified to S_0. The asymmetry between the brighter and fainter populations changes the distribution of magnifications for objects with a *fixed observed* flux of S_0. The steeper the source count slope, the more the *observed* magnification distribution is skewed towards higher magnifications.

If the lens does not have circular symmetry, the magnification calculation is a little more complicated. The mapping from source position $\boldsymbol{\beta} = (\beta_x, \beta_y)$ to image position $\boldsymbol{\theta} = (\theta_x, \theta_y)$ is in general done with a matrix A: a small change in $\boldsymbol{\beta}$, $\mathrm{d}\boldsymbol{\beta} = (\mathrm{d}\beta_x, \mathrm{d}\beta_y)$, relates to $\mathrm{d}\boldsymbol{\theta}$ via $\mathrm{d}\boldsymbol{\beta} = A\,\mathrm{d}\boldsymbol{\theta}$, where

$$A = \frac{\partial \boldsymbol{\beta}}{\partial \boldsymbol{\theta}} = \begin{pmatrix} \partial\beta_x/\partial\theta_x & \partial\beta_x/\partial\theta_y \\ \partial\beta_y/\partial\theta_x & \partial\beta_y/\partial\theta_y \end{pmatrix}. \tag{7.18}$$

(Equation 7.16 is a special case of Equation 7.18 for circular symmetry.)

To calculate the magnification, we want to know how much the background image area (proportional to $\mathrm{d}\beta^2$) relates to the observed image area (proportional to $\mathrm{d}\theta^2$). This comes out as

$$\frac{\mathrm{d}\theta^2}{\mathrm{d}\beta^2} = \frac{1}{\det A},$$

where $\det A$ means the determinant of the matrix A, sometimes written using modulus signs:

$$\det A = \det \begin{pmatrix} a & b \\ c & d \end{pmatrix} = \begin{vmatrix} a & b \\ c & d \end{vmatrix} = ad - bc. \tag{7.19}$$

For this reason the matrix A is sometimes called the inverse magnification tensor. (If you need a reminder about what a tensor is, see the box below.) The magnification tensor is $M = A^{-1}$, so $1/\det A = \det M$.

What is a tensor?

Suppose that you have two springs connected to wires as shown in Figure 7.10a. Both springs have the same spring constant k. What is the force on the object of mass M in this figure? By Hooke's law, the force from each spring is proportional to the displacement, so we have

$$\boldsymbol{F} = (F_x, F_y) = (-kx, ky) = -k(x, y) = -k\boldsymbol{r}, \tag{7.20}$$

where $\boldsymbol{r}$ is the displacement vector.

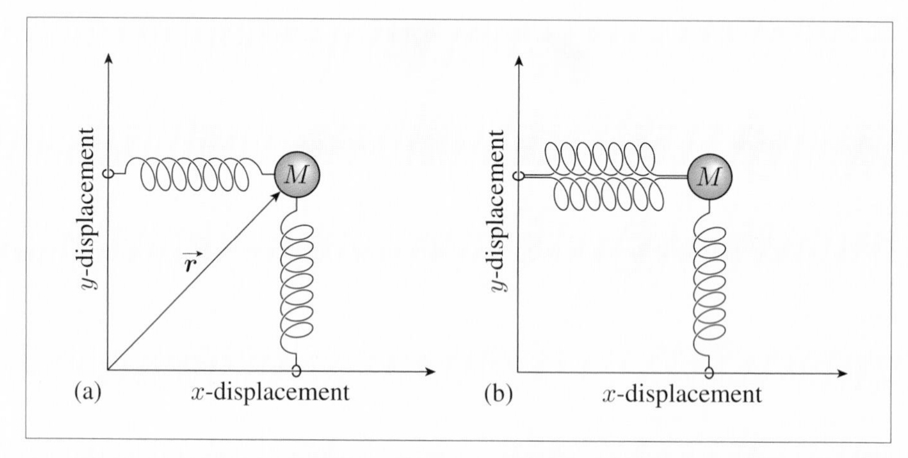

Figure 7.10 (a) A mass M pulled by two springs. The springs have frictionless rings that slide along bars that follow the x- and y-axes. The displacement vector $\boldsymbol{r}$ is also shown. The resulting force vector $\boldsymbol{F}$ is aligned with $\boldsymbol{r}$ (though in the opposite direction). (b) Now the mass M is pulled by one spring along the y-axis direction but two along the x-axis direction. The resulting force vector $\boldsymbol{F}$ is no longer aligned with the displacement vector $\boldsymbol{r}$.

Now let's put a second spring on the x-axis, as shown in Figure 7.10b. What's the force now? The force $\boldsymbol{F}$ is in a different direction to the displacement $\boldsymbol{r}$, and we can't pull out the factor of k as we did in Equation 7.20. But we could write it as a matrix:

$$\begin{aligned}\boldsymbol{F} &= (F_x, F_y)\\ &= (-2kx, -ky)\\ &= -\begin{pmatrix}2k & 0\\ 0 & k\end{pmatrix}(x, y) = -\begin{pmatrix}2k & 0\\ 0 & k\end{pmatrix}\boldsymbol{r} = -K\boldsymbol{r},\end{aligned}$$

where K could be called, say, the 'spring tensor' by analogy to the spring constant. So we can think of this tensor as a matrix that operates on the displacement vector $\boldsymbol{r}$ to give us the force vector $\boldsymbol{F}$.

This is nearly sufficient to define this type of tensor, but not quite. Not every matrix can be a tensor, because a tensor must obey certain transformation rules. You may not have been aware of this, but the *definition* of a vector includes the fact that it obeys the right transformation laws. In Galilean relativity, any spatial three-vector must *by definition* obey the Galilean transformation, so its length and direction are observer-independent. If they aren't, it's not a vector. Similarly, in special relativity, a four-vector must *by definition* obey the Lorentz transformation and have a Lorentz-invariant 'length' (such as the interval Δs in the case of the position four-vector). The definition of a tensor is that it must obey similar transformation laws. This takes us beyond the scope of this book, but it's one of the key ideas underpinning the beautiful theory of general relativity.

We've discussed only two-dimensional tensors here, but there can be higher-order ones too, e.g. cubical arrays or hypercubes of numbers.

7.4 The singular isothermal sphere model

Obviously a point mass isn't a good model for a galaxy lens. Can we come up with something more realistic? Many galaxies are observed to have fairly flat rotation curves, i.e. the one-dimensional velocity dispersion $\sigma_{\rm v}$ is independent of or only weakly dependent on radius r from the centre over much of the radius.

One approach is to imagine that stars or other clumps of matter are like particles in a gas. This 'gas' is imagined to obey an ideal gas law, $p = \rho kT/m$, where ρ is the density and m the typical mass of the star or clump. The temperature T is related to the one-dimensional velocity dispersion of the stars or clumps $\sigma_{\rm v}$ by $m\sigma_{\rm v}^2 = kT$.

To solve this using the ideal gas law, we need to relate the density ρ and the pressure p. Imagine a shell of thickness $\mathrm{d}r$. It must have volume $4\pi r^2\,\mathrm{d}r$ and mass $\mathrm{d}M = 4\pi r^2 \rho\,\mathrm{d}r$. The gravitational force on this shell must be $\mathrm{d}F = -GM(r)\,\mathrm{d}M/r^2$, where $M(r)$ is the mass enclosed by the radius r, and the minus sign accounts for the direction. The pressure exerted by the shell will be this force divided by the area, which is $\mathrm{d}F/(4\pi r^2)$. Alternatively, we could think of this as the pressure drop $\mathrm{d}p$ from going from r to $r + \mathrm{d}r$:

$$\mathrm{d}p = \frac{\mathrm{d}F}{4\pi r^2} = \frac{-GM(r)}{r^2}\,\frac{\mathrm{d}M}{4\pi r^2} = \frac{-GM(r)}{r^2}\,\frac{4\pi r^2 \rho\,\mathrm{d}r}{4\pi r^2} = \frac{-GM(r)}{r^2}\rho\,\mathrm{d}r,$$

so

$$\frac{1}{\rho}\frac{\mathrm{d}p}{\mathrm{d}r} = \frac{-GM(r)}{r^2}.$$

The solution of these equations turns out to be

$$\rho(r) = \frac{\sigma_{\rm v}^2}{2\pi G}\,\frac{1}{r^2}. \tag{7.21}$$

In other words, $\rho(r) \propto r^{-2}$, so $M(r)$ must be $\propto r$ (because $M(r) = \int \rho(r)\,4\pi r^2\,\mathrm{d}r = \text{constant} \times \int \mathrm{d}r$). Therefore the circular velocity of a star or clump in this galaxy would satisfy $v^2/r = GM(r)/r^2$, i.e. $v^2 = GM(r)/r = \text{constant} = 2\sigma_{\rm v}^2$. Projecting along the line of sight, the observed **surface mass density** Σ (we shall spare you the algebra) comes out as

$$\Sigma(\xi) = \frac{\sigma_{\rm v}^2}{2G}\,\frac{1}{\xi}. \tag{7.22}$$

(The distance ξ was shown in Figure 7.7.) This mass distribution is known as the **singular isothermal sphere**. You've seen already why this spherically-symmetric distribution is isothermal. It's called 'singular' because the mass density and surface density tend to infinity as r and ξ respectively tend to zero. There are various modifications that can be made to the model to avoid this singularity. The total mass enclosed within a projected distance ξ is just

$$M(\xi) = \int_0^{\xi} \Sigma(\xi')\, 2\pi\xi'\, \mathrm{d}\xi' = \frac{\pi\sigma_{\mathrm{v}}^2}{G}\xi. \qquad (7.23)$$

- Can the singular isothermal sphere model be extended to infinity?

- $M(r) \propto r$, so the mass would tend to infinity. Therefore in practice this model has to be truncated at some radius (typically $> \theta_{\mathrm{E}}$) for it to be physical.

What about gravitational lensing by a singular isothermal sphere? By Birkhoff's theorem (Chapters 4 and 6), the deflection by any spherically-symmetric mass distribution will depend only on the mass *within* the angular distance ξ, i.e. $M(\xi)$:

$$\widehat{\alpha} = \frac{4GM(\xi)}{c^2\xi}, \qquad (7.24)$$

which comes out as

$$\widehat{\alpha} = 4\pi\frac{\sigma_{\mathrm{v}}^2}{c^2} \simeq (1.4'') \left(\frac{\sigma_{\mathrm{v}}}{220\ \mathrm{km\,s^{-1}}}\right)^2$$

(compare Equation 7.5). Similarly, the Einstein radius is

$$\theta_{\mathrm{E}} = \sqrt{\frac{4GM(\theta_{\mathrm{E}})}{c^2}\frac{D_{\mathrm{LS}}}{D_{\mathrm{L}}D_{\mathrm{S}}}} \qquad (7.25)$$

(compare Equation 7.9), so

$$\theta_{\mathrm{E}}^2 = \frac{4GM(\theta_{\mathrm{E}})}{c^2}\frac{D_{\mathrm{LS}}}{D_{\mathrm{L}}D_{\mathrm{S}}} = \frac{4GM(\theta_{\mathrm{E}})}{c^2}\frac{D_{\mathrm{LS}}}{D_{\mathrm{S}}}\frac{\theta_{\mathrm{E}}}{\xi} = \frac{4G}{c^2}\frac{\pi\sigma_{\mathrm{v}}^2\xi}{G}\frac{D_{\mathrm{LS}}}{D_{\mathrm{S}}}\frac{\theta_{\mathrm{E}}}{\xi}$$

thus

$$\theta_{\mathrm{E}} = \frac{4\pi\sigma_{\mathrm{v}}^2}{c^2}\frac{D_{\mathrm{LS}}}{D_{\mathrm{S}}}. \qquad (7.26)$$

We can use the scalar lens equation because this lens is circularly symmetric:

$$\beta = \theta - \alpha \qquad \text{(Eqn 7.3)}$$

(see Figure 7.7). Remember that α can be positive or negative.

If $\beta = 0$, then $\theta = \theta_{\mathrm{E}}$, i.e. the source is directly behind the lens. The lens equation is therefore

$$\beta = \theta \pm \theta_{\mathrm{E}}. \qquad (7.27)$$

If $\beta > \theta_{\mathrm{E}}$, this gives only one possible solution, $\theta = \beta + \theta_{\mathrm{E}}$. However, if $\beta < \theta_{\mathrm{E}}$, there is also a negative solution for θ, i.e. on the other side:

$$\theta = \beta \pm \theta_{\mathrm{E}}. \qquad (7.28)$$

This solution is shown in Figure 7.11a.

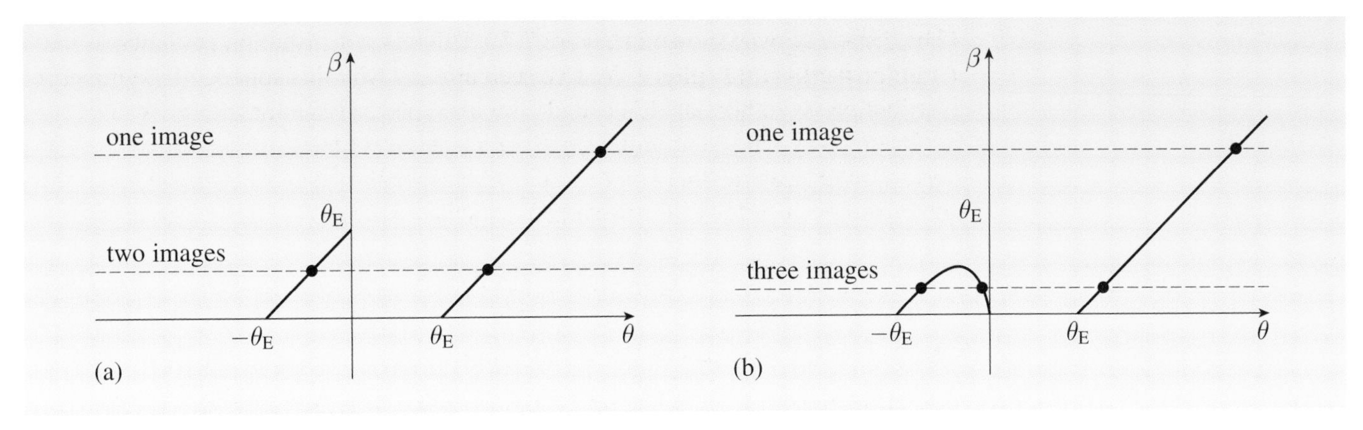

Figure 7.11 Graphical representation of the gravitational lens solution for (a) a singular isothermal sphere in Equation 7.27, (b) an isothermal sphere with a smoothed-out density profile in the core, sometimes called a 'softened isothermal sphere'.

If we write $\theta_\pm$ for the two images, then the magnifications from Equation 7.16 come out as

$$\mu_\pm = \frac{\theta_\pm}{\beta} = 1 \pm \frac{\theta_E}{\beta} = \left(1 \mp \frac{\theta_E}{\theta_\pm}\right)^{-1}. \tag{7.29}$$

Strictly speaking, there would be a third image at $\theta = 0$, because a single photon shot straight through the middle could not be deviated (by symmetry). However, this can only come from a zero-sized point in the background source, and the flux of this central image comes out at zero. There could be a non-zero central image if the central density cusp in the singular isothermal sphere mass distribution is smoothed out somehow, which must be the case if the density profile is physical. An example is shown in Figure 7.11b. This faint central image then appears even if $\beta \neq 0$. The magnification of this image is typically $|\mu| < 1$, i.e. it's demagnified. In general, more complicated density profiles also have faint central images that depend on the central mass distribution. One of the aims of the new eMERLIN array of radio telescopes in the UK (see Figure 7.12) is to detect faint central images in order to determine the density profiles at the centres of galaxies.

Figure 7.12 The Lovell telescope at Jodrell Bank, near Manchester in England. This telescope is part of the eMERLIN array of radio telescopes.

Exercise 7.8 Suppose that the lens is an infinite sheet of matter with a constant surface density Σ. Show that the deflection angle α is given by

$$\alpha(\theta) = \frac{4\pi G\Sigma}{c^2}\frac{D_L D_{LS}}{D_S}\,\theta.$$

Next suppose that Σ takes the critical value

$$\Sigma_{cr} = \frac{c^2}{4\pi G}\frac{D_S}{D_L D_{LS}}. \tag{7.30}$$

What will happen? Do gravitational lenses in general focus light? ■

7.5 Time delays and the Hubble parameter

Gravitational lenses also give us a beautiful geometric way of finding the Hubble parameter. To see how, we'll need to return to the deflection of light by a single mass, which curiously turned out to be exactly a factor of two more than the Newtonian prediction. We'll shed a little more light on that here.

In Section 7.1 we found the Newtonian deflection as

$$\phi_{\text{Newtonian}} = \frac{v_y}{c} = \int_{\theta=-\pi/2}^{+\pi/2} \frac{GM}{bc^2} \cos\theta \, \mathrm{d}\theta$$
$$= \frac{1}{c} \int_{t=-\infty}^{\infty} \frac{GM}{r^2} \cos\theta \, \mathrm{d}t.$$

Now, $(GM/r^2)\cos\theta$ is the gradient of the gravitational potential, $\Phi = -GM/r$, in the y-axis direction in Figure 7.6. We can write this as $\nabla_\perp \Phi$, where $\perp$ refers to differentiation being made along a direction perpendicular to the direction of motion of the particle. (Again, we're treating this as effectively the same thing as the y-axis direction, because the change in direction is small.) The deflection is therefore

$$\phi_{\text{Newtonian}} = \frac{1}{c} \int_{-\infty}^{\infty} \nabla_\perp \Phi \, \mathrm{d}t. \tag{7.31}$$

(Φ is negative, but that sign is absorbed into the definition of $\nabla_\perp$.) If we take 'perpendicular' to mean at right angles to the direction of motion, rather than strictly parallel to the y-axis, then this equation is the correct general Newtonian expression without approximations. Making the approximation that $\mathrm{d}t = \mathrm{d}x/c$, we get

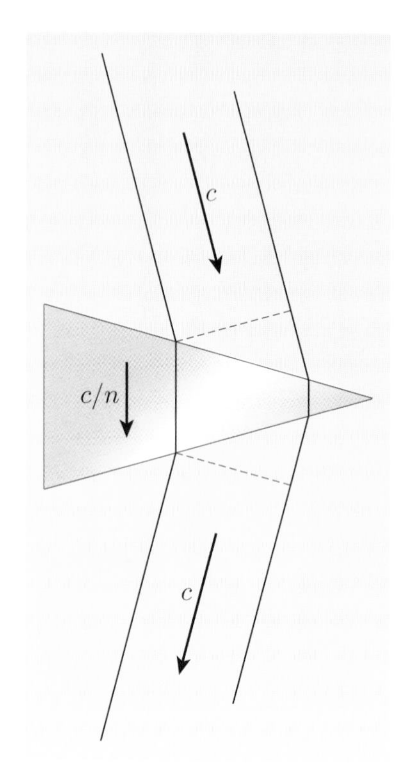

Figure 7.13 An incoming light wavefront with speed c meets a prism with refractive index n and is deflected. The effective speed of light within the prism is c/n. The dashed lines mark lines of constant light-travel time.

$$\phi_{\text{Newtonian}} = \frac{1}{c^2} \int_{-\infty}^{\infty} \nabla_\perp \Phi \, \mathrm{d}x. \tag{7.32}$$

We could also choose to think of the light encountering an effective refractive index n, which varies with position and so veers the light ray off course. The analogous situation for a glass prism is shown schematically in Figure 7.13. Here we have

$$\phi_{\text{Newtonian}} = -\int \nabla_\perp n \, \mathrm{d}x,$$

so we can identify $\nabla_\perp n = (1/c^2)\, \nabla_\perp \Phi$ in the Newtonian case.

The general relativistic equivalent can be found by making a weak-field approximation to the Schwarzschild metric (Chapter 6), using the approximation

$$\left(1 - \frac{2GM}{c^2 r}\right)^{-1} = \left(1 + \frac{2\Phi}{c^2}\right)^{-1} \simeq 1 - \frac{2\Phi}{c^2}$$

to give

$$\mathrm{d}s^2 = \left(1 + \frac{2\Phi}{c^2}\right) c^2\, \mathrm{d}t^2 - \left(1 - \frac{2\Phi}{c^2}\right) \mathrm{d}r^2 - r^2(\mathrm{d}\theta^2 + \sin^2\theta \, \mathrm{d}\phi^2). \tag{7.33}$$

A light ray has $\mathrm{d}s = 0$, and a radial light ray will also have $\mathrm{d}\theta = \mathrm{d}\phi = 0$. In this situation we have

$$\left(\frac{\mathrm{d}r}{\mathrm{d}t}\right)^2 = c^2 \frac{1 + 2\Phi/c^2}{1 - 2\Phi/c^2}. \tag{7.34}$$

So the effective (radial) speed of light is

$$c\sqrt{\frac{1 + 2\Phi/c^2}{1 - 2\Phi/c^2}},$$

i.e. a bit less than c (remember that Φ is negative). We could again think of the lens as having an effective refractive index

$$n = \sqrt{\frac{1 + 2\Phi/c^2}{1 - 2\Phi/c^2}} \simeq 1 - \frac{2\Phi}{c^2}$$

(using the first terms in a Taylor series expansion). This time, however, $\boldsymbol{\nabla}_\perp n = (2/c^2)\,\boldsymbol{\nabla}_\perp \Phi$, explaining the extra factor of two back in Section 7.1.

A photon takes time $(1/c)\,\mathrm{d}\ell$ to travel distance $\mathrm{d}\ell$ in empty flat space. If there is a refractive index n, the time spent is $(n/c)\,\mathrm{d}\ell$. Therefore putting a gravitational lens in between the source and the observer will induce a total time delay of

$$\begin{aligned}\Delta t &= \int_{\text{source}}^{\text{observer}} \frac{1}{c}\,\mathrm{d}\ell - \int_{\text{source}}^{\text{observer}} \frac{n}{c}\,\mathrm{d}\ell \\ &= \int_{\text{source}}^{\text{observer}} \frac{1-n}{c}\,\mathrm{d}\ell \\ &= \int_{\text{source}}^{\text{observer}} \frac{2\Phi}{c^3}\,\mathrm{d}\ell, \end{aligned} \tag{7.35}$$

where the integrations are done over the light path from the source to the observer. This is known as the **Shapiro delay**, after its discoverer. Two different images of a background source would have two different path lengths and experience different potentials, so in general we should expect there to be a relative time delay between different images of a background source.

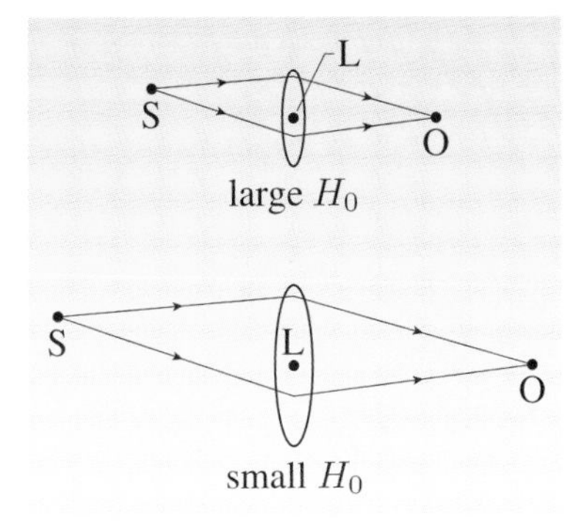

Figure 7.14 Schematic view of how the geometry of a gravitational lens depends on the Hubble parameter H_0. The lens is marked as L, while the source and observer are S and O, respectively. It's not possible to tell from the positions of images alone what the absolute size scale of the system is, but the time delay between two different images can give an absolute scale and hence H_0.

This leads to an ingenious method of finding the Hubble parameter H_0. Most of the lensing equations that we've derived up to now have been dimensionless. For example, angles are dimensionless, and $D_{\mathrm{LS}}/D_{\mathrm{S}}$ is dimensionless. Therefore there's no way to use the lens configuration or arrangement of images to determine the absolute size scale of the lens system (see, for example, Figure 7.14). However, the Shapiro delay is proportional to the path length from the source to the observer. Cosmological distances are proportional to c/H_0 (see Chapter 1) so the time delay between two images will be $\Delta t \propto (1/H_0) \times$ a number that depends on the lens mass model. So if we can find a mass model of the lens that reproduces the lens geometry (e.g. image configurations, lens redshift and source redshift), we can predict the value of $H_0\,\Delta t$; then by measuring Δt we can infer the Hubble parameter!

This has been done in several lenses, such as the quasar QSO 0957+561. The main uncertainty in this experiment is the mass model. (This uncertainty is much larger than the effect that varying Ω_Λ or Ω_{m} would have on the lens geometry.) Also, the time delay itself can sometimes be hard to discern from the data. A recent compilation of time delays from 10 different gravitational lens systems found an average Hubble parameter of $H_0 = 72^{+8}_{-11}\ \mathrm{km\,s^{-1}\,Mpc^{-1}}$.

Saha, P. et al., 2006, *Astrophysical Journal Letters*, **650**, L15.

7.6 Caustics and multiple images

There are many varied and beautiful patterns in gravitational lensing. To understand them, we'll use Fermat's principle, and that's most easily done if we slightly reformulate and simplify the equations that we've found so far.

It follows from Section 7.5 that the deflection from a gravitational lens is

$$\phi = \frac{2}{c^2} \int_{-\infty}^{\infty} \nabla_\perp \Phi \, dx, \tag{7.36}$$

where Φ is the Newtonian potential. This deflection is also the angle $\widehat{\alpha}$ in Figure 7.7. The observed deflection $\boldsymbol{\alpha}$ will therefore be

$$\boldsymbol{\alpha} = \frac{2}{c^2} \frac{D_{\rm LS}}{D_{\rm S}} \int \nabla_\perp \Phi \, dx, \tag{7.37}$$

where we have switched to the more general vector notation. The lens equation is therefore

$$\boldsymbol{\beta} = \boldsymbol{\theta} - \boldsymbol{\alpha} = \boldsymbol{\theta} - \frac{2}{c^2} \frac{D_{\rm LS}}{D_{\rm S}} \int \nabla_\perp \Phi \, dx. \tag{7.38}$$

We could rewrite this in a simpler-looking form as

$$\boldsymbol{\beta} = \boldsymbol{\theta} - \nabla_\theta \psi \tag{7.39}$$

if we can find a suitable new function ψ. Here ∇_θ means derivatives with respect to $\boldsymbol{\theta}$, i.e.

Note that we're not equating two *numbers* or *variables*, but rather two *operators*. This is a subtle but radical change in the use of the = sign.

$$\nabla_\theta = \left(\frac{\partial}{\partial \theta_x}, \frac{\partial}{\partial \theta_y} \right). \tag{7.40}$$

The simplest choice of ψ that works is

$$\psi(\boldsymbol{\theta}) = \frac{D_{\rm LS}}{D_{\rm L} D_{\rm S}} \frac{2}{c^2} \int \Phi \, dx. \tag{7.41}$$

This is sometimes called the scaled projected Newtonian potential. It's related to the deflection angle α through

$$\nabla_\theta \psi = \boldsymbol{\alpha}. \tag{7.42}$$

We can then rewrite the lens equation as

$$\mathbf{0} = \boldsymbol{\theta} - \boldsymbol{\beta} - \nabla_\theta \psi = \nabla_\theta \left[\tfrac{1}{2} (\boldsymbol{\theta} - \boldsymbol{\beta})^2 - \psi \right]. \tag{7.43}$$

To see what the term in square brackets means, here is the corresponding equation for the time delay:

$$\Delta t(\boldsymbol{\theta}) = \frac{(1 + z_{\rm L})}{c} \frac{D_{\rm L} D_{\rm S}}{D_{\rm LS}} \left[\tfrac{1}{2} (\boldsymbol{\theta} - \boldsymbol{\beta})^2 - \psi \right] = \Delta t_{\rm geom} + \Delta t_{\rm grav}, \tag{7.44}$$

where $z_{\rm L}$ is the redshift of the lens. We won't prove this directly (it would take us too far off-topic); instead, we'll point out some general features. The two terms in the square brackets correspond to a gravitational Shapiro time delay ($\Delta t_{\rm grav}$) involving the projected potential ψ, and a geometrical time delay ($\Delta t_{\rm geom}$) involving the angular offset between $\boldsymbol{\beta}$ and $\boldsymbol{\theta}$. The geometrical term is caused by the fact that the light ray is simply travelling further in getting around the lens. The factor of $(1 + z_{\rm L})$ is necessary because a time delay of Δt as the light passes the lens will be time dilated by an additional factor of $(1 + z_{\rm L})$ by the time it's received on the Earth.

Together, Equations 7.43 and 7.44 imply that $\nabla_\theta t(\boldsymbol{\theta}) = \mathbf{0}$. This means that we find images at *stationary points* in the time delay. This is a cosmological version of Fermat's principle. We'll see in Section 7.9 that this projected potential ψ can also be related to the projected mass density Σ.

The time delay is sometimes called the **time delay surface** since it varies in general with both θ_x and θ_y on the sky. Images will form at the stationary points of this surface (minima, maxima, saddle points and points of inflection). However, if the lens is circularly symmetric, we need to consider only one axis. Figure 7.15 shows how the two components of the time delay vary with position for a particular circularly-symmetric lens. Note the images at the three stationary points in the time delay. If we move the position of the background source, the geometric time delay component moves (see Figure 7.16), which changes the shape of the total time delay.

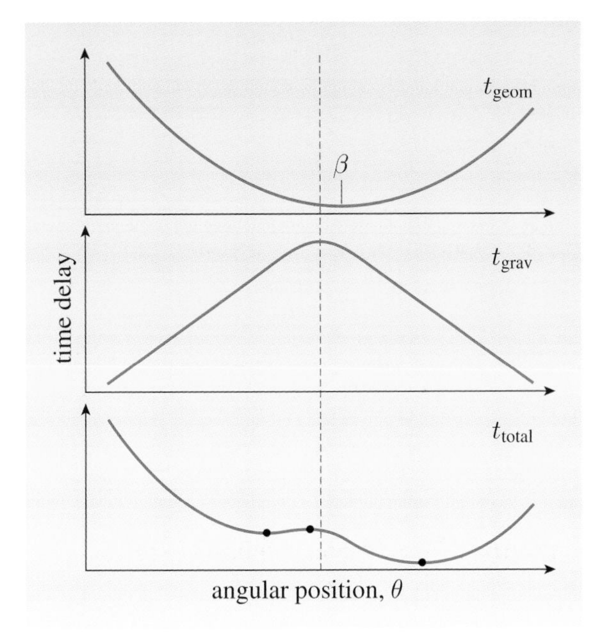

Figure 7.15 The geometric and gravitational time delays for a particular circularly-symmetric lens. The position of the source is marked as β, while the gravitational component peaks at the centre of the lens (marked with a dotted line). There are three images marked as black dots that occur at stationary points in the total time delay curve.

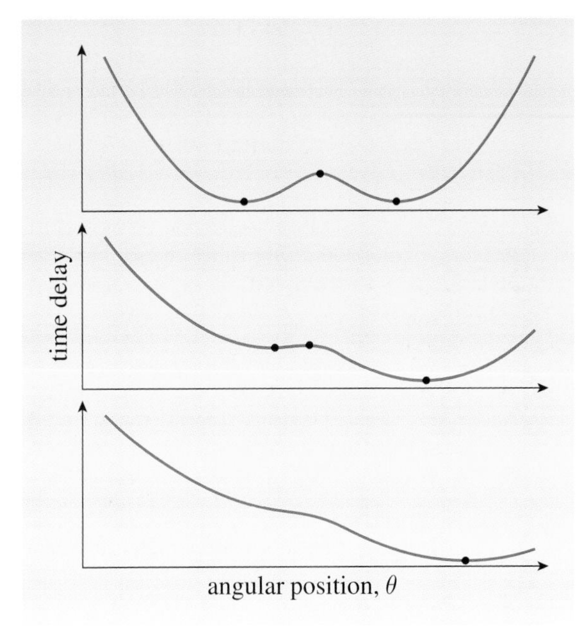

Figure 7.16 The variation of the total time delay and the positions of the images, as the position of the background source is changed. The lens is closely aligned with the background source in the top panel, offset in the central panel, and offset by more in the bottom panel. Note how the leftmost image merges with the central image, and the combined image disappears.

Notice in Figure 7.15 how the image in the left-hand minimum point merges with the image at the maximum, then they vanish. Images can only be created and destroyed in *pairs*, because creating a new minimum means that we must also create a new maximum. Therefore, provided that the lens is non-singular, there must always be an *odd* number of images (sometimes called the *odd-number theorem*). This is also true in the general non-circularly-symmetric case.

Another nice feature of these time delay curves is that the time delay between two images is the vertical distance between them in these plots. In Figure 7.15, for example, the image furthest from the lens will vary first. This is often the case in cosmological lens configurations.

Exercise 7.9 Classify each of the images in Figure 7.15 as a maximum, a minimum or a saddle point. (*Hint*: Don't forget the axis coming out of the paper.)

Exercise 7.10 Suppose that you have a softened isothermal sphere potential, like the one in Figure 7.11b, and you gradually let the potential in the centre get deeper, so it looks more and more like the singular isothermal sphere model in Figure 7.11a. What happens to the time delay of an image seen right through the centre? And where does the image go when the lens potential becomes exactly a singular isothermal sphere? ■

The images that form at maxima, minima and saddle points are each quite different in character. How can we find whether images are minima or maxima? In one-dimensional calculus, a function $y(x)$ with a stationary point at $x = x_0$ has $\mathrm{d}y(x_0)/\mathrm{d}x = 0$. This point is a minimum if $\mathrm{d}^2y/\mathrm{d}x^2 > 0$ there, a maximum if $\mathrm{d}^2y/\mathrm{d}x^2 < 0$, and a point of inflection if $\mathrm{d}^2y/\mathrm{d}x^2 = 0$. The two-dimensional equivalent is to consider the matrix

$$T = \begin{pmatrix} \mathrm{d}^2t/\mathrm{d}\theta_x\mathrm{d}\theta_x & \mathrm{d}^2t/\mathrm{d}\theta_x\mathrm{d}\theta_y \\ \mathrm{d}^2t/\mathrm{d}\theta_y\mathrm{d}\theta_x & \mathrm{d}^2t/\mathrm{d}\theta_y\mathrm{d}\theta_y \end{pmatrix}. \tag{7.45}$$

The criteria are more complicated than in the one-dimensional case. They rely on the determinant and the trace of the matrix. We defined the determinant of a 2×2 matrix in Equation 7.19, while the trace of a 2×2 matrix is defined as

$$\mathrm{tr}\, A = \mathrm{tr} \begin{pmatrix} a & b \\ c & d \end{pmatrix} = ad. \tag{7.46}$$

The criteria are given in Table 7.1.

We have already met something like the T matrix in a different form. If we differentiate Equation 7.44 twice, we find that

$$T \propto \begin{pmatrix} 1 & 0 \\ 0 & 1 \end{pmatrix} - \begin{pmatrix} \mathrm{d}^2\psi/\mathrm{d}\theta_x\mathrm{d}\theta_x & \mathrm{d}^2\psi/\mathrm{d}\theta_x\mathrm{d}\theta_y \\ \mathrm{d}^2\psi/\mathrm{d}\theta_y\mathrm{d}\theta_x & \mathrm{d}^2\psi/\mathrm{d}\theta_y\mathrm{d}\theta_y \end{pmatrix}. \tag{7.47}$$

Back in Section 7.3 we met the inverse magnification tensor, which we defined as $A = \partial\boldsymbol{\beta}/\partial\boldsymbol{\theta}$ (Equation 7.18). If we use the lens equation to expand this (Equation 7.4, $\boldsymbol{\beta} = \boldsymbol{\theta} - \boldsymbol{\alpha}$), we find that

$$\begin{aligned} A &= \begin{pmatrix} \partial\beta_x/\partial\theta_x & \partial\beta_x/\partial\theta_y \\ \partial\beta_y/\partial\theta_x & \partial\beta_y/\partial\theta_y \end{pmatrix} \\ &= \begin{pmatrix} \partial(\theta_x - \alpha_x)/\partial\theta_x & \partial(\theta_x - \alpha_x)/\partial\theta_y \\ \partial(\theta_y - \alpha_y)/\partial\theta_x & \partial(\theta_y - \alpha_y)/\partial\theta_y \end{pmatrix} \\ &= \begin{pmatrix} 1 & 0 \\ 0 & 1 \end{pmatrix} - \begin{pmatrix} \partial\alpha_x/\partial\theta_x & \partial\alpha_x/\partial\theta_y \\ \partial\alpha_y/\partial\theta_x & \partial\alpha_y/\partial\theta_y \end{pmatrix} \\ &= \begin{pmatrix} 1 & 0 \\ 0 & 1 \end{pmatrix} - \begin{pmatrix} \mathrm{d}^2\psi/\mathrm{d}\theta_x\mathrm{d}\theta_x & \mathrm{d}^2\psi/\mathrm{d}\theta_x\mathrm{d}\theta_y \\ \mathrm{d}^2\psi/\mathrm{d}\theta_y\mathrm{d}\theta_x & \mathrm{d}^2\psi/\mathrm{d}\theta_y\mathrm{d}\theta_y \end{pmatrix}, \end{aligned} \tag{7.48}$$

where we've used $\boldsymbol{\alpha} = \boldsymbol{\nabla}_\theta\psi$ in the last step. Therefore the matrix T is just proportional to the inverse magnification tensor A.

One consequence of $T \propto A = M^{-1}$ is that we can immediately say what the magnifications of the different types of images are, because $\mu = 1/\det A$. These magnifications are listed in Table 7.1. The saddle point images also have the curious property of having negative parity, i.e. being mirror-reversed.

Another consequence of $T \propto A = M^{-1}$ is that the curvature of the time delay surface is proportional to inverse magnification, so if the surface is more curved, the image is less magnified.

Table 7.1 The types and properties of gravitational lens images, and how to identify them from the matrix A or T. (We show in the text that T is proportional to A.)

$t(\boldsymbol{\theta})$ shape	Local minimum	Saddle point	Local maximum
Determinant	$\det A > 0$	$\det A < 0$	$\det A > 0$
Trace	$\operatorname{tr} A > 0$	Anything	$\operatorname{tr} A < 0$
Magnification	$\mu > 0$	$\mu < 0$	$\mu > 0$
Parity	$+$	$-$	$+$

- When two images merge, what happens to the magnification?
- The curvature would have to be low in that region (see, for example, Figure 7.16), so the magnification would be high.

We therefore expect that images that are close to each other on the sky would tend to have high magnifications.

The positions on the sky where images merge are known as **critical lines**. The corresponding background source positions are known as **caustics**. In gravitational lensing we tend to refer to the **image plane** (which is what we see) and the **source plane** (which is what's going in the background plane of the source). Figure 7.17 shows an example of a source being moved around in the source plane, and the resulting effects in the image plane.

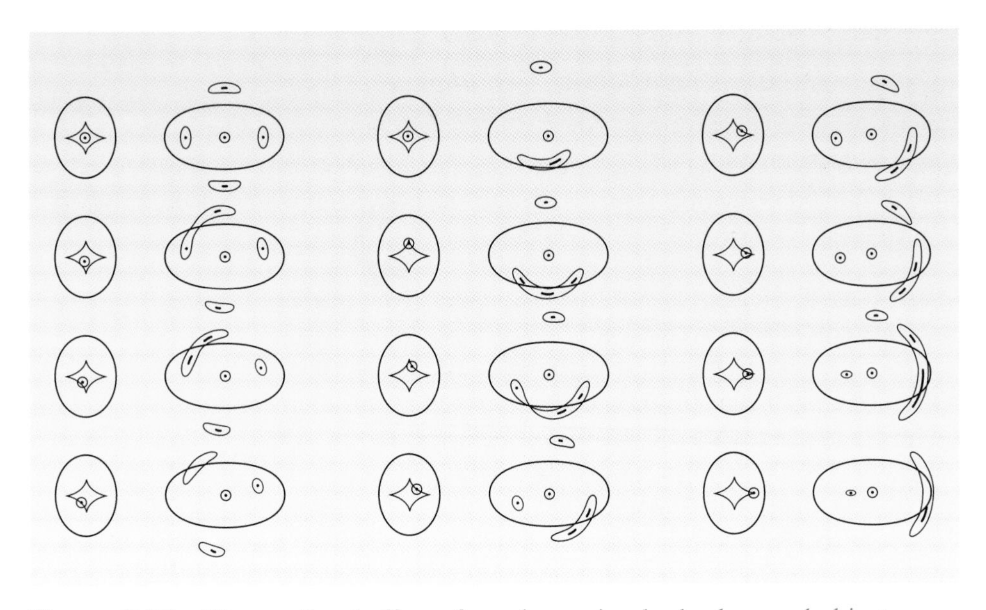

Figure 7.17 The predicted effect of moving a circular background object through the lens caustics (left figures) caused by a simulated elliptical galaxy lens. The images merge at the corresponding critical curves (right figures). The outer caustics and critical curves mark the boundary between one image and three images. The inner caustics and critical curves mark the boundary between three and five images.

7.7 Other lens models

The singular isothermal sphere model is not the only lens model in wide circulation; we'll describe some alternatives briefly here. Though we'll only use them briefly in this book, they are widely used in the lensing community so this is terminology with which you should be familiar.

The Navarro–Frenk–White model[14] is based on predictions from N-body simulations of dark matter haloes (Chapter 4). A generalization of this model is

$$\rho(r) = \frac{\rho_0}{\left(\frac{r}{r_0}\right)^{\alpha}\left(1 + \frac{r}{r_0}\right)^{3-\alpha}}, \tag{7.49}$$

where ρ_0 and r_0 are constants, and $\alpha = 1$ for the Navarro–Frenk–White profile. The best expression for describing galaxy and cluster haloes is still a matter of debate. For example, other groups[15] have found that $\alpha = 1.5$ provides a better fit to their own N-body simulations. If dark matter is self-interacting, this would produce a shallower density profile and reduce the density of the central cusp (Chapter 4).

Elliptical galaxies can also be modelled by generalization of the isothermal sphere:

$$\Sigma(\theta_1, \theta_2) = \frac{\Sigma_0}{\sqrt{\theta_\mathrm{c}^2 + (1-\varepsilon)\theta_1^2 + (1+\varepsilon)\theta_2^2}}, \tag{7.50}$$

where Σ_0 is a constant, θ_1 and θ_2 are angular positions along the major and minor axes, ε is the ellipticity, and θ_c is a core radius. Setting $\theta_\mathrm{c} = 0$ and $\varepsilon = 0$ reduces this to the singular isothermal sphere. The effect of setting $\theta_\mathrm{c} \neq 0$ is to smooth out the density spike in the centre. Alternatively, the Blandford and Kochanek elliptical density profile[16] is often used. For their 'isothermal' lens this is

$$\psi(\theta_1, \theta_2) = \frac{D_\mathrm{LS}}{D_\mathrm{S}}\, 4\pi\, \frac{\sigma_\mathrm{v}^2}{c^2} \left[\theta_\mathrm{c}^2 + (1-\varepsilon)\theta_1^2 + (1+\varepsilon)\theta_2^2\right]^{1/2}, \tag{7.51}$$

where σ_v is the isothermal velocity dispersion. (Non-isothermal lenses have the term in square brackets raised to a positive power less than $1/2$, and have the normalization expressed as a different constant.) Unlike the Navarro–Frenk–White profile and its variants, this functional form is motivated by simplicity of calculation for gravitational lensing, though when ε is small it turns out that it nevertheless is a reasonable approximation (for most uses) to the isothermal ellipse. The demonstration of caustics in Figure 7.17 was made using a Blandford and Kochanek profile with $\theta_\mathrm{E} = 1''$, $\theta_\mathrm{c} = 0.05''$ and $\varepsilon = 0.2$.

Some amount of external shear is always present in gravitational lensing, so one cannot rely on circular symmetry in modelling gravitational lenses. In individual lens models this is sometimes characterized as an additional potential of

$$\psi(\theta_1, \theta_2) = \frac{\kappa}{2}(\theta_1^2 + \theta_2^2) + \frac{\gamma}{2}(\theta_1^2 - \theta_2^2);$$

the convergence κ and shear γ will be described in more detail in Section 7.9.

[14]Navarro, J.F., Frenk, C.S. and White, S.D., 1996, *Astrophysical Journal*, **462**, 563.
[15]Moore, B. et al., 1999, *Monthly Notices of the Royal Astronomical Society*, **310**, 1147.
[16]Blandford, R.D. and Kochanek, C.S., 1987, *Astrophysical Journal*, **321**, 658.

7.8 Microlensing

In 1936 Einstein published a short note about the gravitational amplification that would occur if two stars happen to appear very close in projection on the sky, which has since been called **microlensing**, for reasons that will become clear. He wrote: 'there is no great chance of observing this phenomenon, even if dazzling by the light of the much nearer star ... is disregarded.' He published this paper after being encouraged to investigate the effect by an amateur named Rudi Mandl (though unknown to both, Eddington and Chwolson had each published little-known papers on related effects). Einstein also wrote a private note to the journal editor saying: 'Let me also thank you for your cooperation with the little publication, which Mister Mandl squeezed out of me. It is of little value, but it makes the poor guy happy.' Einstein reckoned without the tremendous advances in optical imaging technology that have happened in the past few decades.

Einstein, A., 1936, *Science*, **84**, 506.

For more on this story, see Renn, J., Sauer, T. and Stachel, J., 1997, *Science*, **275**, 5297.

We can get a rough idea of the probability of one star gravitationally lensing another from the Einstein radius. We found this for star–star lensing in Equation 7.11, with the result that it would be typically measured in milliarcseconds (10^{-3} of an arcsecond, which itself is $1/3600$th of a degree). The number of stars per unit area on the sky varies, with higher densities closer to the Galactic plane. In crowded fields (for example, towards the Galactic bulge), it turns out that we'd expect of the order of one faint foreground star per square arcsecond. It may be too faint to detect on its own, but it might nevertheless be a potential lens. The probability of this foreground star lensing a background one would be of the order of θ_E^2/ρ, where ρ is the number of potential lenses per unit area on the sky, which comes out around 10^{-6}. So, to detect this type of lensing, one would need to monitor millions of stars simultaneously. (A more careful calculation takes into account the fact that lenses close to the source or close to the Earth have smaller θ_E than ones more centrally placed.)

See, for example, Griest et al., 1991, *Astrophysical Journal Letters*, **372**, L79.

In Einstein's time, wide-field optical astronomy could be done only with photographic plates. Wide-field CCD arrays have now made microlensing searches possible. Figure 7.18 shows one of the first discoveries of gravitational microlensing, made with a long-term monitoring campaign of the Large Magellanic Cloud. As the foreground lens passes in front of the background star, the background star is gravitationally lensed and magnified. Note the similar profiles in the red and blue filters: achromaticity is an important test that it is gravitational lensing, and not some unknown type of variable star.

The original aim of microlensing searches was to detect clumps of dark matter, which were given the acronym **MACHOs** (massive compact halo objects). These clumps could be black holes, clumps of non-baryonic elementary particles, or dark baryonic matter such as planetary-sized objects or cometary nuclei such as are found in the Oort cloud of our Solar System. The team that made the early detection in Figure 7.18 also named their survey 'The MACHO Project'. For this experiment one wants to avoid star–star lensing, so surveys for MACHOs have been done outside the Galactic plane, e.g. towards the Large Magellanic Cloud. Microlensing events are rarer outside the plane of the Galaxy. The initial results suggested a large population of $\sim 0.5\,M_\odot$ lenses in the Galactic halo, but with larger surveys the current best limit is that $< 8\%$ of the dark matter halo of the Galaxy is made up of compact objects.

See, for example, Tisserand et al., 2007, *Astronomy and Astrophysics*, **469**, 387.

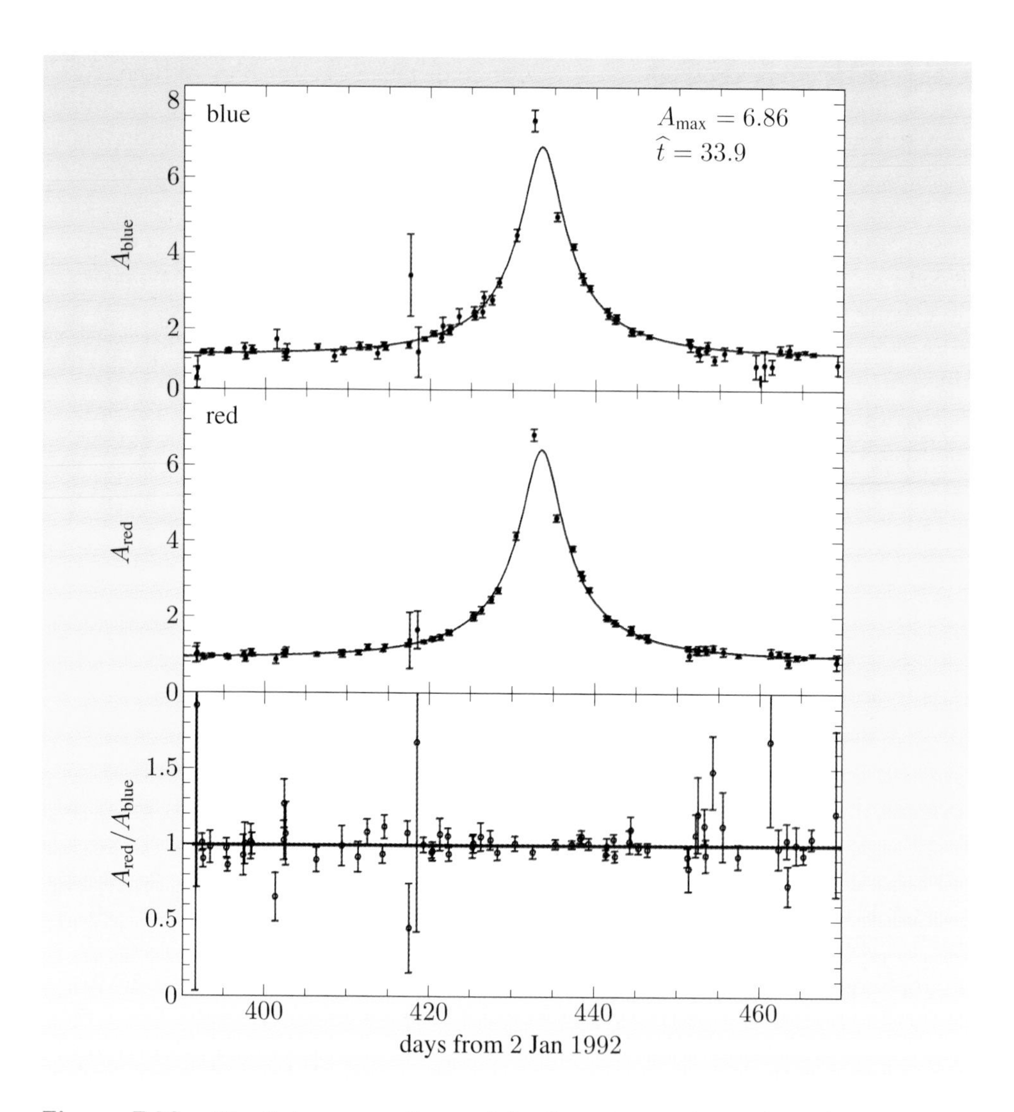

Figure 7.18 The light curve of one of the first observations of gravitational microlensing events, also showing the best fit to the data. The best-fit maximum magnification and timescales are quoted in the figure. Note that the amplification is achromatic, as expected for lensing.

- Could *all* of the dark matter in the Universe be clumps of baryonic matter, like free-floating Jupiters?

- No, because this would violate the Big Bang nucleosynthesis constraint on Ω_b (Chapter 2).

To describe gravitational microlensing, one ideally takes into account the finite source size and limb darkening (stars not being uniformly bright circles), but a good approximation is a point mass lens magnification (Equation 7.17). The distances in this case are not cosmological, so we can just use Euclidean distances in which D_{LS} does equal $D_S - D_L$. Despite the fact that the *lens* is moving across our line of sight to a background star, mathematical descriptions of microlensing are simplest from the lens's point of view, in which the lens is stationary but the background source is moving. This is shown schematically in Figure 7.19.

Pythagoras's theorem gives us the lens–source distance as a function of time:

$$\frac{\beta}{\theta_E} = \sqrt{\left(\frac{b}{\theta_E}\right)^2 + \left(\frac{v}{\theta_E} \times (t - t_0)\right)^2}, \tag{7.52}$$

where b is the impact parameter on the sky, t_0 is the time when the source appears closest to the lens, and v is an angular speed (measured, for example, in microarcseconds per day). Plugging this expression for β/θ_E into Equation 7.17 gives us a rather messy expression for the magnification as a function of time. We won't write this out in full, but it's worth noticing what it depends on and what it doesn't. If we have some microlensing data like those in Figure 7.18, we would use the expression for the magnification as a function of time, and we would vary b/θ_E, v/θ_E and t_0 to find the best fit to the data. The time t_0 just gives us the time of closest approach, which is the time when the curve peaks. The overall normalization of the light curve will depend on b/θ_E, while the width of the curve depends on v/θ_E. We can therefore find the parameters b/θ_E and v/θ_E, but not b or v on their own. There's no way of using a point mass lens microlensing light curve on its own (e.g. Figure 7.18) to find θ_E. Therefore we can't use the light curves to find out how far away the lenses are, how fast they're moving, or how massive they are.

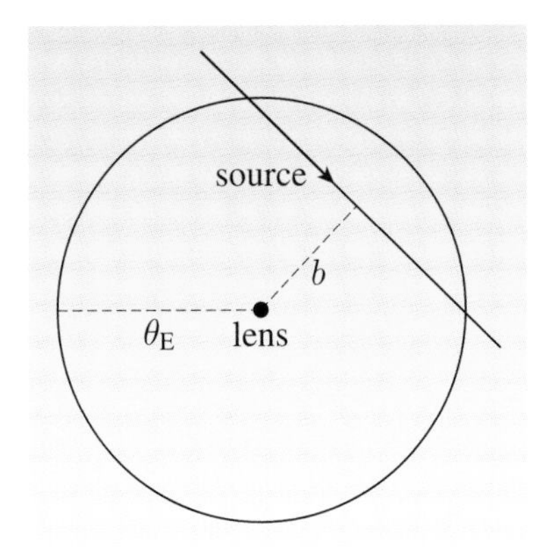

Figure 7.19 Schematic view of a microlensing event, seen from the point of view of the lens. The background source passes through the Einstein radius of the lens θ_E, and its closest approach is an angular separation of b. The source's angular speed is v. In practice it's the lens that moves across our line of sight to the background source, but it's sometimes easier to visualize from the lens's point of view. As this is the *source* plane, the position in this figure corresponds to the angle β in Figure 7.7.

One solution to this problem is to *assume* that the lens has a typical transverse velocity within the galaxy of around $200\,\mathrm{km\,s^{-1}}$, which is typical of stars in the Galaxy. With a large enough sample of microlensing events, plus some assumptions about the spatial distribution of lenses, one could infer a lens mass distribution. This statistical method was the original approach of the MACHO project team.

Another approach is to use parallax. If a microlensing event takes long enough, there could be a measurable change in the lens geometry caused by the Earth's motion around the Sun. These events are rare, because most microlensing events last tens of days, not hundreds. (To calculate a typical duration of a microlensing event, you would assume a lens distance D_L, use it to convert the Einstein radius to a physical distance $\xi_E = \theta_E D_L$, then estimate how long it would take a star to cross it at the typical Galactocentric speed.) It turns out that the parallax supplies enough extra information to derive θ_E and find the lens mass and distance. Alternatively, the background object could be a binary — indeed, about half of the stars in the Galaxy are in binary systems. This superimposes a slight periodic variation on the light curve signal, which depends on the stars' orbital distance in units of θ_E. If the orbital parameters of the binary can be determined by other means, θ_E can be inferred. This is, in some sense, an inverse to the parallax effect on microlensing. It has sometimes been called **Xallarap**.

See, for example, Alcock et al., 1995, *Astrophysical Journal*, **454**, 125.

Griest, K. and Hu, W., 1992, *Astrophysical Journal*, **397**, 362.

Many collaborations have sought to find microlenses, including the MACHO project mentioned above, EROS (Expérience pour la Recherche d'Objets Sombres), OGLE (Optical Gravitational Lens Experiment) and MOA (Microlensing Observations in Astrophysics). The consortia typically arrange for rapid worldwide follow-ups of newly discovered microlensing events, often using robotic telescopes such as the Faulkes telescopes. Much of the current interest in microlensing is in planet discovery; a planet around a lens can create sudden characteristic changes in the light curves, if the source passes through the appropriate caustic from the planet. Microlensing is complementary to other methods of exoplanet discovery; 2005 saw the microlensing discovery of a 5.5 Earth mass planet. The anomalous data point in Figure 7.18 near the maximum magnification may be the result of a binary companion to the lens. This is a very exciting area of research, but it takes us beyond the cosmological theme of this book. There are some suggestions for further reading on this at the end of this chapter.

One final cosmological example of microlensing is worth mentioning. When quasars are gravitationally lensed by foreground galaxies, the small angular sizes of the regions emitting the quasar's continuum and broad emission lines make the quasar susceptible to microlensing by stars in the lensing galaxy. This could in principle give clues about the internal structure of quasars, and attempts have been made to pursue this, though in practice it is limited by the relative rarity of lensed quasars and the microlensing timescales of (in some cases) multiple years.

7.9 Cosmic shear

Gravitational lensing can also trace the large-scale structure of the Universe, through patterns of weak lensing. This directly measures the large-scale matter distribution of the cosmic web and probes the matter power spectrum in the linear regime (Chapter 4). This makes it an excellent test of hierarchical CDM structure formation models. The method appears to be so promising that it could become a powerful route to estimating the equation of state of dark energy.

In this case, the lensing of individual galaxies is weak, so there are not likely to be multiple images, but one can still detect the effect statistically from the tendency of galaxy ellipticities to align, as shown in Figure 7.20. Galaxies are not themselves round, so alignments will not be perfect, but their intrinsic orientations will be random compared to the foreground structure, so the effects of intrinsic ellipticities should average out to zero.

Figure 7.20 Exaggerated view of weak lensing by the cosmic large-scale structure of matter. The shear component of the gravitational magnification will tend to be aligned with the nearby large-scale structure (red), so measured galaxy ellipticities (blue) on average will trace the foreground large-scale matter distribution.

To calculate how much ellipticity is induced by a gravitational lens, we'll show that we can rewrite the inverse magnification tensor as

$$A = (1-\kappa)\begin{pmatrix}1 & 0\\ 0 & 1\end{pmatrix} - \gamma\begin{pmatrix}\cos 2\phi & \sin 2\phi\\ \sin 2\phi & -\cos 2\phi\end{pmatrix}, \tag{7.53}$$

where the κ in the first term is called the **convergence**, the γ in the second is called the **shear**, and ϕ measures the orientation angle of the shear. Figure 7.21 illustrates the different effects of these two terms: convergence is isotropic, so the images are just rescaled by a factor, while shear stretches the shapes in a particular direction. The weak lensing by large-scale structure is often called **cosmic shear**. Figure 7.22 shows a schematic simulated image without shear and another with shear at the level typical for lensing by large-scale structure, though in a fixed direction uniformly over the field. These are subtle effects!

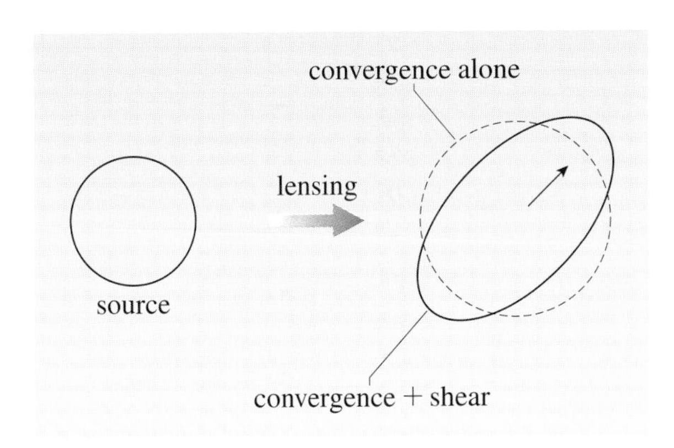

Figure 7.21 Demonstration of the different effects of convergence and shear in gravitational lensing. The arrow to the right is the direction of the shear.

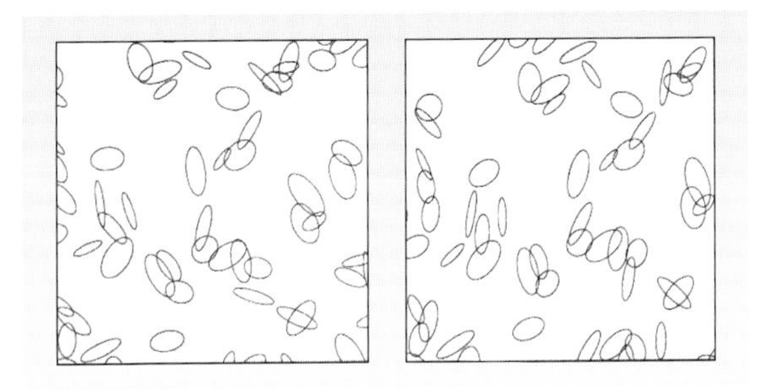

Figure 7.22 Demonstration of an image without shear (left) and with a constant shear applied across the whole image (right). The polarization is 0.1 in the right-hand image (in the image distortion sense rather than the polarized light sense — see page 242).

To get to Equation 7.53, we'll first write A as

$$A = A - \tfrac{1}{2}(\text{tr}\,A)\cdot I + \tfrac{1}{2}(\text{tr}\,A)\cdot I, \tag{7.54}$$

where tr A means the trace of the matrix A (Equation 7.46), and I is the identity matrix:

$$I = \begin{pmatrix}1 & 0\\ 0 & 1\end{pmatrix}. \tag{7.55}$$

To avoid a big mess of partial differentials, we'll write $\partial^2\psi/\partial x^2$ as ψ_{11}, $\partial^2\psi/\partial x\partial y$ as ψ_{12}, and so on.

With this notation, the trace of A is just

$$\text{tr}\,A = 1 - \psi_{11} + 1 - \psi_{22} = 2 - (\psi_{11} + \psi_{22}),$$

so the term $\frac{1}{2}(\text{tr}\,A)\cdot I$ comes out as

$$\tfrac{1}{2}(\text{tr}\,A)\cdot I = \begin{pmatrix}1 - \frac{1}{2}(\psi_{11}+\psi_{22}) & 0\\ 0 & 1 - \frac{1}{2}(\psi_{11}+\psi_{22})\end{pmatrix}. \tag{7.56}$$

The convergence κ is defined via

$$1 - \kappa = \tfrac{1}{2}(\text{tr}\,A) = \tfrac{1}{2}(\psi_{11} + \psi_{22}), \tag{7.57}$$

so with a bit of algebra you can see that the $+\frac{1}{2}(\text{tr}\,A) \cdot I$ term in Equation 7.54 gives rise to the convergence term in Equation 7.53. It also turns out that $\kappa = \Sigma/\Sigma_{\text{cr}}$, where Σ is the surface mass density and Σ_{cr} is the critical surface mass density (see Exercise 7.8).

The next trick is to write

$$\gamma_1 = \tfrac{1}{2}(\psi_{11} - \psi_{22}) = \gamma(\boldsymbol{\theta})\cos(2\phi(\boldsymbol{\theta})), \tag{7.58}$$

$$\gamma_2 = \psi_{12} = \psi_{21} = \gamma(\boldsymbol{\theta})\sin(2\phi(\boldsymbol{\theta})). \tag{7.59}$$

We won't prove here that such a substitution is always possible, but we have three free parameters in the inverse magnification matrix (because $\partial^2\psi/\partial\theta_x\partial\theta_y = \partial^2\psi/\partial\theta_y\partial\theta_x$) and three proposed new parameters κ, γ and ϕ. Therefore we might expect that some substitution of this form should be possible, and we can use it to define γ and ϕ. Plugging these definitions in gets us to Equation 7.53.

In practice, measuring cosmic shear is a difficult experiment that needs a stable and well-characterized point spread function. (An unresolved object in an image takes the shape of the point spread function.) This can be difficult from ground-based astronomy. Also, we've assumed that galaxies don't have intrinsic alignments: galaxies will tend to align themselves with their local large-scale structure. For example, in the vicinity of the lens, this could produce apparent shears perpendicular to the gravitational lensing shear. The current thinking is that intrinsic alignments don't cause a fatal problem for cosmic shear detection measurements, because galaxies that are close on the sky are mostly well-separated in redshift. The shear measurements γ for each galaxy can be estimated from their ellipticities, taking into account the point spread function shape. A circular galaxy would appear as an ellipse with a major axis of $a = (1 - \kappa - \gamma)^{-1}$ and a minor axis of $b = (1 - \kappa + \gamma)^{-1}$. (In the weak lensing regime, it's usually reasonable to assume that the magnification is constant across the image of the lensed galaxy.) Sometimes the shear is expressed as a complex number

$$\varepsilon = \frac{a^2 - b^2}{a^2 + b^2}\,\mathrm{e}^{2\mathrm{i}\phi} = \varepsilon_1 + \mathrm{i}\varepsilon_2.$$

Somewhat confusingly, this complex ellipticity is sometimes referred to as **image polarization**, even though it has nothing to do with polarized light. Similarly, the shear is sometimes expressed as a complex number $\gamma_1 + \mathrm{i}\gamma_2$.

One way of estimating the strength of cosmic shear is to separate the galaxies into foreground and background populations on the basis of, for example, photometric redshifts or even just apparent magnitudes, then measure the tangential component of the shear for every foreground+background pair of galaxies, then finally plot the average tangential shear as a function of the pair separation. An example is shown in Figure 7.23. It's tempting to think of this as the tangential shear *caused by* the foreground galaxies, but because these foreground galaxies cluster, there will also be lensing contributions from neighbours. The theoretical predictions of this signal are complicated because one must estimate all the weak deflections experienced by the light ray in its passage through the intervening inhomogeneous Universe. A good test of whether there are systematic errors lurking in the data analysis is to rotate the background galaxies by $45°$. If the measured tangential shear is due to gravitational lensing, then this $45°$-rotated signal should be consistent with zero. This is indeed what's seen in Figure 7.23.

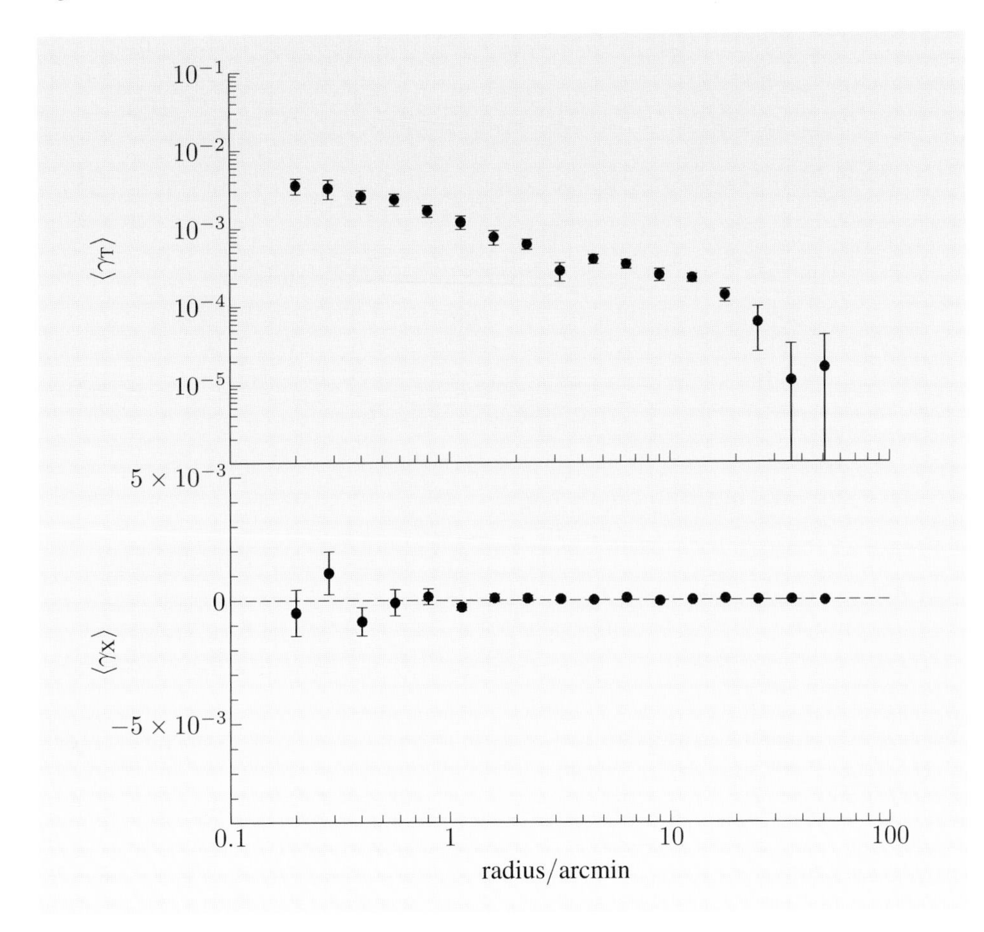

Figure 7.23 Tangential (top) and rotated (bottom) shear components as a function of angular separation of foreground and background galaxies. Galaxies were treated as lenses or background sources on the basis of apparent R-band (660 nm) magnitude.

If there is redshift information available, then it's possible to trace the evolving structure of the cosmic web. This has been achieved by the COSMOS (Cosmological Evolution Survey) project using a wide-area HST survey. Figure 7.24 shows the three-dimensional recovered dark matter distribution as a function of position on the sky (longitude and latitude known as right ascension and declination) and as a function of redshift. Figure 7.25 compares the distribution of total mass (dominated by dark matter) to the distributions of the stellar mass of galaxies, of the numbers of optically-selected galaxies, and of the X-ray-luminous gas. The X-ray emission is proportional to the square of the electron density n_e^2, and since cosmological plasma is overall electrically neutral, the X-ray flux will tend to highlight the higher-density regions of the baryon distribution. (Gravitational lensing sensitivity, however, is linearly proportional to mass.) There is a $z = 0.73$ galaxy cluster in Figure 7.25 at coordinates $149°55'$ and $2°31'$, around which the weak lensing finds filamentary dark matter structures.

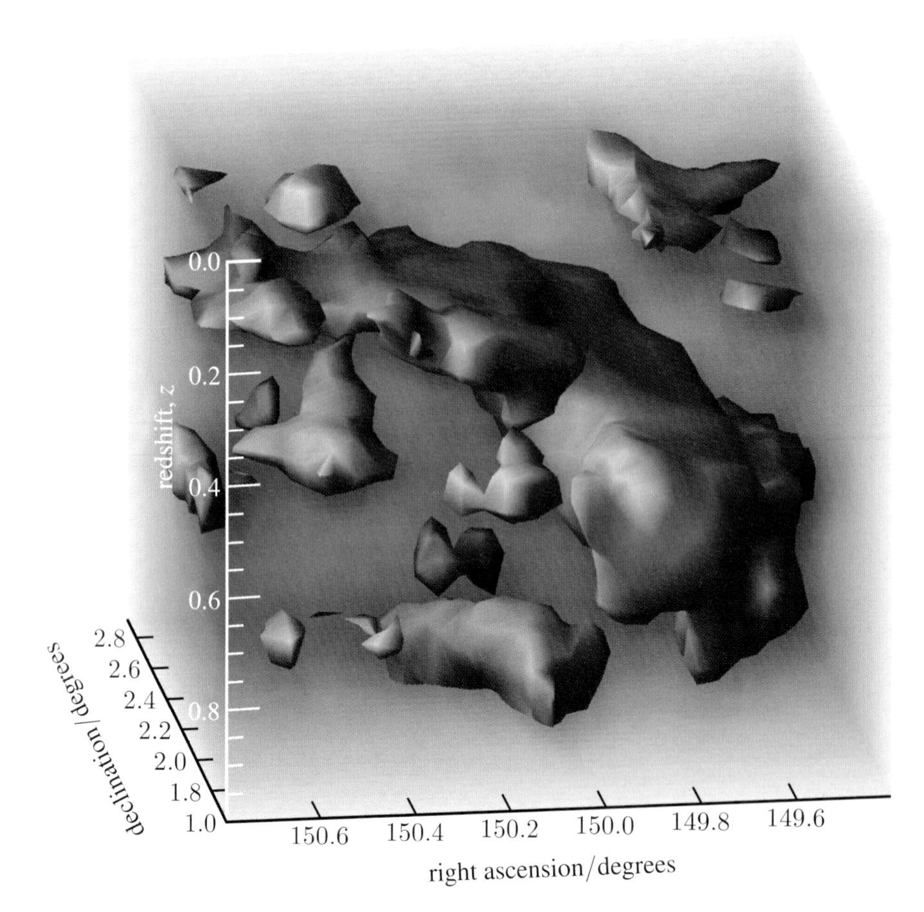

Figure 7.24 Dark matter distribution inferred from weak lensing in the COSMOS survey. The redshift axis is compressed — the survey geometry is really an elongated cone. Regions are marked as opaque where the density is greater than $1.4 \times 10^{13}\,M_\odot$, with a circle of radius 700 kpc on the sky and a redshift interval of $\Delta z = 0.05$. The darkness of the faint greyscale background traces the full density distribution.

The evolution of large-scale structure is a (known) function of the cosmological parameters, including the dark energy equation of state parameter w. Cosmic shear could therefore be a route to constraining cosmological parameters. However, while the signal from cosmic shear itself is much larger than that of intrinsic galaxy alignments, the intrinsic alignments are much larger than the effect of changing w by (say) 1%. It turns out that the redshift-dependence of intrinsic alignments can be used to distinguish the intrinsic alignments from cosmic shear, and this is a subject of much ongoing research.

The prospects for improving the cosmic shear detections are excellent over the next ten years or so. The Pan-STARRS survey (Panoramic Survey Telescope And Rapid Response System) plans to use four dedicated 1.8 m optical telescopes in Hawaii to survey all the sky visible from that site (about three-quarters of the whole sky) in six filters to a typical optical magnitude of $R = 26$ (5σ detection limit for a point source). The primary goal is to detect Solar System moving objects, but the project has great potential for the detection of cosmic shear. The start of operations with its first telescope is imminent at the time of writing. Meanwhile, the Dark Energy Survey (DES) plans to use 30% of the time on the 4 m Cerro Tololo Inter-American Observatory (CTIO) to conduct an optical survey of 5000 deg^2 of the sky in four filters. One of its key objectives is the detection of weak lensing using photometric redshifts to trace the evolution of $w(z)$. The Large Synoptic Survey Telescope (LSST) has similar science objectives and sky area to Pan-STARRS but with a much larger (and hence more sensitive) 8.4 m telescope.

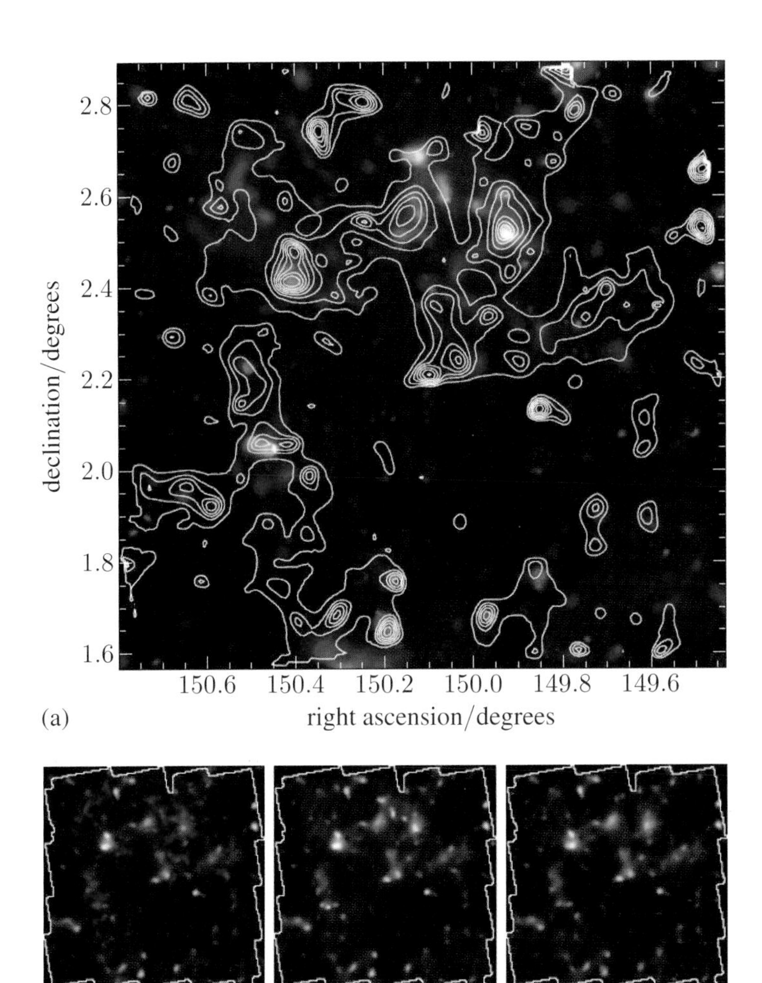

Figure 7.25 The total dark matter distribution in COSMOS, projected onto the sky. Panel (a) has the dark matter marked as contours, while panels (b), (c) and (d) mark the dark matter distribution in greyscale. Also shown are various independent tracers of the baryonic matter distribution: the stellar mass (blue) as traced by near-infrared photometry of galaxies, the density of galaxies (yellow) as traced by optical galaxy counts, and the hot gas (red) as traced by X-ray imaging of the field after removal of X-ray point sources.

At the time of writing, LSST survey operations are planned to begin around 2016. Figure 7.26 shows the projected sensitivity for the dark energy equation of state parameters for the LSST, assuming that the systematics from intrinsic alignments and point spread function variations can be well-characterized. All these future and imminent surveys also seek to measure baryon wiggles, high-redshift supernovae and the evolution of galaxy clustering (Chapter 3).

Besides intrinsic alignments, the main difficulty with ground-based optical measurements of weak lensing is the characterization and stability of the point spread function. There are two quite different solutions that other forthcoming cosmic shear experiments will (or may) use. One solution is to move the telescope above the Earth's turbulent atmosphere. At the time of writing there are two major space missions proposed to do this: the European Space Agency EUCLID mission, and the NASA Joint Dark Energy Mission (JDEM). Both missions are ambitious wide-field optical/near-infrared imaging and spectroscopy surveys using a $\sim$1.2–1.5 m space telescope.

It has been proposed that the missions should merge and form a joint ESA/NASA project. The other option is to use radio telescopes, because the angular resolution of radio interferometry is not subject to the seeing limitations of ground-based optical astronomy. Getting enough galaxies over a large enough sky area is challenging for the current generation of radio telescopes, but the Square

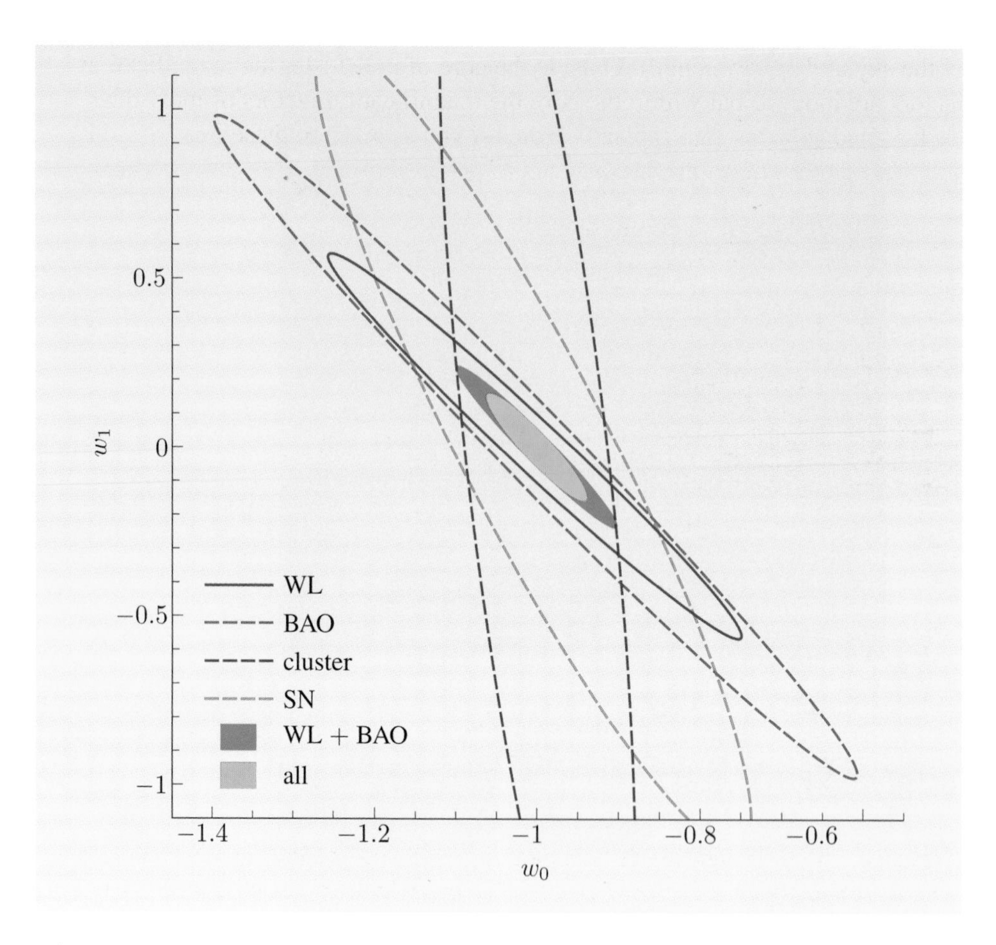

Figure 7.26 Projected dark energy equation of state constraints for the Large Synoptic Survey Telescope. As well as the weak lensing constraints, there are also constraints from baryonic acoustic oscillations (BAOs), supernovae (SN) and galaxy cluster number counts (Chapters 3 and 4).

Kilometre Array (SKA; see also Chapter 8) will revolutionize this field. The SKA should be completed around the year 2020, though early science observations with a subset of the array will happen in the preceding few years. These future projects aimed at measuring cosmic shear may also be useful for finding new strong gravitational lenses (Section 7.11).

7.10 Galaxy cluster lenses

Gravitational lens magnification is one of the best ways of finding the most distant objects in the Universe. Cosmological lenses are rare on the sky, but one place where gravitational lensing can be reliably expected to happen is the cores of the most massive galaxy clusters. We saw an example early on in this chapter in Figure 7.3. This galaxy cluster, Abell 2218, has one of the best-constrained mass models of any galaxy cluster. To make such a model, one first finds the lensed arcs and multiple images. A common procedure is then to approximate the cluster with a smooth model (e.g. isothermal or Navarro–Frenk–White, Sections 7.4 and 7.7), then add in the additional mass from the galaxies, assuming a constant mass-to-light ratio. The parameters of the model are iterated until a good fit to the pattern of arcs and multiple images is found.

The mass model of Abell 2218 has two main condensations, suggesting that the cluster is the product of an ongoing merger. The cluster has been studied very comprehensively at many wavelengths. Figure 7.27 shows some of the deep images of the core of Abell 2218. Many of these are the deepest images ever taken of the sky at that wavelength. Outside the core of the cluster the magnification factors are modest, but within the core the magnification factors of individual background galaxies vary typically from around 2 to 10, so these images are in addition up to 10 times deeper than can be achieved in unlensed parts of the sky.

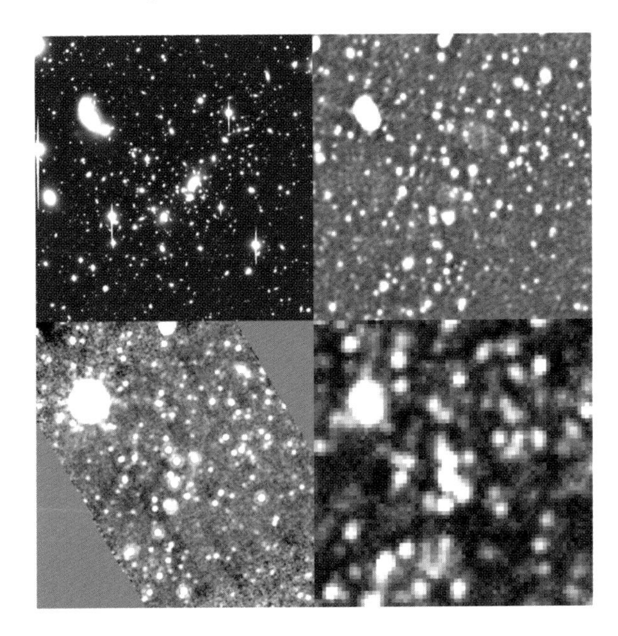

Figure 7.27 Images of Abell 2218 taken with various space telescopes. Top left: I-band HST image. Top right: 15 μm image with the AKARI space telescope. Bottom left: 24 μm image from the Spitzer Space Telescope. Bottom right: 250 μm image from the Herschel Space Observatory.

Exercise 7.11 Suppose that at a particular redshift the background galaxies have a luminosity function $\mathrm{d}\Phi/\mathrm{d}L \propto L^{-\alpha}$. For which values of α would lens magnification increase the number of these background sources? Don't forget that increasing the angular size of a distant region also decreases the comoving volume that's sampled, for a fixed observed area on the sky. ■

In blank-field galaxy surveys (i.e. mapping of blank areas of sky — see Chapter 4), the blending and overlapping of objects limits the depth at which objects can be found. This confusion limit (see Chapter 4) can be circumvented using gravitational lensing by targeting foreground galaxy clusters instead of blank fields.

Galaxy cluster lenses have been used to find ultra-high-redshift galaxy candidates. (At the time of writing this means $z > 6$, but this changes!) These are rare on the sky but the magnification assists. The candidates tend to be selected on the basis of photometric redshifts which at these redshifts are dominated by the Lyman break. An important test of the proposed redshift is the position(s) of multiple images. Another important test is optical spectroscopy, looking for emission lines at the estimated redshift. It's not obvious that emission lines will be visible if the galaxy is very dusty (see, for example, Chapters 4 and 5), but even many submm-selected galaxies have Lyman α lines, so it is certainly worth trying. Two emission lines are needed to confirm a redshift (Chapter 4), but if an emission line is seen at the expected position of Lyman α, it may be taken as confirmation of the redshift. Nevertheless, a claimed $z \simeq 10$ gravitationally-lensed galaxy with an apparent Lyman α emission line in cluster Abell 1835 later turned out to be a red galaxy at a much lower redshift; the emission line appeared to have been an artefact caused by bad pixels in the original data. The danger of these false positive claims is quite high because low-redshift faint red galaxies are far more common on the sky than ultra-high-redshift galaxies. The existence of dark matter has very wide though not universal acceptance within the research community. Some baulk at the prospect of a component of matter for which there have been no direct observations from particle physics, but which has been inferred in astronomy to dominate the matter density of the Universe. The proposed alternative is to modify Newtonian gravitation to explain galaxy rotation curves. This theory, known as 'modified Newtonian dynamics' or MOND, can be given a relativistic context in an alternative to general relativity known as 'Tensor–Vector–Scalar' theory or TeVeS. It can be unwise to make predictions of future discoveries (many expected $\Lambda = 0$ — see Chapter 1); nevertheless, since this is a minority position within the community, we shall not dwell on this theory in this book. One key prediction of MOND/TeVeS is that the gravitational lensing deflection should follow the visible baryonic matter in galaxy clusters. The Bullet cluster turned out to be an excellent place to test this prediction. Like Abell 2218, the cluster is being seen in the process of a merger, though the Bullet cluster is at an earlier stage. The cluster baryonic mass is traced by the X-ray luminous gas, so the mass inferred from gravitational lensing should match the distribution of X-ray-emitting gas. However, while the self-interaction of dark matter is very weak or non-existent, a pocket of gas can interact very strongly with another through pressure, shocks, and so on. When two galaxy clusters collide, we'd therefore expect the two 'clouds' of dark matter to fall towards each other, pass through each other and out the other side, oscillate and eventually settle through tidal forces. Meanwhile, the two 'clouds' of gas would interact strongly and settle into the centre more quickly. The Bullet cluster has a clear separation of the masses inferred from X-ray gas and from gravitational lensing (Figure 7.28). This is particularly challenging for MOND/TeVeS if it is to have no dissipationless dark matter component.

Figure 7.28 Optical image of the Bullet cluster with X-ray image from the Chandra X-ray Observatory superimposed in pink and the mass inferred from gravitational lensing superimposed in blue. The baryonic matter in galaxy clusters is traced by the X-ray-luminous gas. Note the clear separation of the baryonic matter (X-rays) from the total matter, implying a spatially separate dark matter component. This is challenging to models that seek to avoid the existence of dark matter by modifying the gravitational force law.

7.11 Finding gravitational lenses

The first gravitational lens discovered was the double quasar QSO 0957+561 (Figure 7.29). (The numbers refer to the right ascension and declination coordinates on the sky, $\alpha = 09^{\rm h}57^{\rm m}$ and $\delta = +56.1^\circ$.) It was found entirely serendipitously in a survey of optical candidates of radio sources aimed at discovering new radio-loud active galaxies.

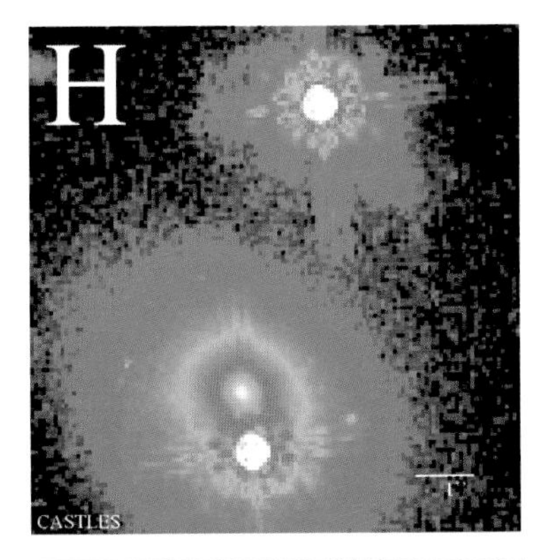

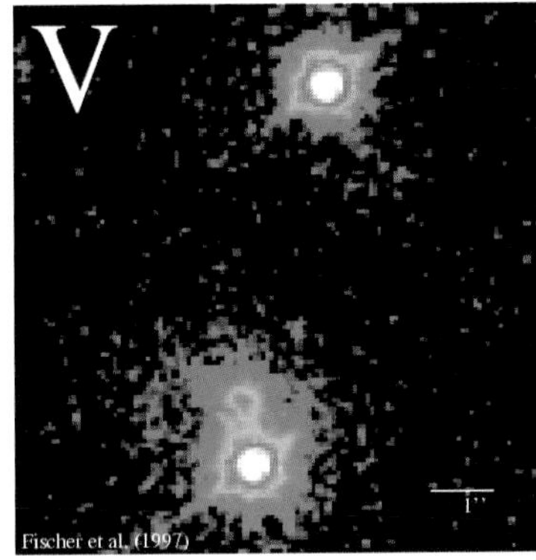

Figure 7.29 1.6 μm image (H-band) and 0.55 μm image (V-band) of the gravitationally-lensed quasar QSO 0957+561. Both images are shown in false colours. The lensing galaxy is much redder than the quasar images, i.e. it is relatively brighter in H than in V. Images from the CASTLES (CfA–Arizona Space Telescope LEns Survey) survey of gravitational lenses.

It turns out that only a small fraction of quasars (radio-loud or otherwise) are gravitationally lensed. Exhaustive efforts were made to follow up radio sources with the UK's MERLIN array to make high-resolution images to find new lenses, by the JVAS (Jodrell/VLA Astrometric Survey) project and later the CLASS (Cosmic Lens All-Sky Survey) project. A total of 16 503 radio sources were surveyed by MERLIN, of which only 22 were found to be strong gravitational lenses.

Meanwhile, other gravitational lenses were being discovered serendipitously. The Einstein Cross, also known more prosaically as QSO 2237+030, was discovered in a spectroscopic redshift survey of nearby galaxies. Figure 7.30 shows an optical image of this lens system. Such systems have become known as 'quadruple lenses' or 'quad lenses', though by the odd-number theorem (Section 7.6) there must be a fifth demagnified image near the centre.

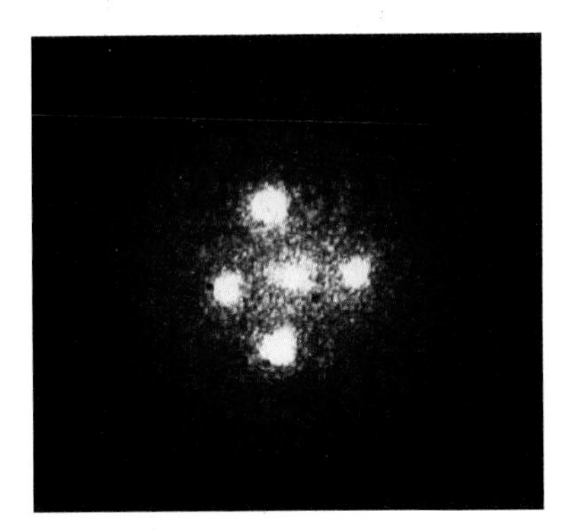

Figure 7.30 Deconvolved image of the Einstein Cross. There are four images of a background quasar at redshift $z = 1.695$, surrounding the core of a foreground galaxy at redshift $z = 0.0394$. The fifth feature at the centre is the core of the foreground galaxy, rather than a fifth image.

The steeper the slope of the number counts or luminosity function, the larger the magnification bias (Section 7.3). For this reason, the search for gravitationally-lensed quasars focused on high-luminosity quasars. One of the first to be discovered in this way was the Cloverleaf lens, so named because it's another quad lens. There is an ongoing search for lenses among the $\simeq 100\,000$ spectroscopically-confirmed quasars in the SLOAN survey (Chapter 3), known as the SLOAN Quasar Lens Survey (SQLS): candidate lenses are selected in the low-resolution ($\simeq 1.3''$ seeing) SLOAN imaging on the basis of morphology and colour, then candidates are followed up at higher angular resolution with other telescopes. At the time of writing, this project has uncovered 32 new lensed quasars.

The SLOAN survey has also been the source of another large catalogue of lenses. The SLOAN Lens ACS survey (SLACS) has found 131 galaxy–galaxy lenses by searching the SLOAN spectra for an absorption-dominated redshift combined with nebular emission lines (e.g. [O II] 372.7 nm or [O III] 500.7 nm) at another, higher, redshift in the same spectrum. These lens candidates were followed up with high-resolution imaging from the HST Advanced Camera for Surveys (ACS). Figure 7.31 shows some of the beautiful lens systems from this survey. The mass profile implied by these lenses is approximately isothermal (Section 7.4), but on average the mass profile is not the same as the light profile. The mass profile does not seem to have evolved since $z = 1$. Most of these lenses are elliptical galaxies, because ellipticals tend to be massive galaxies, so their cross section for lensing is higher. Typical Einstein radii are about an arcsecond, with lens masses roughly in the range 10^{10}–$10^{12}\,M_{\odot}$. The SLACS lenses also follow a fundamental plane (Chapter 3) that is consistent with the local fundamental plane once luminosity evolution is accounted for.

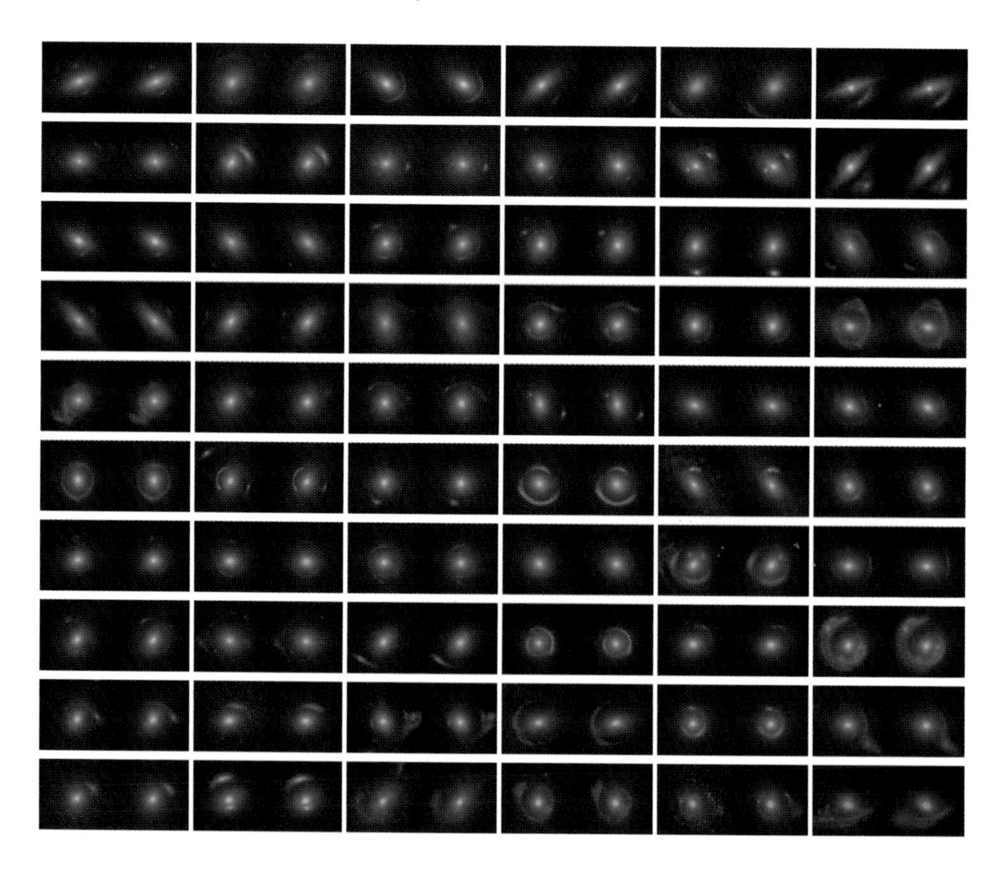

Figure 7.31 Sixty gravitational lenses from the SLACS survey from the HST I-band (814 nm) ACS imaging. Although the HST imaging is intrinsically monochrome, the colours of the foreground lenses in these images have been set using the SLOAN g–r colours, while the background lensed galaxies have been artificially enhanced in blue. In each pair, the left-hand panel is the observed data, while the right-hand panel is the mathematical model used to describe that lens.

Submm-wave surveys also have steep number counts (Chapter 5), particularly at bright fluxes, so bright submm-wave galaxies should be more prone to magnification bias (Section 7.3). At the time of writing, there are two forthcoming surveys that may find many new lenses: the SCUBA-2 All-Sky Survey (SASSy) and the Herschel ATLAS key project (Astrophysical Terahertz Large Area Survey). Both projects aim to scan the sky quickly to a shallow sensitivity in order to find the rare bright objects that may be lensed. Nearby galaxies that make up the Euclidean slope of the counts will probably be easily excludable by their cross-identifications with obvious nearby galaxies in optical surveys such as SLOAN. Similarly, radio-loud active galaxies will have obvious cross-identifications in radio surveys. As a result, the gravitational lens selection efficiency is expected to be around 100%, a far cry from the 22/16 503 lenses found in the early MERLIN surveys.

Another approach is to visually examine galaxies by hand. In an extraordinary solo effort, Neal Jackson from Jodrell Bank checked all 285 000 individual HST galaxy images from the COSMOS survey (see Section 7.9 and Chapter 4) that are brighter than an I-band magnitude of I = 25, finding two definite new strong lenses, one probable strong lens and a list of over a hundred candidates. Some progress has been made in using a computer to find gravitational lenses in high-resolution images automatically, but for the time being at least, human beings still perform better than software, so Neal Jackson's effort cannot easily be superseded.

See, for example, Marshall et al., 2009, *Astrophysical Journal*, **694**, 924.

Summary of Chapter 7

1. Gravitational lensing conserves surface brightness. Images that are magnified in area are therefore also magnified by the same factor in flux.
2. Gravitational lensing is achromatic, though the magnification can vary across an extended background object.
3. Gravitational lensing does *not* focus light rays (except in the case of a mass sheet with the critical mass density). In general a gravitational lens will have every variety of aberration in geometrical optics, except chromatic aberration (because gravitational lensing is achromatic).
4. The gravitational lens deflection by a point mass is exactly twice the Newtonian prediction:
$$\phi = \frac{4GM}{bc^2}. \qquad \text{(Eqn 7.2)}$$
5. The magnification of a point mass is always larger than 1. This does not violate energy conservation.
6. A point exactly behind a point mass is gravitationally lensed into a ring, the radius of which is known as the Einstein radius:
$$\theta_{\rm E} = \sqrt{\frac{4GM}{c^2}\frac{D_{\rm LS}}{D_{\rm L}D_{\rm S}}}. \qquad \text{(Eqn 7.9)}$$
This radius is useful for many lens models.
7. We define the source plane as the plane of the background object, and the image plane as the plane of the lens, i.e. as it appears to the observer.
8. Non-singular lenses have an odd number of images, including a central (demagnified) image whose flux is dependent on the density profile of matter in the core of the lensing galaxy.
9. As a source moves in the source plane through a caustic, a new pair of images is created on the corresponding critical line in the image plane.
10. Magnification bias affects the source counts of background objects and modifies the observed distribution of magnifications.
11. Microlensing of stars can be used to search for dark matter in the halo of the Galaxy and a few nearby galaxies.

12. Microlensing of background quasars by stars in a lensing galaxy causes long timescale drifts (lasting in some cases several years) in the fluxes of individual images.
13. Gravitational lensing by the cores of galaxy clusters has been used to find high-redshift galaxies at many wavelengths, in some cases deeper than the blank-field confusion limit. Although the fluxes are magnified, the comoving volume sampled decreases.
14. Gravitational lensing by galaxy clusters can also be used to map the distribution of dark matter in the clusters.
15. Weak lensing by the large-scale structure of the Universe is sensitive to the matter power spectrum in the linear regime. If the systematic errors can be well-characterized, the prospects are excellent for using this to constrain the dark energy equation of state.
16. The inverse magnification tensor can be used to calculate magnifications, parities, convergences and shears of images.

Further reading

- Adobe Photoshop currently has a downloadable gravitational lensing plug-in.
- More detail on the Schwarzschild solution and the gravitational lensing in Eddington's solar eclipse experiment that confirmed Einstein's theory of general relativity can be found in Lambourne, R., 2010, *Theoretical Cosmology*, Cambridge University Press.
- For more information on lensing theory, see Schneider, P., Ehlers, J. and Falco, E.E., 1992, *Gravitational Lenses*, Springer.
- Alternatively, see Narayan, R. and Bartelmann, M., 1995, 'Lectures on gravitational lensing', in Dekel, A. and Ostriker, J.P. (eds) *Formation of Structure in the Universe*, Proceedings of the 1995 Jerusalem Winter School, Cambridge University Press (available at astro-ph/9606001). We have followed many of the same formalisms and arguments in this chapter.
- For more recent and lengthy reviews of observational lensing, see Meylan, G., Jetzer, P. and North, P. (eds) *Gravitational Lensing: Strong, Weak and Micro*, Saas-Fee Advanced Course 33, Springer. The four lectures are 'Introduction to gravitational lensing and cosmology' by Peter Schneider, 'Strong gravitational lensing' by Christopher S. Kochanek, 'Weak gravitational lensing' by Peter Schneider (available at astro-ph/0407232), and 'Gravitational microlensing' by Joachim Wambsganss.
- More information on exoplanet discovery can be found in Haswell, C.A., 2010, *Transiting Exoplanets*, Cambridge University Press.
- Mellier, A., 1999, 'Probing the Universe with weak lensing', *Annual Review of Astronomy and Astrophysics*, **37**, 127.
- Brainerd, T.G., 'Constraint on field galaxy haloes from weak lensing and satellite dynamics', invited review in Allen, R.E., Nanopoulos, D.V. and Pope, C.N. (eds) *The New Cosmology*, AIP Conference Proceedings vol. 743, available at astro-ph/0411244.
- Albrecht, A. et al., 2006, *Report of the Dark Energy Task Force*, available at astro-ph/0609591.

Chapter 8 The intervening Universe

Birth: the first and direst of all disasters.

Ambrose Bierce

Introduction

After the CMB, what made the first light in the Universe — early stars, or black hole accretion? When were these first things created? In this final chapter, we shall explore what we know of the very earliest objects in the Universe. Much of this evidence comes from absorption lines, which we shall meet first. These absorbers also usefully track the cosmic consumption of gas in star formation, and give us a wonderful method of counting the total number of baryons in the observable Universe.

8.1 The Lyman α forest

One of the possible answers to Olbers' paradox in Chapter 1 was that the Universe is not transparent. As we've seen with the Hubble Deep Fields, the Universe turns out to be surprisingly transparent at optical wavelengths, so this is not the solution (and in any case, the energy absorbed by dust is still re-radiated in the far-infrared as thermal radiation). The answer to Olbers' profound paradox involves a combination of the finite age of the Universe and cosmological $(1+z)^4$ surface brightness dimming.

However, there are wavelength ranges where the Universe is not transparent, at least along certain lines of sight. Figure 8.1 shows a spectrum of the high-redshift quasar QSO 1422+23. The broad peak is the Lyman α emission line, caused by electrons jumping from the $n = 2$ to $n = 1$ quantized energy states in hydrogen (Figure 8.2).

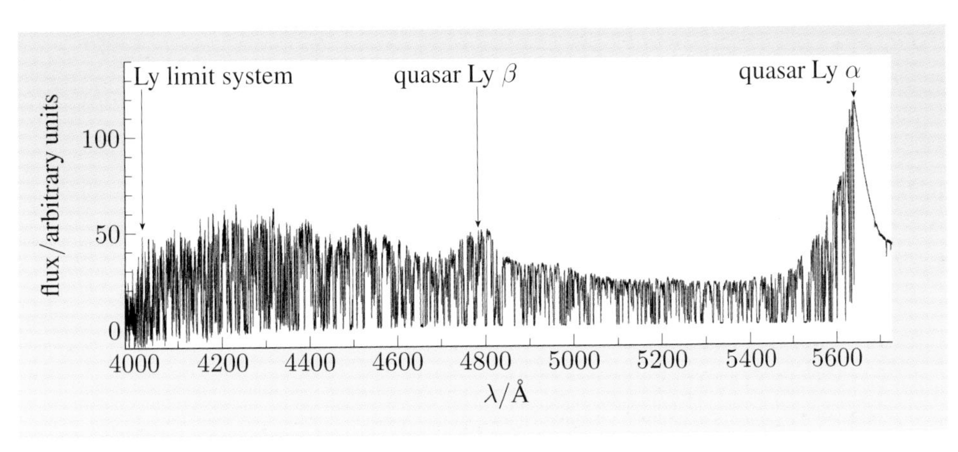

Figure 8.1 The Lyman α forest in the spectrum of a quasar.

- Why is the Lyman α line broad in quasars?

❍ The hydrogen gas is moving quickly close to the central supermassive black hole. The large velocity dispersion gives rise to large Doppler shifts, which in turn give rise to the large emission line width.

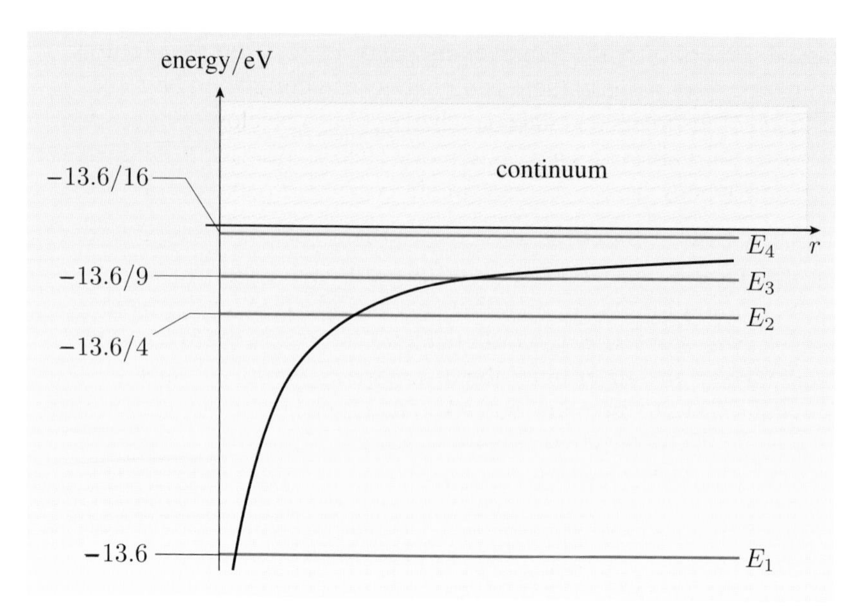

Figure 8.2 Hydrogen energy levels. The energy of an energy level is given by $E_n = -13.6\,\mathrm{eV}/n^2$. The potential energy from the nucleus is shown as a black curve.

Note that $n = 1 \to 2$ is absorption of a photon and a promotion of the election, while $n = 2 \to 1$ is emission of a photon and demotion of the electron.

On the left-hand side of the Lyman α line, i.e. at shorter wavelengths, the spectrum seems much noisier. This is not noise; it is the Lyman α absorption lines from neutral hydrogen clouds between us and the quasar. The absorbing atoms each have an electron in the $n = 1$ energy level that is promoted to $n = 2$ using an absorbed photon's energy. Since the clouds are at lower redshift than the quasar, their Lyman α absorption is less redshifted, so appears at shorter wavelengths. Figure 8.3 shows this schematically. (Note that 'Lyman α' is sometimes abbreviated as Ly α.)

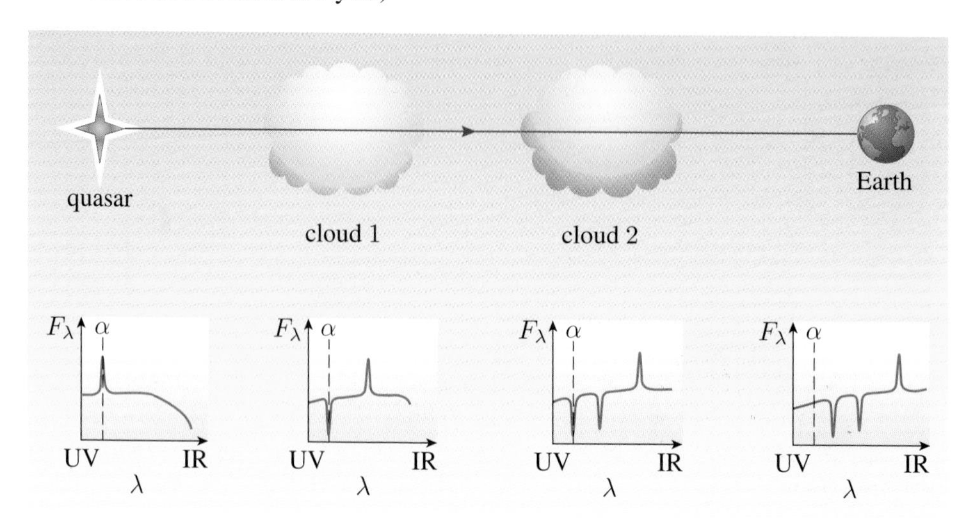

Figure 8.3 Schematic representation of the Lyman α forest.

These Lyman α clouds, collectively called the **Lyman α forest**, are another of the few ways that astronomers can view the *non*-luminous Universe. From a knowledge of the **cross section** of the absorption (see the box in Chapter 3), which one can measure in a laboratory, one can calculate the projected number of hydrogen atoms along this line of sight, per unit area. This is known as the **column density** of the absorption, $N_{\mathrm{H\,I}}$.

What are these absorbers? Are they intervening galaxies, for example? It turns out that they are not — or at least, there is no one-to-one correlation between intervening Lyman α absorbers and galaxies that appear to be close in projection on the sky. Low-redshift absorbers are more likely close to gas-rich local galaxies; nonetheless, galaxy haloes cannot account for all Lyman α clouds.

Instead, it seems that these absorbers are clumps of intergalactic material that (for the most part) have not yet condensed to form galaxies. They would be undetectable, were it not for the fact that they absorb light from background quasars. The distribution of these primordial clumps is not subject to most of the complicated physics that determines the distribution of galaxies, such as non-linear gravitational collapse and feedback. The Lyman α forest can therefore be used as a tracer of the underlying matter distribution. This is very useful for testing cosmological models, as we shall see in the next section.

It may surprise you to read that Lyman α clouds exist even in the present-day Universe. It was once imagined that galaxy formation was something that happened only early in the history of the Universe, but more recently galaxy formation has been seen as an ongoing process. There are even nearby galaxies that seem to have formed all their stars very recently, such as the galaxy I Zw 18 (Figure 8.4). We imagine that a pre-existing puddle of neutral hydrogen has been disturbed or interacted with in some way that has triggered the formation of stars within it. It's not clear what the triggers were for I Zw 18, however.

Figure 8.4 The dwarf galaxy I Zw 18 (read as 'one Zwicky eighteen').

Exercise 8.1 When a Lyman α photon is absorbed by a Lyman α forest cloud, the hydrogen atom is left in an excited state. This state isn't stable, and energy is eventually re-emitted as another Lyman α photon. Why doesn't this fill out the absorption line?

Exercise 8.2 Suppose that you have a spherical cloud of neutral hydrogen with a radius of 1 Mpc and a density of one hydrogen atom per cubic cm. Calculate the column density as seen through the centre of the cloud. ■

Lyman α absorption ($n = 1 \rightarrow 2$) at 121.6 nm in the rest frame of the hydrogen atom isn't the only absorption in these clouds. A sufficiently energetic photon could be absorbed using the $n = 1 \rightarrow 3$ transition, or $n = 1 \rightarrow 4$, and so on.

● What is the rest-frame wavelength of the $n = 1 \rightarrow 3$ absorption?

○ The wavelength of the $n = 1 \rightarrow 2$ transition is 121.6 nm, which corresponds to an energy difference of $(-13.6/2^2 - (-13.6/1^2))$ eV, or 10.2 eV. The $n = 1 \rightarrow 3$ transition is similarly $(-13.6/3^2 - (-13.6/1^2))\,\mathrm{eV} = 12.0\dot{8}\,\mathrm{eV}$. Since $E = h\nu = hc/\lambda$ (where h is Planck's constant, c is the speed of light, and λ is the wavelength),

$$\frac{E(n = 1 \rightarrow 3)}{E(n = 1 \rightarrow 2)} = \frac{\lambda(n = 1 \rightarrow 2)}{\lambda(n = 1 \rightarrow 3)},$$

so $\lambda(n = 1 \rightarrow 3) = 121.6\,\mathrm{nm} \times 10.2/12.0\dot{8} = 102.6\,\mathrm{nm}$.

If the photons are sufficiently energetic, they may ionize the hydrogen atoms entirely. This can happen only at energies > 13.6 eV, or rest-frame wavelengths below 91.2 nm. In the spectrum in Figure 8.1, the quasar's Lyman β emission line ($n = 3 \rightarrow 1$) is marked. At observed wavelengths shorter than those for this emission line, the spectrum will contain absorption from both intervening Lyman α and Lyman β lines. At wavelengths below $91.2\,\mathrm{nm} \times (1 + z_{\mathrm{QSO}})$, where z_{QSO} is the quasar redshift, the atoms along the line of sight could be ionized completely, and the quasar spectrum is noticeably suppressed. This is known as **Lyman limit absorption**.

Just because a photon *can* excite a Lyman β transition, doesn't mean that it *must*. The cross section for Lyman β absorption turns out to be a factor of about five smaller than that for Lyman α absorption. Similarly, the higher transitions of $n = 1 \rightarrow 4$, $n = 1 \rightarrow 5$, and so on are increasingly less probable. Only intervening neutral hydrogen clouds with the sufficiently high column densities cause Lyman limit absorption ($N_{\mathrm{H\,I}} > 10^{21}\,\mathrm{m}^{-2}$ or so — see Exercise 8.4). This is partly why the suppression in the quasar spectrum in Figure 8.1 doesn't start immediately below $91.2\,\mathrm{nm} \times (1 + z_{\mathrm{QSO}})$ — the light from a distant quasar doesn't necessarily immediately encounter a sufficiently opaque absorber.

Exercise 8.3 The Lyman series of hydrogen absorption lines is $n = 1 \rightarrow 2$ (Lyman α), $n = 1 \rightarrow 3$ (Lyman β), $n = 1 \rightarrow 4$ (Lyman γ), and so on. There is also a Balmer series of hydrogen absorption lines: $n = 2 \rightarrow 3$ is known as Hα, $n = 2 \rightarrow 4$ is Hβ, $n = 2 \rightarrow 5$ is Hγ, and so on. But why is there no Hα forest in the spectra of quasars? ■

When are clouds thick enough to be opaque? To answer this, and quantify what we mean by 'opaque', we can write the fraction of radiation passing through the cloud as $\mathrm{e}^{-\tau}$, where τ is known as the **optical depth**. This varies from 0 to 1, and we can regard $\tau > 1$ as opaque (i.e. $< \mathrm{e}^{-1} \simeq 37\%$ of the photons pass through). The optical depth of a Lyman α cloud to ionizing photons is

$$\tau = N_{\mathrm{H\,I}} \frac{\int (\sigma J_\nu/(h\nu))\,\mathrm{d}\nu}{\int (J_\nu/(h\nu))\,\mathrm{d}\nu}, \tag{8.1}$$

where J_ν is the spectrum of the ionizing background, and σ is the cross section for hydrogen ionization. The factor of $h\nu$ converts from energy flux J_ν to photon flux $J_\nu/(h\nu)$. The cross section is $\sigma = 7.88 \times 10^{-22}\,\mathrm{m}^{-2}$ at the Lyman limit, and scales as ν^{-3} at higher frequencies.

Exercise 8.4 Assuming that $J_\nu \propto \nu^{-\alpha}$, show that the optical depth is $\tau > 1$ when $N_{\mathrm{H\,I}} > 1.3\,((\alpha + 3)/\alpha) \times 10^{21}\,\mathrm{m}^{-2}$. This is known as **self-shielding**. ■

For reference, quasar spectra have α around 0.5–1 in the ultraviolet, while galaxies have redder spectra. Taking account of intervening absorption, a Lyman α cloud might experience an ambient light spectrum of $\alpha \simeq 2$, depending on redshift.

8.2 Comparison with cosmological simulations

The density fluctuations $\delta\rho/\rho$ in the Lyman α forest are not too far from the linear regime, so the Lyman α forest is not affected by virialization or dissipation, nor by the complicated biasing physics that affects galaxy clustering. Having said that, the Lyman α forest *is* sensitive to the thermal state of the intergalactic medium (by affecting absorption line widths), so there have been many attempts to numerically simulate the evolving large-scale structure of the Lyman α forest and its ionization history using N-body codes with hydrodynamic inputs and approximations. These COBE-normalized CDM simulations reproduce both the shape and the amplitude of the spatial power spectrum (due to clustering) of the Lyman α forest (Figure 4.1). It's remarkable that Inflation+CDM is more or less enough to reproduce the spatial clustering of the Lyman α forest, since the latter covers a very different part of the Universe's history, and is based on very different physical phenomena. The clustering of the Lyman α forest can be combined with the CMB and galaxy clustering to constrain the power spectrum from $\sim$1 Mpc scales right up to the horizon scale.

8.3 Ω_b and the cosmic deuterium abundance

We've already seen in Chapter 2 how the conditions in the early Universe were hot and dense enough for nuclear reactions, and that the hot Big Bang model makes very specific predictions for the resulting abundances of elements. The nuclear reaction rates don't depend linearly on the density of baryons: for example, a hypothetical universe with double the baryon density Ω_b of our Universe would produce much more than double the lithium (see Figure 2.3). $\Omega_{b,0}\,h^2$ can be determined from the Doppler peaks in the CMB, as we've seen, but its measurement is not independent of other parameters that need to be determined from the same data, such as the primordial spectral index n_s or the Thomson scattering optical depth to reionization. If we can measure the elemental and isotopic abundances in primordial (or nearly primordial) gas, we can derive an independent measurement of the baryon density of the Universe.

Deuterium, also called heavy hydrogen, is the reaction product that depends most sensitively on $\Omega_{b,0}\,h^2$. This isotope of hydrogen has a proton and a neutron as its nucleus. This subtly changes the electronic energy levels. (In the Bohr model, the electron's energy levels change because the increased mass of the nucleus subtly alters the atom's centre of mass). If we have a strong Lyman α absorber, we might hope to detect the corresponding deuterium absorption lines, slightly shifted in wavelength from the hydrogen absorption. From this, one could calculate the deuterium abundance relative to hydrogen, often written as $[D/H]$, and then derive $\Omega_{b,0}\,h^2$ independently of the CMB.

This is a difficult experiment that requires a confirmed low-metallicity Lyman α absorber. Figure 8.5 shows a Lyman α absorber in the quasar Q 0913+072, which is strong enough to be almost completely opaque in the centre.

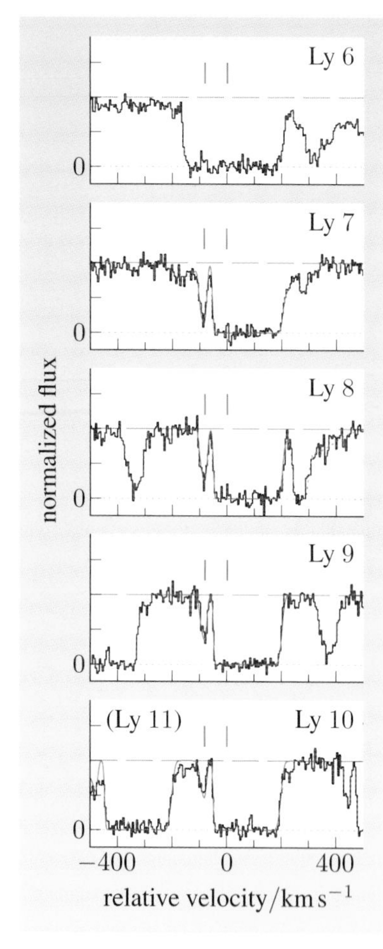

Figure 8.6 Hydrogen and deuterium absorption in the quasar QSO 0913+072. The spectrum has been divided by the expected quasar flux, so a flux of 1.0 means no absorption. In this notation, Lyman α is Ly1, Lyman β is Ly2, and so on. The x-axis units are km s^{-1} relative to the quasar hydrogen Lyman series absorption. The red lines mark the expected positions of the hydrogen and deuterium absorption. Ly11 is off to the left of the bottom panel.

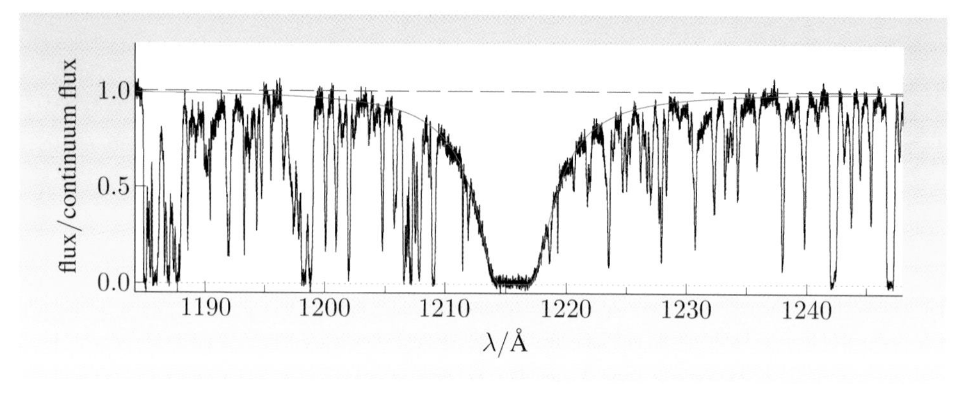

Figure 8.5 Damped Lyman α system in the quasar QSO 0913+072. The spectrum has been divided by the expected quasar flux, so a flux of 1.0 means no absorption. The red line shows the best-fit Voigt profile (see Section 8.5).

This absorber was chosen to have very few associated absorption lines from heavier elements. Higher Lyman transitions have lower optical depth, and in Figure 8.6 the companion deuterium Lyman lines can be seen, slightly blueward of the hydrogen absorption. The [D/H] abundance depends on the hydrogen column density determined from the Lyman α profile in Figure 8.5, which is difficult to fit to given the presence of other intervening Lyman α absorbers. (This is the principal source of systematic uncertainties.) The deuterium abundance is $\log_{10}[\mathrm{D/H}] = -4.56 \pm 0.04$, which when combined with other similar measurements gives $\Omega_{\mathrm{b},0}\, h^2 = 0.0213 \pm 0.0010$. Later in this chapter we shall see how little of this baryonic content of the Universe is stars and planets, and how much is still in its primordial state.

Figure 8.7 compares the WMAP cosmological parameter constraints with the constraint from the [D/H] abundance. This figure shows that combining the WMAP data with the measured baryon density requires that $n_\mathrm{s} < 1$. Several lines of evidence now appear to disfavour an $n_\mathrm{s} = 1$ scale-invariant spectral index. In inflationary models, n_s depends on the shape of the inflation potential. Is this measurement a hint of the new physics of the inflation potential? Inflation models with $n_\mathrm{s} \neq 1$ also predict a gravitational wave background that might eventually be detectable directly in future gravitational wave observatories, or whose effects may be measurable in the polarized CMB with the recently launched Planck space telescope or other later CMB missions. This will be a critical consistency test for inflation.

8.4 The column density distribution

We've seen that the Universe is awash with primordial hydrogen that follows the filaments and clumps of the underlying matter distribution. But is most of this hydrogen still lurking in wispy filaments, or is it already in galaxy-sized clumps waiting to be turned into galaxies?

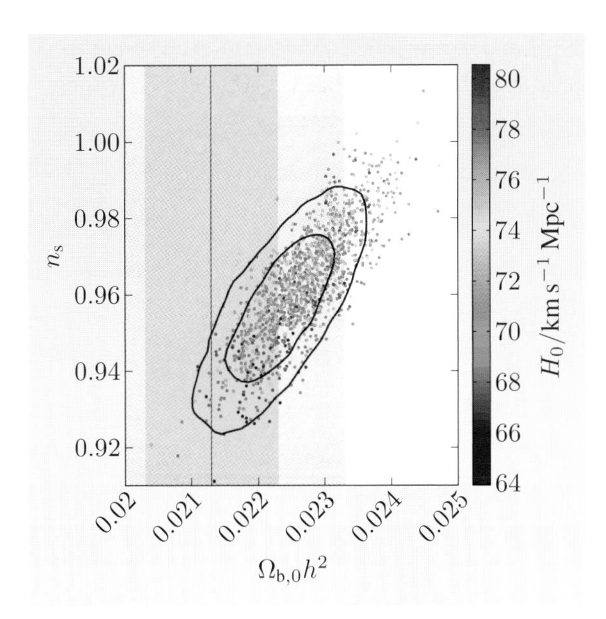

gure 8.7 Constraints on the baryon density of the Universe, $\Omega_{b,0}\,h^2$, and on the imordial spectral index of scalar density perturbations n_s. The points sample the owed distribution from the WMAP data, coloured according to the Hubble rameter H_0 in units $\mathrm{km\,s^{-1}\,Mpc^{-1}}$. The shaded regions are the 1σ and 2σ bounds on $_{,0}\,h^2$ based on the deuterium abundance. (1σ means that there is an $\simeq 68\%$ chance of e true value lying in that range; 2σ corresponds to 95%.) The curves are the 1σ and 2σ nstraints from combining the WMAP measurements with the deuterium abundance.

In Chapter 4, we used the luminosity function $\phi(L)$ of galaxies to find which galaxies contribute most of the luminosity in the Universe: they were around the peak of the $L\,\phi(L)$ distribution. In Chapter 5, we also used the source counts $\mathrm{d}N/\mathrm{d}S$ to find which galaxies dominate the extragalactic background light: they were around the peak of the $S\,\mathrm{d}N/\mathrm{d}S$ distribution. We shall use a similar trick with the column density distribution to find out where most of the neutral hydrogen is in the Universe.

The numbers of Lyman α clouds change strongly with cosmic time. Figure 8.8 shows the spectrum of a quasar at low redshift. Comparison with Figure 8.1 shows that the low-redshift quasar clearly has far fewer Lyman α absorbers than the high-redshift quasar. It's tempting to suppose that this is exactly the emptying out of the voids, and filling up of the overdensities, that the cosmological simulations predict. However, that supposes that we're sampling the same comoving volume in the two spectra. For example, could the 1250–1350 Å observed wavelength range in the low-redshift quasar just be sampling much less volume than the 4600–4700 Å observed wavelength range in the high-redshift quasar? This might explain why there are fewer Lyman α lines at low redshift. To find out, we shall calculate the number of absorbers that a photon would encounter along its travel from the quasar to us.

Unfortunately there's an annoying collision of notation: $N_{\mathrm{H\,I}}$ is conventionally used to mean column density, while N is conventionally used to mean numbers in source counts $\mathrm{d}N/\mathrm{d}S$. The column density distribution (i.e. the number of absorbers per unit column density) would then be $\mathrm{d}N/\mathrm{d}N_{\mathrm{H\,I}}$. To avoid this clumsy notation, it's conventional to use $\mathcal{N}$ to mean the number of absorbers.

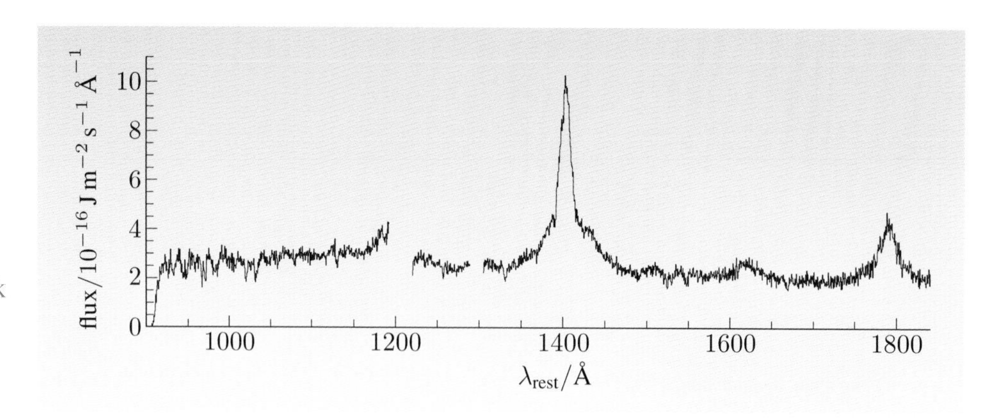

Figure 8.8 The neighbourhood of the Lyman α emission line of the $z = 0.158$ quasar 3C273. Compare the lack of Lyman α forest lines with the high-redshift quasar in Figure 8.1.

The number of absorbers that a photon might encounter will be proportional to the path length that it travels, $d\ell$, and proportional to the density of absorbers ρ, and to the average geometrical cross section A of any single cloud. (Don't confuse this with the absorption cross section σ of a single atom.) Since $\rho = n_{\rm co}(1+z)^3$, where $n_{\rm co}$ is the comoving density of the absorbers, and $d\ell = c\,dt$, we can write the number of absorbers encountered by a photon in a cosmic time interval $t \to t + dt$ as $d\mathcal{N} = n_{\rm co}(1+z)^3 Ac\,dt$. It'll be useful to have the number of absorbers per unit column density per unit redshift, which we can write as

We've written $d^2\mathcal{N}$ as a double differential, which it is, but be warned that some texts use $d\mathcal{N}$ when referring to $d^2\mathcal{N}$.

$$d^2\mathcal{N} = n_{\rm co}(N_{\rm H\,I}, z)\, A(N_{\rm H\,I}, z)\, (1+z)^3 c \left|\frac{dt}{dz}\right| dN_{\rm H\,I}\, dz. \tag{8.2}$$

Exercise 8.5 Suppose that the absorbers have a constant *comoving* space density, and constant *proper* sizes and cross sections. Show that the number of absorbers per unit X would be constant, where $dX/dz = (1+z)^2 H_0/H(z)$. ■

The quantity $X(z)$ in Exercise 8.5 is sometimes known as the **absorption distance**.

It turns out that the number of lines $\mathcal{N}$ per unit absorption distance still evolves strongly:

$$\frac{d\mathcal{N}}{dX} = \int \frac{d^2\mathcal{N}}{dX\, dN_{\rm H\,I}}\, dN_{\rm H\,I} \propto (1+z)^{2.47\pm0.18} \tag{8.3}$$

at $1.5 < z < 4$ and column densities $> 10^{18}\,\rm m^{-2}$. So the increase in the number of lines at high redshifts is not only about the increase in comoving volume sampled.

Figure 8.9 shows the column density distribution of Lyman α absorbers at $z > 1.5$. We've used $d^2\mathcal{N}/dX\, dN_{\rm H\,I}$ to mean the number of absorbers per unit absorption distance X, per unit column density $N_{\rm H\,I}$. It's also conventional to use the symbol f to mean $d^2\mathcal{N}/dX\, dN_{\rm H\,I}$. From Equation 8.2, we have that

$$f(N_{\rm H\,I}, X) = \frac{d^2\mathcal{N}}{dX\, dN_{\rm H\,I}} = n_{\rm co}\, A \frac{c}{H_0}, \tag{8.4}$$

where $n_{\rm co}$ and A are both functions of $N_{\rm H\,I}$ and redshift. In practice, in a redshift interval Δz corresponding to an absorption distance $\Delta X(z)$, we would count the number of absorbers $\Delta\mathcal{N}$ in a column density range $\Delta N_{\rm H\,I}$, to estimate $f(N_{\rm H\,I}, z) \simeq (1/\Delta X)\, \Delta\mathcal{N}/\Delta N_{\rm H\,I}$.

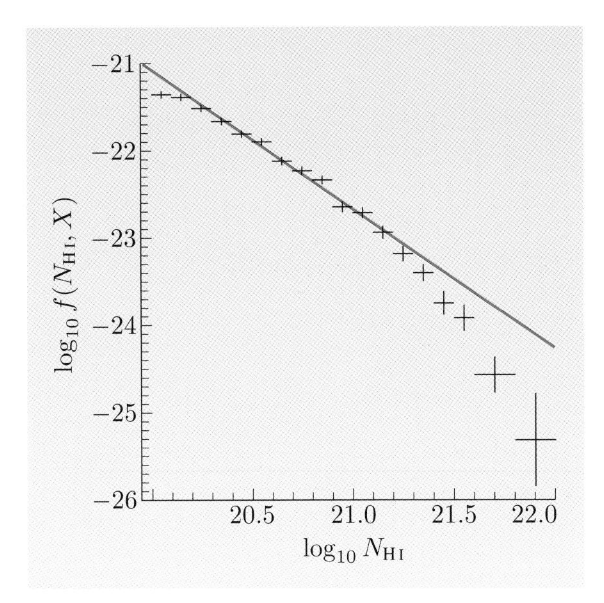

Figure 8.9 The column density distribution of Lyman α clouds with column densities measured in cm^{-2}. Also shown is the best-fit single power-law model; the data, however, show a clear steepening at the high column density end, and clearly depart from the simple single power-law model.

The dotted line in Figure 8.9 is $f \propto N_{\mathrm{H\,I}}^{-1.3}$. The total column density in an absorption distance ΔX, assuming that power-law model, is just

$$\begin{aligned} N_{\mathrm{HI,total}}\,\Delta X &= \Delta X \int N_{\mathrm{H\,I}}\, f(N_{\mathrm{H\,I}})\,\mathrm{d}N_{\mathrm{H\,I}} \\ &= \Delta X \int N_{\mathrm{H\,I}} \times N_{\mathrm{H\,I}}^{-1.3}\,\mathrm{d}N_{\mathrm{H\,I}} \propto N_{\mathrm{H\,I}}^{0.7}, \end{aligned} \tag{8.5}$$

which diverges as $N_{\mathrm{H\,I}}$ tends to infinity. There is evidence that the power-law index of the column density distribution steepens at the highest measured column densities, as it must for $N_{\mathrm{HI,total}}$ to remain finite. Therefore most of the mass of neutral hydrogen in the Universe at $1.5 < z < 5$ was concentrated in the higher column density absorbers. We shall find out more about these in the next section.

8.5 Damped Lyman α systems

The strongest Lyman α absorbers may have played a key role in the birth of galaxies. The objects with the biggest neutral hydrogen column densities locally are spiral discs. Could the strongest high-redshift Lyman α absorbers be their primordial precursors? A primordial galaxy that's gravitationally bound, but which is being seen before it has formed stars, might look very much like a deep Lyman α absorber.

The incidence of strong absorbers evolves intriguingly. Figure 8.10 shows the line density $\mathrm{d}\mathcal{N}/\mathrm{d}X$ for absorbers with $N_{\mathrm{H\,I}} > 2 \times 10^{24}\,\mathrm{m}^{-2}$. There is no evidence for evolution from $z = 2$ to the present, but the incidence of strong absorbers dropped by a factor of 2 from $z = 4$ to $z = 2$. Could this reflect primordial neutral hydrogen clumps merging, which would reduce n_{co} with time in Equation 8.2? Or could this represent the consumption of gas by star formation, which would reduce A with time? Simulations favour the latter option, but it's frustratingly impossible to tell from these data. We can only make inferences about the product $n_{\mathrm{co}} \times A$.

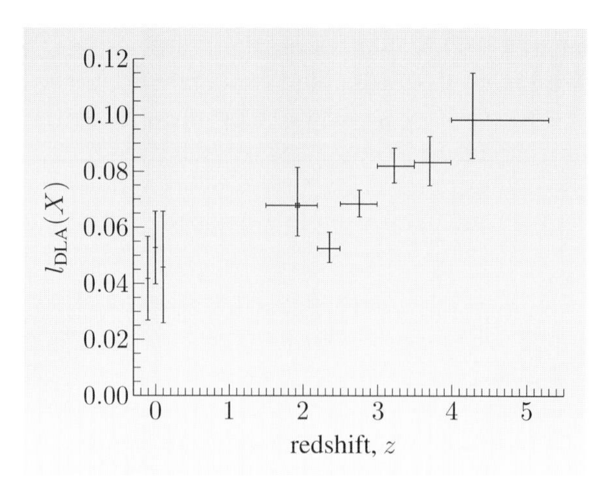

Figure 8.10 The evolution of the line density of damped Lyman α clouds.

The highest column density absorbers, such as the one in Figure 8.5, are almost completely opaque in their centres. The shape of the absorption profile can be predicted from quantum mechanical considerations to follow a Lorentzian profile

$$L(\lambda) \propto \frac{\alpha}{(\lambda - \lambda_0)^2 + \alpha^2},$$

where λ_0 is the line centre, and α is some constant.

Hendrik Antoon Lorentz, who derived this theoretical profile for absorption and emission lines, is also famous today for his contributions to special relativity.

Why are absorption lines not infinitely narrow? The cross section for absorption near — but not equal to — the transition frequency $\nu = E(1 \rightarrow 2)/h$ is not zero. Heisenberg's uncertainty principle makes the required energy E uncertain by $\Delta E\, t_{1/2} \simeq h/(2\pi)$, where $t_{1/2}$ is the half-life of the excited state. The frequency width $\Delta\nu$ would therefore give $h\,\Delta\nu = h/(2\pi t_{1/2})$. Since $\Delta\lambda \simeq \Delta\nu \times |{\rm d}\lambda/{\rm d}\nu|$, and $\nu = c/\lambda$ where c is the speed of light, the wavelength width is

$$\Delta\lambda = \frac{\lambda^2}{2\pi c}\frac{1}{t_{1/2}}. \tag{8.6}$$

In general, if the final state is not stable and has its own half-life $t_{\rm final}$, the wavelength width is

$$\Delta\lambda = \frac{\lambda^2}{2\pi c}\left(\frac{1}{t_{\rm initial}} + \frac{1}{t_{\rm final}}\right), \tag{8.7}$$

where $t_{\rm initial}$ is the half-life of the initial state. (This holds whether the transition is emission or absorption.)

This gives us a width, but not the detailed shape of the Lorentzian profile. Deriving the Lorentzian profile is beyond the scope of this text, but Lorentz himself derived it prior to the invention (or discovery) of quantum mechanics, using a classical argument: he imagined atoms as oscillators, which are being forced by the external electromagnetic field, and which lose energy via radiation (which in quantum mechanics is spontaneous emission). This last effect appears as a damping term in the equation of motion, and is responsible for the α term in the Lorentzian profile expression above. Expressions similar to the Lorentzian profile occur in physical systems with damped, forced harmonic oscillation.

There are also larger-scale reasons why the absorption is not infinitely narrow. The gas will have some motion (e.g. turbulence or coherent rotation), which will cause Doppler shifts. There may also be Doppler shifts from the atoms' thermal motion, known as **thermal broadening**. (A third possibility, relevant in stars but not at the expected densities of Lyman α systems, is **pressure broadening**: the presence of nearby atoms affects the photon emission of any particular atom.) The Doppler broadening in Lyman α clouds is generally treated as a Gaussian distribution, because a Gaussian form occurs in the expected Maxwell–Boltzmann thermal velocity distribution: $\Pr(v) \propto \mathrm{e}^{-mv^2/(2kT)}$.

To find the total effect on the absorption profile, we convolve this Gaussian distribution with the Lorentzian profile. The resulting curve is known as the **Voigt profile**. Figure 8.5 shows the best-fit Voigt profile for this damped Lyman α system. By analogy with the classical case, the shape of the wings of the profile depends on the damping term in the oscillator equation of motion. Since the centres of the profiles in these absorbers are essentially black (i.e. essentially completely opaque), these damped wings dominate the profile shape, which is why these Lyman α absorbers are known as 'damped'. This happens typically at column densities $> 10^{24}\,\mathrm{m}^{-2}$ or so.

The depth of the absorption can also be expressed as **equivalent width**, illustrated in Figure 8.11. This is defined by imagining another absorption line, which removes the same energy but is completely opaque and has a rectangular shape. The width of this line (W in Figure 8.11) is the equivalent width. Note that this is just a measure of the intensity of the absorption, and has nothing to do with velocity widths. Mathematically the equivalent width is

$$W = \int_{-\infty}^{\infty} \frac{C(\lambda) - S(\lambda)}{C(\lambda)}\,\mathrm{d}\lambda, \tag{8.8}$$

where $C(\lambda)$ is the continuum level without the absorption, and $S(\lambda)$ is the observed spectrum with the absorption. In terms of optical depth, equivalent width can be written as

$$W = \int_{-\infty}^{\infty} \left(1 - \mathrm{e}^{-\tau(\lambda)}\right) \mathrm{d}\lambda. \tag{8.9}$$

How can we use the observed equivalent widths to derive the column densities?

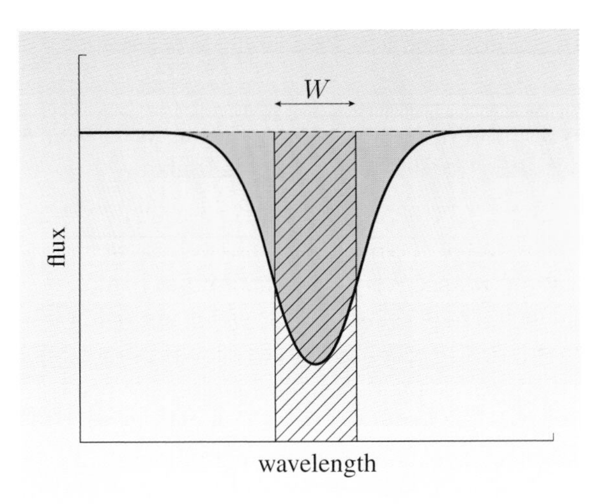

Figure 8.11 The equivalent width, marked as W, is the width of the box that has an area (hatched) the same as the area of the absorption line (in yellow).

Figure 8.12 shows the 'curve of growth' for damped Lyman α absorption, meaning a curve of how the width depends on the optical depth. The optical depth to ionizing photons τ is related to the column density:

$$\tau(\lambda) = \sigma(\lambda)\, N_{\mathrm{H\,I}}, \tag{8.10}$$

where σ is the cross section for absorption. The equivalent width increases linearly with optical depth: $W \propto \tau$ for small τ. This regime corresponds to overdensities of $\delta\rho/\rho \simeq 0$–15, corresponding to the linear or mildly non-linear regime of cosmological structure formation. Once τ is around unity, the absorber is essentially black, and there is little change to the equivalent width with increasing optical depth until column density is high enough for the damping wings to start affecting the equivalent width. Once $\tau > 10^5$ or so, the equivalent width increases as the square root of τ.

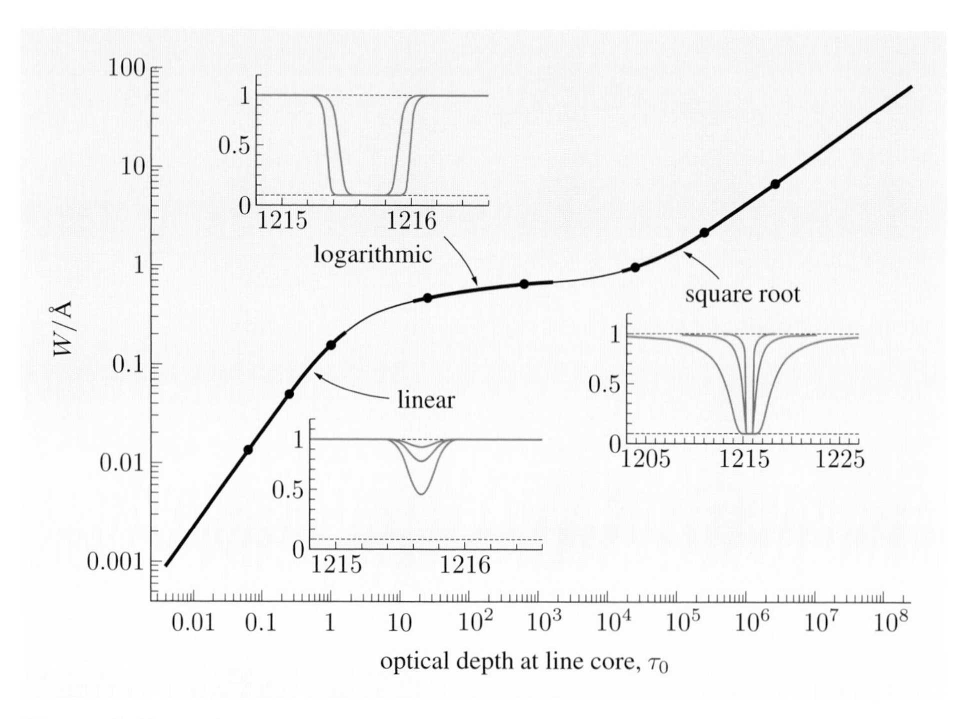

Figure 8.12 The variation of equivalent width with optical depth at the line core.

In Exercise 8.4 we found that Lyman α absorbers above the threshold for Lyman limit absorption are self-shielded, i.e. they are dense enough that ionizing radiation has an optical depth > 1 and does not penetrate the cloud. This implies that the gas in these higher column density absorbers must be mostly neutral, particularly in damped Lyman α systems. This cannot be said of the Lyman α forest in general, as we shall see later in this chapter.

In the nearby Universe, the objects with the biggest neutral hydrogen column densities are spiral discs, and this continues to about $z \simeq 1.6$. Could the higher-redshift damped Lyman α systems be the primordial progenitors of these spiral discs? Even if they are, how can we separate the faint light of these primordial galaxies from the glare of the background quasar?

One approach is to take high-resolution spectra to try to detect narrow emission lines from star formation in the galaxy causing the damped Lyman α system.

Astronomers have looked for Hα, redshifted into the infrared, or Lyman α in the centre of the damped absorption trough. Similarly, one can take an image with a narrow filter, chosen to cover the dark central region of the damped absorption trough (see Figure 8.13); such an image may also detect faint, narrow Lyman α emission from star formation. Only three damped Lyman α systems at $z > 1.6$ have any star formation detected so far using emission lines, though several have upper limits. It seems that we are seeing a key stage in the assembly of galaxies, before they are luminous.

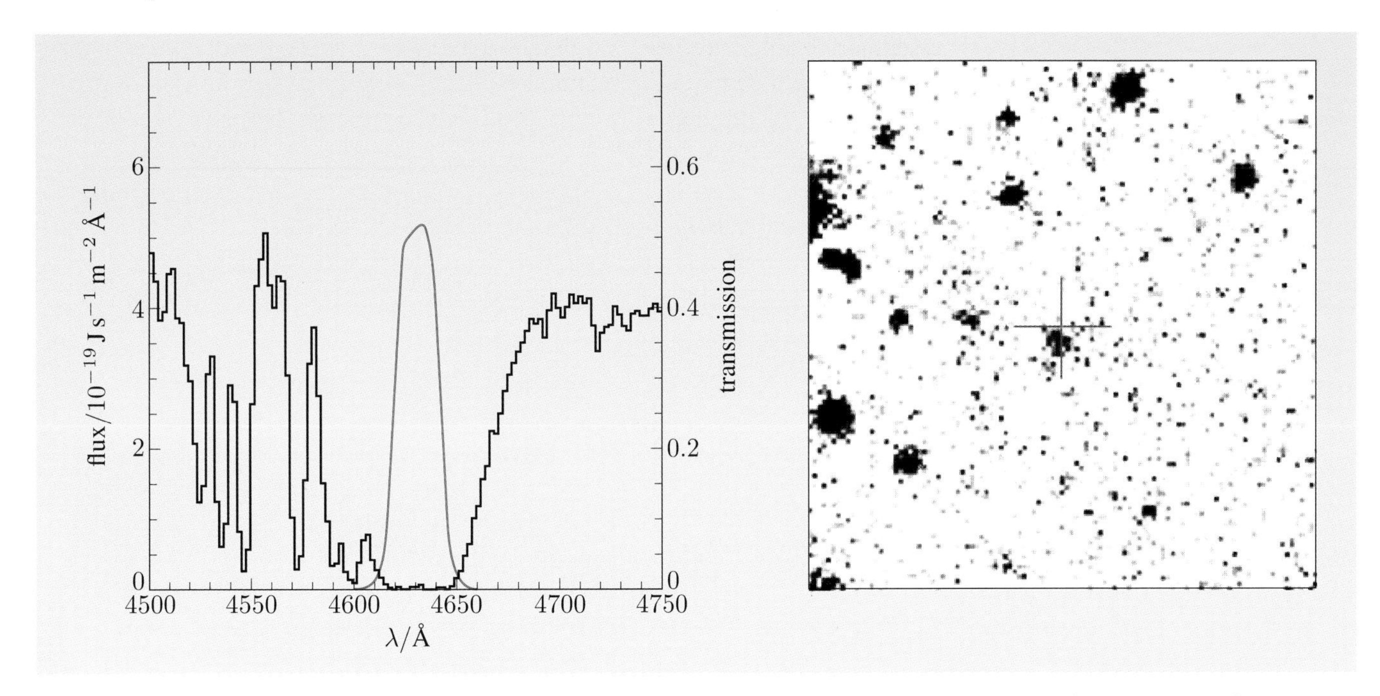

Figure 8.13 The left panel shows the transmission of a narrow-band filter, compared to the spectrum of a damped Lyman α system in the quasar PKS 0528-250. Light that passes through this filter should have little or no contribution from the background quasar. The right image ($65'' \times 65''$ in size) is taken through this narrow-band filter. The position of the quasar is marked as a red cross. Nearby, there is a galaxy that is ostensibly responsible for the damped Lyman α absorption.

Another possibility is to use other absorption lines in the quasar spectrum. If the interstellar medium of the galaxy causing the damped Lyman α absorption has been enriched by star formation, there should be metal absorption lines in the quasar spectrum, and these have been detected in many systems. Also, the C II* 133.57 nm absorption line has been argued to correlate well with the [C II] 158 μm emission line, which in turn is an indirect star formation rate indicator. From this it's possible to estimate the star formation rate in projection, in units of $M_\odot$ per year per kpc^2, in damped Lyman α systems. However, it turns out that the mean metal content of damped Lyman α systems is about 10 times lower than expected from their inferred cosmic star formation history! Could rapid star formation use up the neutral hydrogen, so damped systems don't stay damped and others take over? At these low star formation rates, the timescales are too slow for this to work. Could the metals be ejected from supernova-driven winds? This would disagree with observations of the metallicity of the intergalactic medium. It's not clear what the solution is, but some approaches that have the neutral gas spatially distinct in the absorbing galaxies from their active star forming regions may be consistent with the data. Whatever the solution, it's clear that damped Lyman α absorption gives us a unique window into otherwise invisible aspects of galaxy formation.

Exercise 8.6 There is some evidence that the highest column density damped Lyman α systems are more common in quasars with bright *apparent* magnitudes. What could cause this?

Exercise 8.7 How could one use observations of the background quasars to investigate the dust content of damped Lyman α systems? ■

8.6 The proximity effect

Lyman limit systems and damped Lyman α systems may be self-shielded, but ionizing photons will penetrate lower column density systems. The Lyman α absorption comes only from neutral hydrogen — how much hydrogen is being missed by absorption line surveys? A great deal, as it turns out.

The equilibrium ionized fraction (conventionally, if bizarrely, given the symbol x) is set by the balance between recombination and photoionization: the number of atoms being ionized must equal the number recombining. The latter is $xn_{\mathrm{H}}\beta n_{\mathrm{e}}$, where $\beta = \beta(T)$ is the temperature-dependent recombination rate, n_{H} is the total hydrogen density, xn_{H} is the number of hydrogen ions (i.e. protons) available, and n_{e} is the number of electrons. The number of atoms being ionized is given by the number of neutral atoms, i.e. $(1-x)n_{\mathrm{H}}$, times the cross section for absorption σ, times the flux density of ionizing photons. These last two terms are frequency-dependent, so we integrate over frequency and find

$$xn_{\mathrm{H}}\beta n_{\mathrm{e}} = (1-x)n_{\mathrm{H}} \int_{\nu_{\mathrm{ion}}}^{\infty} \sigma(\nu)\, \frac{4\pi I_\nu}{h\nu}\, \mathrm{d}\nu, \tag{8.11}$$

where I_ν is the ionizing background and $h\,\nu_{\mathrm{ion}} = 13.6\,\mathrm{eV}$ is the minimum energy needed to ionize hydrogen. The factor of $h\nu$ converts from energy flux I_ν to photon flux $I_\nu/(h\nu)$. Cancelling the n_{H} term gives

$$x\beta n_{\mathrm{e}} = (1-x) \int_{\nu_{\mathrm{ion}}}^{\infty} \sigma(\nu)\, \frac{4\pi I_\nu}{h\nu}\, \mathrm{d}\nu. \tag{8.12}$$

It turns out that the temperature dependence of β is not important, since the gas tends to be at $kT \simeq 13.6\,\mathrm{eV}$. Therefore the ionization fraction is mainly determined by the balance between the numbers of ionizing photons and free electrons. It's useful to define the dimensionless **ionization parameter** U to be the number of photons per electron:

$$U = \frac{n_\gamma}{n_{\mathrm{e}}} = \frac{1}{n_{\mathrm{e}}} \int_{\nu_{\mathrm{ion}}}^{\infty} \sigma(\nu)\, \frac{4\pi I_\nu}{ch\nu}\, \mathrm{d}\nu. \tag{8.13}$$

The hydrogen ionization then comes out as about

$$\frac{x}{1-x} = \frac{U}{10^{-5.2}}. \tag{8.14}$$

So the amount of ionized (and invisible) hydrogen depends on the ambient ionizing light that the Lyman α clouds experience. But how can we find out what ambient light they experience?

The vital clue has come from the **proximity effect** in quasar spectra: as we approach the redshift of any quasar, the numbers of Lyman α clouds in that quasar's spectrum decreases. This is caused by the ionizing radiation from the quasar, which can be estimated independently from extrapolating the quasar spectrum. When the quasar's ionization equals that from the ambient background, the number of Lyman α clouds $\mathrm{d}\mathcal{N}/\mathrm{d}X$ will be half the number that there are elsewhere (e.g. along other lines of sight to other quasars, far from a quasar). At $z \simeq 2.5$ the background turns out to be around $10^{-24}\,\mathrm{J\,m^{-2}}$. As with the Hubble parameter, this is sometimes expressed as a dimensionless quantity: $I_\nu = J_{-21} \times 10^{-21}(\nu_{\rm ion}/\nu)^\alpha\,\mathrm{erg\,cm^{-2}\,s^{-1}\,Hz^{-1}\,sr^{-1}}$, where α is the slope of the spectrum. (Note: $1\,\mathrm{erg\,cm^{-2}} = 10^{-3}\,\mathrm{J\,m^{-2}}$.) In other words, J_{-21} is the background at $\nu_{\rm ion}$ in units of $10^{-21}\,\mathrm{erg\,cm^{-2}\,s^{-1}\,Hz^{-1}\,sr^{-1}}$. If $\alpha = 1$, then $J_{-21} = 1$ corresponds to a proper photon density of $63\,\mathrm{photons\,m^{-3}}$.

The value of J_{-21} comes curiously close (within a factor of a few) to the total ionizing background estimated from integrating the quasar luminosity function. Do star-forming galaxies provide the rest of this ionizing background? The similarity of the quasar contribution to the total would then just be a coincidence. Or are there errors or inaccuracies in the calculations, and quasars provide it all? The jury is still out. In case 'coincidence' is read as pejorative, remember that there are other coincidences in astronomy (indeed, there *must* be): for example, the similar angular sizes of the Sun and the Moon are a coincidence that makes total solar eclipses possible. In any case, this 'coincidence' may reflect some underlying physical connection between quasar activity and star formation, already hinted at in the Magorrian relation.

The ionizing background is a fairly constant $J_{-21} \simeq 1$ at $1.6 < z < 4$, but there is a very quick decline in the ionizing background at $z < 1.6$. At $z \simeq 0.5$, J_{-21} is only 6×10^{-3}, as the epochs of cosmic quasar activity and star formation draw to a close. At the earliest cosmic epochs, the ability of primordial galaxies to ionize their environments will depend on the escape fraction of ionizing photons from these galaxies, of which we shall hear more later in this chapter.

Finally, a creative way to constrain the lifetimes of quasars and test the isotropy of their emission is the **transverse proximity effect**: if you have two quasars that have different redshifts but appear close on the sky, then you can use the Lyman α forest in the spectrum of the more distant quasar to measure the ionization effect of the nearer quasar. If quasars are found in rich environments on average, this will complicate the interpretation, since the richer environment might compensate for the loss of Lyman α clouds from ionization. (A similar bias may be present in the proximity effect measurements of J_{-21}.)

See Goncalves, Steidel and Pettini, 2008, *Astrophysical Journal*, **676**, 816.

8.7 $\Omega_{\rm H\,I}$, the neutral hydrogen density parameter

We've seen that it's frustratingly impossible to constrain the comoving number density of absorbers $n_{\rm co}$ separately from their absorption cross section A, along any one line of sight. However, it turns out to be possible to estimate the comoving density of neutral hydrogen.

The total neutral hydrogen mass of any single absorber must be $\mu m_{\rm H}\, N_{\rm H\,I}\, A$, where μ is the mean molecular mass, and $m_{\rm H}$ is the mass of a hydrogen atom. We include the μ term to account for the contribution of helium to the neutral gas mass. The comoving neutral hydrogen matter density must therefore be

$$\begin{aligned}\rho_{\mathrm{H\,I}}(z) &= \mu m_{\mathrm{H}} \int n_{\mathrm{co}}\, N_{\mathrm{H\,I}}\, A(N_{\mathrm{H\,I}}, z)\, \mathrm{d}N_{\mathrm{H\,I}} \\ &= \frac{H_0 \mu m_{\mathrm{H}}}{c} \int N_{\mathrm{H\,I}}\, f(N_{\mathrm{H\,I}}, z)\, \mathrm{d}N_{\mathrm{H\,I}},\end{aligned} \tag{8.15}$$

using Equation 8.4. It's conventional to measure $\rho_{\mathrm{H\,I}}$ in units of the critical (matter) density, i.e.

$$\Omega_{\mathrm{H\,I}} = \frac{8\pi G\, \rho_{\mathrm{H\,I}}}{3H^2} \tag{8.16}$$

(compare Equation 1.15). In practice, the total HI is estimated over an absorption distance ΔX by summing the column densities in the interval ΔX:

$$\int N\, f(N_{\mathrm{H\,I}}, z)\, \mathrm{d}N_{\mathrm{H\,I}} = \frac{\sum N_{\mathrm{H\,I},i}}{\Delta X}.$$

Here, $N_{\mathrm{H\,I},i}$ refers to the column density of the ith absorber.

Figure 8.14 shows the evolution in $\Omega_{\mathrm{H\,I}}$ measured by quasar absorption lines. At the time of writing, the picture is somewhat confusing. Over the redshift interval $2 < z < 6$, there seems to be significant evolution, consistent with the consumption of gas by star formation. The data point at $z = 0$ is consistent with this broad trend. However, at $0.16 < z < 2$ there are marginally discrepant data points. It's not yet clear what are the causes of the discrepancy, or whether this represents a genuine effect.

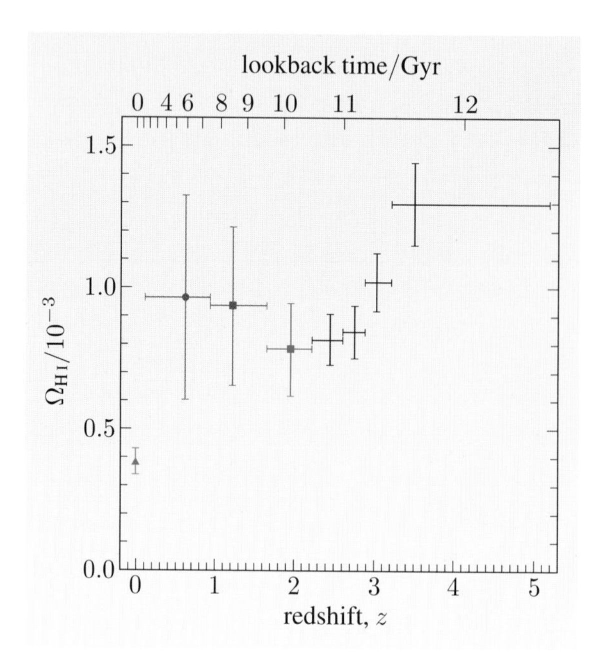

Figure 8.14 The redshift evolution of $\Omega_{\mathrm{H\,I}}$. Data from different sources are shown in different colours.

It was originally thought that the high-redshift $\Omega_{\mathrm{H\,I}}$ matched the current comoving stellar mass density, often written as Ω_*. The decline in $\Omega_{\mathrm{H\,I}}$ could then be attributed to the consumption of gas by star formation in galactic discs. This interpretation, in which damped Lyman α systems (which dominate $\Omega_{\mathrm{H\,I}}$) do not interact much with their environment, is sometimes called the 'closed box' model. Curiously, closed box models consistently overestimated the number of low-metallicity stars in the solar neighbourhood, known as the **G-dwarf problem**.

However, this interpretation of the evolution in $\Omega_{\rm H\,I}$ rested on the assumption of an $\Omega_{\rm m} = 1$, $\Lambda = 0$ universe. The advent of the concordance cosmology (Chapter 2) changed this neat picture. It seems now that the present-day Ω_* is significantly larger than $\Omega_{\rm H\,I}$ at any epoch (Figure 8.14). Hand in hand with these observational changes, numerical simulations changed the theoretical picture. It's now thought that the neutral gas used up by star formation in damped Lyman α systems can be replenished by accretion of hydrogen from the intergalactic medium. This would naturally explain why the *observed* $\Omega_{\rm H\,I}$ (which is dominated by damped Lyman α clouds) is at all times less than the present-day Ω_*. In hierarchical structure formation, matter overdensities often accrete neighbouring clumps (Chapter 4), or merge with larger neighbours. Damped Lyman α systems are seen as dynamic neutral gas reservoirs, in this picture.

Even before the influential Madau diagram (Chapters 4 and 5), Pei and Fall presciently modelled the consumption of gas in damped Lyman α systems and broadly correctly predicted the shape of the cosmic star formation history.

Pei, Y.C. and Fall, S.M., 1995, *Astrophysical Journal*, **454**, 69.

8.8 How big are Lyman α clouds?

As we've seen, it's frustratingly impossible to tell from one Lyman α forest what the sizes are of the Lyman α clouds. However, there are important clues from quasar pairs and gravitationally-lensed quasars. If Lyman α clouds are large enough, they will intersect more than one quasar line of sight, so we can constrain the sizes of Lyman α clouds by comparing the Lyman α forests of nearby quasars or lensed images of a quasar. Lensed quasars have strikingly similar Lyman α forests (e.g. Figure 8.15), which turn out to give stringent lower limits to Lyman α cloud sizes of $> 25\,h^{-1}$ kpc.

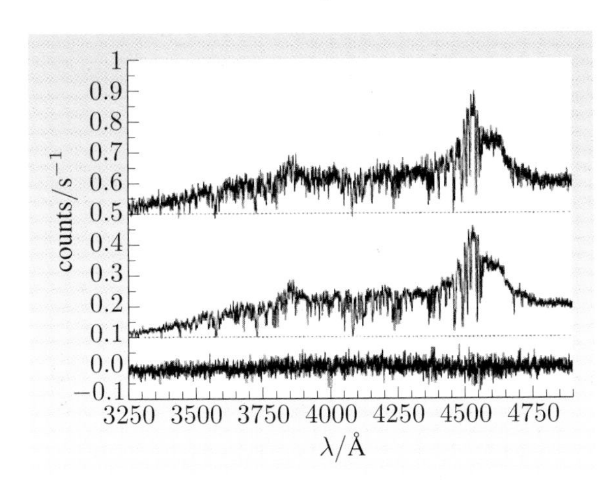

Figure 8.15 The Lyman α forests in two images of the gravitationally lensed quasar UM 673. The upper spectrum has been scaled by 7.9 and offset by 0.5, and the lower spectrum offset by 0.1. The bottom panel shows the difference of the spectra.

Quasar pairs, on the other hand, are more widely separated and probe separations of 1–2 h^{-1} Mpc. The line-of-sight comparisons are much less striking. From statistical cross-correlations, there do appear to be some coherent structures on Mpc scales, but it's less clear that one is taking two lines of sight through a single, coherent object — one might be just tracing the same large-scale structure.

These size constraints are already enough to constrain the physics of Lyman α clouds. One early suggestion was that the Lyman α clouds are neutral clumps in pressure equilibrium with a tenuous ionized medium, but the predicted range of sizes of 0.03–30 kpc in this model is inconsistent with these size observations. However, as we've seen, the sizes are consistent with cosmological simulations in which the Lyman α forest is the neutral 'tip of the iceberg' of the predominantly ionized hydrogen gas, which follows the bottom-up gravitational collapse of matter perturbations.

8.9 Reionization and the Gunn–Peterson test

After the epoch of recombination at $z \simeq 1000$, about the time of the CMB, the Universe entered what's sometimes called its dark ages. The first luminous objects in the Universe formed long after the epoch of recombination, and reionized the Universe. We've already seen in Chapter 2 that the free electrons generated in reionization can Thomson scatter the microwave background photons. To a good approximation, this reduces the temperature fluctuations on all scales, i.e. $\delta T/T \rightarrow (\delta T/T)\exp(-\tau_{\mathrm{T}})$. The effect of this can therefore be seen in the C_l anisotropies of the microwave background. The optical depth to Thomson scattering (which we write here as τ_{T} to distinguish it from other optical depths in this chapter) is one of the free parameters in the CMB fits, and the reionization epoch has a 95% confidence constraint from the CMB of $z_{\mathrm{reion}} = 11 \pm 3.5$, assuming a single instantaneous reionization redshift. Can we do better, and find when and how the first objects lit up the Universe?

As the first luminous objects formed, they ionized their immediate surroundings, then more distant regions. These ionized bubbles or regions, known as **Strömgren spheres**, eventually overlapped, and the Universe was almost entirely reionized, apart from the self-shielded clumps. A simulation of this is shown in Figure 8.16.

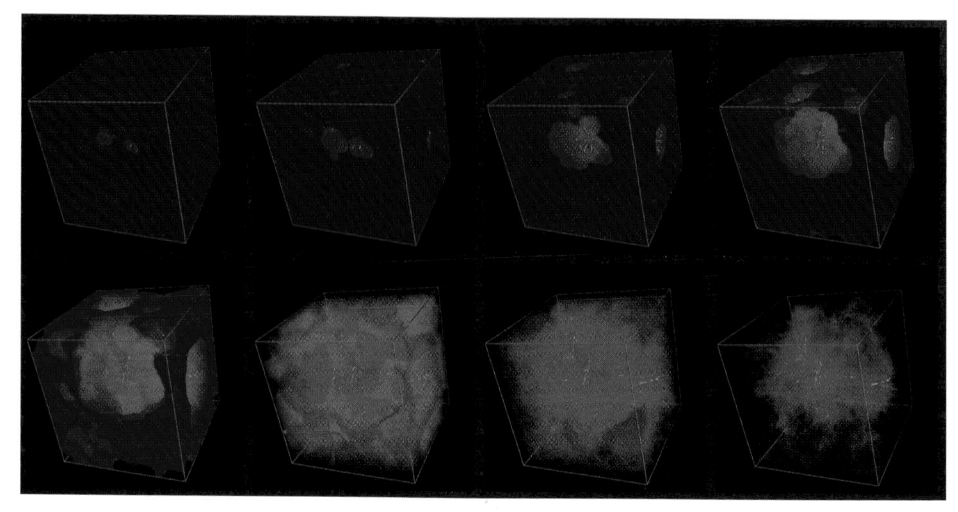

Figure 8.16 Numerical simulation of reionization in a $(2h^{-1})^3$ comoving Mpc^3 volume by Nick Gnedin. The brown opaque fog symbolizes neutral hydrogen. Glowing blue gas is dense ionized hydrogen, and less dense ionized hydrogen is rendered as transparent. Yellow dots represent galaxies. The redshifts shown are $z = 12.1$, 10.4, 9.1, 8.1, 7.3, 6.6, 6.3, 6.0.

It's easy to show that most of the Universe at $1.6 < z < 4$ is, on average, ionized. The present-day density of the Universe is $\rho_0 = 1.8789 \times 10^{-26}\Omega_0 h^2\,\mathrm{kg\,m^{-3}}$ (Chapter 1). Putting in the nucleosynthesis value of $\Omega_{\mathrm{b},0}\,h^2 \simeq 0.015$, and remembering that density scales as $(1+z)^3$, we find the baryon density of the Universe to be $\rho_{\mathrm{b}}(z) \simeq 2.8(1+z)^3 \times 10^{-28}\,\mathrm{kg\,m^{-3}}$. About 75% is hydrogen, as we've seen, and the mass of a proton is 1.67×10^{-27} kg, so there are on average about $0.13(1+z)^3$ hydrogen ions or atoms per cubic metre. The average free electron density is therefore $n_{\mathrm{e}} = 0.13(1+z)^3 x$ per cubic metre, where x is the average hydrogen ionization fraction. We've already seen that the estimated $J_{-21} \simeq 1$ at $z \simeq 3$ implies about $n_\gamma = 63$ photons per cubic metre, so the ionization parameter is $U = n_\gamma/n_{\mathrm{e}} = 500x^{-1}(1+z)^{-3}$. Using Equation 8.14 we can find a quadratic equation for x:

$$x^2 = (1-x)\frac{10^{7.9}}{(1+z)^3},$$

for which the only positive solution is $(1-x) \simeq 10^{-8}(1+z)^3$ or $x \simeq 1$.

Therefore the $z \simeq 3$ Universe should on average be highly transparent to Lyman α, and it's only because of density inhomogeneities that any Lyman α absorbers can be seen. If we assume that a Lyman α cloud is 25 kpc in size (Section 8.8), the neutral hydrogen density must be around $\rho_{\mathrm{H\,I}} \simeq N_{\mathrm{H\,I}}/25\,\mathrm{kpc}$, which comes out as $\rho_{\mathrm{H\,I}} \simeq (N_{\mathrm{H\,I}}/10^{19}\,\mathrm{m^{-2}}) \times 0.013$ atoms per cubic metre. This is much lower than the total hydrogen density of the Universe from primordial nucleosynthesis (see above), so again we see that most of the hydrogen must be ionized.

What's more, an ionizing flux of $J_{-21} = 1$ is clearly enough to ionize the Universe at any redshift for which we are likely to observe an object. However, if we see the highest-redshift Universe becoming opaque on average to Lyman α photons, then J_{-21} must have dropped sharply, and the Universe will be predominantly neutral. This was first proposed by Gunn and Peterson in 1965 and is now known as the Gunn–Peterson test. The transition between opaque and transparent would then be probing the epoch of reionization in which the first Strömgren spheres expand around the very first luminous objects in the Universe.

We had to wait several decades for the first thrilling hints of reionization in quasar spectra from a Gunn–Peterson absorption trough at the highest redshifts. Figure 8.17 shows the spectra of the highest redshift quasars — note the rapidly decreasing lack of flux in the Lyman α forest at redshifts $z > 6$. We can convert this to a Lyman α optical depth, shown in Figure 8.18. Whether this represents a transition to the epoch when the Strömgren spheres were just beginning to overlap is still a matter of debate. The Lyman α opacity is sensitive to the presence of rare voids in the intergalactic medium, so measurements of the Gunn–Peterson trough are sensitive to assumptions about the distribution of gas. Known quasars are also biased tracers of the underlying matter distribution, and the quasars or starbursts responsible for reionization may well also have been strongly biased, so reionization is likely to have been inhomogeneous. Nevertheless, these high-redshift quasars are the first to give useful constraints on reionization simulations.

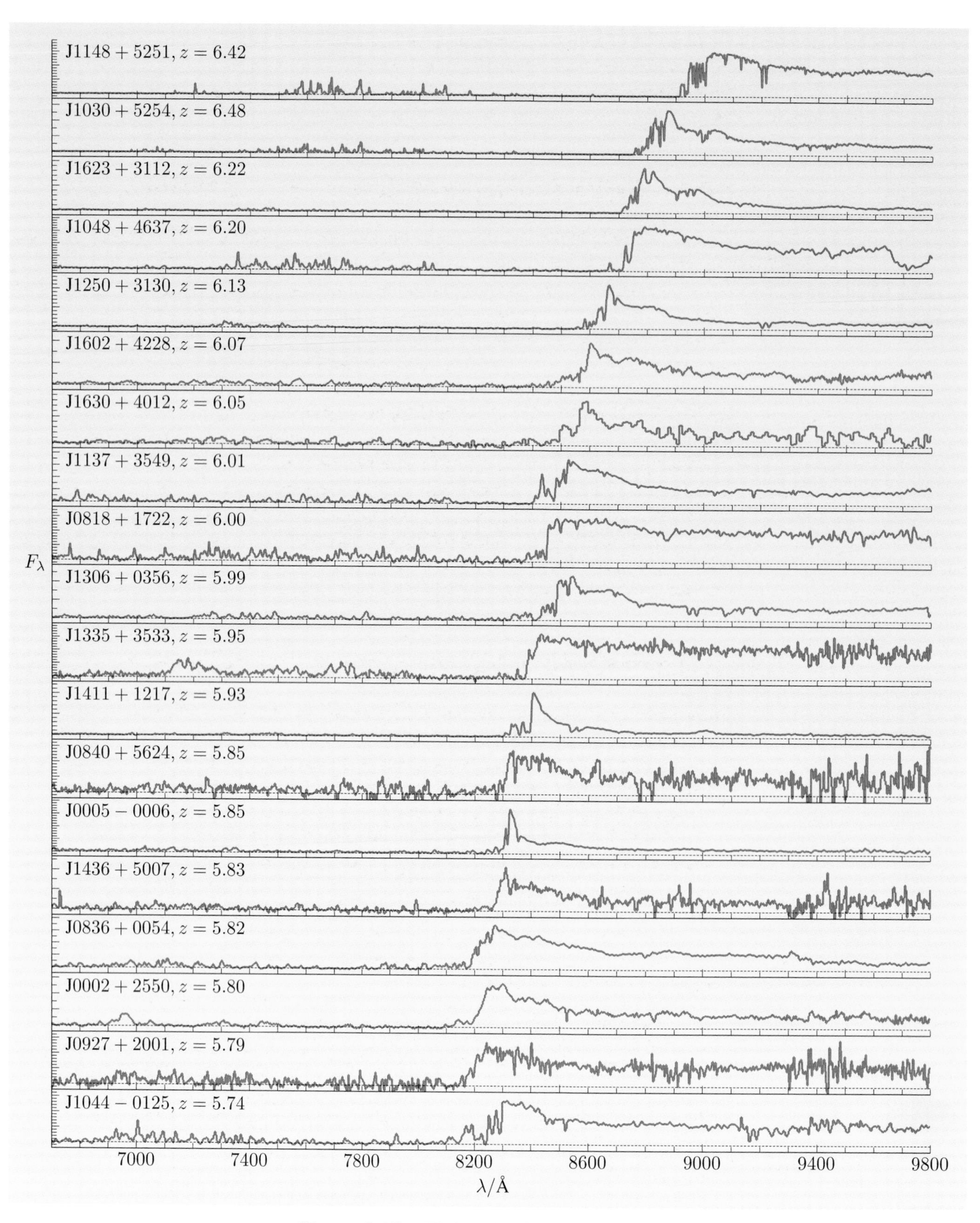

Figure 8.17 High-redshift quasars in the Sloan Digital Sky Survey (SDSS). Note the increasing Gunn–Peterson opacity at the highest redshifts.

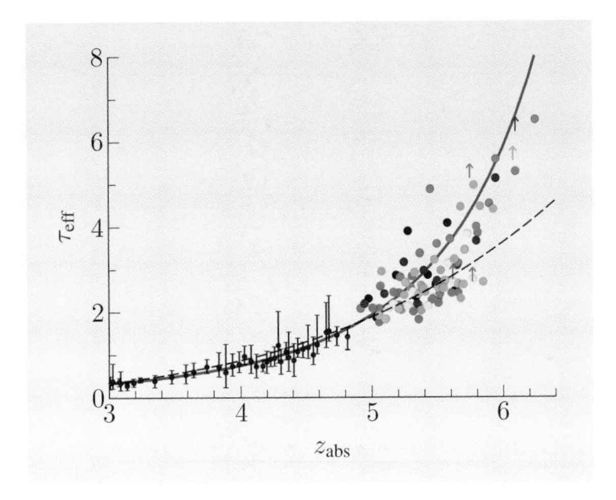

Figure 8.18 The opacity of the Universe to Lyman α photons. Data on different quasars are plotted in different colours, with lower limits shown as arrows.

Was the first light in the Universe from star formation, or from black hole accretion in quasars? At redshifts $z > 3$ the comoving number density of the most luminous quasars drops quickly (Chapter 4). The slope of the quasar luminosity function tells us whether fainter quasars could be important. It turns out that the slope at $z > 4$ is shallower than at low redshift, which implies that quasars did not contribute the majority of J_{-21} during the tail-end of reionization at $z \simeq 6$. Another possibility for the origin of the first light is star-forming galaxies, since young massive O and B stars are prodigious emitters of ionizing radiation. But what fraction of this ionizing radiation escapes star-forming galaxies? It's difficult to measure Lyman continuum photons from high-redshift galaxies because of the presence of intervening Lyman α absorbers and Lyman limit systems; measurements of escape fractions so far range from $f_{\rm esc} < 0.1$ to $f_{\rm esc} > 0.5$. However, even assuming $f_{\rm esc} = 1$, the luminosity function of $z > 6$ optically-selected galaxies suggests, as for quasars, that they are insufficient to reionize the Universe.

Perhaps a new population of objects — such as intermediate mass accreting black holes — reionized the Universe, but this mini-quasar population could easily exceed the unresolved soft X-ray background. Perhaps the luminosity function of $z > 5$ star-forming galaxies steepens at luminosities much fainter than probed so far, invoking a new population of dwarf star-forming galaxies. At the time of writing, the objects that reionized the Universe remain a mystery.

We can't yet rule out more than one reionization epoch. Figure 8.19 shows the constraints on the neutral fraction x as a function of redshift. Two reionization epochs might happen if there is an initial flurry of star formation that creates predominantly massive stars because of the low metallicity (known as **population III** stars), but subsequent stars (population II) are less massive so less able to ionize their surroundings. The intergalactic medium in this model would then recombine until enough stars have formed to ionize it again.

Radiation pressure limits the maximum luminosity of stars, but the primordial gas from which population III stars formed lacked metal absorption lines, reducing the radiation pressure on the gas.

8.10 The Lyman α forest of He II

The epoch of hydrogen reionization is tantalizingly just beyond our grasp, but 25% of the baryons in the Universe are in helium, and helium reionization is already within our grasp. Helium is harder to ionize than He II: 54.4 eV are needed, as opposed to 13.6 eV with hydrogen. (He I is neutral helium, while He II refers to He^+ ions, i.e. helium with one electron missing; similarly, H I means neutral hydrogen, and the Strömgren sphere of ionized hydrogen is an H II region.) This means that a much harder radiation field (i.e. more photons at higher energies) is needed for complete helium reionization than for hydrogen. Also, helium reionization will have happened later than hydrogen reionization, because of the availability of these high-energy ambient photons.

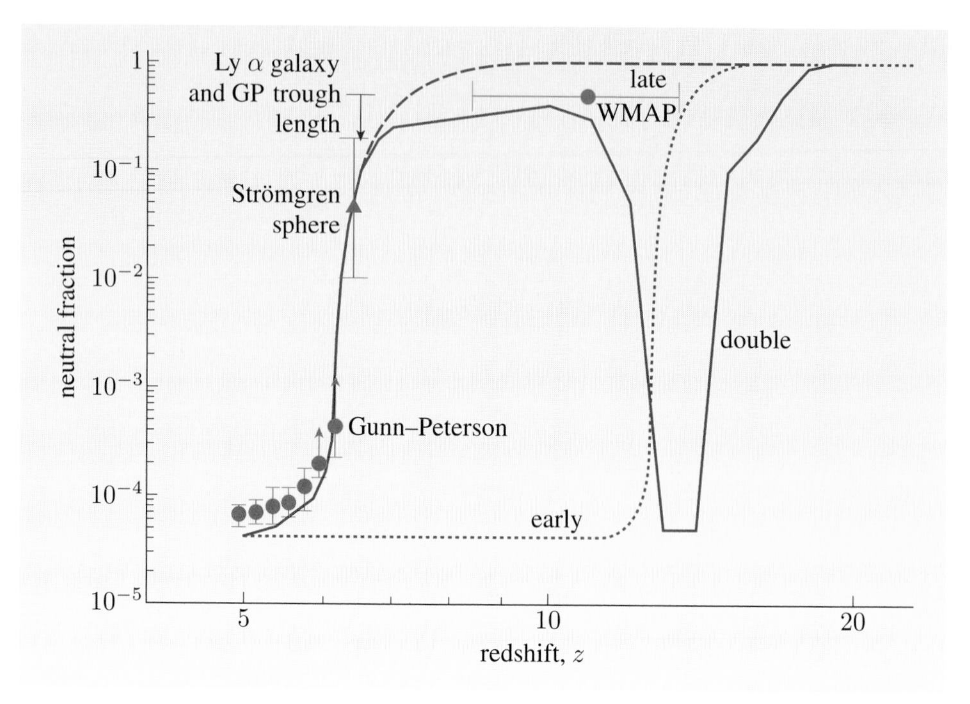

Figure 8.19 Experimental constraints on the reionization history of the Universe. The lines are two models that are consistent with the Gunn–Peterson data, and one that is not (but which still is marginally consistent with WMAP). The Strömgren sphere point is a constraint on the sizes of ionized regions around high-redshift quasars.

Exercise 8.8 Calculate the rest-frame wavelengths of the hydrogen and helium Lyman limits, and identify the redshifted hydrogen Lyman limit in Figure 8.20. ■

The helium Lyman α forest and the Gunn–Peterson test could tell us about the spectrum of objects that are keeping the Universe ionized. However, the helium lines are a factor of four shorter in wavelength (because the energy levels are $\propto Z^2$, where Z is the atomic number), so the rest wavelength of He II Lyman α is $121.6/4\,\text{nm} = 30.38\,\text{nm}$. The He II Lyman α forest will therefore all lie below the hydrogen Lyman limit (rest wavelength 91.2 nm).

This doesn't mean that He II Lyman α lines are unobservable, but rather means that they need a quasar that happens to have not much total hydrogen column density. Figure 8.20 shows the simulated typical opacity in the spectrum of a quasar at redshift $z = 3.2$. Below the H I Lyman limit, the spectrum is heavily attenuated, but at shorter wavelengths the attenuation reduces (because the cross section for hydrogen ionization varies as ν^{-3} at frequencies above the Lyman limit), and there is a possibility of the He II Lyman α lines being observable in some quasars.

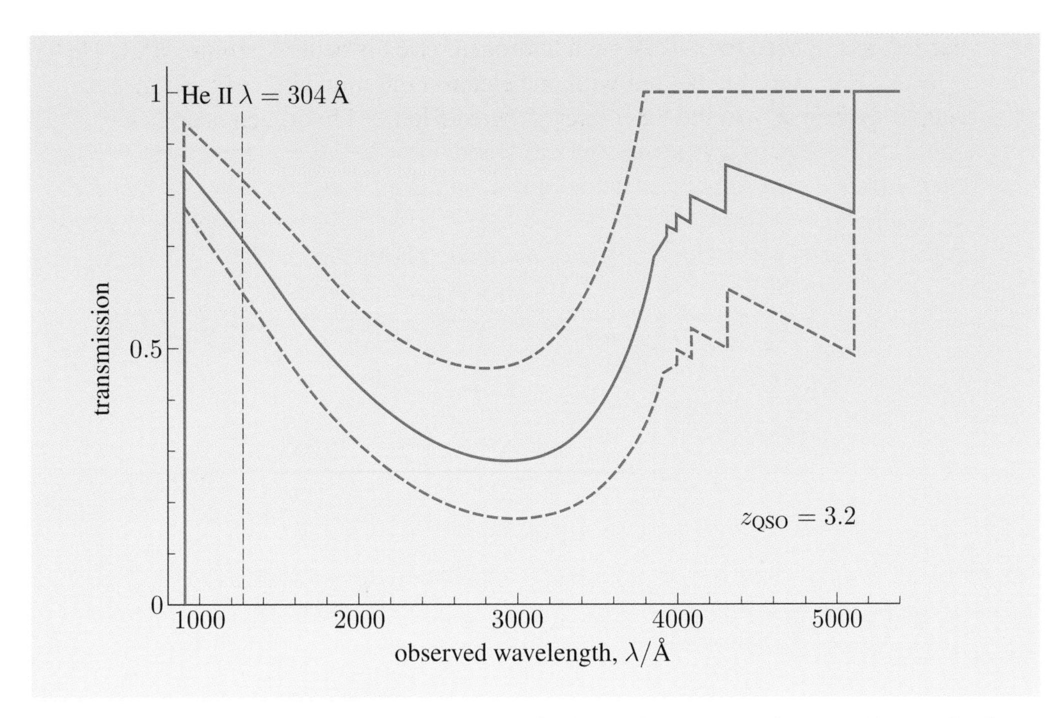

Figure 8.20 The expected average transmission to Lyman continuum photons in the spectrum of a $z = 3.2$ quasar. The dashed lines show the $\pm 1\sigma$ range expected in the opacity. Also shown is the location of the He II Lyman α line at $z = 3.2$. For some quasars, we might expect enough transparency to be able to detect this line.

Figure 8.21 shows the He II Gunn–Peterson trough in the quasar HE 2347-4342. The He II opacity is strikingly different to the H I opacity, at 4 times longer wavelengths. Some regions lacking H I Lyman α lines are also relatively transparent to He II, but for the most part the spectrum is opaque to He II. (There is a region with high He II opacity but no obvious H I Lyman α absorbers, possibly caused by thermal broadening or instrumental noise, or possibly related to variations in the hardness of the ionizing radiation.)

Taking all available observations, He II reionization is measured to have happened at a redshift of $z = 2.8 \pm 0.2$. Despite the decline in quasar comoving number density at $z > 2$, quasars are more than enough to reionize He II. There is some tentative evidence, from comparing the H I and He II opacities, that the spectrum of ionizing radiation is softer at high redshifts, i.e. a smaller proportion of high-energy photons, consistent with star-forming galaxies providing a bigger proportion.

8.11 The first light in the Universe and gamma-ray blazars

Will it ever be possible to directly detect the objects that reionized the hydrogen in the Universe? If they are (for example) dwarf galaxies, then they are far beyond the reach of current telescopes, but in the future, the James Webb Space Telescope (JWST) may be able to detect them.

At the time of writing, the JWST's launch is several years away, but the integrated light from the reionization population may already have been detected. This light, heavily redshifted, should appear as an all-sky background light in the near-

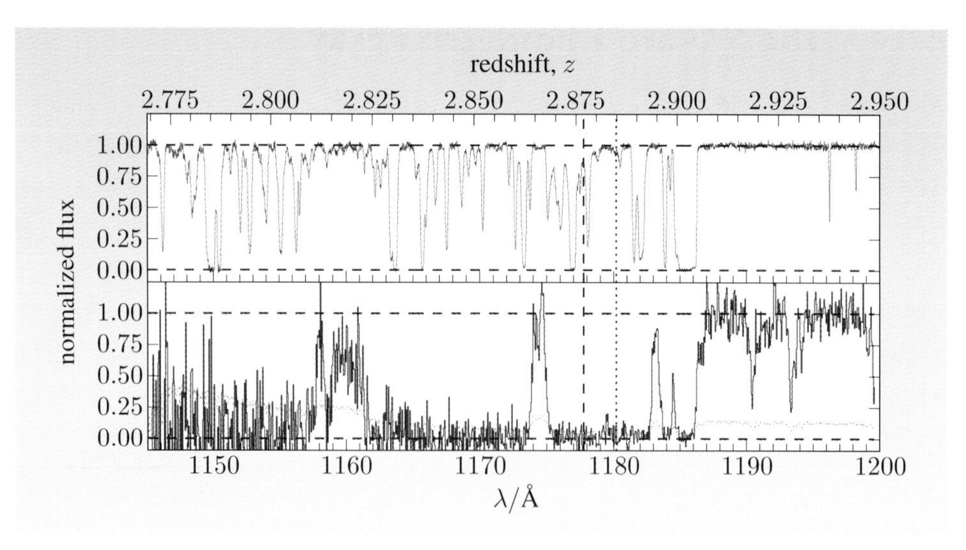

Figure 8.21 Signatures of He II reionization in the quasar HE 2347-4342. The top panel shows the optical spectrum, normalized to the quasar spectrum (so a flux of 1.0 is no absorption). The lower panel shows the ultraviolet spectrum. The wavelengths in the top panel have been divided by approximately four, to match the wavelengths of the H I and He II Lyman α forests. The thin dotted, roughly horizontal line is the 1σ uncertainty in the ultraviolet measurements. The thick vertical dotted line marks the expected position of He II Lyman α (no emission line is detected), and data redward (i.e. rightward) of the dashed vertical line are affected by absorbers within the quasar itself or its host galaxy. The quasar redshift is $z = 2.885$, and redward of the dashed line the quasar is known to have absorption lines associated with the quasar itself, i.e. $z_{\text{abs}} \simeq z_{\text{QSO}}$.

infrared. There have been several claims of detections of this cosmic near-infrared background, independently from the Infrared Telescope in Space (IRTS) and the Diffuse Infrared Background Experiment (DIRBE) on the COBE CMB mission. This would be a ground-breaking discovery, but this faint background ($\sim$10–50 $\text{nW}\,\text{m}^{-2}\,\text{sr}^{-1}$ at wavelengths of 1–4 μm) is around a hundred times fainter than the reflected sunlight from the zodiacal dust in our own Solar System. This is a very delicate experiment that requires careful control of the systematic uncertainties, and opinion is still divided as to whether genuinely cosmic infrared background signals have been detected. Another approach is to take the DIRBE maps, subtract the infrared fluxes of known stars and galaxies, and/or mask them out, then look for the *clustering* of the residuals (analogously to the CMB). This is independent of the absolute cosmic infrared background level. The clustering measurements in the cosmic near-infrared have been argued to be consistent with reionization population predictions, but opinion is again divided because this is again an experiment that needs careful treatment of systematic uncertainties. Nevertheless, the potential reward of discovering the reionization population makes this a hot topic in current cosmology.

A completely independent approach to constraining the cosmic near-infrared background comes from gamma-ray observations of quasars. If the Universe is filled with a homogeneous cosmic background of near-infrared photons, they should interact with the gamma rays through the pair production reaction ($\gamma + \gamma \rightarrow \text{e}^- + \text{e}^+$, the inverse reaction of electron–positron annihilation, where one γ is a gamma-ray photon and the other γ is a near-infrared photon), which results in a measurable gamma-ray opacity. This opacity has not been seen, which places important limits on the cosmic near-infrared background.

8.12 The Square Kilometre Array

A powerful new telescope, or rather array of telescopes, will eventually join the race to find the epoch of reionization. The Square Kilometre Array (SKA) will be a radio-wave observatory that will have a combined effective collecting area (factoring in telescope efficiencies) more than 30 times bigger than the largest telescope ever built. Although full SKA science operations are still some way off (not before 2020) at the time of writing, the science promises to be revolutionary when it comes.

Cosmological neutral hydrogen doesn't just absorb photons at the Lyman α wavelength ($n = 1 \rightarrow 2$). The energy of an $n = 1$ electron is slightly different, depending on whether its spin is parallel to the nucleus spin, or antiparallel (Figure 8.22). The energy difference is very small in comparison to the Lyman α transition: $\Delta E = 5.87 \times 10^{-6}$ eV, compared to $\Delta E = 10.2$ eV for Lyman α. This corresponds to photon energies of $\lambda = hc/\Delta E \simeq 21$ cm for this $n = 1$ spin-flip transition. This is in the radio wavelengths and is accessible to radio telescopes even when redshifted to $z > 10$ (though conditions in the Earth's ionosphere limit performance at the longest wavelengths). The 21 cm transition during reionization could be seen in absorption or emission, depending on the thermal history of the gas. In only a few hours' exposure time, the SKA is expected to detect many tens or even hundreds of thousands of galaxies in H I out to $z \sim 1.5$, from which redshifts can immediately be calculated.

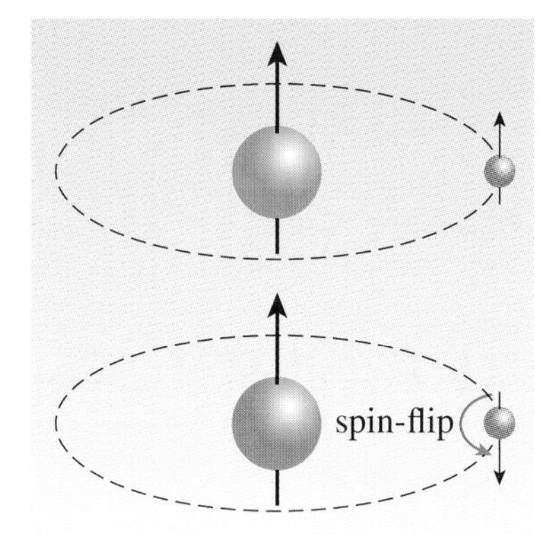

Figure 8.22 The hydrogen spin-flip transition. The electron changes its spin from parallel to antiparallel relative to the nucleus, which releases an energy of $\Delta E = 5.87 \times 10^{-6}$ eV as a photon with wavelength 21 cm.

There is also expected to be a 21 cm forest , analogous to the Lyman α forest, and a 21 cm Gunn–Peterson test. Unlike Lyman α, however, the 21 cm line doesn't saturate when the neutral fraction is large. A simulated SKA radio spectrum of a $z = 10$ radio-loud quasar is shown in Figure 8.23. We have yet to find any suitable $z \simeq 10$ background radio source, but it is not unreasonable to expect the SKA to be able to find one.

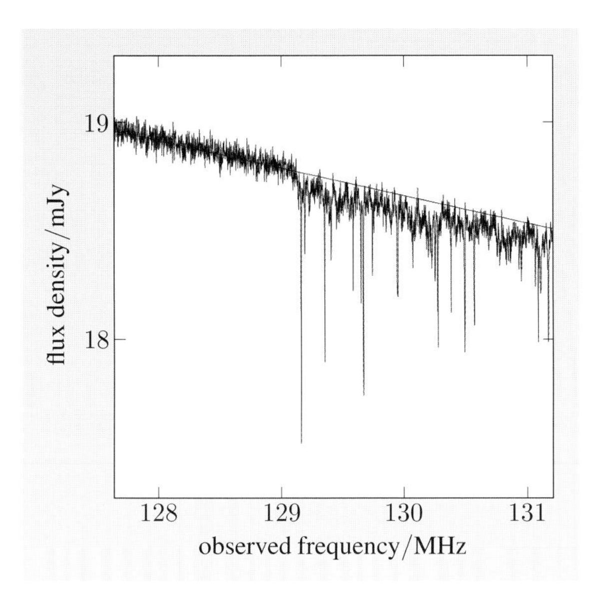

Figure 8.23 A numerical simulation of the 21 cm forest by the Square Kilometre Array, by Chris Carilli.

The SKA also promises to revolutionize many of the topics discussed in this book: the SKA team aim to measure the dark energy equation of state (from cosmic shear and baryon wiggles), test whether dark energy clusters (using the Integrated Sachs–Wolfe effect), measure the power spectrum of primordial density fluctuations, and measure the Hubble parameter to 1% accuracy (from observations of extragalactic water masers), and the SKA could even work as a gravitational wave detector by comparing the timings of millisecond pulsars.

8.13 The CODEX experiment

We shall end this book with the story of another extraordinary and ambitious experiment. Obtaining the results will take some decades, but the astonishing goal is to detect the *real-time* expansion of the Universe. Broadly, the aim is to watch distant objects over decades, and see their redshifts slowly change because of the expansion of the Universe.

This is an extremely small effect. The quantity $\mathrm{d}z/\mathrm{d}t_{\mathrm{observed}}$ has dimensions of $1/\mathrm{time}$, i.e. the same as the Hubble parameter H_0. Broadly speaking, we would expect $\mathrm{d}z/\mathrm{d}t_{\mathrm{observed}} \propto H_0$, with the constant of proportionality depending on the cosmological parameters, and with some dependence on redshift. (Note: do not confuse this with Equation 1.34, which links redshift z with lookback time.)

A Hubble parameter of $H_0 = 72\,\mathrm{km\,s^{-1}\,Mpc^{-1}}$ is equivalent to $H_0 = 2.3 \times 10^{-18}\,\mathrm{s^{-1}}$. If the constant of proportionality mentioned above is of order 1, then in ten years we would expect a change in redshift of only $\sim 10^{-9}$. Measuring this minute change would require extremely accurate redshifts. Even Doppler shifts within the objects become significant: to first order, $v/c \simeq z_{\mathrm{Doppler}}$, where c is the speed of light, so a change of just $\delta v = 0.3\,\mathrm{m\,s^{-1}}$ would result in a redshift change of $\delta z \simeq 10^{-9}$. If you compare this with typical velocity dispersions of galaxies, $\sim 230\,\mathrm{km\,s^{-1}}$, or the velocity widths of narrow emission lines of active galaxies, $\sim 1000\,\mathrm{km\,s^{-1}}$, you will appreciate the enormous difficulty of obtaining such a measurement.

We might hope that the constant of proportionality is much bigger than 1, but alas it is not. If we differentiate $1 + z = R_{\mathrm{obs}}/R_{\mathrm{em}}$ (writing 'obs' for observed and 'em' for emitted), we find that

$$\frac{\mathrm{d}z}{\mathrm{d}t_{\mathrm{obs}}} = \frac{\mathrm{d}R_{\mathrm{obs}}/\mathrm{d}t_{\mathrm{obs}}}{R_{\mathrm{em}}} - \frac{\mathrm{d}R_{\mathrm{em}}}{\mathrm{d}t_{\mathrm{obs}}}\frac{R_{\mathrm{obs}}}{R_{\mathrm{em}}^2},$$

using the quotient rule for differentiation. The chain rule then gives

$$\frac{\mathrm{d}z}{\mathrm{d}t_{\mathrm{obs}}} = \frac{\mathrm{d}R_{\mathrm{obs}}/\mathrm{d}t_{\mathrm{obs}}}{R_{\mathrm{em}}} - \frac{\mathrm{d}R_{\mathrm{em}}}{\mathrm{d}t_{\mathrm{em}}}\frac{R_{\mathrm{obs}}}{R_{\mathrm{em}}^2}\frac{\mathrm{d}t_{\mathrm{em}}}{\mathrm{d}t_{\mathrm{obs}}}.$$

Now, we know that the Hubble parameter is $H = (1/R)\mathrm{d}R/\mathrm{d}t$ in general, so the value now must be $H_0 = (1/R_{\mathrm{obs}})\,\mathrm{d}R_{\mathrm{obs}}/\mathrm{d}t_{\mathrm{obs}}$. Meanwhile, at the time of redshift z, the Hubble parameter was $H(z) = (1/R_{\mathrm{em}})\,\mathrm{d}R_{\mathrm{em}}/\mathrm{d}t_{\mathrm{em}}$. We can use these values to simplify the terms in the equations above:

$$\frac{\mathrm{d}R_{\mathrm{obs}}/\mathrm{d}t_{\mathrm{obs}}}{R_{\mathrm{em}}} = H_0\frac{R_{\mathrm{obs}}}{R_{\mathrm{em}}}$$

and

$$\frac{\mathrm{d}R_{\mathrm{em}}}{\mathrm{d}t_{\mathrm{em}}}\frac{R_{\mathrm{obs}}}{R_{\mathrm{em}}^2} = H(z)\frac{R_{\mathrm{obs}}}{R_{\mathrm{em}}},$$

which gives

$$\frac{\mathrm{d}z}{\mathrm{d}t_{\mathrm{obs}}} = H_0\frac{R_{\mathrm{obs}}}{R_{\mathrm{em}}} - H(z)\frac{\mathrm{d}t_{\mathrm{em}}}{\mathrm{d}t_{\mathrm{obs}}}\frac{R_{\mathrm{obs}}}{R_{\mathrm{em}}}.$$

But we know that $R_{\rm obs}/R_{\rm em} = {\rm d}t_{\rm obs}/{\rm d}t_{\rm em} = (1+z)$, so this just becomes

$$\frac{{\rm d}z}{{\rm d}t_{\rm obs}} = (1+z)H_0 - H(z).$$

Without being too disingenuous we could write this as

$$\frac{{\rm d}z}{{\rm d}t_{\rm obs}} = \dot{z} = H_0\left\{(1+z) - \frac{H(z)}{H_0}\right\}, \tag{8.17}$$

where $H(z)/H_0$ is the factor by which the Hubble parameter has changed. The rate of change of redshift, $\dot{z}$, is plotted in Figure 8.24. So the constant of proportionality is generally a bit smaller than 1, at least in the concordance cosmology.

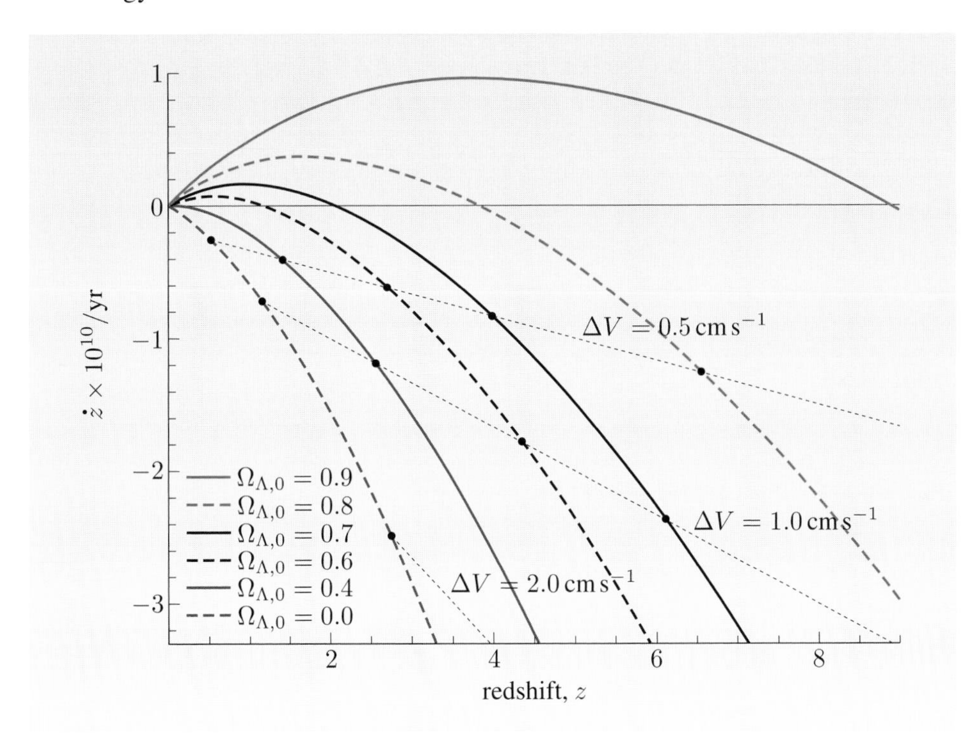

Figure 8.24 The real-time rate of change of redshift for various cosmological models. Some reference velocity changes are shown as thin black dashed lines.

So the prospects for measuring the real-time expansion of the Universe seem to look grim. However, one approach that might work is the Cosmical Dynamics Experiment, or CODEX. The objective is to take very-high-resolution spectra of the Lyman α forest with an extremely careful and stable wavelength calibration. By cross-correlating the spectrum with a second spectrum at least 10 years later, the shifts in redshifts should be detectable. Figure 8.25 shows simulated spectra separated in time; note that it's only by averaging the shifts of many Lyman α absorbers that the expansion is detectable. This averaging also washes out any peculiar acceleration in individual objects. CODEX is currently a proposed experiment for the proposed European Extremely Large Telescope, and is still many years from taking its first data. Another approach could be to use the SKA to monitor the H I 21 cm forest in a $z > 10$ radio-loud active galaxy, if we can find one that is bright enough (see Figure 8.23). Either way, it is possible that within our lifetimes we shall have detected the real-time expansion of the Universe.

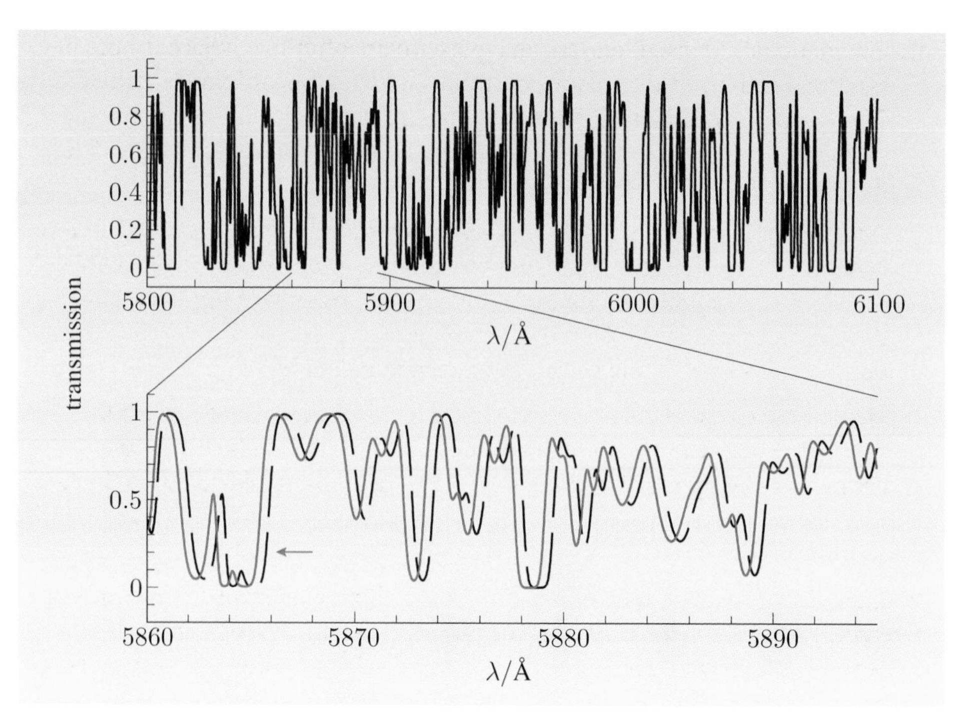

Figure 8.25 The change in redshift in the Lyman α forest expected in five million years. The shift in ten years will be much smaller, and detectable only statistically.

Summary of Chapter 8

1. The Lyman α forest of absorption lines blueward of the Lyman α emission line in quasars and galaxies is caused by intervening, lower-redshift Lyman α absorbers.
2. Most of the electrons in hydrogen atoms in the Lyman α clouds are in the ground state, implying no Hα absorption.
3. If sufficiently dense, the clouds will be self-shielding, i.e. will have an optical depth > 1 to Lyman α photons.
4. The large-scale structure of the Lyman α forest traces the power spectrum of baryonic matter on scales that are close to the linear regime for the growth of perturbations.
5. The largest contribution to $\Omega_{\text{H I}}$ comes from damped Lyman α systems. The term 'damped' refers to the damping wings of the Lorentzian absorption profile.
6. The column density and optical depth of an absorption line can be related to the equivalent width via the curve of growth.
7. The comoving number density of absorbers along a line of sight is measured with the use of absorption distance.
8. Ionizing photons near a quasar reduce the comoving number density of Lyman α clouds. This is known as the proximity effect.

9. Gravitationally-lensed quasars and quasar pairs offer two adjacent lines of sight that can be used to place constraints on the sizes of Lyman α clouds. The cloud sizes cannot be determined from a single line of sight; only the product $n_{co} \times A$ (where n_{co} is the comoving density and A is the area) can be found from a single line of sight.
10. After recombination, the Universe was neutral (i.e. unionized) and therefore opaque to Lyman α photons. The first luminous objects later reionized the Universe. The epoch of reionization is (at the time of writing) uncertain, but the constraints from the Gunn–Peterson test (the opacity of the Universe derived from Lyman α absorption) place it at redshifts $z > 6$.
11. He II reionization occurs later than hydrogen reionization, at a redshift of $z = 2.8 \pm 0.2$. The He II Lyman α lines occur at about a factor of four shorter wavelength than hydrogen, making them observable along only those few lines of sight without Lyman limit systems over a sufficiently wide redshift range.
12. Other constraints on the reionization population come from the gamma-ray opacity of the Universe and the cosmic near-infrared background. The Square Kilometre Array aims to map the reionization of the Universe using the 21 cm spin-flip transition of neutral hydrogen.
13. There are proposals to monitor the real-time expansion of the Universe using the Lyman α forest or the 21 cm forest.

Further reading

- Fan, X., 2006, 'Observational constraints on cosmic reionization', *Annual Review of Astronomy and Astrophysics*, **44**, 415.
- Loeb, A. and Barkana, R., 2001, 'The reionization of the Universe by the first stars and quasars', *Annual Review of Astronomy and Astrophysics*, **39**, 19.
- Rauch, M., 1998, 'The Lyman alpha forest in the spectra of QSOs', *Annual Review of Astronomy and Astrophysics*, **36**, 267.
- Wolfe, A.M., Gawiser, E. and Prochaska, J.X., 2005, 'Damped Ly α systems', *Annual Review of Astronomy and Astrophysics*, **43**, 861.
- Cen, R., 2003, 'The Universe was reionized twice', *Astrophysical Journal*, **591**, 12.
- Faucher-Giguère, C.A., Lidz, A., Hernquist, L. and Zaldarriaga, M., 2008, 'Evolution of the intergalactic opacity: implications for the ionizing background, cosmic star formation, and quasar activity', *Astrophysical Journal*, **688**, 85.
- Hauser, M.G. and Dwek, E., 2001, 'The cosmic infrared background: measurements and implications', *Annual Review of Astronomy and Astrophysics*, **39**, 249.
- Rybicki, G.B. and Lightman, A.P., 1979, *Radiative Processes in Astrophysics*, Wiley.
- More on the Square Kilometre Array can be found at its website, currently http://www.skatelescope.org.

Epilogue

How wonderful it would be to become wise.

Genesis 3, 6

Where will the next big changes in thinking in cosmology come from? Many of the previous big changes have come from unexpected observational discoveries, which makes it difficult to foresee the next leaps. As we've seen, the population of high-redshift submm-luminous galaxies seemed unremarkable to optical telescopes, yet were found to be convulsed in violent star formation by submm-wave imaging. This led in part to the new model of galaxy downsizing. As I write this, the submm-wave Herschel Space Observatory will be launched in six days, and the submm-wave SCUBA-2 camera will shortly be commissioned at the James Clerk Maxwell Telescope in Hawaii. Both have tremendous scope for new discoveries. Cosmology has also seen a change from small teams and lone scientists, to large international consortia using many different astronomical facilities and techniques. Despite that, it's still possible for individual scientists to make a mark, whether on their own or as part of a small or large team. As a result of these large-scale international efforts and developments in survey technology at all wavelengths, we are in a very data-rich phase of astronomy.

So what's next? Large CCD arrays are just making time-domain optical astronomy possible. Projects such as PAN-STARRS and Gaia will repeatedly survey large areas of sky. These will almost certainly uncover many new gravitational microlens events and many nearby supernovae. Gamma-ray monitoring of the sky led to the completely unexpected discovery of gamma-ray bursts, which themselves have optical transients, so what else lies in wait to be discovered in time-domain optical astronomy? Perhaps the new generation of radio telescopes such as LOFAR and the SKA, or the HST's successor the JWST, or the next generation of $\simeq 50$ m-diameter optical/near-infrared telescopes, will detect unexpected reionization populations that generated the first light in the Universe after the Big Bang. Perhaps the new gravitational wave observatories LIGO and LISA will detect inspiralling black holes and confront us with irreconcilable inconsistencies with general relativity. Perhaps the anisotropies in the CMB will eventually be found inconsistent with inflation, or the LHC could fail to find the Higgs boson, either of which would force big changes in fundamental physics. Perhaps the signatures of dark matter particles will be found in terrestrial direct detection experiments or at the LHC, or their annihilation signatures will be inferred from cosmic rays, which will tell us what dominates most of the matter content of the Universe. We know very little indeed about the dark sector in general, whether dark matter or dark energy. We assume, perhaps blithely, that dark matter only responds to gravity, but perhaps it has its own intricate suite of dark physics. Perhaps the delicate measurements of cosmic shear or baryon wiggles or the expansion of the Universe will constrain the phenomenological parameters of dark energy, and give some insight on the physical causes of what dominates the current expansion of the Universe, or even challenge our assumptions of the size scales at which the Universe is homogeneous. There has surely never been a more exciting time in observational cosmology.

Appendix A

Table A.1 Common SI unit conversions and derived units.

Quantity	Unit	Conversion
speed	$\mathrm{m\,s^{-1}}$	
acceleration	$\mathrm{m\,s^{-2}}$	
angular speed	$\mathrm{rad\,s^{-1}}$	
angular acceleration	$\mathrm{rad\,s^{-2}}$	
linear momentum	$\mathrm{kg\,m\,s^{-1}}$	
angular momentum	$\mathrm{kg\,m^2\,s^{-1}}$	
force	newton (N)	$1\,\mathrm{N} = 1\,\mathrm{kg\,m\,s^{-2}}$
energy	joule (J)	$1\,\mathrm{J} = 1\,\mathrm{N\,m} = 1\,\mathrm{kg\,m^2\,s^{-2}}$
power	watt (W)	$1\,\mathrm{W} = 1\,\mathrm{J\,s^{-1}} = 1\,\mathrm{kg\,m^2\,s^{-3}}$
pressure	pascal (Pa)	$1\,\mathrm{Pa} = 1\,\mathrm{N\,m^{-2}} = 1\,\mathrm{kg\,m^{-1}\,s^{-2}}$
frequency	hertz (Hz)	$1\,\mathrm{Hz} = 1\,\mathrm{s^{-1}}$
charge	coulomb (C)	$1\,\mathrm{C} = 1\,\mathrm{A\,s}$
potential difference	volt (V)	$1\,\mathrm{V} = 1\,\mathrm{J\,C^{-1}} = 1\,\mathrm{kg\,m^2\,s^{-3}\,A^{-1}}$
electric field	$\mathrm{N\,C^{-1}}$	$1\,\mathrm{N\,C^{-1}} = 1\,\mathrm{V\,m^{-1}} = 1\,\mathrm{kg\,m\,s^{-3}\,A^{-1}}$
magnetic field	tesla (T)	$1\,\mathrm{T} = 1\,\mathrm{N\,s\,m^{-1}\,C^{-1}} = 1\,\mathrm{kg\,s^{-2}\,A^{-1}}$

Table A.2 Other unit conversions.

wavelength
1 nanometre (nm) = 10 Å = 10^{-9} m
1 ångstrom = 0.1 nm = 10^{-10} m

angular measure
1° = 60 arcmin = 3600 arcsec
1° = 0.017 45 radian
1 radian = 57.30°

temperature
absolute zero: 0 K = −273.15 °C
0 °C = 273.15 K

spectral flux density
1 jansky (Jy) = $10^{-26}\,\mathrm{W\,m^{-2}\,Hz^{-1}}$
$1\,\mathrm{W\,m^{-2}\,Hz^{-1}} = 10^{26}$ Jy

cgs units
1 erg = 10^{-7} J
1 dyne = 10^{-5} N
1 gauss = 10^{-4} T
1 emu = 10 C

mass–energy equivalence
1 kg = $8.99 \times 10^{16}\,\mathrm{J}/c^2$ (c in $\mathrm{m\,s^{-1}}$)
1 kg = $5.61 \times 10^{35}\,\mathrm{eV}/c^2$ (c in $\mathrm{m\,s^{-1}}$)

distance
1 astronomical unit (AU) = 1.496×10^{11} m
1 light-year (ly) = 9.461×10^{15} m = 0.307 pc
1 parsec (pc) = 3.086×10^{16} m = 3.26 ly

energy
1 eV = 1.602×10^{-19} J
1 J = 6.242×10^{18} eV

cross-sectional area
1 barn = $10^{-28}\,\mathrm{m^2}$
$1\,\mathrm{m^2} = 10^{28}$ barn

pressure
1 bar = 10^5 Pa
1 Pa = 10^{-5} bar
1 atmosphere = 1.013 25 bar
1 atmosphere = $1.013\,25 \times 10^5$ Pa

Table A.3 Constants.

Name of constant	Symbol	SI value
Fundamental constants		
gravitational constant	G	$6.673 \times 10^{-11}\,\mathrm{N\,m^2\,kg^{-2}}$
Boltzmann's constant	k	$1.381 \times 10^{-23}\,\mathrm{J\,K^{-1}}$
speed of light in vacuum	c	$2.998 \times 10^{8}\,\mathrm{m\,s^{-1}}$
Planck's constant	h	$6.626 \times 10^{-34}\,\mathrm{J\,s}$
	$\hbar = h/2\pi$	$1.055 \times 10^{-34}\,\mathrm{J\,s}$
fine structure constant	$\alpha = e^2/4\pi\varepsilon_0\hbar c$	$1/137.0$
Stefan–Boltzmann constant	σ	$5.671 \times 10^{-8}\,\mathrm{J\,m^{-2}\,K^{-4}\,s^{-1}}$
Thomson cross section	σ_T	$6.652 \times 10^{-29}\,\mathrm{m^2}$
permittivity of free space	ε_0	$8.854 \times 10^{-12}\,\mathrm{C^2\,N^{-1}\,m^{-2}}$
permeability of free space	μ_0	$4\pi \times 10^{-7}\,\mathrm{T\,m\,A^{-1}}$
Particle constants		
charge of proton	e	$1.602 \times 10^{-19}\,\mathrm{C}$
charge of electron	$-e$	$-1.602 \times 10^{-19}\,\mathrm{C}$
electron rest mass	m_e	$9.109 \times 10^{-31}\,\mathrm{kg}$ $= 0.511\,\mathrm{MeV}/c^2$
proton rest mass	m_p	$1.673 \times 10^{-27}\,\mathrm{kg}$ $= 938.3\,\mathrm{MeV}/c^2$
neutron rest mass	m_n	$1.675 \times 10^{-27}\,\mathrm{kg}$ $= 939.6\,\mathrm{MeV}/c^2$
atomic mass unit	u	$1.661 \times 10^{-27}\,\mathrm{kg}$
Astronomical constants		
mass of the Sun	$\mathrm{M}_\odot$	$1.99 \times 10^{30}\,\mathrm{kg}$
radius of the Sun	$\mathrm{R}_\odot$	$6.96 \times 10^{8}\,\mathrm{m}$
luminosity of the sun	$\mathrm{L}_\odot$	$3.83 \times 10^{26}\,\mathrm{W}$
mass of the Earth	$\mathrm{M}_\oplus$	$5.97 \times 10^{24}\,\mathrm{kg}$
radius of the Earth	$\mathrm{R}_\oplus$	$6.37 \times 10^{6}\,\mathrm{m}$
mass of Jupiter	$\mathrm{M_J}$	$1.90 \times 10^{27}\,\mathrm{kg}$
radius of Jupiter	$\mathrm{R_J}$	$7.15 \times 10^{7}\,\mathrm{m}$
astronomical unit	AU	$1.496 \times 10^{11}\,\mathrm{m}$
light-year	ly	$9.461 \times 10^{15}\,\mathrm{m}$
parsec	pc	$3.086 \times 10^{16}\,\mathrm{m}$
Hubble parameter	H_0	$(70.4 \pm 1.5)\,\mathrm{km\,s^{-1}\,Mpc^{-1}}$ $(2.28 \pm 0.05) \times 10^{-18}\,\mathrm{s^{-1}}$
age of Universe	t_0	$(13.73 \pm 0.15) \times 10^{9}$ years
current critical density	$\rho_{\mathrm{c},0}$	$(9.30 \pm 0.40) \times 10^{-27}\,\mathrm{kg\,m^{-}}$
current dark energy density	$\Omega_{\Lambda,0}$	$(73.2 \pm 1.8)\%$
current matter density	$\Omega_{\mathrm{m},0}$	$(26.8 \pm 1.8)\%$
current baryonic matter density	$\Omega_{\mathrm{b},0}$	$(4.4 \pm 0.2)\%$
current non-baryonic matter density	$\Omega_{\mathrm{c},0}$	$(22.3 \pm 0.9)\%$
current curvature density	$\Omega_{\mathrm{k},0}$	$(-1.4 \pm 1.7)\%$
current deceleration	q_0	-0.595 ± 0.025

Appendix B

Introduction

In this appendix, we shall take you on a very quick revision of special relativity. You will need the Lorentz transformation, time dilation and Lorentz contraction in this book, as well as to be able to use Einstein's mass–energy equivalence. Proving $E = mc^2$ will take us into discussions of four-vectors in this appendix, though four-vectors are not needed in themselves for this book. Where algebraic steps have been left out for brevity, enough information should be given for you to fill them in, should you wish to.

There isn't space to describe Einstein's ingenious thought experiments that led him to this theory, nor the many astonishing relativistic paradoxes. For these and more, consult a specialist text, such as Lambourne's *Relativity, Gravitation and Cosmology* published by Cambridge University Press.

B.1 Principles

The principles of special relativity are:

- There is no universal standard of rest.
- The speed of light (c) is invariant.

B.2 Feynman light clock

Consider a light clock (two mirrors with a light ray bouncing between them), shown in Figure B.1. The clock is moving with velocity v. One can use Pythagoras's theorem to show that

$$\delta t_1 = \gamma\, \delta t_0, \tag{B.1}$$

where

$$\gamma = \frac{1}{\sqrt{1 - v^2/c^2}}, \tag{B.2}$$

δt_1 is the time between reflections in the moving frame, and δt_0 is the time in the stationary frame. This can be remembered as 'moving clocks run slowly'. Sometimes the notation $\beta = v/c$ is used.

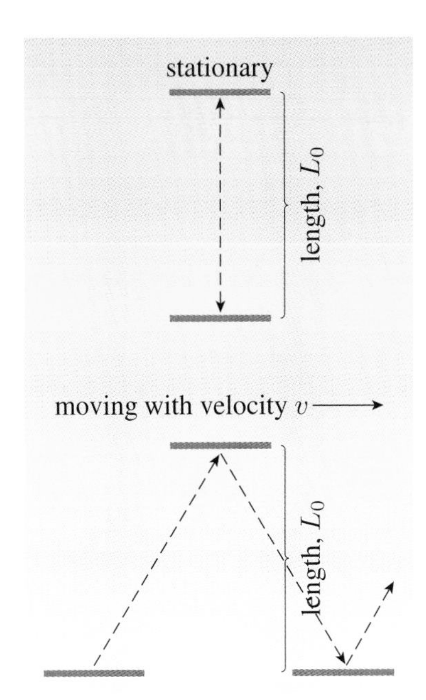

Figure B.1 A Feynman light clock, made of two mirrors at a fixed distance L_0, between which a light pulse bounces. In a stationary clock, bounces occur at intervals of $\delta t_0 = L_0/c$. In a moving clock, the intervals δt_1 between bounces are longer, because the light travels a distance $\sqrt{L_0^2 + (v\,\delta t_1)^2} > L_0$. Setting this distance equal to $c\,\delta t_1$ and rearranging, one obtains Equation B.1.

B.3 Lorentz contraction and simultaneity

In Figure B.2, the light pulses leave the corner simultaneously. The impacts at the top and side mirrors are simultaneous in the stationary frame, but not in the moving frame because the outward journey is longer than the return journey for the light ray moving parallel to the direction of motion. In general, there is no universal standard of simultaneity in special relativity.

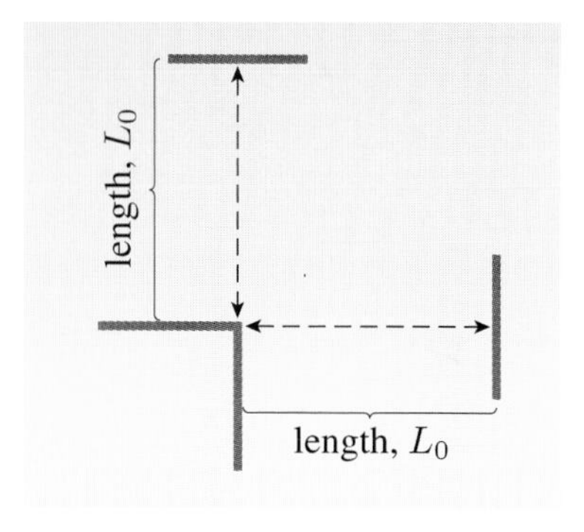

Figure B.2 A modified Feynman light clock, made of two sets of mirrors, both at a fixed rest-frame distance L_0, between which light pulses bounce (shown as dashed lines).

It can be shown that the length in the stationary frame L_0 (measured along the direction of motion) and the length in the moving frame L are related by

$$L = \frac{1}{\gamma} L_0. \tag{B.3}$$

Note that this is contraction, not dilation: 'moving rulers are short'. There is *no* contraction perpendicular to the motion. To prove this, imagine two circular hoops with the same rest-frame radius, both aligned to be perpendicular to the x-axis. One hoop moves along the x-axis. If one hoop passed inside the other, then there would be a preferred standard of rest, in contradiction with the first principle.

B.4 Lorentz transformation

The transformation of ct, x, y, z coordinates from one reference frame to another (which we denote as primed and unprimed coordinates) can be expressed as

$$\begin{pmatrix} ct' \\ x' \\ y' \\ z' \end{pmatrix} = \Lambda \begin{pmatrix} ct \\ x \\ y \\ z \end{pmatrix}, \tag{B.4}$$

where Λ is a 4×4 matrix (*not* to be confused with the cosmological constant). We assume that the origins of the coordinate systems coincide. For x-axis motion with velocity v,

$$\Lambda = \begin{pmatrix} \gamma & (-v/c)\gamma & 0 & 0 \\ (-v/c)\gamma & \gamma & 0 & 0 \\ 0 & 0 & 1 & 0 \\ 0 & 0 & 0 & 1 \end{pmatrix}. \tag{B.5}$$

This can be proved elegantly using only symmetries.

First, $y' = y$ and $z' = z$, because there is no Lorentz contraction perpendicular to the motion. Suppose that

$$\begin{pmatrix} ct' \\ x' \end{pmatrix} = \Lambda \begin{pmatrix} ct \\ x \end{pmatrix} = \begin{pmatrix} A & B \\ C & D \end{pmatrix} \begin{pmatrix} ct \\ x \end{pmatrix}.$$

(We neglect the y- and z-components for reasons of space.) Consider light rays in the positive and negative x-directions. From the principles of special relativity, the line $x = ct$ must transform to $x' = ct'$, i.e.

$$\begin{pmatrix} A & B \\ C & D \end{pmatrix} \begin{pmatrix} 1 \\ 1 \end{pmatrix} = a \begin{pmatrix} 1 \\ 1 \end{pmatrix},$$

where a is a non-zero constant, which implies that $A + B = C + D$. Similarly, $x = -ct$ must transform to $x' = -ct'$, i.e.

$$\begin{pmatrix} A & B \\ C & D \end{pmatrix} \begin{pmatrix} 1 \\ -1 \end{pmatrix} = b \begin{pmatrix} 1 \\ -1 \end{pmatrix},$$

where $b \neq 0$ is another constant, implying that $A - B = -(C - D)$. Together, these imply that $A = D$ and $B = C$, i.e. Λ is a symmetric matrix:

$$\Lambda = \begin{pmatrix} A & B \\ B & A \end{pmatrix}.$$

Clearly, if $x' = 0$, then $x = vt$, so

$$\begin{pmatrix} ct' \\ 0 \end{pmatrix} = \begin{pmatrix} A & B \\ B & A \end{pmatrix} \begin{pmatrix} ct \\ x \end{pmatrix}$$

gives $\quad 0 = Bct + Ax \quad$ and so $\quad B = (-v/c)A$. Hence

$$\Lambda = \begin{pmatrix} A & (-v/c)A \\ (-v/c)A & A \end{pmatrix}.$$

Finally, we must have

$$\begin{pmatrix} ct \\ x \end{pmatrix} = \Lambda^{-1} \begin{pmatrix} ct' \\ x' \end{pmatrix},$$

where Λ^{-1} is the inverse matrix to Λ. By symmetry, we must also have

$$\Lambda^{-1} = \begin{pmatrix} A & (+v/c)A \\ (+v/c)A & A \end{pmatrix}.$$

The equation $\Lambda\Lambda^{-1} = I$ (where I is the identity matrix) can be solved to show that $A = \gamma$, as required.

B.5 Invariants

Using the Lorentz transformation, one can show that the **interval** δs is invariant (i.e. the same in all reference frames) under Lorentz transformations, where

$$(\delta s)^2 = c^2(\delta t)^2 - (\delta x)^2 - (\delta y)^2 - (\delta z)^2.$$

Note that $\delta s = c\,\delta\tau$, where τ is the proper time. Note also that $\delta s = 0$ for photons.

We can write this as

$$(\delta s)^2 = \sum_{\alpha,\beta} \eta_{\alpha\beta}\, \delta x^\alpha\, \delta x^\beta,$$

where x^α does not mean 'x to the power of α', but rather in this context refers to the components of the four-vector (ct, x, y, z). The convention is for this to count from zero, i.e. $x^0 = ct$, $x^1 = x$, $x^2 = y$, $x^3 = z$. One can only apologize for the obvious inadequacies of this very common notation. It is usually clear from the context whether superscripts refer to components, or mean 'to the power of'.

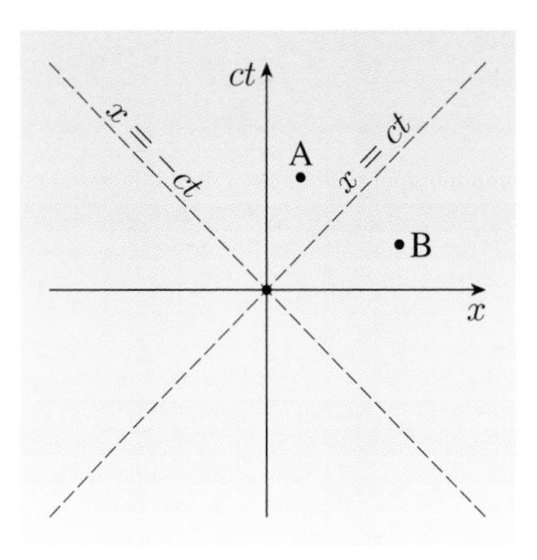

Figure B.3 Special relativistic lightcone diagram. The origin and point A have a time-like separation for all observers, while the origin and point B have a space-like separation for all observers.

The matrix

$$\eta = \begin{pmatrix} 1 & 0 & 0 & 0 \\ 0 & -1 & 0 & 0 \\ 0 & 0 & -1 & 0 \\ 0 & 0 & 0 & -1 \end{pmatrix}$$

is called the **metric tensor**. Real intervals are known as time-like, and imaginary intervals as space-like (see Figure B.3). (Note that some textbooks use $\text{diag}(-1, 1, 1, 1)$, resulting in the opposite convention for δs for space-like and time-like intervals.) This metric is sometimes known as **Minkowski spacetime**.

In general, if A^α and B^α are the components of four-vectors, then

$$\sum_{\alpha,\beta} \eta_{\alpha\beta}\, A^\alpha B^\beta$$

is invariant. We define the scalar product of four-vectors as

$$\boldsymbol{A} \cdot \boldsymbol{B} = \sum_{\alpha,\beta} \eta_{\alpha\beta}\, A^{\alpha} B^{\beta},$$

so $s = \sqrt{\boldsymbol{x} \cdot \boldsymbol{x}}$ is the invariant 'length' of the position four-vector $\boldsymbol{x} = (ct, x, y, z)$.

Other four-vectors Lorentz transform in the same way as the position four-vector. (This is, in fact, the definition of a four-vector.) In the following sections we shall introduce some useful four-vectors and some useful invariant lengths.

B.6 Position four-vector

As mentioned above, the position four-vector is (ct, x, y, z), or x^{α} with $\alpha = 0, 1, 2, 3$ in component notation. The invariant length (squared) is $\boldsymbol{x} \cdot \boldsymbol{x} = s^2$.

B.7 Velocity and acceleration four-vectors

We differentiate the position four-vector x^{α} with respect to proper time τ (an invariant scalar) to obtain another four-vector, the velocity four-vector, the components of which are:

$$U^{\alpha} = \frac{\mathrm{d}x^{\alpha}}{\mathrm{d}\tau}$$

(except for photons).

The invariant length (squared) of any velocity four-vector can be shown to be $\boldsymbol{U} \cdot \boldsymbol{U} = -c^2$.

If one differentiates again with respect to τ, one can show that $\boldsymbol{U} \cdot \boldsymbol{a} = 0$, where $\boldsymbol{a}$ is the four-acceleration, the components of which are:

$$a^{\alpha} = \frac{\mathrm{d}U^{\alpha}}{\mathrm{d}\tau}.$$

Hence four-acceleration is 'orthogonal' to four-velocity.

B.8 Relationship between four- and three-velocities

We consider x-axis motion only, for simplicity. Writing u for the three-velocity, the components of the four-velocity are

$$U^0 = c\frac{\mathrm{d}t}{\mathrm{d}\tau} = \frac{c}{\sqrt{1 - u^2/c^2}} = \gamma(u)\, c,$$

$$U^1 = \frac{\mathrm{d}x}{\mathrm{d}\tau} = \frac{\mathrm{d}x/\mathrm{d}t}{\mathrm{d}t/\mathrm{d}\tau} = \gamma(u)\, u,$$

$$U^2 = \frac{\mathrm{d}y}{\mathrm{d}\tau} = 0,$$

$$U^3 = \frac{\mathrm{d}z}{\mathrm{d}\tau} = 0.$$

In general, if $u^i = \mathrm{d}x^i/\mathrm{d}t$ is the three-velocity (where $i = 1, 2, 3$), then

$$U^\alpha = \gamma \times (c, u^i)$$

is the four-velocity, where γ is a function of $|\boldsymbol{u}|$.

The Lorentz transformation of the velocity four-vector can be used to derive the addition law for three-velocities:

$$u_{\text{total}} = \frac{u + v}{1 + uv/c^2}. \tag{B.6}$$

This also applies if one of the velocities is c. For example, the headlights on a car moving on a road at speed u can send photons out at $v = c$ in the car's reference frame. Nevertheless, according to a stationary observer beside the road, the photons' speed is still c, not $v + c$. (Try substituting $v = c$ into Equation B.6.)

B.9 Momentum four-vector

This is defined for a massive particle as $P^\alpha = mU^\alpha$, where m is the rest mass. Therefore

$$P^\alpha = m\,\frac{\mathrm{d}x^\alpha}{\mathrm{d}\tau},$$

so

$$P^\alpha = m\gamma \times (c, u^i).$$

B.10 Force four-vector

Non-relativistically, force is the rate of change of momentum, so we define the four-force vector components as

$$F^\alpha = \frac{\mathrm{d}P^\alpha}{\mathrm{d}\tau} = \frac{\mathrm{d}}{\mathrm{d}\tau}\left(m\,\frac{\mathrm{d}x^\alpha}{\mathrm{d}\tau}\right) = ma^\alpha.$$

From this one can show that

$$F^\alpha = \left(\frac{\mathrm{d}P^\alpha}{\mathrm{d}\tau}, \gamma f^i\right),$$

where f^i is the relativistic force three-vector:

$$f^i = \frac{\mathrm{d}}{\mathrm{d}\tau}\left(\gamma m\,\frac{\mathrm{d}x^i}{\mathrm{d}\tau}\right) = \frac{\mathrm{d}}{\mathrm{d}\tau}\left(\gamma m u^i\right).$$

B.11 $E = mc^2$

To reach Einstein's famous equation, we start from $\boldsymbol{U} \cdot \boldsymbol{a} = 0$:

$$\begin{aligned} 0 &= \sum_{\alpha,\beta} U^\alpha m a^\beta = \sum_{\alpha,\beta} U^\alpha F^\beta \\ &= -U^0\,\frac{\mathrm{d}P^0}{\mathrm{d}\tau} + U^i \gamma f^i = -c\,\frac{\mathrm{d}t}{\mathrm{d}\tau}\,\frac{\mathrm{d}(\gamma mc)}{\mathrm{d}\tau} + \frac{\mathrm{d}x^i}{\mathrm{d}\tau}\,\gamma f^i. \end{aligned}$$

Using the chain rule, we get

$$0 = -c\frac{dt}{d\tau}\frac{dt}{d\tau}\frac{d(\gamma mc)}{dt} + \frac{dt}{d\tau}\frac{dx^i}{dt}\gamma f^i.$$

But $dt/d\tau = \gamma$, so

$$0 = -c\gamma^2\frac{d(\gamma mc)}{dt} + \gamma\frac{dx^i}{dt}\gamma f^i$$

thus

$$f^i\frac{dx^i}{dt} = \frac{d}{dt}(\gamma mc^2).$$

In non-relativistic physics, we have

$$f^i\frac{dx^i}{dt} = \frac{dE}{dt},$$

where E is the energy. Therefore we choose to identify energy with $E = \gamma mc^2 +$ a constant, and we can assume that the constant is zero without loss of generality. One can show using a Taylor series that $E(u) = mc^2 + \frac{1}{2}mu^2 +$ higher-order terms starting with the order (mu^4/c^2). The second term here is the non-relativistic kinetic energy, $\frac{1}{2}mu^2$. For a stationary object, $u = 0$, so it has energy $E = mc^2$. Rest mass therefore has an equivalent energy. Also, $P^\alpha = (E/c, m\gamma u^i)$. By considering the invariant length of the momentum four-vector, one can show that

$$m^2c^4 = -E^2 + p^2c^2, \tag{B.7}$$

where $p = m\gamma u^i$ is the relativistic three-momentum.

B.12 Photons

Photons have zero rest mass, but nevertheless carry momentum and energy consistent with Equation B.7: $E = pc$. The four-velocity is not defined for a photon, but the four-momentum is $(E/c, p_xc, p_yc, p_zc)$, where p_x is the x-component of the relativistic three-momentum, and so on. The invariant interval δs along any two points on a light ray is always zero.

One curious and little-known aspect of special relativity is that it implies Planck's famous formula $E = h\nu$, but doesn't give a value for h. If we consider a photon with energy E moving along the x-axis, and Lorentz transform to the frame of an observer also moving along the x-axis with speed v, one can show that $E'/E = \sqrt{(c+v)/(c-v)}$. This is the relativistic Doppler shift (but don't confuse it with cosmological redshift in Chapter 1).

Alternatively, we could consider a monochromatic plane wave. We can define a wave four-vector as $\boldsymbol{k} = (\omega/c, k_x, k_y, k_z)$, where the three-vector (k_x, k_y, k_z) points along the direction of the wave, and $\omega = 2\pi\nu$, where ν is the frequency. (To see why $\boldsymbol{k}$ must be a four-vector, note that the phase ϕ must be an invariant scalar, and that $\boldsymbol{k}\cdot\boldsymbol{x} = \phi$.) The wavelength is $\lambda = 2\pi/\sqrt{k_x^2 + k_y^2 + k_z^2}$. A Lorentz transformation of the wave four-vector of a wave moving along the x-axis leads eventually to $\nu'/\nu = \sqrt{(c+v)/(c-v)}$. Therefore $E'/E = \nu'/\nu$, or $E \propto \nu$.

Solutions to exercises

Exercise 1.1 When we calculated that the sky is as bright as the Sun, we assumed that the line of sight stopped on the star, i.e. stars are opaque. When we calculated the brightness of the sky for a $S^{-5/2}$ power law, we integrated down to zero flux, which (for any particular type of star) means integrating to $r = \infty$. So the lines of sight don't stop on stars in the latter case; stars are treated as transparent.

Exercise 1.2 We use $E = \gamma m_0 c^2$, where E is the energy, m_0 is the rest mass, $\gamma = (1 - v^2/c^2)^{-1/2}$, and c is the speed of light. We have that $10^{20}\,\mathrm{eV} = \gamma c^2 \times 938.28\,\mathrm{MeV}/c^2 \simeq \gamma \times 10^9\,\mathrm{eV}$. The quoted accuracy of the energy does not justify carrying more than just the first significant figure on the proton's rest mass. The γ factor is then just $\gamma = 10^{20}/10^9 = 10^{11}$. The cosmic ray is moving at very close to the speed of light, so it would take about 100 000 years for the proton to cross the Galaxy in the Galaxy's rest frame. But moving clocks run slow, so it would take $100\,000/\gamma$ years in the proton's rest frame, i.e. $10^5/10^{11}$ years, or 10^{-6} years, or about 30 seconds!

Exercise 1.3 First we differentiate Equation 1.7 with respect to time t to get

$$2\dot{R}\ddot{R} = \frac{8\pi G}{3}\left(\dot{\rho}R^2 + 2\rho R\dot{R}\right) + \frac{2\Lambda c^2 R\dot{R}}{3}, \tag{S1.1}$$

where we write $\rho = \rho_\mathrm{m} + \rho_\mathrm{r}$ for brevity and the 'dot' notation is used to indicate differentiation with respect to time, i.e. $\dot{R} = \mathrm{d}R/\mathrm{d}t$ and $\ddot{R} = \mathrm{d}^2R/\mathrm{d}t^2$. The conservation of matter energy gives

$$\begin{aligned}\frac{\mathrm{d}}{\mathrm{d}t}\left(\rho c^2 R^3\right) &= \dot{\rho}c^2R^3 + 3\rho c^2 R^2\dot{R}\\ &= -p\,\frac{\mathrm{d}(R^3)}{\mathrm{d}t}\\ &= -3pR^2\dot{R},\end{aligned}$$

so

$$\dot{\rho}c^2R^3 = -3pR^2\dot{R} - 3\rho c^2R^2\dot{R}.$$

Equation S1.1 has a term $\dot{\rho}R^2$, so we rearrange the above to find

$$\begin{aligned}\dot{\rho}R^2 &= \frac{-3pR\dot{R}}{c^2} - 3\rho R\dot{R}\\ &= -R\dot{R}\left(\frac{3p}{c^2} + 3\rho\right).\end{aligned}$$

Substituting this into Equation S1.1 gives

$$\begin{aligned}2\dot{R}\ddot{R} &= \frac{8\pi G}{3}\left\{2\rho R\dot{R} - R\dot{R}\left(\frac{3p}{c^2} + 3\rho\right)\right\} + \frac{2\Lambda c^2 R\dot{R}}{3}\\ &= \frac{8\pi G R\dot{R}}{3}\left(2\rho - \frac{3p}{c^2} - 3\rho\right) + \frac{2\Lambda c^2}{3}R\dot{R}\\ &= \frac{-8\pi G R\dot{R}}{3}\left(\rho + \frac{3p}{c^2}\right) + \frac{2\Lambda c^2}{3}R\dot{R}\\ &= -8\pi G\left(\rho + \frac{3p}{c^2}\right)\frac{R\dot{R}}{3} + \frac{2\Lambda c^2}{3}R\dot{R}.\end{aligned}$$

Dividing this by $2\dot{R}$ gives

$$\begin{aligned}\ddot{R} &= -4\pi G\left(\rho + \frac{3p}{c^2}\right)\frac{R}{3} + \frac{\Lambda c^2 R}{3}\\ &= -4\pi G\left(\rho_m + \rho_r + \frac{3p}{c^2}\right)\frac{R}{3} + \frac{\Lambda c^2 R}{3},\end{aligned}$$

as required.

Exercise 1.4 If $\Lambda = 0$, then Ω_Λ is always zero (Equation 1.17). From Equation 1.33, we therefore have that $(H/H_0)^2 = (1+z)^3$ when $\Omega_m = 1$ and $\Lambda = 0$. Now, Equation 1.28 tells us that

$$H = \frac{-1}{1+z}\frac{dz}{dt},$$

so

$$\frac{1}{H_0^2}\frac{1}{(1+z)^2}\left(\frac{dz}{dt}\right)^2 = (1+z)^3,$$

which we may write more simply as $dz/dt \propto (1+z)^{5/2}$, or $dt/dz \propto (1+z)^{-5/2}$. Integrating this with respect to z, we get $t \propto (1+z)^{-3/2}$. But $1 + z = R_0/R$, so $t \propto R^{3/2}$, or

$$R = \alpha t^{2/3}, \tag{S1.2}$$

where α is some constant. In particular, at the current time $t = t_0$ we have

$$R_0 = \alpha t_0^{2/3}, \tag{S1.3}$$

and dividing Equation S1.2 by Equation S1.3 gives $R/R_0 = (t/t_0)^{2/3}$.

Exercise 1.5 We can rearrange Equation 1.35 to read

$$R = R_0\left(\frac{t}{t_0}\right)^{2/3}. \tag{S1.4}$$

From Equation 1.12, we have that $H = (1/R)\,dR/dt$. Differentiating Equation S1.4, we get

$$\frac{dR}{dt} = \frac{2}{3}\frac{R_0}{t_0^{2/3}}t^{-1/3}.$$

At a time $t = t_0$, this is just

$$\left.\frac{dR}{dt}\right|_{t=t_0} = \frac{2}{3}\frac{R_0}{t_0}.$$

Therefore the Hubble parameter at a time $t = t_0$ in this model Universe is

$$H_0 = \frac{1}{R_0}\left.\frac{dR}{dt}\right|_{t=t_0} = \frac{1}{R_0}\frac{2}{3}\frac{R_0}{t_0} = \frac{2}{3t_0},$$

or $t_0 = 2/(3H_0)$ as required. Putting in $H_0 = 72 \pm 3\,\text{km}\,\text{s}^{-1}\,\text{Mpc}^{-1}$, we find $t_0 = 9.1 \pm 0.4\,\text{Gyr}$.

Exercise 1.6 The angular diameter in degrees will be inversely proportional to d_A (Equation 1.47), so the angular area (e.g. in square degrees) will vary as

$\theta^2 \propto d_{\rm A}^{-2}$. The flux will be inversely proportional to $d_{\rm L}^2$ (Equation 1.49), i.e. $S \propto d_{\rm L}^{-2}$. The surface brightness will therefore vary as $S/\theta^2 \propto d_{\rm A}^2/d_{\rm L}^2$. But $d_{\rm L} = (1+z)^2 d_{\rm A}$ (Equation 1.50), so surface brightness must vary as $(1+z)^{-4}$.

Exercise 1.7 In Section 1.5 we are given that $H_0 = 72 \pm 3\,{\rm km\,s^{-1}\,Mpc^{-1}}$ and $\Omega_{\Lambda,0} = 0.742 \pm 0.030$. One parsec is 3.09×10^{16} m, so in SI units, $H_0 = (2.3 \pm 0.1) \times 10^{-18}\,{\rm s^{-1}}$. Equation 1.17 relates these two quantities to Λ: $\Omega_{\Lambda,0} = \Lambda c^2/(3H_0^2)$, so $\Lambda = 3\Omega_{\Lambda,0}\, H_0^2/c^2$. Putting in the numbers, we get $\Lambda = (1.3 \pm 0.2) \times 10^{-52}\,{\rm m^{-2}}$. The horizon size will be $\sqrt{3/\Lambda} = (1.5 \pm 0.1) \times 10^{26}$ m, or 4900 ± 300 Mpc. This cosmological event horizon will be exceedingly distant; for comparison, the current radius of the observable Universe in Section 1.9 is about $3.53c/H_0 = 14\,900$ Mpc.

Exercise 2.1 The 13.6 eV photon *does* ionize another atom. However, the process of recombination needn't result in the emission of just *one* photon. Sometimes the electron will bind first in a high energy state (releasing one photon with an energy < 13.6 eV), then release the remaining energy in stages as the electron drops down the energy levels of the hydrogen atom. Each of these stages will involve the release of a photon, but none of these photons will have enough energy on its own to ionize hydrogen atoms.

Exercise 2.2 We are given that $T = 2.725 \pm 0.001$ K, so the energy density must be $\rho_{\rm r,0}\, c^2 = 4\sigma T^4/c = 4 \times 5.67 \times 10^{-8} \times 2.725^4/(3.00 \times 10^8)$ joules per cubic metre, i.e. $\rho_{\rm r,0}\, c^2 = 4.17 \times 10^{-14}\,{\rm J\,m^{-3}}$, or mass-equivalent density of $\rho_{\rm r,0} = 4.64 \times 10^{-31}\,{\rm kg\,m^{-3}}$. Applying Equation 1.16, and remembering that $H_0 = 100h\,{\rm km\,s^{-1}\,Mpc^{-1}} = 3.24 \times 10^{-18} h\,{\rm s^{-1}}$, we find that

$$\Omega_{\rm r,0} = \frac{8\pi G\,\rho_{\rm r,0}}{3H_0^2} = 2.47h^{-2} \times 10^{-5}.$$

So

$$\Omega_{\rm r,0}\, h^2 \simeq 2.5 \times 10^{-5},$$

as required.

Exercise 2.3 The matter energy density scales as R^{-3}, while the photon/neutrino energy density scales as R^{-4}. Therefore from Equations 1.15 and 1.16, $\Omega_{\rm r}/\Omega_{\rm m} = (1+z)\,\Omega_{\rm r,0}/\Omega_{\rm m,0}$. From Exercise 2.2 and the text following it, we have that $\Omega_{\rm r,0}\, h^2 \simeq 4.2 \times 10^{-5} (T_{\rm CMB,0}/2.725\,{\rm K})^4$. The epoch of matter–radiation equality must by definition satisfy $\Omega_{\rm r}/\Omega_{\rm m} = 1$, so

$$\begin{aligned} 1 + z_{\rm eq} &= \frac{\Omega_{\rm m,0}}{\Omega_{\rm r,0}} \\ &= \frac{h^2}{4.2 \times 10^{-5}}\, \Omega_{\rm m,0}\, (T_{\rm CMB,0}/2.725\,{\rm K})^{-4} \\ &\simeq 23\,800\, \Omega_{\rm m,0}\, h^2 (T_{\rm CMB,0}/2.725\,{\rm K})^{-4}, \end{aligned}$$

as required.

Exercise 2.4 The analysis is the same up to Equation 1.30, where ρ this time is $\rho_{\rm r}$. However, instead of $\rho = \rho_0 \times R_0^3/R^3$, we must also take into account the fact that photons lose energy from redshifting, so $\rho_{\rm r} = \rho_0 \times R_0^4/R^4$. With Λ set to zero, the equivalent of Equation 1.32 comes out as

$$\left(\frac{H}{H_0}\right)^2 = (1+z)^2 \left(1 - \Omega_{\rm r,0} + \Omega_{\rm r,0}\,(1+z)^2\right),$$

and inserting $\Omega_{\rm r,0} = 1$ and using $H^2 = (1+z)^{-2}\,({\rm d}z/{\rm d}t)^2$, we find that

$$\left(\frac{{\rm d}z}{{\rm d}t}\right)^2 = H_0^2(1+z)^6$$

so

$$\frac{{\rm d}z}{{\rm d}t} = \frac{{\rm d}(1+z)}{{\rm d}t} = H_0(1+z)^3.$$

Now the dimensionless scale factor a is related to redshift via $a = 1/(1+z)$, so we could write this as

$$\frac{{\rm d}(a^{-1})}{{\rm d}t} = H_0 a^{-3}$$

thus

$$-a^{-2}\frac{{\rm d}a}{{\rm d}t} = H_0 a^{-3}$$

hence

$$a\,{\rm d}a \propto {\rm d}t.$$

Integrating this gives $a^2 \propto t$, or $a \propto t^{1/2}$ as required.

Exercise 2.5 $\hbar$ is measured in J s. A Joule has dimensions of energy (like $\frac{1}{2}mv^2$) so it has dimensions $\rm ML^2T^{-2}$, where we write M for the dimension of mass, L for length, and T for time. (Note that numerical constants are ignored in dimensional analysis.) Therefore we can write the dimensions of $\hbar$ as $[\hbar] = \rm ML^2T^{-1}$. Similarly, the dimensions of c are $[c] = \rm LT^{-1}$. To find the dimensions of G, we can start with the familiar equation $F = GMm/r^2$, and note that force is mass times acceleration, so $ma = GMm/r^2$ or $G = ar^2/M$, so the dimensions of G are $[G] = \rm LT^{-2}L^2/M = M^{-1}L^3T^{-2}$. Now let's suppose that the Planck time is given by a formula of the form $\hbar^x c^y G^z$, where the constants x, y and z are to be determined. The result must have the dimensions of time, so

$$\rm T = \left(ML^2T^{-1}\right)^{\mathit{x}} \left(LT^{-1}\right)^{\mathit{y}} \left(M^{-1}L^3T^{-2}\right)^{\mathit{z}}.$$

Multiplying this out and rearranging gives

$$\rm T = M^{\mathit{x-z}} L^{2\mathit{x}+\mathit{y}+3\mathit{z}} T^{-\mathit{x}-\mathit{y}-2\mathit{z}}.$$

The left-hand side has no mass M, so $x - z$ must equal zero, i.e. $x = z$. The left-hand side also has no length L, so $2x + y + 3z = 0$. The left-hand side has exactly one power of T, so $-x - y - 2z = 1$. We have three simultaneous equations for three unknowns. Substituting in $x = z$ into the other two equations gives $5x + y = 0$ and $-3x - y = 1$. Therefore $y = -1 - 3x = -5x$, or $x = 1/2$. Since $x = z$, we have $z = 1/2$. Finally, any of the equations involving y imply that $y = -5/2$. Therefore the characteristic time must be of the form $\hbar^x c^y G^z = \hbar^{1/2}c^{-5/2}G^{1/2} = \sqrt{\hbar G/c^5}$, as required.

Exercise 2.6 We have already that $(1/R)\,{\rm d}^2R/{\rm d}t^2 = \alpha(\alpha-1)t^{-2}$. Since t is positive and $\alpha > 1$, the right-hand side must be positive. Therefore the left-hand side must also be positive. Since R is also positive, ${\rm d}^2R/{\rm d}t^2 > 0$.

Exercise 2.7 We start with

$$3H\dot{\phi} = -V' \qquad \text{(Eqn 2.23)}$$

and then use

$$H^2 = \frac{8\pi}{3m_{\mathrm{Pl}}^2}V. \qquad \text{(Eqn 2.24)}$$

Now the $H\,\mathrm{d}t$ term in the integral in the question can also be expressed as

$$H\,\mathrm{d}t = H\,\frac{\mathrm{d}t}{\mathrm{d}\phi}\,\mathrm{d}\phi = H\,\frac{\mathrm{d}\phi}{\dot{\phi}}.$$

Next we use Equation 2.23 to get

$$H\,\mathrm{d}t = H\,\frac{\mathrm{d}\phi}{(-V'/3H)} = -3H^2\,\frac{\mathrm{d}\phi}{V'}.$$

Finally, using Equation 2.24 this comes out as

$$H\,\mathrm{d}t = \frac{-8\pi}{m_{\mathrm{Pl}}^2}\left(\frac{V}{V'}\right)\mathrm{d}\phi,$$

so we reach the required integral:

$$N = \frac{-8\pi}{m_{\mathrm{Pl}}^2}\int_{\phi_2}^{\phi_1}\frac{V}{V'}\,\mathrm{d}\phi.$$

For the next part, we set $V' \simeq V/\phi$ and $\phi_1 = 0$ (as advised in the question) to write this as

$$N = \frac{-8\pi}{m_{\mathrm{Pl}}^2}\int_{\phi_2}^{0}\frac{V}{V}\phi\,\mathrm{d}\phi.$$

Evaluating this integral gives

$$N = \frac{4\pi}{m_{\mathrm{Pl}}^2}\phi_2^2 = \left(\frac{2\sqrt{\pi}\phi_2}{m_{\mathrm{Pl}}}\right)^2.$$

Thus to have $N > 60$ we need $\phi_2 > m_{\mathrm{Pl}}\sqrt{60}/(2\sqrt{\pi})$, or in other words, $\phi_2 > 2.2m_{\mathrm{Pl}}$.

Exercise 2.8 No, not immediately. At first the CMB will appear very uniform, as you receive light from only your immediate neighbourhood. As time progresses you will receive light from larger and more distant parts of the Universe. You'll only be able to see the structures with wavelength λ once light has had time to travel the distance λ, i.e. after a time $\delta t = \lambda/c$, where c is the speed of light. The size of the largest acoustic peak is set by the sound horizon after inflation. Once light has had time to travel this distance, all the acoustics will start to become visible. Also, the acoustic peaks will have a different *angular* size on the sky, because the surface of last scattering was closer. Finally, the CMB wouldn't have peaked at microwave wavelengths then, so perhaps we shouldn't call it the CMB then!

Exercise 2.9 We found in Section 2.7 that the particle horizon radius at recombination was $2c/H = 0.46\,\mathrm{Mpc}$. The sound speed is $c_{\mathrm{s}} = c/\sqrt{3}$, so the sound horizon will be $2c_{\mathrm{s}}/H = (2c/H) \times (c_{\mathrm{s}}/c) = 0.46/\sqrt{3}\,\mathrm{Mpc} = 0.27\,\mathrm{Mpc}$.

Exercise 2.10 Dark matter clumps through gravitation, while dark energy appears to be smoothly distributed through space. Dark matter is also essentially pressureless, with Ω_{m} dominated by the rest mass of the dark matter particles, while dark energy has a strong negative pressure. Dark matter makes up about 20% of the total energy density of the Universe, and at recombination made up about 70%. Dark energy, meanwhile, was negligible at recombination and yet dominates the present-day energy density of the Universe. (One hopes that it will soon be possible to add that the dark matter particle has been directly detected, though that is not yet true at the time of writing; certainly, the proposed particle physics mechanisms for generating dark matter and dark energy are very different.)

Exercise 2.11 One parsec is about 3.09×10^{16} m, so $H_0 = 72 \times 10^3/(10^6 \times 3.09 \times 10^{16}) \simeq 2.33 \times 10^{-18}\,\mathrm{s}^{-1}$. In Chapter 1 we saw that $\Omega_{\Lambda,0} = \Lambda c^2/(3H_0^2)$, so $\Lambda = 3\Omega_{\Lambda,0}\,H_0^2/c^2$. Putting in the numbers gives $\Lambda = 1.3 \times 10^{-52}\,\mathrm{m}^{-2}$.

Exercise 3.1 The luminosity contributed by a shell of radius $r \to r + \mathrm{d}r$ will be $I(r)$ times the area of the shell, $2\pi r\,\mathrm{d}r$. Summing these shells, the total luminosity will be $L = \int_0^\infty I(r)\,2\pi r\,\mathrm{d}r$. Let's define L_0 to be the luminosity with $I_0 = r_0 = 1$, i.e.

$$L_0 = \int_0^\infty f(r)\,2\pi r\,\mathrm{d}r.$$

Now let's calculate the luminosity in the more general case:

$$\begin{aligned} L &= \int_0^\infty I_0\,f\left(\frac{r}{r_0}\right)\,2\pi r\,\mathrm{d}r \\ &= I_0 r_0^2 \int_0^\infty f\left(\frac{r}{r_0}\right)\,2\pi\frac{r}{r_0}\,\mathrm{d}\left(\frac{r}{r_0}\right). \end{aligned}$$

But this integral has the same form as the integral defining L_0, which also integrates from 0 to ∞, so $L = I_0 r_0^2 L_0$.

Exercise 3.2 A shell of thickness $\mathrm{d}r$ and radius r will have mass $\mathrm{d}M = 4\pi r^2 \rho\,\mathrm{d}r$. The gravitational potential energy of this shell will be

$$\mathrm{d}E_{\mathrm{GR}} = \frac{-G\,M(<r)\,\mathrm{d}M}{r}, \qquad \text{(S3.1)}$$

where $M(<r)$ is the mass enclosed within a radius r, i.e.

$$M(<r) = \tfrac{4}{3}\pi r^3 \rho,$$

and the mass of the shell is

$$\mathrm{d}M = 4\pi r^2 \rho\,\mathrm{d}r.$$

Substituting this into Equation S3.1 gives

$$\mathrm{d}E_{\mathrm{GR}} = \frac{-G\frac{4}{3}\pi r^3 \rho}{r}\,\mathrm{d}M = -G\tfrac{4}{3}\pi r^2 \rho \times 4\pi r^2 \rho\,\mathrm{d}r$$

so

$$\mathrm{d}E_{\mathrm{GR}} = -3G \times \left(\tfrac{4}{3}\pi r^2 \rho\right)^2 \mathrm{d}r.$$

Integrating this from radius 0 to radius R gives

$$\begin{aligned}
E_{\rm GR} &= -3G \int_0^R \left(\tfrac{4}{3}\pi r^2 \rho\right)^2 \, dr = -3G \left(\tfrac{4}{3}\pi\rho\right)^2 \frac{R^5}{5} \\
&= \frac{-3G}{5R} \left(\tfrac{4}{3}\pi R^3 \rho\right)^2 \\
&= -\frac{3GM^2}{5R},
\end{aligned}$$

where $M = \frac{4}{3}\pi R^3 \rho$ is the total mass of the sphere.

Exercise 3.3 The kinetic energy will be $E_{\rm K} = \frac{3}{2}NkT$, where N is the number of gas particles. Virial equilibrium is $2E_{\rm K} = -E_{\rm GR}$, i.e.

$$3NkT = \frac{3}{5}\frac{GM^2}{R}.$$

The requirement for gravitational collapse is therefore

$$3NkT < \frac{3}{5}\frac{GM^2}{R}.$$

To reach Equation 3.7, we need to eliminate N and R. To a good approximation, at recombination we can assume that the gas particle masses are the proton mass $m_{\rm p}$, so the number of particles must be $N = M/m_{\rm p}$. We can also use $M = \frac{4}{3}\pi R^3$ to eliminate R, since $R = (3M/4\pi\rho)^{1/3}$. Inserting these substitutions gives

$$3\frac{M}{m_{\rm p}}kT < \frac{3}{5}GM^2 \left(\frac{4\pi\rho}{3M}\right)^{1/3},$$

which when rearranged in terms of M gives the required equation.

The current temperature of the CMB is about 2.7 K, and the redshift of recombination is about $z = 1000$, so the photon temperature at recombination must be $T = 2.7(1+z) \simeq 3000\,{\rm K}$. Matter and radiation will just have been in thermal equilibrium, so this will have been the matter temperature too. The baryonic density will be proportional to $(1+z)^3$, and using Equation 1.26 and $\rho_{\rm b} = \Omega_{\rm b}\,\rho_{\rm crit}$ (Equation 1.22), we have that the baryonic density at $z = 1000$ will be

$$\begin{aligned}
\rho_{\rm b} &= \rho_{\rm b,0}(1+z)^3 \\
&= \rho_{\rm crit} \times \Omega_{\rm b,0}\,(1+z)^3 \\
&= 1.8789 \times 10^{-26} \times \Omega_{\rm b,0}\,h^2 (1+z)^3\,{\rm kg\,m^{-3}} \\
&= 1.8789 \times 10^{-26} \times 2.273 \times 10^{-2} \times (1+1000)^3\,{\rm kg\,m^{-3}} \\
&\simeq 4.3 \times 10^{-19}\,{\rm kg\,m^{-3}}.
\end{aligned}$$

Putting in the numbers gives

$$\begin{aligned}
M &> \left(\frac{5 \times (1.381 \times 10^{-23}\,{\rm J\,K^{-1}}) \times 3000\,{\rm K}}{(6.673 \times 10^{-11}\,{\rm N\,m^2\,kg^{-2}}) \times (1.673 \times 10^{-27}\,{\rm kg})}\right)^{3/2} \times \left(\frac{3}{4\pi \times 4.3 \times 10^{-19}\,{\rm kg\,m^{-3}}}\right)^{1/2} \\
&\simeq 2 \times 10^{36}\,{\rm kg},
\end{aligned}$$

or $M > 10^6\,{\rm M_\odot}$, as required.

Exercise 3.4 For a *flat* universe, the comoving distance is the same as the proper motion distance (Equation 1.56). This isn't true in general (watch out!) but it's true in a flat universe. The proper motion distance is related to the angular diameter distance d_A by Equation 1.50, which gives $d_A = d_{\text{comoving}}/(1+z)$. The definition of angular diameter distance in Equation 1.47 gives us a relationship between the size of an object *as it was at the time of redshift* z and the angular size as it appears today. The proper size of the BAO wiggles is just the comoving size divided by $(1+z)$, i.e. $L_{\text{BAO}}/(1+z)$. The angular diameter distance to redshift z is therefore $d_A = (L_{\text{BAO}}/(1+z))/\theta_{\text{BAO}}$. The comoving distance to redshift z must therefore be $d_{\text{comoving}} = d_A \times (1+z) = L_{\text{BAO}}/\theta_{\text{BAO}}$, as required.

Exercise 3.5 Here the trick is to use Equation 1.43. It follows from that relation that a small comoving interval along the redshift axis must equal $\delta d_{\text{comoving}} = c\,\delta z/H(z)$. Setting this comoving interval to L_{BAO} gives us $L_{\text{BAO}} = c\,\delta z/H(z)$, so $H(z) = c\,\delta z/L_{\text{BAO}}$, as required.

Exercise 3.6 No. The amplitude of the fluctuations could depend on the bias, but the scale length itself is bias-independent.

Exercise 4.1 First, we need to get Equation 1.7 into a form where the only time-dependent parameter is R. The density ρ is time-dependent and varies as $\rho = \rho_0(R_0/R)^3$ (where subscript 0 indicates present-day values), so we have

$$\left(\frac{\mathrm{d}R}{\mathrm{d}t}\right)^2 = \frac{8\pi G}{3}\rho_0\left(\frac{R_0}{R}\right)^3 R^2 - c^2 = \frac{8\pi G\rho_0 R_0^3}{3}R^{-1} - c^2$$

(where we've used $k = +1$). If we set $\mathrm{d}R/\mathrm{d}t = 0$ and solve, we find that $R_{\max} = R = 8\pi G\rho_0 R_0^3/(3c^2)$. Therefore

$$\left(\frac{\mathrm{d}R}{\mathrm{d}t}\right)^2 = \frac{R_{\max}}{R}c^2 - c^2.$$

Using the chain rule we have that

$$\left(\frac{\mathrm{d}R}{\mathrm{d}\theta}\right)^2 = \left(\frac{\mathrm{d}R}{\mathrm{d}t}\right)^2\left(\frac{\mathrm{d}t}{\mathrm{d}\theta}\right)^2 = \left(\frac{\mathrm{d}R}{\mathrm{d}t}\right)^2\left(\frac{R}{c}\right)^2$$

and so

$$\left(\frac{\mathrm{d}R}{\mathrm{d}\theta}\right)^2 = \left(\frac{R}{c}\right)^2\left(\frac{R_{\max}}{R}c^2 - c^2\right) = R_{\max}\,R - R^2,$$

as required.

We're asked to verify that Equation 4.2 works rather than proving it, so all we have to do is substitute it in. Differentiating Equation 4.2 with respect to θ gives

$$\frac{\mathrm{d}R}{\mathrm{d}\theta} = \frac{R_{\max}}{2}\sin\theta$$

so

$$\left(\frac{\mathrm{d}R}{\mathrm{d}\theta}\right)^2 = \frac{R_{\max}^2}{4}\sin^2\theta = \frac{R_{\max}^2}{4}\left(1-\cos^2\theta\right).$$

Meanwhile,

$$\begin{aligned} R_{\max} R - R^2 &= \frac{R_{\max}^2}{2}(1-\cos\theta) - \frac{R_{\max}^2}{4}(1-\cos\theta)^2 \\ &= \frac{R_{\max}^2}{4}(2-2\cos\theta) - \frac{R_{\max}^2}{4}\left(1+\cos^2\theta - 2\cos\theta\right) \\ &= \frac{R_{\max}^2}{4}\left(2-2\cos\theta-1-\cos^2\theta+2\cos\theta\right) \\ &= \frac{R_{\max}^2}{4}\left(1-\cos^2\theta\right), \end{aligned}$$

which equals $(\mathrm{d}R/\mathrm{d}\theta)^2$ as above.

Finally, we just need to differentiate Equation 4.3, which gives

$$\frac{\mathrm{d}t}{\mathrm{d}\theta} = \frac{R_{\max}}{2c}(1-\cos\theta) = \frac{R}{c},$$

as required.

Therefore Equations 4.2 and 4.3 are a solution.

Exercise 4.2 To show this, we'll first get things in terms of H. It's a flat matter-dominated universe, so $\Omega_\mathrm{m} = 1 = 8\pi G\rho_\mathrm{m}/(3H^2)$, thus $4\pi G\rho_\mathrm{m} = 3H^2/2$. We also know that $H(t) = \dot{a}/a$. Substituting this into Equation 4.9, we have

$$\ddot{\delta} + 2H(t)\,\dot{\delta} = 3H^2(t)\,\delta/2.$$

Next we use $H(t) = 2/(3t)$ to reformulate this in terms of a differential equation involving just δ and time:

$$\ddot{\delta} + \frac{4}{3t}\dot{\delta} = \frac{3}{2}\left(\frac{2}{3t}\right)^2\delta = \frac{2}{3t^2}\delta.$$

Next, let's try power law solutions $\delta = bt^c$ where b and c are constants. Then $\dot{\delta} = bct^{c-1}$ and $\ddot{\delta} = bc(c-1)t^{c-2}$. Substituting in, we find

$$bc(c-1)t^{c-2} + \frac{4}{3t}bct^{c-1} = \frac{2}{3t^2}bt^c.$$

Collecting the terms together, we find that

$$bc(c-1)t^{c-2} + \tfrac{4}{3}bct^{c-2} = \tfrac{2}{3}bt^{c-2},$$

and dividing through by bt^{c-2} gives

$$c(c-1) + \tfrac{4}{3}c = \tfrac{2}{3}.$$

The solution to this quadratic equation is $c = 2/3$ or $c = -1$. The -1 solution is known as the decaying mode, and is not physically relevant in this universe (it decays more rapidly than the growing mode grows and is quickly negligible). The $2/3$ power law time-dependence (which we found ultimately from linearized fluid dynamic equations) is identical to Equation 4.8, which is why the latter is known as the linear theory.

Exercise 4.3 The redder colour will be the one with the larger V-band to B-band flux ratio $S_\mathrm{V}/S_\mathrm{B}$. The fluxes are related to the magnitudes by

$V = -2.5 \log_{10} S_V + c_V$ and $B = -2.5 \log_{10} S_B + c_B$, where c_V and c_B are constants (not necessarily identical). Therefore

$$\begin{aligned}(\text{B–V}) &= -2.5 \log_{10} S_B + c_B + 2.5 \log_{10} S_V - c_V \\ &= -2.5(\log_{10} S_B - \log_{10} S_V) + (c_B - c_V) \\ &= -2.5 \log_{10}(S_B/S_V) + (c_B - c_V) \\ &= 2.5 \log_{10}(S_V/S_B) + (c_B - c_V),\end{aligned}$$

which gives

$$2.5 \log_{10}(S_V/S_B) = (\text{B–V}) - (c_B - c_V)$$

so

$$\log_{10}(S_V/S_B) = (\text{B–V})/2.5 - (c_B - c_V)/2.5$$

thus

$$\begin{aligned}(S_V/S_B) &= 10^{(\text{B–V})/2.5-(c_B-c_V)/2.5} \\ &= 10^{(\text{B–V})/2.5} \times 10^{-(c_B-c_V)/2.5} \\ &= 10^{(\text{B–V})/2.5} \times \text{constant.}\end{aligned}$$

Therefore the larger the value of (B–V), the larger the value of S_V/S_B. Therefore $(\text{B–V}) = 1$ is redder than $(\text{B–V}) = 0$.

Exercise 4.4 We haven't specified the geometry yet, so let's keep things simple. Let's take the dust and stars to be in a cylinder facing us, with cross-sectional area A. Let's set the length of the cylinder to be h, and measure distances along this length with the variable x. An infinitesimal layer would have thickness $\mathrm{d}x$ and volume $A\,\mathrm{d}x$. The bigger the volume, the more stars it will contain, so let's set the luminosity of the shell to be $\mathrm{d}L = \rho A\,\mathrm{d}x$, where ρ is a constant (the luminosity density). By the time the light emerges from the end of cylinder, it will have been extinguished by a factor of $\mathrm{e}^{\tau(x)}$, where $\tau(x)$ is the optical depth at a distance x into the cylinder. This optical depth must be proportional to x, because each increment δx will suppress the light by the same factor, which we could write as $\mathrm{e}^{\delta\tau}$, so let's write that as $\tau = kx$. We could, for example, write the optical depth from one end of the cloud to the other as $\tau_{\text{total}} = kh$. The light that emerges from the shell at $x \to x + \mathrm{d}x$ will therefore be $\mathrm{d}L_{\text{out}} = L \times \mathrm{e}^{-\tau(x)} = \rho A\,\mathrm{d}x \times \mathrm{e}^{-kx}$. If we integrate that from $x = 0$ to $x = h$, we get

$$L_{\text{out}} = \int_{x=0}^{h} \rho A\,\mathrm{e}^{-kx}\,\mathrm{d}x = \frac{\rho A}{k}\left(1 - \mathrm{e}^{-kh}\right).$$

Some quick checks: note that k has dimensions of one over length (because $\tau = kx$ and τ is dimensionless), so A/k has dimensions of volume, and so $\rho A/k$ is luminosity density times volume, which is a luminosity. Note also that kh is dimensionless.

Now, what would happen if there were no dust? The luminosity would just be $L_{\text{no dust}} = \rho A h$. The dust has therefore reduced the output luminosity by a factor

$$\frac{L_{\text{out}}}{L_{\text{no dust}}} = \frac{\rho A/k}{\rho A h}\left(1 - \mathrm{e}^{-kh}\right) = \frac{1}{kh}\left(1 - \mathrm{e}^{-kh}\right).$$

This ratio is independent of the geometrical cross section A and of the luminosity density ρ. If the cloud is deep enough, then the term in brackets is $\simeq 1$, so we just have $L_{\rm out}/L_{\rm no\ dust} = 1/(kh)$. We can now write this for Hα light:

$$\frac{L_{\rm out}({\rm H}\alpha)}{L_{\rm no\ dust}({\rm H}\alpha)} = \frac{1}{k_{{\rm H}\alpha}\,h}.$$

For Hβ, we have that $\tau_{{\rm H}\beta} \simeq 1.45\,\tau_{{\rm H}\alpha}$, so $k_{{\rm H}\beta} = 1.45\,k_{{\rm H}\alpha}$, thus

$$\frac{L_{\rm out}({\rm H}\beta)}{L_{\rm no\ dust}({\rm H}\beta)} = \frac{1}{k_{{\rm H}\beta}\,h} = \frac{1}{1.45\,k_{{\rm H}\alpha}\,h} = \frac{1}{1.45}\,\frac{L_{\rm out}({\rm H}\alpha)}{L_{\rm no\ dust}({\rm H}\alpha)}.$$

Therefore

$$\frac{L_{\rm out}({\rm H}\alpha)}{L_{\rm out}({\rm H}\beta)} = 1.45\,\frac{L_{\rm no\ dust}({\rm H}\alpha)}{L_{\rm no\ dust}({\rm H}\beta)}. \tag{S4.1}$$

This is independent of h, so we've now removed all dependence on the geometry. So even if kh is enormous and $L_{\rm out} \ll L_{\rm no\ dust}$, the luminosity ratio of Hα and Hβ is only ever 1.45 times the ratio that you get with no dust, when enough dust is evenly mixed with the gas emitting the emission lines.

Now suppose that you wrongly assumed that it's a simple dust screen with an optical depth of $\tau_{{\rm H}\alpha}$ for Hα and $\tau_{{\rm H}\beta} = 1.45\,\tau_{{\rm H}\alpha}$ for Hβ. Your luminosities would be

$$L_{\rm out}({\rm H}\alpha) = L_{\rm no\ dust}({\rm H}\alpha) \times {\rm e}^{-\tau_{{\rm H}\alpha}},$$
$$L_{\rm out}({\rm H}\beta) = L_{\rm no\ dust}({\rm H}\beta) \times {\rm e}^{-1.45\,\tau_{{\rm H}\alpha}},$$

so the luminosity ratio would be

$$\frac{L_{\rm out}({\rm H}\alpha)}{L_{\rm out}({\rm H}\beta)} = \frac{L_{\rm no\ dust}({\rm H}\alpha)}{L_{\rm no\ dust}({\rm H}\beta)}\,{\rm e}^{0.45\,\tau_{{\rm H}\alpha}}. \tag{S4.2}$$

Comparing this to Equation S4.1, we have $1.45 = {\rm e}^{0.45\,\tau_{{\rm H}\alpha}}$, or $\tau_{{\rm H}\alpha} = \ln(1.45)/0.45 \simeq 0.83$. Since $\tau_{{\rm H}\alpha} \simeq 0.7A_{\rm V}$, we have $A_{\rm V} \simeq 1.2$. So, if you have an optically-thick cloud in which the dust is well-mixed with the gas, but you wrongly assumed a foreground dust screen, you'd infer a V-band extinction of just 1.2 magnitudes, regardless of what the real extinction $\tau_{\rm total}$ is from one end of the cloud to the other.

Exercise 4.5 Astronomical absolute magnitudes are defined as $m = -2.5\log_{10} L + \text{constant}$, so

$$dm = -2.5\,d(\log_{10} L) = -2.5\,\frac{d(\ln L)}{\ln 10} = \frac{-2.5}{\ln 10}\,\frac{1}{L}\,dL. \tag{S4.3}$$

Therefore

$$\frac{dN}{dm} = \frac{-\ln 10}{2.5}\,L\,\frac{dN}{dL}. \tag{S4.4}$$

The $-$ sign just indicates that the magnitude increment dm is in the opposite sense to the luminosity increment dL, and is usually neglected.

Exercise 4.6 The variance of a probability distribution $p(x)$ is the mean of the squares minus the square of the mean, i.e.

$${\rm Var}(x) = \int_0^1 x^2\,p(x)\,dx - \left(\int_0^1 x\,p(x)\,dx\right)^2.$$

Now, our probability distribution is uniform, so $p(x) = 1$ for all x from 0 to 1, hence this is just

$$\begin{aligned}\mathrm{Var}(x) &= \int_0^1 x^2\,\mathrm{d}x - \left(\int_0^1 x\,\mathrm{d}x\right)^2\\ &= \left[\frac{x^3}{3}\right]_{x=0}^{x=1} - \left[\left(\frac{x^2}{2}\right)^2\right]_{x=0}^{x=1}\\ &= \tfrac{1}{3} - \tfrac{1}{4} = \tfrac{1}{12},\end{aligned}$$

as required. The standard deviation is the square root of the variance, so the standard deviation of the uniform distribution is $1/\sqrt{12}$. The central limit theorem states that if you have N measurements, each with an uncertainty σ (i.e. taken from the same distribution with standard deviation σ), then the standard deviation of the mean average of these measurements is $\sigma/\sqrt{N}$. Now, if our null hypothesis holds, then V/V_{max} is uniformly distributed, so each measurement of V/V_{max} is taken from a distribution with standard deviation $1/\sqrt{12}$. Therefore the standard deviation of the average N measurements of V/V_{max} must be $1/\sqrt{12N}$, as required.

Exercise 4.7 Yes, provided that the selection function has been correctly stated.

Exercise 4.8 No, not necessarily. Suppose that you had a volume-limited sample with $V_{\mathrm{max}} = V(z_{\mathrm{max}})$ for all galaxies. Now suppose that half the galaxies exist at exactly $z = 0$, half are at $z = z_{\mathrm{max}}$, and there are none in between. Clearly, the numbers of galaxies are evolving very strongly and discontinuously, but $\langle V/V_{\mathrm{max}}\rangle = 1/2$.

Exercise 4.9 The amount of light emitted per unit volume will be given by the number density of galaxies multiplied by their luminosity, i.e. $L \times \phi(L)$. At luminosities far below the break, $\phi(L) \propto L^{-\alpha}$, so $L\,\phi(L) \propto L^{1-\alpha}$. Since we're given that the faint-end slope α satisfies $\alpha < 1$, this must be increasing with luminosity. At the bright end we have that $\phi \propto \exp(-L/L_*)$, which tends to zero faster than $1/L$, so $L\,\phi(L)$ (which is proportional to $L\exp(-L/L_*)$) must also tend to zero. We'd expect one turning point — but where? We can differentiate $L\,\phi(L)$, set the result equal to zero and rearrange. This gives

$$\frac{\mathrm{d}(L\phi)}{\mathrm{d}L} = \phi_*(-\mathrm{e}^{-L/L_*}(L/L_*)^{-\alpha+1} + (1-\alpha)\mathrm{e}^{-L/L_*}(L/L_*)^{-\alpha}) = 0.$$

Dividing by $\phi_*\mathrm{e}^{-L/L_*}$ gives $(L/L_*)^{1-\alpha} = (1-\alpha)(L/L_*)^{-\alpha}$. Further dividing by $(L/L_*)^{-\alpha}$ gives $L/L_* = 1-\alpha$, or $L = (1-\alpha)L_*$. The galaxies that dominate the cosmic luminosity density are therefore those with luminosities of $(1-\alpha)L_*$.

Exercise 4.10 PDE is vertical translations, while PLE is horizontal translations.

Exercise 4.11 Active galaxies can be seen to much higher redshifts than the elliptical galaxies used in the Tolman test in Chapter 3, and as the predicted redshift-dependence of surface brightness is strong, i.e. $(1+z)^4$, it might appear that the radio lobes of radiogalaxies have a strong advantage. The attraction of the Tolman test is that the $(1+z)^4$ surface brightness prediction is independent of the cosmological parameters. In order to apply it, we need a population of objects whose luminosity per unit area (in, for example, square parsecs) is constant. In this case, rearranging the relation in the question gives us $L/r^2 \propto Q^{7/6} r^{-4/3} \rho^{7/12}$. We might hope to find active galaxies with the same Q on average if we match other properties of the central engine (e.g. optical emission lines and continuum) on average. We might also be able to calibrate out any variations in density through other observations as indicated in the question, but we're still left with a surface brightness that depends on the linear size of the system. Without additionally having a standard rod as a comparison, we can't apply the Tolman test as it stands.

Exercise 4.12 There are $60 \times 60 = 3600$ arcseconds in a degree, so there are $3600^2 \simeq 1.30 \times 10^7$ square arcseconds in a square degree. Therefore the number of random 5σ noise spikes in one square degree would be $(1.30 \times 10^7)/(3.5 \times 10^6) \simeq 3.7$. So we'd expect one 5σ noise spike in $1/3.7$ square degrees, or about 0.27 square degrees. In practice, noise spikes can occur more frequently than this for a variety of reasons (including instrumental effects).

Exercise 4.13 Suppose that your camera or detector covers an area A on the sky. Let's say that you invest all your time in a pencil-beam survey, and it reaches a flux S. The number counts are Euclidean, so $N(>S) = kS^{-1.5}$, where k is some constant. Therefore the number of galaxies seen in the pencil-beam survey is

$$n_{\text{pencil}} = A \times N(>S) = AkS^{-1.5}.$$

Now suppose that instead of doing a pencil-beam survey, you spread your integration time over m fields of view, each of which has area A. The total area that you cover is $m \times A$, but the images would be shallower by a factor of $\sqrt{m}$, so the total number of galaxies in the wide-field survey would be

$$n_{\text{wide}} = mA(\sqrt{m}S)^{-1.5} = mAm^{-0.75}S^{-1.5} = m^{0.25}AS^{-1.5}.$$

Comparing this to n_{pencil}, we see that $n_{\text{wide}} = m^{0.25}\, n_{\text{pencil}}$, so the wide-field survey finds more galaxies by a factor of $m^{0.25}$.

A similar calculation shows that if the source counts are steeper than $N(>S) \propto S^{-2}$, then the pencil-beam survey would see more. However, only rarely are source counts that steep (we'll see an example in Chapter 5). In the vast majority of cases, wide-field surveys find more objects in a given observing time than pencil-beam surveys. In practice, though, there's often a limit to how wide you can make a survey, because the time spent simply moving the telescope or reading out the detector becomes significant (we've neglected both effects here).

Exercise 5.1 $I_\nu\, \mathrm{d}\nu$ is the background intensity in an interval $\nu \to \nu + \mathrm{d}\nu$. The background per decade is the background in a logarithmic interval, $\nu \to \nu + \mathrm{d}\log_{10}\nu$. Let's write this as $B\,\mathrm{d}\log_{10}\nu$. If we can set this equal to something times $\mathrm{d}\nu$, then that something must be I_ν. Now, $\mathrm{d}\log_{10}\nu = (\mathrm{d}\ln\nu)/\ln(10)$, so $B\,\mathrm{d}\log_{10}\nu = (B/\ln(10))\,\mathrm{d}\ln\nu$. But $\mathrm{d}\ln\nu = (1/\nu)\,\mathrm{d}\nu$, so

$$B\,\mathrm{d}\log_{10}\nu = B\frac{1}{\nu\ln(10)}\,\mathrm{d}\nu.$$

Therefore

$$I_\nu = B\frac{1}{\nu\ln(10)},$$

so $B = \ln(10)\,\nu I_\nu$. Therefore the background intensity per decade of frequency is proportional to νI_ν. Looking at Figure 5.1, we see that the far-infrared bump has a similar height and area to the optical/near-infrared bump, each over roughly the same logarithmic frequency interval of about $\Delta \log_{10} \nu = 1.5$. Therefore there's about the same energy output in the far-infrared bump as in the optical/near-infrared bump.

Exercise 5.2 This will be the one in which $S_\nu\,\mathrm{d}N/\mathrm{d}\ln S_\nu$ is a maximum, and since $\mathrm{d}\ln S_\nu = S_\nu^{-1}\,\mathrm{d}S_\nu$, we can also express this as $S_\nu^2\,\mathrm{d}N/\mathrm{d}S_\nu$. This is similar (though not quite identical) to Figure 5.2.

Exercise 5.3 The angular resolution in radians is $1.22\lambda/D = 1.22 \times 500 \times 10^{-6}/3.5 = 1.7429 \times 10^{-4}$ (we'll carry some extra significant figures until the end of the calculation). In degrees this is $1.7429 \times 10^{-4} \times 360^\circ/(2\pi) = 0.009\,985\,8^\circ$. In arcseconds this is $0.009\,985\,8 \times 3600 = 35.95''$, or $36.0''$ to the accuracy of the initial numbers.

Exercise 5.4 (a) The fractional range would be $0.09/0.15 = 0.6$ or 60%, which we could also quote as a possible variation of a factor of $1/0.6 = 1.7$.

(b) The variation in β changes the extrapolation from the $800\,\mu$m quoted to the rest frame, which is $850/(1+z)\,\mu\mathrm{m} = 850/4\,\mu\mathrm{m} = 212.5\,\mu\mathrm{m}$. The wavelength dependence is $\lambda^{-\beta}$, so

$$\frac{k_\mathrm{d}(800\,\mu\mathrm{m})}{k_\mathrm{d}(212.5\,\mu\mathrm{m})} = \left(\frac{800}{212.5}\right)^{-\beta} = 3.765^{-\beta},$$

i.e. 0.0705–0.2656 when $\beta = 1$–2, or a further variation of a factor of 3.8. The total variation so far is $1.7 \times 3.8 \simeq 6.5$.

(c) Using the black body spectrum given in Equation 2.2 and putting in the numbers for a wavelength of $212.5\,\mu$m (i.e. $\nu = c/212.5\,\mu\mathrm{m} = 2.998 \times 10^8\,\mathrm{m\,s^{-1}}/(212.5 \times 10^{-6}\,\mathrm{m}) = 1.411 \times 10^{12}\,\mathrm{Hz}$) and temperatures of $T = 20$ K and 40 K, we find that

$$\begin{aligned}\frac{B(1.411\,\mathrm{THz}, 40\,\mathrm{K})}{B(1.411\,\mathrm{THz}, 20\,\mathrm{K})} &= \frac{\exp(h\nu/kT_1) - 1}{\exp(h\nu/kT_2) - 1} \\ &= \frac{\exp(6.626 \times 10^{-34}\,\mathrm{J\,s} \times 1.411 \times 10^{12}\,\mathrm{Hz}/1.381 \times 10^{-23}\,\mathrm{J\,K^{-1}} \times 20\,\mathrm{K}) - 1}{\exp(6.626 \times 10^{-34}\,\mathrm{J\,s} \times 1.411 \times 10^{12}\,\mathrm{Hz}/1.381 \times 10^{-23}\,\mathrm{J\,K^{-1}} \times 40\,\mathrm{K}) - 1} \\ &= 6.435.\end{aligned}$$

The range of allowed temperatures therefore gives an additional fractional range of 6.4, so the total fractional range is $1.7 \times 3.8 \times 6.4 \simeq 41$, i.e. we cannot even quote a dust mass to within an order of magnitude!

However, if we measure fluxes at more wavelengths, we might be able to reduce these uncertainties by constraining the value of β on the Rayleigh–Jeans tail, and determining the temperature from the wavelength λ_max of the location of the peak of the spectral energy distribution. This is quantified with the Wien displacement law, which can be expressed in astrophysically-useful quantities as

$$\frac{\lambda_\mathrm{max}}{100\,\mu\mathrm{m}} = 1.45\frac{20\,\mathrm{K}}{T}.$$

There is, however, still the issue that galaxies do not have single temperatures.

Exercise 5.5 Suppose that there were no background. In some fixed observing time, suppose that we collect N photons from a distant object. Using Poisson statistics, the variance on this number will also be N, so the standard deviation (i.e. the noise) will be $\sqrt{N}$. The signal-to-noise ratio will therefore be $N/\sqrt{N} = \sqrt{N}$. Now suppose that there's a strong background, so we observe $N + N_{\text{back}}$ photons, with $N_{\text{back}} \gg N$. The noise on this will be $\sqrt{N + N_{\text{back}}} \simeq \sqrt{N_{\text{back}}} \gg \sqrt{N}$. What we want is N and not $N + N_{\text{back}}$, so we have to observe an additional blank bit of sky to estimate N_{back}. This can be done if we have a small object in our camera, so we can use blank bits of the image, but if our detector has only one or a small number of pixels, we have to spend extra time observing blank sky. However, even neglecting the uncertainty on our N_{back} estimate, we still have a signal-to-noise ratio of $N/\sqrt{N_{\text{back}}}$, which is much less than the $N/\sqrt{N}$ that we'd have in the case of no background. So once $N_{\text{back}} \geq N$ we enter the *background-limited* regime where good signal-to-noise is harder to get. In the case of the SCUBA camera, the faintest objects are $\simeq 10^5$–10^6 times fainter than the sky background. Worse, the background varies on timescales of less than a second, so observing techniques at submm wavelengths are often geared towards making the best background subtraction.

Exercise 5.6 See Figure S5.1.

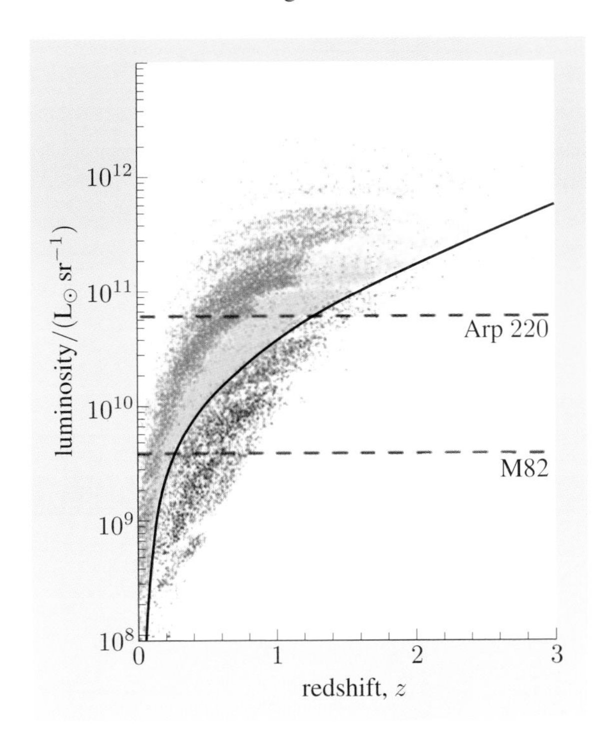

Figure S5.1 This is the same as Figure 5.15, but with the approximate location of one possible flux limit marked as a thick black line.

Exercise 6.1 Suppose that we wanted to separate a human being into protons and electrons, then hold them one metre apart. For a 60 kg mass, the force required would be $F = (ne)^2/(4\pi\varepsilon_0 r^2)$, where $r = 1\,\text{m}$, $n = 60\,\text{kg}/m_\text{p}$ and ε_0 is the vacuum permittivity of free space. This comes out as a gigantic $F \simeq 3 \times 10^{29}\,\text{kg}\,\text{m}\,\text{s}^{-2}$. The luminosity of the Sun is $\text{L}_\odot = 3.83 \times 10^{26}\,\text{W}$, so the momentum flux from the Sun is $\text{L}_\odot/c = 1.28 \times 10^{18}\,\text{kg}\,\text{m}\,\text{s}^{-2}$. If we could employ all the momentum flux from all the $\simeq 10^{11}$ stars in the Galaxy in keeping the positive and negative parts separate, it would be just sufficient to maintain a 1 m separation for just 60 kg. The potential barrier for separating the charged components of a plasma accreting around a black hole is clearly insuperable for radiation pressure.

Exercise 6.2 Putting the numbers into Equation 6.6 gives

$$L_{\mathrm{E}} = \frac{4\pi \times (6.67\times10^{-11}\,\mathrm{N\,m^2\,kg^{-2}}) \times (3.00\times10^{8}\,\mathrm{m\,s^{-1}}) \times (1.99\times10^{30}\,\mathrm{kg}) \times (1.67\times10^{-27}\,\mathrm{kg})}{6.65\times10^{-29}\,\mathrm{m^2}}$$
$$= 1.26\times10^{31}\,\mathrm{W}.$$

The luminosity of the Sun is 3.83×10^{26} W, which is far below the Eddington limit.

Exercise 6.3 Assuming that the mass of a 100 W light bulb is (say) about 50 g, we get an Eddington limit of just 0.2 W . Clearly, a light bulb radiates at much more than the Eddington limit. Light bulbs don't blow themselves apart because they are not gravitationally bound.

Exercise 6.4 To obtain Equation 6.26 we start with Equation 1.53, then use Equation 1.41. It immediately follows that

$$\mathrm{d}V = d_{\mathrm{A}}^2(1+z)^3 \frac{4\pi c\,\mathrm{d}z}{(1+z)\,H(z)} = 4\pi d_{\mathrm{A}}^2(1+z)^2 \frac{c\,\mathrm{d}z}{H(z)}.$$

(We ignore the $-$ sign, which just refers to the directions in which the infinitesimal increments are measured.) Next, putting in the relationship between angular diameter and luminosity distance, $d_{\mathrm{L}} = (1+z)^2 d_{\mathrm{A}}$ (Equation 1.50), gives

$$\mathrm{d}V = \frac{4\pi d_{\mathrm{L}}^2}{(1+z)^4}(1+z)^2 \frac{c\,\mathrm{d}z}{H(z)} = \frac{4\pi d_{\mathrm{L}}^2}{(1+z)^2}\frac{c\,\mathrm{d}z}{H(z)}.$$

Dividing by $\mathrm{d}z$ and multiplying by H_0/H_0 gives

$$\frac{\mathrm{d}V}{\mathrm{d}z} = \frac{4\pi c d_{\mathrm{L}}^2}{(1+z)^2 H(z)} = \frac{c}{H_0}\frac{4\pi d_{\mathrm{L}}^2}{(1+z)^2 H(z)/H_0},$$

as required.

We can rearrange this as

$$\frac{4\pi d_{\mathrm{L}}^2}{\mathrm{d}V/\mathrm{d}z} = (1+z)^2 \frac{H(z)}{c}.$$

Finally, we use Equation 1.28: $|\mathrm{d}z/\mathrm{d}t| = (1+z)\,H(z)$ (again we'll not worry about the sign). Therefore

$$\frac{4\pi d_{\mathrm{L}}^2}{\mathrm{d}V/\mathrm{d}z}\,\mathrm{d}t = \frac{1}{c}(1+z)\,\mathrm{d}z,$$

which is Equation 6.24, as required.

Exercise 6.5 The angular size θ will satisfy $\theta \simeq \tan\theta = r_h/D$, where $D = 10$ Mpc and r_h is given by Equation 6.29: $r_h = (10^8/10^8) \times (220/200)^{-2}\,\mathrm{pc} = 0.83\,\mathrm{pc}$. Plugging in the numbers, we have $\theta \simeq r/D = 0.83\,\mathrm{pc}/10\,\mathrm{Mpc} = 8.3\times10^{-8}$ radians. In arcseconds this is

$\theta = 8.3 \times 10^{-8} \times (360°/2\pi) \times 60 \times 60 = 0.017''$. This is clearly a lot smaller than the seeing limit of ground-based telescopes.

Exercise 6.6 The e-folding timescale for Eddington-limited black hole growth is the Salpeter timescale t_E divided by the efficiency η, i.e. $t_{\text{e-fold}} = 4 \times 10^8/\eta$ yr. There have been $3 \times 10^9/t_{\text{e-fold}}$ e-foldings since the start of the Universe, or $0.75(\eta/0.1)$ e-foldings. To reach $10^6\,M_\odot$, one needs $\log_e(10^6/10^1) = 11.5$ e-foldings. Even if $\eta = 1$, you have only 7.5 e-foldings, so 3 Gyr is not long enough.

Exercise 7.1 Comoving distances add, so $r_S = r_L + r_{LS}$. Therefore $r_{LS} = r_S - r_L$. In flat space, angular diameter distance is simply comoving distance divided by $(1+z)$ (Chapter 1), but in this case we need the redshift of the background source *as seen from the lens*. We could write this factor as $(1+z_{LS})$. This is the factor by which the Universe expanded between the source redshift and the lens redshift, i.e. R_L/R_S, where R is the scale factor. But

$$\frac{R_L}{R_S} = \frac{R_L/R_0}{R_S/R_0} = \frac{R_0/R_S}{R_0/R_L}$$

(where the subscript 0 refers to the present day), so $(1+z_{LS}) = (1+z_S)/(1+z_L)$. Therefore our final expression for the angular diameter distance D_{LS} is

$$D_{LS} = (r_S - r_L) \times \frac{(1+z_L)}{(1+z_S)}.$$

Exercise 7.2 First, matching distances along the top of Figure 7.7 shows that $\boldsymbol{\theta} D_S = \boldsymbol{\beta} D_S + \hat{\boldsymbol{\alpha}} D_{LS}$. But $\boldsymbol{\alpha} = \hat{\boldsymbol{\alpha}} D_{LS}/D_S$, so $\boldsymbol{\theta} D_S = \boldsymbol{\beta} D_S + \boldsymbol{\alpha} D_S$. Dividing out the scalar D_S gives $\boldsymbol{\theta} = \boldsymbol{\beta} + \boldsymbol{\alpha}$, which we can rearrange to $\boldsymbol{\beta} = \boldsymbol{\theta} - \boldsymbol{\alpha}$, as required.

Exercise 7.3 We set $\beta = 0$ in Equation 7.8. We can rearrange this to show that

$$\theta = \sqrt{\frac{4GM}{c^2}\frac{D_{LS}}{D_L D_S}}.$$

But what would this look like? The background object is exactly behind the lens and it's deflected by an angle θ. Is it deflected to the left or right or up or down? In fact, there is nothing to give the deflection any particular direction, so the background source is lensed into a *ring*. These are very rare, but an example is shown in Figure S7.1.

Exercise 7.4 $\beta^2 + 4\theta_E^2$ is always positive, but the square root of it can be positive or negative. $\sqrt{\beta^2 + 4\theta_E^2} > \beta$ unless $\theta_E = 0$, so the negative root must always give a negative θ. This is indeed a physical solution and represents an angle measured in the opposite direction: as shown in Figure 7.7, the image is on the other side of the lens. Note that one image is at $\theta > \theta_E$ and the other is at $\theta < \theta_E$, unless $\theta = \theta_E$ and the system is an Einstein ring.

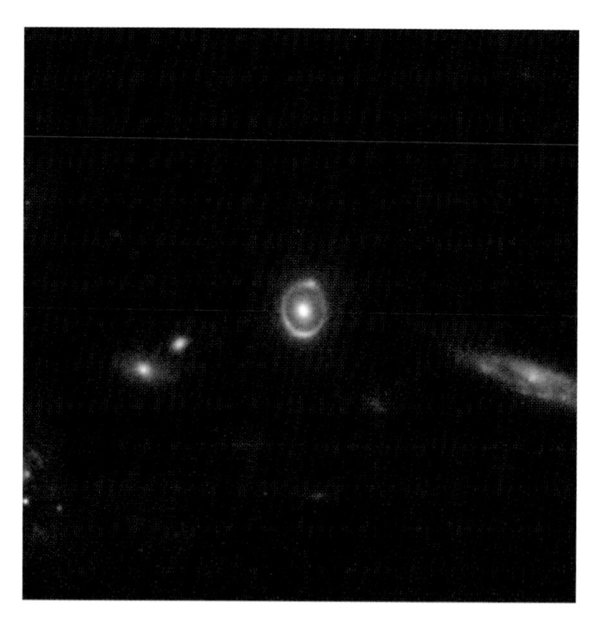

Figure S7.1 The gravitational lens 0038+4133 (an Einstein ring) from the COSMOS survey, taken by the HST . The image is 15″ by 15″.

Exercise 7.5 From the previous exercise, a source can have multiple images, so there is *not* necessarily a unique image position $\boldsymbol{\theta}$ for a given source position $\boldsymbol{\beta}$. In mathematical terms, we would speak of the mapping $\boldsymbol{\beta} \rightarrow \boldsymbol{\theta}$ as being one-to-many. However, each image position $\boldsymbol{\theta}$ *does* map in a one-to-one way onto a source position $\boldsymbol{\beta}$, i.e. each image position can correspond to only one position in the background source. To see why, consider Equation 7.4. The function $\boldsymbol{\alpha}(\boldsymbol{\theta})$ must be a single-valued function, i.e. any particular input $\boldsymbol{\theta}$ can give only one possible output $\boldsymbol{\alpha}$. Therefore there can be only one value of $\boldsymbol{\beta}$ for a given input $\boldsymbol{\theta}$.

Exercise 7.6 We're asked to differentiate Equation 7.12, which gives $\mathrm{d}\beta/\mathrm{d}\theta = 1 + (\theta_\mathrm{E}^2/\theta^2)$. This gives us one of the fractions in Equation 7.16. The magnification is therefore

$$\begin{aligned}\frac{\theta}{\beta}\frac{\mathrm{d}\theta}{\mathrm{d}\beta} &= \frac{\theta}{\beta}\left(1+\frac{\theta_\mathrm{E}^2}{\theta^2}\right)^{-1} = \theta\left(\theta-\frac{\theta_\mathrm{E}^2}{\theta}\right)^{-1}\left(1+\frac{\theta_\mathrm{E}^2}{\theta^2}\right)^{-1}\\ &= \left(1-\frac{\theta_\mathrm{E}^2}{\theta^2}\right)^{-1}\left(1+\frac{\theta_\mathrm{E}^2}{\theta^2}\right)^{-1} = \left(1+\frac{\theta_\mathrm{E}^2}{\theta^2}-\frac{\theta_\mathrm{E}^2}{\theta^2}-\frac{\theta_\mathrm{E}^4}{\theta^4}\right)^{-1}\\ &= \left(1-\frac{\theta_\mathrm{E}^4}{\theta^4}\right)^{-1},\end{aligned}$$

as required.

Exercise 7.7 A negative magnification means that the image is mirror-reversed. For example, a positive change $\mathrm{d}\beta$ would have a corresponding $\mathrm{d}\theta$ in the opposite direction, so $\mathrm{d}\theta$ is negative. Therefore $\mathrm{d}\theta/\mathrm{d}\beta$ is negative in Equation 7.16.

Exercise 7.8 We start from Equation 7.24. The mass enclosed is $\Sigma\pi\xi^2$ and we set $\xi = D_\mathrm{L}\theta$:

$$\widehat{\alpha} = \frac{4GM(\xi)}{c^2\xi} = \frac{4G}{c^2\xi} \times \Sigma\pi\xi^2 = \frac{4G}{c^2} \times \Sigma\pi \times D_\mathrm{L}\theta.$$

Now,

$$\alpha = \frac{D_{\rm LS}}{D_{\rm S}}\widehat{\alpha}, \qquad \text{(Eqn 7.6)}$$

so

$$\alpha = \frac{4\pi G\Sigma}{c^2}\,\frac{D_{\rm L}D_{\rm LS}}{D_{\rm S}}\,\theta,$$

as required.

If we then set $\Sigma = \Sigma_{\rm cr}$, we find that $\alpha(\theta) = \theta$ for any θ, so $\beta = 0$. This means that the gravitational lens is acting as a perfect focusing lens! However, this is a very special case — gravitational lenses in general do *not* focus light. As 'lenses' in the optical sense, they have all forms of aberration, except of course chromatic aberration since gravitational lensing is strictly achromatic.

Exercise 7.9 From left to right, they are a saddle point, a maximum and a minimum.

Exercise 7.10 The time delay of the image at the centre increases. In a diagram like Figure 7.15, the central panel showing the gravitational time delay would be acquiring a sharper and higher point in the centre. When the lens potential becomes a singular isothermal sphere, the time delay becomes infinite, so the image disappears. Photons would take an infinite amount of time to climb out of the infinitely-deep potential well, and (by symmetry) spend another infinite amount of time falling in beforehand. But a more thoughtful answer is that this deep potential well would form a black hole. Right from Equation 7.1, we've been assuming a weak-field limit, so a better answer is that these simple assumptions break down as the potential becomes more extreme.

Exercise 7.11 The background objects have the same redshift, so we could think of the luminosity function as differential source counts, thus $\mathrm{d}N/\mathrm{d}S \propto S^{-\alpha}$. Therefore the number of objects per unit area brighter than a flux S_0 will be $N(> S_0) \propto S_0^{1-\alpha}$, which we could write as

$$N(> S_0) = kS_0^{1-\alpha}.$$

If the background galaxies are gravitationally magnified by a factor of μ, the intrinsic fluxes will be $S_{\rm intrinsic} = S/\mu$, while the comoving volume sampled will be smaller by a factor of $1/\mu$. Therefore the number of galaxies brighter than an *observed* flux S_0 will be

$$N_{\rm lensed}(> S_0) = \frac{k}{\mu}\left(\frac{S_0}{\mu}\right)^{1-\alpha} = k\mu^{-1}S_0^{1-\alpha}\mu^{\alpha-1} = kS_0^{1-\alpha}\mu^{\alpha-2} = N(> S_0)\,\mu^{\alpha-2}.$$

Therefore for a magnification of μ (where $\mu > 1$), the lensing changes the number of background galaxies per unit area by a factor of $\mu^{\alpha-2}$. For this factor to be bigger than 1 we need

$$\mu^{\alpha-2} > 1,$$

so $\quad \log(\mu^{\alpha-2}) > \log(1) = 0,$
thus $\quad (\alpha - 2)\log(\mu) > 0.$
We already know that $\log(\mu) > 0$ (because $\mu > 1$), so this can happen only if $\alpha > 2$. For example, if the source counts have a Euclidean slope ($\alpha = 2.5$), then

lensing would increase the number of objects. The effect of sampling less volumes due to lensing, and so finding fewer objects than the flux magnification on its own would suggest, is known as the *Broadhurst effect*. (See Broadhurst, T.J., Taylor, A.N. and Peacock, J.A., 1995, *Astrophysical Journal*, **438**, 49.)

Exercise 8.1 There's no guarantee that the re-emitted photon comes out in the same direction — in fact, it probably won't. A corollary is that any Lyman α cloud should glow faintly in Lyman α light in all directions from these re-emitted photons, even if the cloud is not intercepting our line of sight to a quasar (because there will always be *some* line of sight that does). This re-emission is in general too faint to detect. However, Lyman α emission can sometimes be seen if there are internal ionizing sources (e.g. star formation) within damped Lyman α systems, which you will meet later in the chapter.

Exercise 8.2 The column density through the centre will be the same as that seen through a cubical cloud with a side 2 Mpc, facing the observer (because the absorption doesn't depend on the distribution of material that the light *doesn't* pass through). One Mpc is about 3×10^{24} cm, so we can write the density as $(3 \times 10^{24})^3\,\mathrm{cm}^{-3} = 2.7 \times 10^{73}\,\mathrm{Mpc}^{-3}$. The total number of neutral hydrogen atoms in the cube must be $2.7 \times 10^{73}\,\mathrm{Mpc}^{-3} \times 8\,\mathrm{Mpc}^3 = 21.6 \times 10^{73}$, which is spread over a projected area of $2 \times 2\,\mathrm{Mpc}^2 = 36 \times 10^{48}\,\mathrm{cm}^2$. Therefore the column density must be $21.6 \times 10^{73}/(36 \times 10^{48})\,\mathrm{cm}^{-2} \simeq 6 \times 10^{24}\,\mathrm{cm}^{-2}$.

Exercise 8.3 In order for a hydrogen atom to absorb an Hα photon, the photon must have the right energy, and there must be an atom with an electron in the $n = 2$ energy level ready to absorb the photon. This energy level is at $E = -13.6/n^2\,\mathrm{eV} = -13.6/4\,\mathrm{eV} = -3.4\,\mathrm{eV}$. In order to be in such a state, the atom must have absorbed a photon of energy $(-3.4\,\mathrm{eV}) - (-13.6\,\mathrm{eV}) = 10.2\,\mathrm{eV}$. Photons of this energy require a black body temperature of order

$$T \simeq E/k = \frac{10.2\,\mathrm{eV} \times 1.602 \times 10^{-19}\,\mathrm{J\,eV^{-1}}}{1.381 \times 10^{-23}\,\mathrm{J\,K^{-1}}} = 120\,000\,\mathrm{K}.$$

This is hotter than the surface of an O star, and is much hotter than the typical temperatures in the intergalactic medium. Lyman α clouds are too cold to have many atoms with electrons already excited to the $n = 2$ level, so the clouds have almost no Hα absorption.

Exercise 8.4 We can write $\sigma(\nu) = \sigma_0(\nu/\nu_{\mathrm{limit}})^{-3}$, where $\sigma_0 = 7.88 \times 10^{-22}\,\mathrm{m}^{-2}$, and ν_{limit} is the frequency of the Lyman limit. Writing $J_\nu = k\nu^{-\alpha}$ and plugging the terms in, we find

$$\begin{aligned}
\tau = N_{\mathrm{H\,I}} \frac{\int_{\nu_{\mathrm{limit}}}^{\infty} (\sigma J_\nu/(h\nu))\,\mathrm{d}\nu}{\int_{\nu_{\mathrm{limit}}}^{\infty} (J_\nu/(h\nu))\,\mathrm{d}\nu} &= N_{\mathrm{H\,I}}\,\sigma_0 \frac{\int_{\nu_{\mathrm{limit}}}^{\infty} (\nu/\nu_{\mathrm{limit}})^{-3} k\nu^{-\alpha-1}\,\mathrm{d}\nu}{\int_{\nu_{\mathrm{limit}}}^{\infty} k\nu^{-\alpha-1}\,\mathrm{d}\nu} \\
= \frac{N_{\mathrm{H\,I}}\,\sigma_0}{\nu_{\mathrm{limit}}^{-3}} \frac{\int_{\nu_{\mathrm{limit}}}^{\infty} \nu^{-\alpha-4}\,\mathrm{d}\nu}{\int_{\nu_{\mathrm{limit}}}^{\infty} \nu^{-\alpha-1}\,\mathrm{d}\nu} &= \frac{N_{\mathrm{H\,I}}\,\sigma_0\,\nu_{\mathrm{limit}}^{-\alpha-3}}{\nu_{\mathrm{limit}}^{-3}\,(\alpha+3)} \frac{\alpha}{\nu_{\mathrm{limit}}^{-\alpha}} \\
= \frac{N_{\mathrm{H\,I}}\,\sigma_0 \alpha}{\alpha + 3},
\end{aligned}$$

where we first cancelled the h terms, then cancelled the k terms. Setting $\tau > 1$, we find $N_{\mathrm{H\,I}} > 1.3\,((\alpha+3)/\alpha) \times 10^{21}\,\mathrm{m}^{-2}$, as required.

Exercise 8.5 Equation 1.28 relates $\mathrm{d}z/\mathrm{d}t$ to $H(z)$. Taking the modulus and reciprocal of that equation gives $(1+z)\,|\mathrm{d}t/\mathrm{d}z| = 1/H(z)$. A population with constant proper sizes has constant A in Equation 8.2, and a constant comoving density is constant n_{co} in the same equation. Therefore $\mathrm{d}^2\mathcal{N} \propto (1+z)^3\,|\mathrm{d}t/\mathrm{d}z| \propto (1+z)^2/H(z)$. If we write $\mathrm{d}X/\mathrm{d}z = (1+z)^2 H_0/H(z)$, then

$$\mathrm{d}^2\mathcal{N} = n_{\mathrm{co}}\,A \times (1+z)^2 c \left|\frac{1}{H(z)}\right| \mathrm{d}N_{\mathrm{H\,I}}\,\mathrm{d}z$$

gives

$$\mathrm{d}^2\mathcal{N} = n_{\mathrm{co}}\,A\frac{c}{H_0}\,\mathrm{d}X\,\mathrm{d}N_{\mathrm{H\,I}},$$

which is constant.

Exercise 8.6 Gravitational lensing of the background quasar by the damped Lyman α system could cause such an effect. The strength of this effect, and the biases that it creates on the measured cosmic evolution of neutral gas, are still the subject of debate. However, it turns out that this is probably only a 10–20% effect on $\Omega_{\mathrm{H\,I}}$ at $z > 2$.

Exercise 8.7 Dust in the damped Lyman α systems should redden the quasar spectra, so one might compare the optical spectral indices or B–V colours of quasars with and without damped Lyman α absorbers. However, if damped systems are very dusty, they may induce so much reddening that the quasars drop out of the parent sample, so bright quasar catalogues would be biased to detecting low-reddening systems. Statistical analyses suggest that this latter effect does not dominate, but direct results on quasar reddening are currently conflicting.

Exercise 8.8 The energy of the hydrogen Lyman limit is $E = 13.6\,\mathrm{eV}$, i.e. $E = 13.6 \times 1.602 \times 10^{-19}\,\mathrm{J} = 2.179 \times 10^{-18}\,\mathrm{J}$. This corresponds to a frequency of $\nu = E/h$, where h is Planck's constant, which comes out as $\nu = 3.289 \times 10^{15}\,\mathrm{Hz}$. The wavelength of this light is $\lambda = c/\nu$, where c is the speed of light, which comes out as $\lambda = 9.116 \times 10^{-8}\,\mathrm{m}$, or 91.2 nm (i.e. 912 Å) to three significant figures. For the helium Lyman limit, $\lambda_{\mathrm{He}} = \lambda \times 13.6/54.4 = 22.8\,\mathrm{nm}$.

The redshifted hydrogen Lyman limit in Figure 8.20 is at a wavelength of $912 \times (1+z)\,\text{Å} = 912 \times (1+3.2)\,\text{Å} = 3830\,\text{Å}$.

Acknowledgements

Grateful acknowledgement is made to the following sources:

Figures

Cover image courtesy of the Spitzer Space Telescope, ©NASA/JPL-Caltech/STScI/CXC/UofA/ESA/AURA/JHU;

Figure 1.9: supernova data taken from Blondin, S. et al. (2008) *The Astrophysical Journal*, **682**, 724; Figure 1.10 top left: http://astrosurf.com, Christian Buil; Figure 1.10 top right: European Southern Observatory (ESO); Figure 1.10 bottom left: Stanford, S. A. et al. (2000) 'The first sample of ultraluminous infrared galaxies at high redshift', *The Astrophysical Journal Supplement Series*, **131**, 185, The American Astronomical Society; Figure 1.10 bottom right: van Dokkum, P. G. et al. (2005) 'Gemini near-infrared spectrograph observations of a red star-forming galaxy at $z = 2.225$: evidence of shock ionization due to a galactic wind', *The Astrophysical Journal*, **622**, L13, The American Astronomical Society; Figure 1.11: Carroll, S. M. (2004), 'Why is the Universe accelerating?', Freedman, W. L. ed. Measuring and Modelling the Universe, Carnegie Observatories Astrophysics Series, **2**, Carnegie Observatories; Figure 1.13: NASA and the Hubble Heritage Team (STScI/AURA); Figure 1.16: Springel, V. et al. (2005) 'Simulations of the formation, evolution and clustering of galaxies and quasars', *Nature*, **435**, 629 ; Figures 1.18 & 1.19: adapted from Carroll, S. M., Press, W. H. and Turner, E. L. (1992) 'The Cosmological Constant', *Annual Review of Astronomy & Astrophysics*, **30**, 499, ©Annual Reviews Inc.; Figure 1.20: Adapted from Knop R. A. et al. (2003), 'New Constraints on Ω_M, Ω_Λ and w from an independent set of 11 high-redshift supernovae observed with the Hubble Space Telescope', *The Astrophysical Journal*, **598**, 102, ©The American Astronomical Society;

Figures 2.1, 2.2 & 2.9: NASA/WMAP Science Team; Figure 2.3: adapted from Coc, A. (2009) 'Big-bang nucleosynthesis: a probe of the early Universe', *Nuclear Instruments & Methods in Physics Research A*, **611**, 224, Elsevier Science BV; Figure 2.4: adapted from a figure by Professor Edward L. Wright, UCLA; Figures 2.7 & 2.8: Peacock, J. A. (1999) *Cosmological Physics*, Cambridge University Press; Figure 2.10: University of Hawaii; Figure 2.11: Granett, B. R. et al. (2008) 'An imprint of super-structures on the microwave background due to the Integrated Sachs–Wolfe effect', *The Astrophysical Journal Letters*, **683**, L99, Institute of Physics Publishing; Figure 2.12: adapted from Dunkley, J. et al. (2009) 'Five year Wilkinson Microwave Anisotropy Probe (WMAP1) observations: likelihoods and parameters from the WMAP data', *Astrophysical Journal Supplement Series*, **180**, 306, Institute of Physics Publishing; Figures 2.13 & 2.15: Hu, W. and Dodelson, S. (2002) 'Cosmic microwave background anisotropies', *Annual Reviews of Astronomy & Astrophysics*, **40**, 171, Annual Reviews; Figures 2.14 & 3.7: adapted from figures by Edward L. Wright, UCLA and based on data from Kowalski, M. et al. (2009) *The Astrophysical Journal Supplement Series*, **686**, 749; Figure 2.16: adapted from Larson, D. et al. (2010) 'Seven year Wilkinson Microwave Anisotropy Probe (WMAP1) observations: power spectra and WMAP-derived parameters', *Astrophysical Journal Supplement Series* (in press, arXiv:1001.4635), Institute of Physics Publishing; Figures 2.17 & 2.18: adapted from Komatsu, E. et al. (2009)

'Five year Wilkinson Microwave Anisotropy Probe (WMAP1) Observations: cosmological interpretation' *Astrophysical Journal Supplement Series*, **180**, 330, Institute of Physics Publishing;

Figure 3.1a: Justin Yaros and Andy Schlei/Flynn Haase/NOAO/AURA/NSF; Figure 3.1b: adapted from Begeman, K. G., Broeils, A. H. and Sanders, R. H. (1991) 'Extended rotation curves of spiral galaxies: dark haloes and modified dynamics', *Monthly Notices of the Royal Astronomical Society*, **249**, 523; Figure 3.2: ESO Online Digital Sky Survey www.eso.org/dss/dss.; Figure 3.4: adapted from a figure of Professor Edward L. Wright, UCLA; Figure 3.5: adapted from Dressler, A. (1980) 'Galaxy morphology in rich clusters: implications for the formation and evolution of galaxies', *The Astrophysical Journal*, **236**, 351, American Astronomical Society; Figure 3.6: adapted from Ciardullo, R. (2004) 'The Planetary Nebula Luminosity Function', A contribution to the ESO International Workshop on *Planetary Nebulae beyond the Milky Way*, Garching (Germany), May 19–21, 2004; Figure 3.8: Chris Schur, www.schursastrophotography.com; Figure 3.9: www.astro.uu.se; Figure 3.10: NASA/Jason Ware; Figure 3.11: Günter Kerschhuber, Gahberg Observatory; Figures 3.12 & 3.13: Richard Powell, www.atlasoftheuniverse.com; Figure 3.14: adapted from de Lapparent, V. et al. (1986) 'A slice of the Universe', *The Astrophysical Journal*, **302**, 1, The American Astronomical Society; Figures 3.15 & 3.18: The 2dF Galaxy Redshift Survey team (http://www2.aao.gov.au/2dFGRS/); Figure 3.16: adapted from Peacock, J. A. et al. (2001) 'A measurement of the cosmological mass density from clustering in the 2dF Galaxy Redshift Survey', *Nature*, **410**, 169, Nature Publishing Group; Figure 3.17: adapted from Peacock, J. A. and Dodds, S. J. (1994), 'Reconstructing the linear power spectrum of cosmological mass fluctuations', *Monthly Notices of the Royal Astronomical Society*, **267**, 1020, The Royal Astronomical Society; Figure 3.19: adapted from Percival, W. J. et al. (2007) 'Measuring the Baryon Acoustic Oscillation Scale using the Sloan Digital Sky Survey and 2df Galaxy Redshift Survey', *Monthly Notices of the Royal Astronomical Society*, **381**, 1053, The Royal Astronomical Society;

Figure 4.1: adapted from Tegmark, M. and Zaldarriaga, M. (2002) 'Separating the Early Universe from the Late Universe: cosmological parameter estimation beyond the black box', *Physical Review D*, **66**(10), 103508, The American Physical Society; Figure 4.2: adapted from Lacey, C. and Cole, S. (1993) 'Merger rates in hierarchical models of galaxy formation', *Monthly Notices of the Royal Astronomical Society*, **262**, 627, The Royal Astronomical Society; Figure 4.5: Moore, B. et al. (1999) 'Dark matter substructure within galactic halos', *The Astrophysical Journal*, **524**, L19, American Astronomical Society; Figure 4.6: adapted from Rocca-Volmerange, B. and Guiderdoni, B. (1988) 'An atlas of synthetic spectra of galaxies', *Astronomy & Astrophysics Supplement Series*, **75**, 93, European Southern Observatory; Figure 4.7: Dr Henner Busemann, School of Earth, Atmospheric and Environmental Sciences (SEAES), The University of Manchester; Figure 4.8: adapted from Gordon, K. D. et al. (2003) 'A quantitative comparison of the Small Magellanic Cloud, Large Magellanic Cloud, and Milky Way ultraviolet to near-infrared extinction curves', *The Astrophysical Journal*, **594**, 279, The American Astronomical Society; Figure 4.9: Brammer, G. B. et al. (2008) 'EAZY: A fast, public photometric redshift code', *The Astrophysical Journal*, **686**, 1503, The American Astronomical Society; Figure 4.10: adapted from Dey, A. et al. (1998) 'A galaxy at $z = 5.34$', *The Astrophysical Journal*, **498**, L93, The American Astronomical Society; Figure 4.11: adapted from Bell, E. F. et al. (2003) 'The optical and near infrared properties of galaxies. 1. luminosity and stellar mass functions', *The Astrophysical Journal Supplement Series*, **149**, 289, The American Astronomical Society; Figure 4.12: NRAO; Figure 4.13: Sloan Digital Sky Survey; Figure 4.14: adapted from Yates, M. G. and Garden R. P. (1989) 'Near-simultaneous optical and infrared spectrophotometry of active galaxies', *Monthly Notices of the Royal Astronomical Society*, **241**, 167, The Royal Astronomical Society; Figure 4.16: adapted from Figure 2.3 of Peterson, B. M. (1997) *An Introduction to Active Galactic Nuclei*, Cambridge University Press; Figure 4.17: adapted from Richards, G. T. et al. (2006), 'The Sloan Digital Sky Survey Quasar Survey: quasar luminosity function from data release 3', *The Astronomical Journal*, **131**, 2766, The American Astronomical Society; Figure 4.18 left: A. Fujii; Figure 4.18 right: R. Williams (STScI), the Hubble Deep Field Team and NASA; Figure 4.19: Robert Williams and the Hubble Deep Field Team (STScI) and NASA; Figure 4.20: NASA/ESA, CXC, JPL-Caltech, STScI, NAOJ, J. E. Greach (Univ Durham) et al.; Figure 4.21: adapted from Gabasch, A. et al. (2004) 'The evolution of the luminosity functions in the FORS deep field from low to high redshift', *Astronomy & Astrophysics*, **421**, 41, ESO; Figure 4.22: NASA, ESA, S. Beckwith (STScI) and the HUDF Team; Figures 4.23 & 4.24: adapted from Bouwens, R. J. et al. (2009) 'Constraints on the first galaxies: z 10 Galaxy Candidates from HST WFC3/IR', Submitted to Nature (arXiv:0912.4263); Figure 4.25: adapted from Cohen, J. G. et al. (1996) 'Redshift clustering in the Hubble Deep Field', *The Astrophysical Journal*, **471**, 5, The American Astronomical Society; Figure 4.26: adapted from Bouwens, R. J. et al. (2004) 'Galaxy size evolution at high redshift and surface brightness selection effects:constraints from the Hubble Ultra Deep Field', *The Astrophysical Journal*, **611**, 1. The American Astronomical Society; Figure 4.27: adapted from van Dokkum, P. G., Kriek, M. and Franx, M. (2009) 'A high stellar velocity dispersion for a compact massive galaxy at redshift $z = 2.186$', *Nature* , **460**, 717, Macmillan Publishers Limited; Figure 4.28: NASA Jet Propulsion Laboratory (NASA-JPL); Figure 4.29: H. Ferguson, M. Dickinson, R. Williams, STScI and NASA; Figure 4.30: adapted from Bell, E. F. et al. (2004) 'Nearly 5000 distant early type galaxies in COMBO-17: a red sequence and its evolution since z 1', *The Astrophysical Journal*, **608**, 752, The American Astronomical Society;

Figure 5.1: adapted from Hauser, M. G. and Dwek, E. (2001) 'The Cosmic Infrared Background: Measurements and Implications', *Annual Review of Astronomy & Astrophysics*, **39**, 249, Annual Reviews Inc; Figure 5.2: adapted from Hopwood, R. H. et al. 'Ultra deep AKARI observations of Abell 2218: resolving the 15 m extragalactic background light', *Astrophysical Journal Letters*, **716**, 45; Figure 5.3: http://alma.asiaa.sinica.edu.tw; Figure 5.4: adapted from Blain, A. W. et al. (2002) 'Submillimeter galaxies', *Physics Reports*, **369**, 111, Elsevier Science B.V.; Figure 5.5: adapted from Hughes D. H. et al. (1998) 'High-redshift star formation in the Hubble Deep Field revealed by a submillimetre-wavelength survey', *Nature*, **394**, 241; Figure 5.6: BLAST Collaboration; Figure 5.7: ESA and SPIRE Consortium; Figure 5.8: adapted from Serjeant, S. et al. (1998) 'A spectroscopic study of IRAS F10214+4724', *Monthly Notices of the Royal Astronomical Society*, **298**, 321, Royal Astronomical Society; Figure 5.9: adapted from Surace, J. A. et al. (1998) 'HST/WFPC2 Observations of warm ultraluminous infrared galaxies', *Astrophysical Journal*, **492**, 116, The American Astronomical Society; Figure 5.10 top: Brad Whitmore (STScI) and NASA; Figure 5.10 bottom: NASA/JPL-Caltech/Z. Wang (Harvard-Smithsonian CfA); Visible: M. Rushing/NOAO; Figure 5.11: Courtesy of JAXA; Figures 5.12 & 5.13: adapted from Condon, J. J. (1992) 'Radio emission from normal galaxies', *Annual Reviews of Astronomy & Astrophysics*, **30**, 575, Annual Reviews Inc; Figure 5.14: adapted from Dole, H. et al. (2006) 'The cosmic infrared background resolved by Spitzer', *Astronomy & Astrophysics*, **451**, 417, EDP Sciences; Figures 5.15 & S5.1: adapted from Griffin, M. et al. (2007) 'The Herschel-SPIRE instrument and its capabilities for extragalactic astronomy', *Advances in Space Research*, **40**, 612, ©COSPAR, Published by Elsevier Ltd; Figures 5.16, 5.17 & 5.18: Pérez-González, P. G. et al. (2008) 'The Stellar Mass Assembly of Galaxies from $z = 0$ to $z = 4$', *The Astrophysical Journal*, **675**, 234, The American Astronomical Society; Figure 5.19: adapted from Le Floc'h, E. et al. (2005) 'Infrared Luminosity Functions from the Chandra Deep Field-South', *The Astrophysical Journal*, **632**, 169, The American Astronomical Society; Figure 5.20: adapted from Di Matteo, T. et al. (2005) 'Energy input from quasars regulates the growth and activity of black holes and their host galaxies', *Nature*, **433**, 604, Nature Publishing Group; Figure 5.21: McNamara, B. R. et al. (2000) 'Chandra X-ray observations of the Hydra A cluster:an interaction between the radio source and the X-ray emitting gas', *Astrophysical Journal*, **534**, L135, The American Astronomical Society; Figures 5.22 & 5.23: Fabian, A. C. et al. (2003) 'A very deep Chandra observation of the Perseus cluster: shocks and ripples', *Monthly Notices of the Royal Astronomical Society*, **344**, L43, The Royal Astronomical Society;

Figure 6.3: Science Photo Library; Figure 6.4: Misner, C. W., Thorne, K. S. and Wheeler, J. A. (1973) *Gravitation*, W. H. Freeman & Co Ltd; Figure 6.5: adapted from Kormendy, J. (1988) 'Evidence for a supermassive black hole in the nucleus of M31', *The Astrophysical Journal*, **325**, 128, American Astronomical Society; Figure 6.6: adapted from Miyoshi, M. et al. (1995) 'Evidence for a black hole from high rotation velocities in a sub parsec region of NGC4258', *Nature*, **373**, 127, Nature Publishing Group; Figure 6.7: adapted from Schdel, R. et al. (2003) 'Stellar dynamics in the central arcsecond of our galaxy', *The Astrophysical Journal*, **596**, 1015, The American Astronomical Society; Figures 6.8 & 6.9: adapted from Peterson, B. M. (2001) 'Variability of active galactic nuclei', Aretxaga, I., Knuth, D. and Mujica, R. eds. Advanced Lectures on the Starburst-AGN Connection, World Scientific; Figure 6.10: Ferraresse, L. (2002) 'Black Hole Demographics', Proceedings of the 2nd KIAS Astrophysics Workshop held in Seoul, Korea (Sep 3–7 2001), Lee, C. H. ed. World Scientific; Figure 6.11: J. Schmitt et al. ROSAT Mission, MPE, ESA; Figure 6.13: Brandt, W. N. and Hasinger, G. (2005) 'Deep Extragalactic X-ray Surveys', *Annual Review of Astronomy & Astrophysics*, **43**, 827, Annual Reviews; Figure 6.14: X-ray: NASA/CXC/U. of Michigan/J. Liu et al.; Optical: NOAO/AURA/NSF/T. Boroson; Figure 6.15: adapted from Alexander, D. M. et al. (2008) 'Weighing the black holes in $z \approx 2$ submillimeter-emitting galaxies hosting active galactic nuclei', *The Astrophysical Journal*, **135**, 1968, The American Astronomical Society; Figure 6.16: adapted from Kauffmann, G. and Heckman, T. M. (2009) 'Feast and famine: regulation of black hole growth in low redshift galaxies', *Monthly Notices of the Royal Astronomical Society*, **397**, 135,

The Royal Astronomical Society; Figure 6.17: Weisberg, J. M. and Taylor, J. H. (2005) 'The relativistic binary pulsar B1913+16: thirty years of observations ad analysis', Rasio, F. A. and Stairs, I. H. (eds) Binary Radio Pulsars, *ASP Conference Series*, **328**, 25, Astronomical Society of the Pacific; Figure 6.18: CalTech; Figure 6.19: adapted from Boroson, T. A. and Lauer, T. R. (2009) 'A candidate sub-parsec supermassive binary black hole system', *Nature*, **458**, 53, Nature Publishing Group;

Figure 7.3: NASA, Andrew Fruchter and the ERO Team [Sylvia Baggett (STScI), Richard Hook (ST-ECF), Zoltan Levay (STScI)] (STScI); Figure 7.4: adapted from Nguyen, H. T. et al. (1999) 'Hubble Space Telescope imaging polarimetry of the gravitational lens FSC 10214+4724', *The Astronomical Journal*, **117**, 671, The American Astronomical Society; Figure 7.5: adapted from Serjeant, S. et al. (1998) 'A spectroscopic study of IRAS F10214+4724', *Monthly Notices of The Royal Astronomical Society*, **298**, 321, The Royal Astronomical Society; Figure 7.12: Dr A. Holloway, University of Manchester; Figure 7.17: Burke, B. et al. (1993) *Sub-Arcsecond Radio Astronomy*, Davis, R. J. and Booth, R. S. eds. Cambridge University Press; Figure 7.18: adapted from Alcock, A. et al. (1993) 'Possible gravitational microlensing of a star in the Large Magellanic Cloud', *Nature*, **365**, 621, Nature Publishing Group; Figure 7.20: Stephane Colombi, International Astronomical Union; Figure 7.22: adapted from Blandford, R. D. et al. (1991) 'The distortion of distant galaxy images by large scale structure', *Monthly Notices of the Royal Astronomical Society*, **251**, 600, The Royal Astronomical Society; Figure 7.23: adapted from Hoekstra, H. et al. (2004) 'Properties of galaxy dark matter halos from weak lensing', *The Astrophysical Journal*, **606**, 67, The American Astronomical Society; Figures 7.24 & 7.25: adapted from Massey, R. et al. (2007) 'Dark matter maps cosmic scaffolding', *Nature*, **445**, 286, Nature Publishing Group; Figure 7.26: Large Synoptic Survey Telescope Corporation; Figure 7.27: top right panel adapted from Hopwood et al. (2010) *Astrophysical Journal Letters*, **716**, 45; bottom left panel adapted from Egami et al., paper in preparation; Figure 7.28: X-ray: NASA/CXC/CfA/M. Markevitch et al. Lensing Map: NASA/STScI; ESO WFI; Magellan/U. Arizona/D. Clowe et al. Optical: NASA/STScI; Magellan/U. Arizona/D. Clowe et al.; Figure 7.29: CASTLES (CfA-Arizona Space Telescope Lens Survey); Figure 7.30: NASA Johnson Space Center Collection; Figure 7.31: A. Bolton (UH IfA) for SLACS and NASA/ESA; Figure S7.1: NASA, ESA, C. Faure (Zentrum für Astronomie, University of Heidelberg) and J. P. Kneib (Laboratoire d'Astrophysique de Marseille);

Figure 8.1: Rauch, M. (1998) 'The Lyman alpha forest in the spectra of quasistellar objects', *Annual Reviews of Astronomy & Astrophysics*, **36**, 267, Annual Reviews; Figure 8.4: NASA, ESA, Y. Izotov (Main Astronomical Observatory, Kyiv, UA) and T. Thuan (University of Virginia); Figures 8.5, 8.6 & 8.7: Pettini, M. et al. (2008) 'Deuterium abundance in the most metal-poor damped Lyman alpha system', *Monthly Notices of the Royal Astronomical Society*, **391**, 1499, The Royal Astronomical Society; Figure 8.8 adapted from Kriss, G. A. et al. (1999) 'The Ultraviolet Peak of the Energy Distribution in 3C 273: Evidence for an Accretion Disk and Hot Corona around a Massive Black Hole', *The Astrophysical Journal*, **527**, 683, The American Astronomical Society; Figures 8.9 & 8.14: adapted from Noterdaeme, P. et al. (2009) 'Evolution of the cosmological mass density of neutral gas from Sloan

Digital Sky Survey II-data release 7', *Astronomy & Astrophysics*, **505**, 1087, European Southern Observatory; Figure 8.10: Prochaska, J. X. et al. (2005) 'The SDSS damped Ly alpha survey: data release 3', *Astrophysical Journal*, **635**, 123, The American Astronomical Society; Figure 8.12: Reynolds, S. C. (2007) 'Quasar Absorbers and the InterGalactic Medium', taken from a pedagogical Seminar at the Royal Observatory, Edinburgh, 8 March 2007, www.roe.ac.uk/ifa/postgrad/pedagogy/2007_reynolds.pdf; Figure 8.13: Möller, P. and Warren, S. J. (1993) 'Emission from a damped Ly alpha absorber at $z = 2.81$', *Astronomy & Astrophysics*, **270**, 43, European Southern Observatory; Figure 8.15: Smette, A. et al. (1992) 'A spectroscopic study of UM 673 A & B: on the size of the Lyman-alpha clouds', *Astrophysical Journal*, **389**, 39, The American Astronomical Society; Figure 8.16: Nick Gnedin, Department of Astronomy & Astrophysics, The University of Chicago; Figures 8.17 & 8.19: Fan, X. et al. (2006) 'Observational constraints on cosmic reionization', *Annual Review of Astronomy & Astrophysics*, **44**, 415 ©2006 by Annual Reviews; Figure 8.18: Becker, G. D. et al. (2007) The evolution of optical depth in the Ly alpha Forest: evidence against reionization at z 6, *The Astrophysical Journal*, **662**, 72, The American Astronomical Society; Figure 8.20: adapted from Möller, P. and Jakobsen, P. (1990) 'The Lyman continuum opacity at high redshifts: through the Lyman forest and beyond the Lyman valley', *Astronomy & Astrophysics*, **228**, 299, European Southern Observatory; Figure 8.21: Smette, A. et al. (2002) 'Hubble Space Telescope Space Telescope Imaging System Observations of the He II Gunn–Peterson effect toward HE 2347-4342', *Astrophysical Journal*, **564**, 542, The American Astronomical Society; Figure 8.23: Carilli, C. L. et al. (2002) 'H I 21 centimeter absorption beyond the epoch of reionization', *The Astrophysical Journal*, **577**, 22, The American Astronomical Society; Figures 8.24 & 8.25: Cristiani, S. et al. (2007) 'The CODEX-ESPRESSO experiment: cosmic dynamics, fundamental physics, planets and much more . . .', *Il Nuovo Cimento*, **122B**, 1165, Societa Italiana di Fisica.

Every effort has been made to contact copyright holders. If any have been inadvertently overlooked the publishers will be pleased to make the necessary arrangements at the first opportunity.

Index

Items that appear in the Glossary have page numbers in **bold type**. Ordinary index items have page numbers in Roman type.